Lecture Notes in Electrical Engineering

Volume 209

For further volumes:
http://www.springer.com/series/7818

Wenjiang Du
Editor

Informatics and Management Science VI

Editor
Wenjiang Du
College of Elementary Education
Chongqing Normal University
Chongqing
People's Republic of China

ISSN 1876-1100 ISSN 1876-1119 (electronic)
ISBN 978-1-4471-4804-3 ISBN 978-1-4471-4805-0 (eBook)
DOI 10.1007/978-1-4471-4805-0
Springer London Heidelberg New York Dordrecht

Library of Congress Control Number: 2012952024

Printed on acid-free paper

Springer is part of Springer Science+Business Media (www.springer.com)

Preface

Welcome to the proceedings of the International Conference on Informatics and Management Science (IMS) 2012, which will be held in December 21–23, 2012, in Kunming, China.

IMS 2012 will be a venue for leading academic and industrial researchers to exchange their views, ideas and research results on innovative technologies, and sustainable solutions leading to Informatics and Management Science. The conference will feature keynote speakers, a panel discussion, and paper presentations.

The objective of IMS 2012 is to facilitate an exchange of information on best practices for the latest research advances in the area of Informatics and Management Science. IMS 2012 will provide a forum for engineers and scientists in academia, industry, and government to address the most innovative research and development including technical challenges, social and economic issues, and to present and discuss their ideas, results, work in progress and experience on all aspects of Informatics and Management Science.

There was a very large number of paper submissions (2351). All submissions were reviewed by at least three Program or Technical Committee members or external reviewers. It was extremely difficult to select the presentations for the conference because there were so many excellent and interesting submissions. In order to allocate as many papers as possible and keep the high quality of the conference, we finally decided to accept 614 papers for presentations, reflecting a 26.1 % acceptance rate. We believe that all of these papers and topics not only provided novel ideas, new results, work in progress, and state-of-the-art techniques in this field, but also stimulated the future research activities in the area of Informatics and Management Science.

The exciting program for this conference was the result of the hard and excellent work of many others, such as Program and Technical Committee members, external reviewers, and Publication Chairs under a very tight schedule. We are also grateful to the members of the Local Organizing Committee for supporting us in handling so many organizational tasks, and to the keynote speakers for accepting to come to the conference with enthusiasm. Last but not least, we hope you enjoy the conference program, and the beautiful attractions of Kunming, China.

With our warmest regards.

December 2012

Wenjiang Du
Guomeng Dong
General and Program Chairs
IMS 2012

Organization

IMS 2012 was organized by Electric Power Research Institute, YNPG, Yunnan Normal University, Wuhan Institute of Technology, Guizhou University, Chongqing Normal University, Chongqing University, Yanshan University, Xiangtan University, Hunan Institute of Engineering, Shanghai Jiao Tong University, Nanyang Technological University, and sponsored by National Natural Science Foundation of China (NSFC). It was held in cooperation with *Lecture Notes in Electrical Engineering* (LNEE) of Springer.

Executive Committee

General Chairs: Maode Ma, Nanyang Technological University, Singapore
Yuhang Yang, Shanghai Jiao Tong University, China

Program Chairs: Yanchun Zhang, University of Victoria, Australia
Rafa Kapelko, Wrocaw University of Technology, Poland
Rongbo Zhu, Virginia Tech, USA
Ming Fan, University of Washington, USA

Local Arrangement Chairs: Qing Xiao, Chongqing University, China
Wenjiang Du, Chongqing Normal University, China

Steering Committee: Maode Ma, Nanyang Technological University, Singapore
Nadia Nedjah, State University of Rio de Janeiro, Brazil
Lorna Uden, Staffordshire University, UK
Dechang Chen, Uniformed Services University of the Health Sciences, USA
Mei-Ching Chen, Tatung University, Taiwan
Rong-Chang Chen, National Taichung Institute of Technology, Taiwan
Chi-Cheng Cheng, National Sun Yat-Sen University, Taiwan
Naohiko Hanajima, Muroran Institute of Technology, Japan
Shumin Fei, Southeast University, China
Yingmin Jia, BeiHang University, China
Weiguo Liu, Northwesten Polytechnic University, China

Program/Technical Committee

Meh Shafiei	Dalhousie University, Canada
Girij Prasad	University of Ulster, UK
Jams Li	University of Birmingham, UK
Liang Li	University of Sheffield, UK
Hai Qi	University of Tennessee, USA
Yuezhi Zhou	Tsinghua University, China
Duolin Liu	ShenYang Ligong University, China
Zengqiang Chen	Nankai University, China
Dumisa Wellington Ngwenya	Illinois State University, USA
Hu Changhua	Xi'an Research Institute of Hi-Tech, China
Juntao Fei	Hohai University, China
Zhao-Hui Jiang	Hiroshima Institute of Technology, Japan
Michael Watts	Lincoln University, New Zealand
Chun Lee	Howon University, Korea
Cent Leung	Victoria University of Technology, Australia
Haining Wang	College of William and Marry, USA
Worap Kreesuradej	King Mongkuts Institute of Technology Ladkrabang, Thailand
Muhammad Khan	Southwest Jiaotong University, China
Stefa Lindstaedt	Division Manager Knowledge Management, Austria
Yiming Chen	Yanshan University, China
Tashi Kuremoto	Yamaguchi University, Japan
Zheng Liu	Nagasaki Institute of Applied Science, Japan
Seong Kong	The University of Tennessee, USA
R. McMenemy	Queens University Belfast, UK
Sunil Maharaj Sentech	University of Pretoria, South Africa
Paolo Li	Polytechnic of Bari, Italy
Cheol Moon	Gwangju University, Korea
Zhanguo Wei	Beijing Forestry University, China
Hao Chen	Hunan University, China
Xiaozhu Liu	Wuhan University, China
Xilong Qu	Hunan Institute of Engineering, China
Lilei Wang	Beijing University of Posts and Telecommunications, China
Liang Zhou	ENSTA-ParisTech, France
Yanbing Sun	Beijing University of Posts and Telecommunications, China
Xiyin Wang	Hebei Polytechnic University, China
Hui Wang	University of Evry in France, France
Uwe Kuger	Queen's University of Belfast, UK
Nin Pang	Auckland University of Technology, New Zealand
Yan Zhang	Simula Research Laboratory and University of Oslo, Norway

Contents

Part I
Web Science, Engineering and Applications II

Chapter 1
Education Training System of Primary and Middle School Teachers in Network Environment

Hui Zhao, Shujuan Fu and Zhihong Qiang

Abstract In order to develop teachers' ability of educational technology, this paper designed the Langfang city primary and secondary school teachers' educational technology training system. The main technical line of teacher's education technical skills training system is to build a platform. This platform can support the wide-area learning resource sharing, support personalized learning and support large-scale learners learning corporately. The training system for primary and middle school teachers' educational technology ability broke through the limitation of time and space, overcome the deficiency of the old training mode. The teacher can make a choice and grasp the study progress based on their own needs; the training system can reduce the burden of teachers and improve the efficiency.

Keywords Network environment · Educational technology · Training system

1.1 Introduction

Years of work for training and teaching reflect that the training of teachers' overall level is different. In order to make the teacher education technology the ability to get comprehensive development, this paper designs a strategic plan for primary and secondary school teacher's education technology training web site to realize the teacher education information resources sharing [1].

H. Zhao (✉) · S. Fu · Z. Qiang
Langfang radio and Television University, Hebei, People's Republic of China
e-mail: zhaohui20011211@sina.com

W. Du (ed.), *Informatics and Management Science VI*, Lecture Notes in Electrical Engineering 209, DOI: 10.1007/978-1-4471-4805-0_1,

Primary and middle school teachers' education technology training website design process is as follows:

1.2 Demand Analysis

1.2.1 Questionnaire Survey

In order to get the teachers' education present situation of Langfang technical training authentically, we investigated Langfang City Department of education, audio-visual education center and other units, get some information from the questionnaire. The questionnaire is divided into two parts, the first part is the basic personal information, and the second part is the investigation content. The investigation content includes several aspects: The person's education technology of consciousness and attitude, skills and knowledge, innovation and application, the needs of the training and social responsibility, and so on. The objects of the investigation are teachers, the managers of the school and technical personnel in six middle schools which has the different level [2].

1.2.2 The Existing Problems in Langfang City, Educational Technology Training

The analysis of questionnaires shows that the teachers in secondary school can not master the educational technology sufficiently, especially the research methods of educational technology. According to the survey, only 46.2 % of the teachers know it. Most of the training schools are lack of modern equipment and application and experience for education technology. If the training still takes the traditional single teaching mode "teach–demonstration—practice", it neither improve the effect of training, nor enhance the operating ability.

1.2.3 The Necessity of Constructing Network Training Platform

The survey showed that the educational structure should designee an environment which adapts to meet the teachers' needs and characteristics of the training situation. It can realize through the activities of the organization. The needs of the training teachers are changing with the flowing time, so it is very important to build the network platform of educational technology training [3].

1.3 The Overall Design

1.3.1 The Overall Structure

Based on the analysis of needs for the teacher education and technical training system in Langfang City, combined with destination and the content of the educational technology training, this paper designed the overall structure of the system. It consists of four main components: (1) The basic theory of education technology module. (2) Learning resources application module. (3) The training activities of the organization module. (4) Learning evaluation module.

1.3.1.1 The Basic Theory of Education Technology Module

Educational technology is an emerging comprehensive discipline. If the teachers want to improve the ability of using the media, they must master the basic theory of educational technology which includes: the concept, characteristics and the function of educational technology, the basic theory of the educational technology, varieties of principles of the teaching media, etc. In addition, educational technology, educational psychology theory and practical training are also need, this can solve the problem: how to use modern ways education to teach?

1.3.1.2 Learning Resources Application Module

The main effect of the module is that it can share the multimedia learning resources, so the teacher can apply them better. The training website in this paper emphasize that the website can push the resources when the people are learning, the teachers who are trained can look for the resources with their destination, the system can feedback the resources which are the teachers' needs. So it not only improves the efficiency of acquiring the resource, but also saves the time to find the resources.

1.3.1.3 The Training Activities of the Organization Module

The teachers should solve the problems in the teaching time through the training. Therefore in the activity arrangements, the module takes the problems which need to be solved and the tasks to be completed as the guidance, takes the activities as the center, takes the updating knowledge as the goal, takes the personalize application and the rich resources as the support, it does not list all of the courses. The design of the training activities ultimately represent as the design of the training tasks. It achieves the purpose of the training through stipulating the aim, the content, the strategy, the results and the methods of the activities, all of these can lead the learner to think deeply [4].

1.3.1.4 Learning Evaluation Module

Reasonable appraisal system is the guarantee of the success for educational technology training. Evaluation of training is not only a summary of the training results, but also the reflection of the training process. It is also one of the important methods which ensure the training to achieve the purpose. This paper gives the actual situation of the education and technical training for teachers in Langfang City three evaluations: process evaluation, performance evaluation, summative evaluation.

1.3.2 The Technical Route and the Application Architecture Analysis

The main technical line of teacher's education technical skills training system is to build a platform. This platform can support the wide-area learning resource sharing, support personalized learning and support large-scale learners learning corporately.

Combined with information sharing and parallelism characteristics of the system, this paper designed the teacher training system which use the B/S mode network architecture, the user will be able to quickly and easily access the system by the client browser. In the B/S structure, the client only need to display the web page on the screen, all the data processing tasks are lead to the server to process. The system uses the B/S system architecture, including the presentation layer, application layer and data layer. The presentation layer is an interface, the client interacts with the system can use the interface. Based on the client's request, the application layer can realize all the tasks in the training process, at the same time, it manages the entire database access rules, and it determines the service management method, so it can realize the security guarantees. The data layer uses the relational database. It can ensure the integrity of the data in the system through the rule defining, constraint, transactions and the concurrency control. The system use the program which based on the JSP technology, the back-end database uses the SQL server 2000 system architecture [5, 6].

1.3.3 System Database Design

The database design is to establish a database and application systems technology, it is the core technology in the information system development and construction, it is also the fundamental basis of the system. Teacher training system is based on teacher training. In this training system, the describe information include registration forms, course training table, teaching table, the training teachers table, work units table, and so on.

As this system is a database-based Web system, the system uses a connection pool technology. The core of the connection pool is connectivity between the cached JDBC and the database, it can reduce the number of the establishment of the connection, when the application requires access to the database, it directly obtain a connection from the cache, the connection is provided to the users, thus it can effectively reduce the time which is need for connection establishment.

1.4 The Operation Condition of the Training System

The interface of the system login module is shown in Fig. 1.1.

After a successful login in the system, the interface turns into the teacher education page, as shown in Fig. 1.2.

Teacher training which under the network environment, uses the computer network as a delivery platform, uses the network nodes as the basis for the resources sharing and services providing. The teacher can enter the teacher

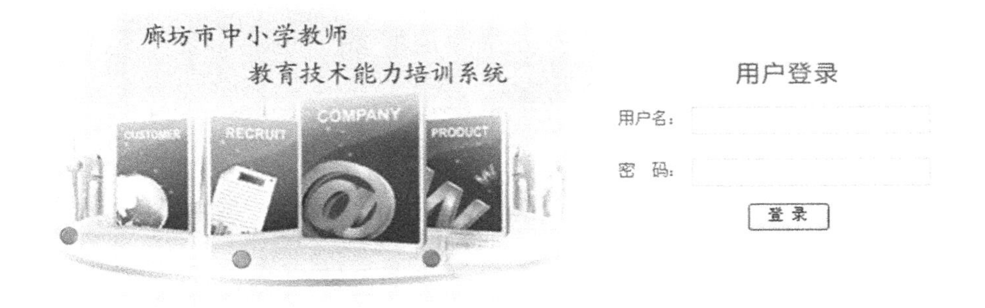

Fig. 1.1 Login module page

Fig. 1.2 Teacher education and training page

training system through the web portal to obtain needed services and carry out the training activities.

1.5 The Significance

Teacher education remote training broke through spatial temporal limitation, overcome the deficiency of the old training model. It provides a strong support for teachers to choose training content according to their own need. It also helps the teacher to grasp the learning progress, to accept lifelong education. At the same time, it reduces the burden of teachers and improves training effectiveness. It finds out a shortcut for the primary and secondary school teachers' Educational Technology ability training. It ensures the school-based training and teachers' continuing education work along the correct path. It provided a powerful guarantee for the professional development, ability promoting, and curriculum reform of teachers.

References

1. Nana C, Haodong Z (2008) Correct understanding of the transition from information technology training to technical skills training in education. Inf Technol Educ Primary Secondary Sch 12(01):550–557
2. Hailong W (2008) Our country middle and primary school teachers distance training present situation analysis. Leg Sys Soc 12:239–240
3. Guangzhen X (2010) "Learner-centered" of "participatory" teacher training practice and exploration. Primary Middle Sch Teach Training 3:22–23
4. Yan-li L (2008) Primary and secondary school teachers' educational technology ability training in experiential learning activity design. China Educ Inf 24:61–63
5. Kunpeng Y, Fanqi M, Caiyan W (2005) ASP.NET + SQL server dynamic web development from basic to the practice. Electron Ind Press 2(5):5–10
6. Haijing L, Zhen L (2007) The.NET framework asynchronous programming. Technol Square 11:194–195

Chapter 2
Classifier Selection Based on Support Vector Technique

Dajin Gao, Xiang-Sheng Rong, Xiang-Yang You, Ming Xu and Fujiang Huo

Abstract This paper presents an ensemble approach consisting of global SVM and local SVM. Global SVM is estimated according to its decision confidence. Local SVM handles the query whose global decision is of low confidence. Local SVM is constructed over query's neighborhood, which is developed under the guidance of an informative metric. And its training is based on a query-based objective function. Global SVM helps to define the new metric. Some heuristics proposed to specify neighborhood size and hyper parameters. We present experimental evidence of classification performance improved by our schema over state of the arts on real datasets.

Keywords Local classifier · Global classifier · Support vector technique · New metric

2.1 Introduction

As a qualified classification algorithm, SVM [1] has proved efficiency in a wide range of applications from pattern recognition to function regression, and time series prediction. Its basic idea of structural risk minimization [2] equips SVM with high generalization. In spite of the success in most cases, SVM can't provide qualified decision for some outliers or some ambiguous data that are located

D. Gao (✉) · X.-S. Rong · X.-Y. You
Training Department, Xuzhou Air Force College of P. L. A, Xuzhou 221000, China
e-mail: rxs12@126.com

M. Xu · F. Huo
Department of Logistic Command, Xuzhou Air Force College of P. L. A,
Xuzhou 221000, China

W. Du (ed.), *Informatics and Management Science VI*, Lecture Notes in Electrical Engineering 209, DOI: 10.1007/978-1-4471-4805-0_2,

around the decision interface. The reason lies in the generalization property of SVM to make it hold an overly high confidence in decisions on all inputs. However that space is not necessarily suitable everywhere, just like a global model is not reared to some local regions. A case of point is the point with $0 < f(x) < \delta$, assuming δ being a small value. Such a point takes much possibility to change its membership if being perturbed with small amount.

Essentially local SVM is in an adaptive and informative space, where the discriminate direction concerned with the query can be revealed. Global SVM helps to modify input space into new space over query's neighborhood, which is achieved by specifying hyper parameters of local space. The local property of new SVM is reflected in the objective function of local SVM. It adopts Kernel affinities as penalty coefficients of slack variables. Another point is the heuristic rules for global SVM hyper parameters which bring computation ease. The proposed approach is experimented on real datasets to find it does a better or competitive job than traditional SVM-based schemas and gives very competitive results compared with the state-of-the-art methods while using less computation cost.

2.2 Related Work

Three techniques used in this paper are given in brief. The first one is SVC. Let $x_i \in X$, $X = \Re^n$ be the input space, and Φ be the nonlinear transformation from X to the feature space. To find a minimum hyper sphere that encloses all data, the optimal objective function with slack variable ξi is designed as:

$$\min_{R,\xi} \quad R^2 + C\Sigma_i\xi_i \text{ s.t } || \Phi(x_i) - a||^2 \leq R^2 + \xi_i, \xi_i \geq 0 \tag{2.1}$$

a and R are the center and radius of sphere, C is the penalty parameter. Transfer its Lagrangian function into Wolf dual, and introduce Kernel trick, leading to:

$$\max_{\beta} \quad \Sigma_i\beta_i K(x_i, x_i) - \Sigma_{i,j}\beta_i\beta_j K(x_i, x_j) \text{ s.t. } \Sigma_i\beta_i = 1, \ 0 \leq \beta_i \leq C \tag{2.2}$$

Gaussian Kernel $K(x_i, x_j) = \exp\left(-q||x_i - x_j||^2\right)$ is used. Points with $\xi_i = 0$ and $0 < \xi_i < C$ are referred as non-bounded Support Vector (nbSV) and they describe cluster contours. Points with $\xi_i > 0$ and $\beta_i = C$ are bounded Support Vector (bSV).

*k*NN is a simple but attractive method. It labels the query as the most frequent class of neighborhood.

SA procedure [3, 4] is the third technique. SA obtains data spectral projections by eigen-decomposing a pairwise matrix H, which is usually the normalized affinity matrix. Select top p eigenvectors and form spectral embedding matrix S by stacking p eigenvectors in columns. Rows of S are data's spectral coordinates. Then it clusters spectrums with a simple method, and assigns point the same label as its spectrum.

2.3 NSVC Algorithm

NSVC uses the objective function of traditional SVC, but modifies its Kernel scale data-dependently so that data representatives (DRs) are extracted. Then SA is conducted on DRs to collect label information. Simultaneously, a new metric is defined, and this metric helps find query's NEI. That NEI is divided into sub neighborhoods (sNEI) in terms of classes. Each sNEI is enriched with convex hull technique. Label assignment is done in sNEI. The steps are:

1. Optimize: min $R^2 + C\Sigma_i\xi_i$, to produce {DRs}.
2. Conduct SA on {DRs}, to obtain labels and new distance definition $\|\cdot\|_{Sd}$.
3. For query $Q \neq SV$
4. Generate Q's NEI according to $\|\cdot\|_{Sd}$;
5. Divide NEI into sub region $sNEI_j$ for involved class j.
6. $esNEI_j$ = convex_hull($sNEI_j$).
7. Formulate weight w_j based on $esNEI_j$.
8. t_j = frequency of class j in NEI.
9. Label (Q) = $\max_j \{t_j \cdot w_j\}$.

Here, the affinity matrix H of SA is normalized in the fashion $H' = D^{-1/2} H D^{-1/2}$, where D is diagonal-shape with $D_{ii} = \sum_{j=1}^{n} H_{ij}$. p controls the number of selected eigenvectors, and it is ups to the max gap in the descending eigenvalue list [5]. K-means method is used to classify data spectrum coordinates.

2.3.1 Self-Tuning Kernel Scale

This paper investigates q in data local context. For point x we set its scale factor as: $\sigma_x = \|x - x_r\|$. To measure affinity between x and y, their scale factors are combined together to develop $q = 1/\sigma^2 = \sigma_x \cdot \sigma_y$. This leads to the tuning Kernel:

$$k(x, y) = \exp\left(-\frac{\|x - y\|^2}{\|x - x_r\| \cdot \|y - y_r\|}\right) \quad (2.3)$$

r is regarded as the size of the neighborhood. It is set as following steps: (a) Sort rows of Euclidean distance matrix $d\ (i,j)$ in an ascending order. (b) Let $gap\ (i) = \max_j \{d\ (i, j) - d(i, j-1)\}$. (c) r = average $\{gap(i)\}$.

Below is visual proof. For dataset nvSVs of tuning SVC. Clearly SVs are located on cluster contours and important positions where sharp changes of distribution density happen. They provide a sketch of dataset, and their *NEI*s are believed to cover the entire dataset. This justifies the feasibility of NSVC's labeling method.

2.3.2 Individual Setting Penalty Coefficient

In this paper C is parameterized by integrating diverse C_i values expected by points individually.

Set N as dataset size, and rows of the Kernel matrix $k\ (i,j)$ has been sorted in a descending order. With $Kgap(i) = \max_j\{k(i,j) - k(i,j+1)\}$, the definition of C_i for x_i is:

$$C_i = \exp(-\frac{aveIn(i)/aveAll(i) - 1}{1 - Kgap(i)/N}) \tag{2.4}$$

Here,

$$aveIn(i) = (\sum\nolimits_{j=1}^{Kgap(i)} k(i,j))/Kgap(i) \tag{2.5}$$

$$aveAll(i) = (\sum\nolimits_{j=1}^{N} k(i,j))/N \tag{2.6}$$

$Kgap(i)$ acts as the size estimate of neighborhood. $aveIn(i)$ is the average affinity of x_i neighborhood. $aveAll(i)$ is the average affinity of x_i to all other points. Both of them tell density information. C_i reflects x_i's individual demand to penalty term. So the global C is defined as the average of all individual C_i:

$$C = average\{C_i\} \tag{2.7}$$

2.4 Local SVM

When global SVM yields output of low confident, a query-based SVM is trained in query's neighborhood. That neighborhood is developed by a locally adaptive and informative metric, which is described next.

2.4.1 New Metric

Decision function of global SVM helps to derive new metric. Viewed under the light of theory, SVM decision function is optimal in the sense of structural risk minimization and therefore it is very desirable for seeking discriminates directions between classes. Viewed from geometry light, to any point x on level curve f(x) = 0, the gradient vector f'(x) reveals the perpendicular orientation along which data can be well separated over x's neighborhood. Integrate that orientation into metric definition and classification can b benefited a lot.

Without loss of generality, given query Q, we find its nearest neighbor P on the interface $f(x) = 0$ by:

$$\min_{P} \quad \|Q - P\| \text{ s.t. f(P)} = 0. \tag{2.8}$$

It is regarded that discriminates information is rich over P's neighborhood. But this optimization asks for extra cost, so we simulate P with Q and use $f'(Q)$ as the guidance to formulate new metric. Let $G_Q = f'(Q) = (G_{Q,1} \ldots G_{Q,\ n})$, then the new distance is defined as

$$||x - y||_{new} = \sqrt{(x-y)^T \mu (x-y)}. \text{ With } \mu_i = \frac{\exp(B \cdot |G_{Q,i}|)}{\Sigma_{j=1}^n \exp(B \cdot |G_{Q,j}|)} \tag{2.9}$$

Therein, the exponential mechanism is added to guarantee value stability. B controls the influence of elements G_Q on the whole weights. It is set as $B = 1/|f(Q)|$. The nearer Q is to $f(x)$, the more effect of G_Q should be strengthened, and the more informative local metric is. On the contrary, if Q is far to $f(x)$, B gets to zero and weight $\mu_i = 1/n$, which results to the original distance definition.

In M-classification, we create 1-vs-r SVMs, so M decision functions f_j and $f'_j(Q)$ are involved. These gradient vectors are combined in a weighted fashion. We design combination weights proportion to decision function values in the sense that the closer Q is to f_j, more influence $f'_j(Q)$ makes in formulating the final discriminant orientation. Let $G_Q^j = f_j'(Q)$, so the comprehensive orientation is defined:

$$G_Q = \sum\nolimits_{j=1}^{M} \quad (1 - \frac{f_j(Q)}{\Sigma_{l=1}^M f_l(Q)}) \cdot G_Q^j \tag{2.10}$$

In the new feature space spanned by $\|\cdot\|_{new}$, the Kernel is updated into:

$$k_{new}(x, y) = \exp(-||x - y||_{new}^2) \tag{2.11}$$

2.5 Experiment Results

2.5.1 Test New Metric

Firstly the quality of new metric is checked by introducing it into the kNN procedure to develop a classifier that probes query's *NEI* based on *Sd* and labels query with weighted voting strategy. That classifier is named WkNN. Six datasets are taken from UCI Machine Learning Repository. In Table 2.2, WkNN are compared on the average of 20 runs with following classifiers: (1) kNN. (2) SVMs of 1-vs-r version (SVM_{1r}). (3) SVMs of 1-vs-1 version (SVM_{11}) (4) C4.5 decision tree. (5) Machete [6]. (6) Scythe. (7) DANN. (8) Adamenn. Here, 30% data are sampled randomly for training.

Generally speaking, Adamenn achieves better performance than other classifiers by giving 4 optimal results of 6 experiments. It collects entire statistics about data distribution to define a globally informative metric, so its work is steady and fine. But its good behaviors are at the price of expensive consumption that is spent on tuning six parameters. W*k*NN achieves the optimal result in Banana, and follows Adamenn in other cases with a gentle distance to the optimal result. That demonstrates the validation of W*k*NN idea and the fine performance of this algorithm. If consider the computation ease brought by the self parameterization, W*k*NN is a more appealing choice in practice. DANN is on the second place. The metric employed by DANN approximates the weighted Chi-squared distance, which causes it fails in datasets of non-Gaussian distribution. Machete and Scythe are rooted from the same spirit, but the latter modifies the greedy nature of the former, so it improves the clustering accuracy in most cases.

Then it proceeds to SVM_{1r} and SVM_{11}. They don't depend on the deriving new metric, but on constructing the wise separating surface. They work well in the scenarios where classes are non-linear separated by mapping data into a feature space and then transferring the non-linear classification into the linear classification. Here, six datasets involved are mostly linear-separated, but with irregular cluster shapes. That data distribution permits their performance, so they can't play all potential power. C4.5 and kNN work poorly due to their greedy idea and unsupervised partition respectively.

2.5.2 Test NSVC

Now NSVC is performed and compared with some popular clustering algorithms: K-means; traditional SVC; Girolami method; and NJW. For each algorithm, the minimum number of incorrectly clustered points is documented. And we also present results of another NSVC version that is encoded with a width searching approach. We find the best clustering result by running over a specified range of $1/\sigma^2$. This method is named as search-NSVC. As to Wine data, the fact that 178 points cover 13 dimensions leads to the wide spreading information and the weak neighborhood information in the local context. So the tuning approach exhibits little help to refine affinities and NSVC produces a high error rate.

Between two versions of NSVC, surely search-NSVC is better that the tuning version. The difference between two versions is not large. Among other four methods, NJW does the best job. Search-NSVC is competitive with it, which verifies the capacity of NSVC algorithm idea. Girolami's performance follows search-NSVC, then SVC and K-means in turn. Their work has apparent gap with the above three methods. Girolami sometimes are affected by the unsteady optimization process. SVC's challenges lie in its expensive labeling process, whose randomness degrades SVC's final results. But the non-linear map hidden by Kernel makes SVC does better than K-means, the method depends only on input space

Table 2.1 Comparisons on classification error (%)

Data	Diabetes	Ionosphere	Banana	Liver	Sonar	Waveform
*k*NN	1514.7	5.92	13.6	31.5	12.65	18.4
SVM_{1r}	8.12	6.72	14.3	28.4	11.12	18.0
SVM_{11}	8.3	6.21	13.9	26.7	12	18.6
C4.5	10.52	6.84	14.6	38.3	23.1	23.95
Machete	9.6	5.63	12.76	25.5	21.2	22.3
Scythe	7.6	5.06	12.15	25	16.3	18.1
DANN	8.55	4.92	11.4	30.1	9.7	19.23
Adamenn	7.8	4.78	11.6	26.2	7.7	17.2
W*k*NN	7.92	4.88	11.3	26.67	8.14	17.8

Table 2.2 Comparisons on classification error on news group (%)

Dataset	K-means	Girolami	SVC	NJW	NSVC	Search-NSVC
(1)	12.0	15.8	12.9	17.7	13.1	14.4
(2)	13.79	15.8	14.2	15.92	11.07	8.03
(3)	7.5	6.4	4.96	8.1	4.92	5.3
(4)	5.95	5.1	5.3	7.3	6.5	5.2
(5)	5.8	3.2	3.98	6.4	5.83	7.74
(6)	4.9	7.8	5.9	6.55	4.72	7.68

information. K-means's behavior is moderate since it is heavily affected by data distribution, and depends on the hard partition classification idea.

The paper considers Facebook-Dataset, which is taken from Max Plank institute for software systems (http://socialnetworks.mpi-sws.org/data-wosn2009.html). This dataset includes two classes and the size of dataset equal to 57 kB. Each line of dataset contains of two unknown user identifiers, where the second user posts on the first user's Facebook wall. The other attributes are the number of frequent message, frequent post, frequent business, and frequent application. The number of samples equal to 2,700, which is divided to 2,430 (Training data = (2,700/10) ×9)) instances as a training data and 270 (Test data = 2,700−2,430) as a test data (Table 2.1). This dataset includes six attributes as listed in Table 2.2. All three-classification methods are implemented in Weka 3.7.4 software. The system requirements are Weka 3.7.4 software on Mac OS X version 10.6.8 with processor 2.4 GHz Intel Core 2 Duo, memory 4 GB, and 1,067 MHz DDR3 for comparing the percentage of accuracy of classifications. We take some data to form the experimental subsets: (1)–(6).

2.6 Conclusion

This paper presents a simple approach to estimate SVM output confidence. Using the confidence value as weights, WSVM schema works well in practical classification problems. AkNN deals with the difficult cases rejected by WSVM. It employs an informative metric to develop neighborhood. The hyper parameters are auto learned which facilitates computation. Experiments on real datasets evidence fine performance and efficiency of WSVM.

References

1. Cristianini N, Shawe-Taylor J (2000) An introduction to support vector machines, vol 10. Cambridge University Press, London, pp 381–388
2. Richard MD, Lippmann RP (1991) Neural network classifiers estimate bayesian a posteriori probabilities. Neural Comput 1(3):461–483
3. Craven M, DiPasquo D, Freitag D (1998) Learning to extract symbolic knowledge from the World Wide Web. In: Proceedings of 15th national conference on artificial intelligence, 8:880–886
4. Milgram J, Mohamed CSR (2005) Estimating accurate multi-class probabilities with support vector machines. In: Proceedings of IEEE international joint conference on neural networks, 3:1906–1911
5. Kwork TJ (1999) Moderate the outputs of support vector machine classifiers. IEEE Trans Neural Networks 10:1018–1032
6. Friedman JH (1994) Flexible metric nearest neighbor classification. In: Technical report, department of statistics, vol 6. Stanford University, pp 95–104

Chapter 3
Implementation of Remote Experiment Platform Based on Web Technology

Xu Liu and Hongyan Zhang

Abstract Distance education is a big leap in the field of education. It breaks through the limit of time and space, and provides very strong support for the universal education and lifelong learning. It has injected new vitality to the traditional way of teaching. Now the major schools' distance education platform also matures gradually, whenever and wherever we can all share curriculum resources of well-known colleges and universities. This paper mainly introduces the meaning of distance education, distance education platform function module, and remote education platform architecture, finally introduced the concrete implementation and testing.

Keywords Internet · Distance education platform

3.1 Introduction

With the continuous development of computer networks, a variety of Internet technologies are emerged. Distance education is an important part of the education system has become an indispensable means of education EFA and lifelong learning [1]. The continuous development of modern distance education subverts the mode and ideas of traditional education. Teachers and students are in the state

X. Liu (✉)
Qiqihaer Medical University, Qiqihaer 161006, China
e-mail: liuxu222@yeah.net

H. Zhang
The Third Affiliated Hospital of Qiqihaer Medical College, Qiqihaer 161006, China

W. Du (ed.), *Informatics and Management Science VI*, Lecture Notes in Electrical Engineering 209, DOI: 10.1007/978-1-4471-4805-0_3,

of separation in distance education, the role of teachers has a clear conversion. Learners have become selectors and users of educational resources from education objectors. This makes learning become independent thing [2–4].

3.2 The Significance of Distance Education

Distance education, also known as distance learning. The name suggests it is a long-distance teaching, the geographical boundaries are most significant features, it is the product of the modernization of network. Specifically, distance education via the Internet, breaking the geographical constraints, disseminates knowledge to remote or non-local region. It is different from traditional education which students don't need learning to the designated location, only need to connect to the network class and learn anytime and anywhere [5, 6].

3.2.1 Open

Distance education platform is open for the group who want to learn the course, regardless of age, regional or even nation. It is by means of the Internet and multimedia technologies to breakthrough learning limitations of space and time, all teaching content on the platform is open. Distance education is not restricted by time, any person at any time can directly open a Web page and carry out the study through distance education. Due to its openness, Distance education can be popularized widely public, provide personalized instruction courses for different people, and provide quality training services, which traditional education can't match.

3.2.2 Resource Sharing

Distance education platform provides a rich variety of teaching and learning resources, makes learners breakthrough restraints of geography, knowledge to enjoy shared learning resources and information. Distance education platform can integrate with existing technology, faculty, Knowledge to all learners, which not only saves the cost of learning and improves the utilization of educational resources, but also allows more people to get a higher level of education.

3.2.3 Independent and Flexibility

Through distance education, on the one hand, learners can choose interested content to learn, learner has more independent and flexibility; on the other hand, teachers arrange different learning programs according to students' different level of knowledge, teachers can independently design personalized curriculum according to their own experience. In addition, regardless of the learners and teachers are not constrained by time, they have flexible choice of teaching and learning time for teaching and learning.

3.2.4 Advanced Technology

The key reason why distance education has the openness, resource sharing, independent and flexibility which is unmatched by traditional education, is that relying on advanced technology, the implementation of distance education is to rely on computer technology, Internet technology, software technology, modern communications technology and so on.

3.3 The Demand for Distance Learning Platform

The distance learning platform has five main modules: resource management module, teaching support module, educational administration module, course management module, background module. We will introduce in the following.

3.3.1 Resource Management Module

The main function of the resource management module is divided into two blocks. First, teachers can upload the data which includes a variety of formats, documentation such as word, excel, txt, pdf, video data such as avi, rmvb, mkv, etc. Second, Students browse the learning materials. Students can view online information, watch instructional videos, download the data. Students and teachers could not be online at the same time.

3.3.2 Teaching Support Module

Teaching support module supports online teachers' teaching and students' lectures, teachers and students need online at the same time when they use the module, teachers explain the video at the network end, students ask questions at the other end in real time, and teachers answer students' questions in real time. Generally, students have a lot of questions need to instant communication with the teachers when exam is approaching; teachers need to release exam-date information to students. Video communication with students, not only can inspire students with confidence, but also can pass more emergency exam information.

3.3.3 Educational Administration Module

Educational management module is the core of the management in distance education platform which can manage teachers and students. Teacher management includes teacher qualification certification management, the approval of teachers applying for teaching support, the approval of the space to create, increase and modify, deleting teacher information management. Student management includes the approval of distance learning for students, the audit of student data.

3.3.4 Course Management Module

Course management module includes courses classification management and publication of course information. Course classification management includes course to add, delete, modify and other operations, management of course instructor information. It supports different courses distance learning according to different departments.

3.3.5 Background Module

The background module is mainly used by administrator who is the platform manager. Administrator logins background module on a regular basis, checks all the information, deletes the malicious files, compresses large video software, deletes unknown user, resets password for forgot password user.

We conduct a detailed description of user actions and teachers operations in the following.

Students operation: When student's login distance learning platform and order to study a course or courses of the department, the course documentation, video

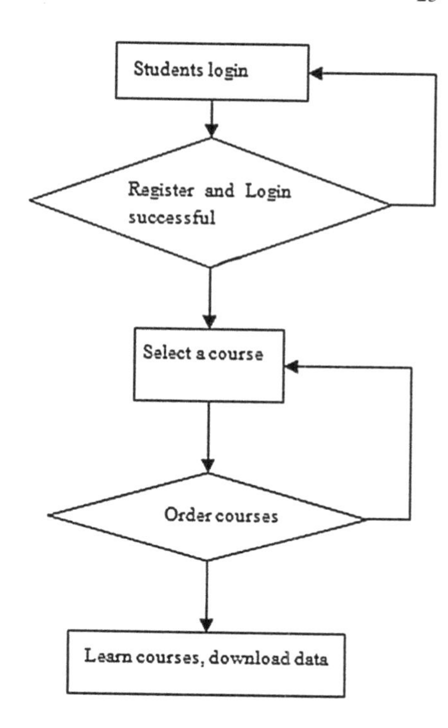

Fig. 3.1 Students operation flowchart

information and teacher information are presented to students, notice information about the course is also shown to students, and student can learn a specific course on demand. Each course has different levels of division, system will record learning courses. Students could learn the premise of this course before learning courses, which suggests students the difficulty of the course and so on.

Teachers operation: Educational administration module have developed the teachers' information in the allocation of curriculum to teachers, when teachers login remote education platform, it will list courses information taught by the teacher, be available to teachers for each course courseware upload, video upload is equal to the course curriculum data upload operation. The following is activities flowchart for students and teachers (Figs. 3.1, 3.2).

3.4 Design Architecture of Distance Education Platform

3.4.1 The Overall Framework

Under normal circumstances, the Internet platform-based design framework is based on B/S architecture. We use J2EE architecture in the platform, the specific architecture design is as shown in Fig. 3.3.

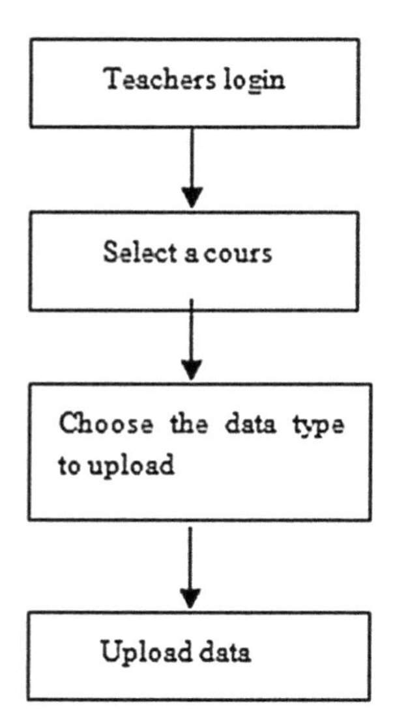

Fig. 3.2 Teachers operation flowchart

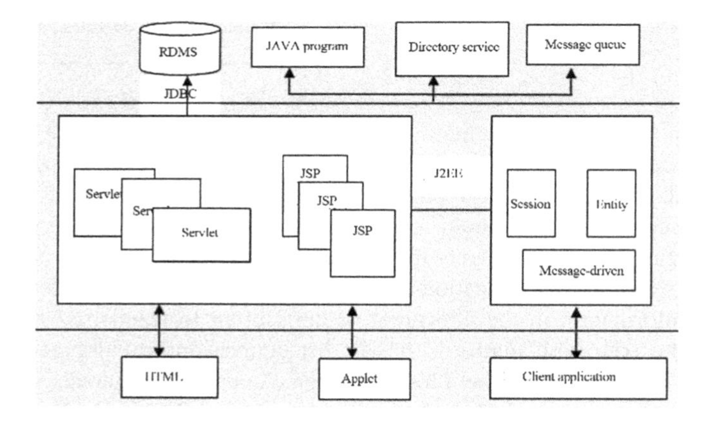

Fig. 3.3 Platform architecture

In the architecture of distance education platform, RDMS is a relational database, we use MySQL database as background database, the background database connection uses JDBC connection. Servlet is a server-side Java application program, has platform and protocol-independent features and could generate dynamic Web pages. JSP is java written web page. Session, entity and message-driven are three types of bean objects which is defined by EJB architecture (Table 3.1).

MyEclipse is J2EE development plug-ins in Eclipse, fully supports Struts, spring and Hibernate framework. It is a relatively completed and very powerful development environment.

Table 3.1 Operating environment

Operating system	Microsoft windows 2003
Support environment	JDK, tomcat6.0
Database	My SQL
Development tools	MyEclipse6.0

Tomcat is a free open source Servlet container, could deal with JSP pages and Servlet in addition to the traditional web server function.

MySQL is open source and free, is a small relational database. Compared with other large databases such as SQL Server, DB2 and Oracle, MySQL has its deficiencies, but the MySQL functionality provided more than enough for small and medium distance education platform. The biggest feature to use MySQL is that you can save costs, it is completely free.

3.5 The Distance Education Platform

3.5.1 Page Design of Distance Education Platform:

3.5.1.1 Interface Design

The user interface is designed to friendly interface, easy operation and reasonable layout. Platform page layout is divided to up, left and right. This layout is generous and concise, the top floor means the title bar which is divided into left and right columns, the left is tree navigation area, and the right is the list of information area which each function data reflected.

3.5.1.2 Page Style

The platform's overall background color is white, normal font is V Arial. The focus of the font is highlighted in blue to distinguish between the courses of play and not play, play courses with a blue font, and not play course with the default black font. The default size of the video is 600 $\times$ 600 pixels which could display full screen.

3.5.1.3 Page Specification

Paging must be standardized, placement of same operating buttons is relatively fixed, pages have different functions try to separate, which makes the page a strong logic. If the operation is an error, it gives the user a variety of error, warning, inquires.

3.5.2 Build Steps of Distance Education Platform:

1. Install and configure the jdk, after installing the jdk, type "java-version" to detect installation and configuration on command line window.
2. Install Tomcat, to detect whether the installation is successful installation, a successful installation page will be displayed.
3. Install MyEclipse6.0; after the installation is complete it is needed to enter the registration code to register. After successful registration, you need to set it up, bind with the tomcat and jdk.
4. Install SQL Server 2005.
5. Install the jtds-1.2.2.jar driver package.

3.5.3 The Design of the Database:

In addition to the two basic tables (Tables 3.2 and 3.3), there are many tables and stored procedures in mysql database. Main tables include course information table, student scores table. It has established referential integrity between tables.

Table 3.2 StudentInfo table

Serial number	Field	Name	Data type	Note
1	StudentID	Student's ID	Int	Primary key
2	Studentname	Student's name	Varchar (6)	Non-empty
3	Password	Password	Varchar (20)	Non-empty
4	Nickname	Nickname	Varchar (20)	
5	Coursenum	Number of selected courses	Int	Non-empty
6	Coursename1	Selected course 1	Varchar (20)	Non-empty
7	Coursename2	Selected course 2	Varchar (20)	Non-empty
...	...	...	...	...

Table 3.3 Teacherinfo table

Serial number	Field	Name	Date type	Note
1	TeacherID	Teacher's ID	Int	Primary key
2	Teachername	Teacher's name	Varchar(6)	Non-empty
3	Password	Password	Varchar(20)	Non-empty
4	TTCoursenum	Number of taught courses	Int	
5	Coursename1	Taught course 1	Varchar(20)	Non-empty
6	Coursename2	Taught course 1	Varchar(20)	Non-empty
...	...	...	...	...

3.6 The Test and Run of Distance Education Platform

We need unit test, stress test and release test after distance education platform is finished. Unit test for each functional module is mainly to click each module to test whether the link correctly, whether the interface deployment, whether page color visible, whether upload information could download and whether video could play, it needs specialized persons to test, and the submitted bug needs to be modified, then does the bug test. Stress test is there are a large number of students log on the website at the same time, which can withstand the load, this test needs to use specialized stress testing software to simulate a large number of user login. The release test refers to the site whether could normally browse in the outside network, and whether the publication is configured correctly.

3.7 Conclusion

The release and operation of the distance education platform, has great significance in the history of education, is a leap of innovation. It could not only solve the problem of remote areas of student to learn, but also make teacher' courses of major schools shared with the students. Education, regardless of rich or poor, regardless of geography, only needs a networked computer, you can learn with more teacher and more students to conduct extensive exchanges. With distance education platform, it brings very convenient learning environment and resources to a variety of learners, is the gospel of the learners.

References

1. Zuodong F, Qing D, Ruilin Z (2006) Design and implementation of distance education platform based on streaming media. Computer 03:132–135
2. Gengzhong Z, Qiumei L (2005) Design and implementation of distance education platform based on J2EE and XML schema. Mod Comput 11:69–71
3. Chen Y, Zhouya, Wang W (2005) Design and implementation of modern distance education network platform based on ASP technology. Comput Modernization 11:280–288
4. Guangrui Y (2003) Design and implementation of distance education teaching management based on J2EE component technology, vol 32(4). Northwest University. pp 92–97
5. Xiaoling H, Gaixue Y (2001) The development trend of modern distance education, vol 05. China Distance Education, pp 38–44
6. Mang W, Huimin M, Zhijing L (2001) The design of web-based distance education platform. Electron Sci Technol 01:44–53

Chapter 4
Design of Shopping Site Based on Struts Framework

Ruihui Mu

Abstract The construction of the shopping site is all this we must first consider its development we must first understand the background of the current information, domestic and international status of site development and popular information technology. The paper is starting from the development of a retail Web site, expand the function of the system and database on the basis of the analysis and design, the use of theoretical knowledge and develop technology based on the MVC Struts framework applied to the system implementation, the final completion of an online shopping website. The site can be achieved commodity browse, search, purchase; customers can give the website a message, view shopping help to achieve the basic requirements of a general shopping site. The paper put forward the development of shopping site based on struts framework.

Keywords Struts · Shopping site · E-commerce · MVC

4.1 Introduction

IT is the rapid development of online shopping to become a reality. The construction of the shopping site is all this we must first consider its development we must first understand the background of the current information, domestic and international status of site development and popular information technology.

R. Mu (✉)
College of Computer and Information Engineering, Xinxiang University, Xinxiang 453003, China
e-mail: mrh112@163.com

W. Du (ed.), *Informatics and Management Science VI*, Lecture Notes in Electrical Engineering 209, DOI: 10.1007/978-1-4471-4805-0_4,

E-commerce activities on the network to expand the company's corporate influence, increase traffic, improve customer satisfaction, has become a key consideration by many companies [1]. Which are built above with a well-functioning shopping site in the most efficient way, the selection of the best tools technology, rapid development, it is deployment, application shopping site as the current focus of the issue.

Mature models and frameworks, you can quickly build a well-structured, post-maintenance online sales system, in order to achieve a convenient, fast online shopping, shopping site is a typical Web application, and it is a distributed application system. Technically speaking, Web development techniques can be divided into two major categories of client-side technologies and server technologies.

Web server development technology have improved the development of complex Web applications become possible, so have the two most important enterprise-class development platform—J2EE and .NET. J2EE is a pure Java platform, including the three core technologies, Servlet, JSP and EJB. Java technology is platform independence, portability, and the support of many IT manufacturers.

Web client, the main task is to show the information content; the HTML language is an effective carrier of the information displayed. Cascading Style Sheets (CSS) and Dynamic HTML (DHTML) Web pages more wealth to the dynamic show. The technology is the first dynamically generated HTML pages CGI technology, now the mainstream server-side technology PHP, ASP, JSP, have produced and widely used. ASP also mainly running on top of Microsoft's server products, the PHP can run on Windows, Linux, Unix, Web server, but some code to make changes in the platform conversion, and the JSP can run on almost all platform. The paper puts forward the development of shopping site based on struts framework.

4.2 J2EE Platform, MVC Architecture and Struts Framework

J2EE as the development platform SUN Company is in order to promote the support of many IT vendors. The MVC architecture is the platform of a development model, as an important separation technology, so that the coupling of the system greatly reduced, while improving the reusability of the system. Struts are an implementation of the MVC architecture is a popular Web development framework, has been widely used [2]. This chapter introduces the J2EE platform, MVC architecture, Struts framework.

Model-View-Controller (MVC) is a classical architecture; the application is divided into three parts of the model, view, and controller. Model is the main part of the application, it is generally said that the business data and business logic to provide data for the view layer. A model can be used for multiple views, improve

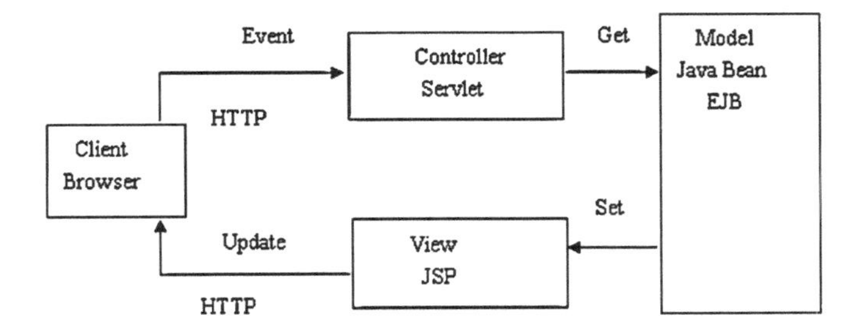

Fig. 4.1 J2EE model based on MVC architecture

the reusability of the system. View is the interaction with the user interface. Can be used to display information and accept user input. It generally does not conduct business processes, but can handle model changes the view changes to update the view synchronization, as is shown by Eq. 4.1.

$$\begin{cases} C_{j+1} = HC_jH' \\ D^k_{j+1} = GC_jH' \\ D^v_{j+1} = HC_jG' \\ D^D_{j+1} = GC_jG' \end{cases} \quad (j = 0, 1, 2, \ldots, J-1) \tag{4.1}$$

Controller accepts user's request, and then calls the appropriate modules and views to complete this request. Servlet in J2EE projects generally take up this task. J2EE application model based on the MVC architecture in Fig. 4.1.

The framework model is software architecture; it focuses on the description of the overall structure. This section briefly describes the components of the Struts framework, principle of operation and its characteristics [3].

The Jakarta-Struts is an open source project provided by the Apache Software Foundation, it provides a model for Web application—View—Controller (MVC) architecture. In the Struts framework MVC architecture from different components, Model constitutes a Java Bean or EJB components that implement business logic. Controller ActionServlet, and other auxiliary objects such as: Action, the Action Forward, Action Mapping to achieve, and view a set of JSP files. Struts concrete realization of the MVC framework in the form shown in Fig. 4.2.

Implement the view of the JSP file, there is no business logic, there is no model, only the label, these labels can be a standard JSP tag or customized labels (such as Struts comes with the label). Under normal circumstances, the Action Form bean in the Struts framework also divided into the View module [4]. The Action Form bean is actually a Java Bean, Java Bean conventional methods, it also contains some special method: such as Validate () method (used for verification of the submitted value), the reset () method (for Form in the value of re-assignment). Struts framework with the Action Form bean form data passed between the view and controller. View of the form is submitted, the controller automatically update and synchronize the Action Form bean.

The model represents the state of the application and business logic. For large Web applications, business logic is usually a Java bean and the EJB component, for a small program or the business logic is not complex processing, can be implemented in the Action class execute () method, as is shown by Eq. 4.2.

$$\begin{aligned} C_{j-1} &= H'C_jH + G'D_j^hH + H'D_j^vG + G'D_j^vG \\ (J &= J, J-1, J-2, \ldots, 1) \end{aligned} \tag{4.2}$$

The controller is composed of the ActionServlet classes and the Action class. In an application using the Struts framework, the ActionServlet object at run time is generally only one, and is automatically provided by the Struts framework.

ActionServlet in the Struts framework plays the role of the core controller, it receives an HTTP request, according to the configuration files Struts-config.xml configuration information, forwards the request to the appropriate Action object. Action object receives a request, call the execute () method. This method is accomplished through the use of the Java bean or EJB business logic. Processing business logic, the view of the user to switch to another view, as is shown by Eq. 4.3.

$$\hat{C}_J = (C_J^1 + C_J^2)/2 \tag{4.3}$$

Struts-config.xml is the form of an XML document storage requests are forwarded to the configuration file mapping information. The configuration information: the data source, the global forwarding, global exception, Action mapping, the Form the bean registered. ActionServlet through this file is in order to get configuration information [5]. Figure 4.2 is the icon of the Struts implementation of MVC. The browser makes a request to the Web container, Web container to transfer the request to the Struts framework controller ActionServlet is responsible for request processing, the controller to obtain configuration information from the Struts_config.xml file, and then by specific Action in the system to perform the business logic which may call for certain business logic processing system model of the Java bean or EJB, has been dealt with then the data to the appropriate JSP page, and finally the Web container Web page sent to the client, this process will be completed under the Struts framework the client and server interaction.

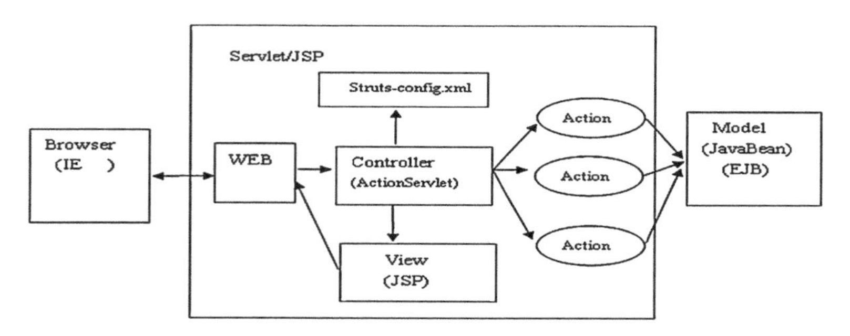

Fig. 4.2 Struts implementation of MVC framework map

Web applications using the Struts framework is first when it starts to load and the initialization AcitonServlet, AcitonServlet read from the Struts-config.xml application configuration information and store the information to a variety of used to the starting role in allocating the object(such as: ActionMapping, the Action-Forward). When the program runs, dynamic call these configuration objects.

A strut is open source software, which allows developers to in-depth understanding of its internal implementation mechanism. Framework of each piece has a strict division of labor, especially in the part of the controller, such as the Action to handle the business logic mapping Action Mapping processing, the ActionForward used to handle the steering configuration. In addition, the advantages of Struts are also reflected in: Taglib tag library and page navigation. Taglib Struts comes with the tag library; flexible use of the tag library can reduce the program code, and make the page simple. Struts page navigation by configuring the Struts-config.xml to achieve this configuration file in source code (i.e., in the form of an XML document) can graphically configure, it makes clearer the context of the system, for future maintenance, updates there are great benefits.

4.3 Description and Design of the System Requirements

After the existing traditional business processes, inspection and observation, and careful consideration, the analysis of business processes that may occur in the future, can get the system functional requirements as follows: (1) Goods browse. Show Web page to highlight the performance of commodities, goods page, basic information of the display of goods, such as the name, date of manufacture of goods, commodities, production companies, and other information. Have a page, at the same time, be able to view the detailed information of the goods, in addition to the basic information of the goods you want to add the detailed description of the goods, the evaluation of other customers for the goods, the goods from different sides, look that can be observed in pictures [6].

Search of merchandise, online store selling the goods can not within an unfolded all at the same time consumers are unwilling to view each category of each commodity, so on the home page should be able to retrieve all goods, the retrieval result is presented in a separate page. For retrieval, fuzzy queries, but also for different product features, keyword combinations queries. Query results, a page can not be fully displayed, paging, and display the total number of entries and results of the search results pages.

Sign on as a way of trading, the two companies know each other's information, at the same time as the mall party, to distinguish between different purchasers, and consumers buy goods records, generate orders, which require the user's information. User registration information stored in the database, this information is called when the user later in this web site to purchase goods is given by Eq. 4.4.

$$STD = \sqrt{\sum_{i=0}^{M-1} \sum_{J=0}^{N-1} (F(x,y) - MEAN)^2/(M \times N)} \quad (4.4)$$

Shopping help take into account that some customers are not familiar with online shopping, to design a page to display the entire shopping process to help users shopping, shopping among the common problems encountered while listed for reference. Feedback, user suggestions and comments on the company after the day-to-day operations will have a huge impact, and promptly collected customer feedback after the mall to provide quality, customer satisfaction is of great significance [7].

The shopping cart, the user site to buy something to temporary storage was Suoyu the purchase of goods. Design a simulated shopping cart, you can save, modify, and delete this temporary shopping information. Site news, site may have some announcements, promotions and other information you want to publish, to design a page to display the information published on the website, as is given by Eq. 4.5.

$$RMSE = \sqrt{\frac{\sum_{i=1}^{M} \sum_{j=1}^{N} [R(x_i, y_j) - F(x_i, y_j)]^2}{M \times N}} \quad (4.5)$$

Consider the system requirements analysis, the modular design of the site are as follows: (1) Users registered login module. Contains display registration agreement, registration information records, user login; Merchandise display module. Include general merchandise display, retrieve, and display detailed information; Shopping help module. Show the average customer shopping help; Press the display module. Display the site daily published notice, news, and so on.

The pros and cons of the system database design directly affect the after site with data connection speed and complexity of an update query [8]. This section describes the design of this system is based on the database the overall structure of the relational database design and data sheets.

The system's database design is named for the shop data write database scripts for later migration of the database. Take into account the initial amount of data for this site, began to run when the database file size to 10, 50 M, increments 5 M. Trigger for some of the operations of the database you want to write the script, to prevent the data does not change, so ensure the safety and accuracy of the data.

The system interface is the user interface; it is the user can directly see. Research in computer ergonomics, it is interface design and more attention. Design the user interface to fully take into account human factors, such as the user's characteristics, how the user interacts with the system; the user how to understand the output generated by the system as well as users of the system have any expectations, as is shown by Eq. 4.6.

$$PSNR = 10 \times \log_{10} \frac{MN[\max(F(x,y) - \min(F(x,y)))]}{\sum_{i=1}^{M} \sum_{j=1}^{N} [R(x_i, y_j) - F(x_i, y_j)]^2} \quad (4.6)$$

Needs relating to commodity price information, the user's personal information page of the site, the information should ensure their accuracy, so in the interface of the system to verify the processing of such information. The same time, the user's private information, to ensure the basic security of the information displayed on the page, such as hide the display to the user's password, the user's question to check before they can perform certain actions.

4.4 Site Specific Implementation of the Struts Framework

Struts framework to develop the introduction of the Struts framework in project development model, and then in accordance with the specification of this model system [9].

The JSP page is the most tedious part of the development, in Myeclipse development tools can drag completed part of the control, but most of the information display to use code written in the form. The JSP page of this site is mainly HTML tags, JSP tags, custom tags (mainly Struts tag), CSS and JavaScript [10]. This system is mainly used HTML tags, bean tags, and the Tiles tag. When the page is the introduction of the java bean, you can use the import attribute of the JSP page tags can also use the jsp label use bean property.

ActionForm is used to submit the Form data synchronization mapping class, if not dynamic Form Generally speaking, each ActionForm corresponds to a JSP page.System ActionForm: loginAgreeForm, noteForm, orderForm Forms 8, respectively, corresponding with the corresponding JSP page, of course, some pages may not need to configure the Forms.

In this system, the ActionServlet is the central controller, it is automatically provided by the Struts framework, so no programming is required to achieve the program. Struts-config.xml configuration file, which controls this site page flow, which set up the database connection information, global forwarding, global exception mapping relations, Action, ActionForm, the resource file location information when the application up and running ActionServlet will read information from the configuration file to determine the next step.

Struts-config.xml there are two editorial formats, one is shown above in source code form, another form of the graphics page. Editor in the graphics page which is more convenient, graphics drag configure. You can also right click, and then generate the associated Action ActionForm, the ActionForward.

In order to make the site pages to form a unified layout, this program use the Tiles tag library, so that the whole site to form a consistent interface. The layout page file the: layout. jsp. It formed the layout of other pages can be inserted into the internal, while reducing the complexity and difficulty of development of the

JSP page. For example, the development of the home page, which involves a lot of encoding, the Tiles framework, it can be split into a functional part of, and then the various parts of the preparation, finish writing the various parts of the rounds, and then insert a unified layout page can it.

The formation of a unified layout, the other page becomes very simple, the other page is essentially the same, simply insert one of the necessary pages to form the final page.(Message of the page) such as index.jsp (home page) and innote.jsp only slight differences in the content part into the local page inserts the contents of the index.jsp indexcontent.jsp, in innote.jsp content inserted note.jspdisplay different pages under the unified framework of the page.

4.5 Conclusions

The main features of this system are reflected in the following aspects: (1) Application configuration file of the Struts framework Struts-config.xml makes the transfer of control is very clear. In the future maintenance, change this configuration file can create a new page flow. (2) Using the Tiles framework development page, reduce the complexity of the page developed to maintain a uniform appearance and the system page. Based on the MVC pattern, the establishment of a reusable class reduces the duplication of development of the system.

The development process of the shopping site, select the appropriate development framework will not only speed up development, improve the quality of development, but also to improve program maintainability, reusability. The three-tier separation of MVC architecture reduces the coupling of the system. Presentation layer, model layer, business logic to assume the task of the layers, reducing the inter-linkages between developments of the complexity of the system greatly reduced. Struts framework, develop the corresponding JSP, ActionForm, Action, the first design page to submit the data, then generate the ActionForm Struts-config.xml graphics development, while allowing the system to automatically generate the JSP page, as long as a slightly modified, it can be used. Finally generate the corresponding Action class, which is a more appropriate order.

References

1. Chen L, Gao Q (2011) Research on framework developing technology based on MVC. AISS 3:25–31
2. Zhen L, Weiwei F, Konggui S, Feng L, Fangnan Y (2011) iTDTS: a SIP-based telephone system for train dispatching. JDCTA 4:88–100
3. Kwangmu S, Sunghwan C, Kidong C (2011) An efficient mode selection exploiting property of region in multi-view video coding. IJIPM 3:44–51
4. Liu G, Lou Z, Hou Y, Shen C (2011) Rural emergency system based on WebGIS. JCIT 2:342–346

5. Dan Z, Xiaoqing Z, Hongming C, Wei H (2011) Evaluation of customer value in E-commerce with 2-tuple linguistic information. JDCTA 11:95–100
6. Kun F (2012) Credit risk comprehensive evaluation method for online trading company. AISS 6:102–110
7. Xianqiang GUO (2011) The using of D-S evidence theory in building the trust model in B–C. JCIT 8:263–269
8. Hongxia R (2011) An approach to evaluating enterprise financial performance with linguistic information. AISS 11:398–404
9. Yiqun L, Cong S, Li X (2011) Information services platform of international trade based on E-commerce. AISS 1:78–86
10. Duolin L (2010) E-commerce system security assessment based on grey relational analysis comprehensive evaluation. JDCTA 10:279–284

Chapter 5
A Household Fetal ECG Monitoring System Based on Cloud Computing

Guo-jun Li, Nai-qian Liu, Linhong Wang and Xiao-na Zhou

Abstract Monitoring the fetal electrocardiogram (FECG) can provide important clinical information to obstetricians for the assessment of the fetal well-being. For the regular check-up of the fetal ECG, pregnant women have to come to hospital 2–3 times a week. It's difficult for the working women and the women with high-risk pregnancies. The great progress in sensor technology, wireless broadband communications and Cloud computing offers the possibility of real-time collection and dissemination of personal healthy data to patients and expert physicians anytime and from anywhere. This study presents a Cloud computing-based medical care network system for at-home FECG monitoring to enables the mother to complete the check-up themselves. This will help facilitate the regular checkups during pregnancy without compromise in the wellbeing of both the mother and the foetus.

Keywords Fetal ECG · Cloud computing · Household healthcare · Mobile monitoring · Wavelet Kalman filtering

G.-j Li (✉) · X.-n Zhou
Chongqing Communication Institute, Shapingba Chongqing 400035, China
e-mail: lgjsw@sina.com

G.-j Li · N.-q Liu
College of Communication Engineering, Chongqing University,
Shapingba Chongqing 400044, China

L. Wang
Chongqing College of Electronic Engineering, Shapingba Chongqing 401331, China

W. Du (ed.), *Informatics and Management Science VI*, Lecture Notes in Electrical Engineering 209, DOI: 10.1007/978-1-4471-4805-0_5,

5.1 Introduction

Nowadays, several technological advances such as Body Area Networks (BAN), pervasive wireless communications and Cloud computing are producing a great impact on our daily life for health take-care support, which benefit both the patient and the health experts. Specially, they enable the development of a system to perform remote real-time medical data collection and analysis [1].

The advantages of a pervasive health information system is evident to those who require long-term monitoring but often reside in remote non-clinical environment and have difficulty taking part in frequent therapy sessions [2]. Usually, pregnant women require attending regular check-up 2–3 times a week to assess the fetal well-being. It becomes a problem for women having a job and women with diabetes or other disease. This need motivates us to design a Cloud computing-based real time fetal monitoring system capable of aiding the mother to evaluate the fetal health condition by fetal ECG without the on-site consultations of obstetric professionals.

The fetal electrocardiogram (FECG) is an important approach to estimate the healthy condition of the fetus. The fetal cardiac electrical activity can be measured by placing electrodes on the mother's abdomen. As a result, FECG is completely noninvasive and can be done over extended periods. On the other hand, the most common Doppler ultrasound–based approach is apparently invasive since an ultrasound wave is directly beamed to the fetal cardiac. Moreover, it needs a hospital setting because a nurse is needed for the repositioning and skillful placement of the ultrasound probe [3]. Obviously, the fetal ECG is more feasible for the long-term at-home monitoring of pregnant women.

However, the FECG is largely distorted by variant sources of noise, where the most important source of noise is the maternal electrocardiogram (MECG) collected on the abdomen [1]. For the reliable extraction of FECG multiple abdominal recording channels are commonly utilized [3]. This multiple channels-based method means complicated electrodes configuration on the mother abdomen. Meanwhile, the recordings from each channel need to be transported to the web server, which requires more time to connect the network. For the facilitation of at-home FECG monitoring and the prolongation of the monitoring instrument's endurance, a single-channel FECG extraction method is proposed. It is great beneficial to aid the mother herself to complete the regular check-up and reduce the difficulty of developing monitoring device as well as the cost of the Cloud computing-based monitoring system.

5.2 System Architecture

The household FECG monitoring system is developed in Browse/Sever (B/S) mode. With Cloud computing the design of household monitoring instrument is much easier, which no longer analyzes the ECG data and stores the results. The

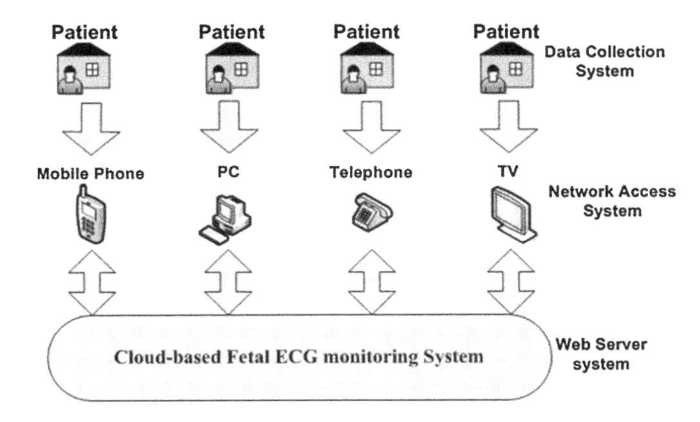

Fig. 5.1 Blockdiagram of the proposed cloud computing-based household fetal ECG monitoring system

client monitoring instrument only needs to collect the abdominal ECG signal and transport the single channel ECG data to the web server.

The extraction, analysis and interpretation of fetal ECG are implemented in the Cloud server. The proposed Cloud computing-based fetal monitoring system consists of three parts: data collection system, network access system and server system with Cloud computing. Figure 5.1 depicts the framework of a Cloud-based FECG monitoring system.

5.2.1 Data Connection System

A wireless broadband data acquisition system with 2.4 GHz frequency is designed to collect the maternal abdominal ECG signal and dissimilate it to local network access equipment, such as mobile phone, desktop and so on. The wireless data collection approach allows the mother can be more comfortable in the monitoring process without to be at rest [4, 5].

A low power-consumption microcontroller (C8051F930) controls the ECG data collection from the abdominal wall. Two effective 2.4 GHz wireless broadband communication modules (24LE1 and 24LU1) are employed to disseminate the ECG signal. Figure 5.2 illustrates the block diagram of the data acquisition system.

5.2.2 Server System with Cloud Computing

This study uses Elastic Compute Cloud (EC2) from Amazon company as the web server. Amazon EC2 is a commercial Cloud computer system that allows clients to elastically perform their enantiomorphism programs by the web service.

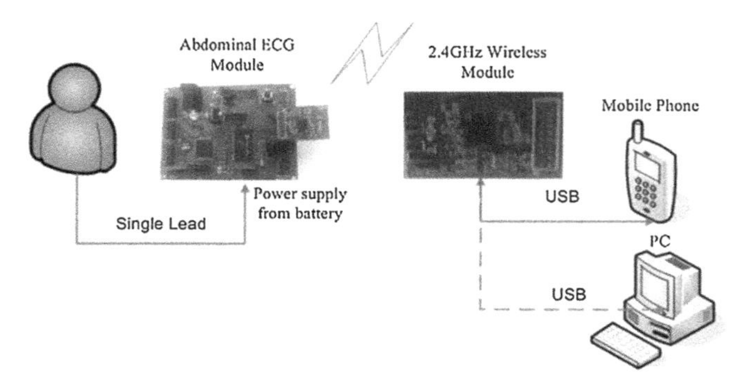

Fig. 5.2 Pictorial view of the abdominal ECG collecting system

All application programs including the FECG extraction, its analysis and the client database management, are implemented in the server with the framework of B/S. Patients are only required to enter an appointed web site by means of a certain network access device such as mobile phone, personal computer and so on. After validating the user name and the password, the patient can read the past fetal health recordings. After uploading the collected abdominal ECG signal, the fetal ECG waveform and fetal heart rate curve are automatically displayed on the web site. Meanwhile, the assessment results of fetal health conditions are also provided and stored in the client database. According to the fetal health situation, the regular check-up of the mother would be redesigned by the obstetric professionals.

5.3 FECG Extraction Algorithms

The extraction of fetal ECG is the most important for the assessment of the fetal well-being in the womb. For the reliable extraction of FECG and its simplicity, the Wavelet decomposition is cooperated into the Kalman Filtering framework. As a result, a Wavelet Kalman filter algorithm like the one proposed in [6] is introduced to extract the single channel fetal ECG signal by subtracting the maternal ECG form the abdominal ECG signal.

A Considering the maternal ECG has evident pseudo-periodical morphology with five characteristic waveforms (PQRST wave), five Gaussian functions are designed to model the maternal ECG signal and its pseudo-periodical characteristic is depicted by the phase of a unit circle.

$$ECG = \sum_{i \in \{P,Q,R,S,T\}} \alpha_i \exp(-\frac{(\varphi_k - \theta_i)^2}{2\beta_i^2}) \tag{5.1}$$

By forward differential of expression (5.1), the recursive representation of maternal ECG signal can be obtained as follows.

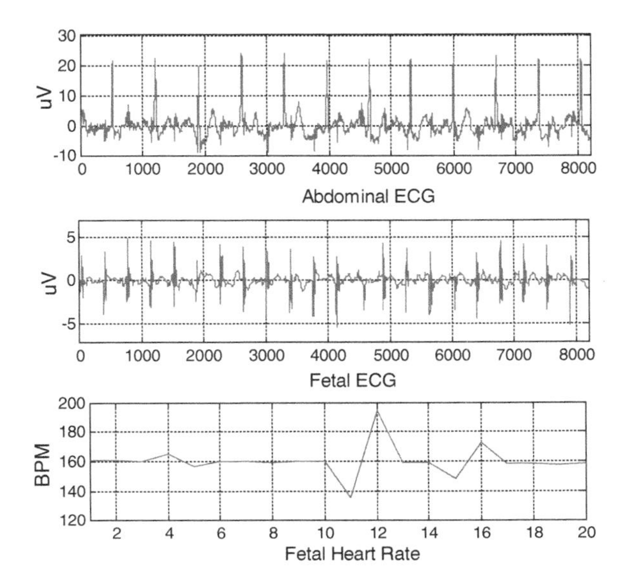

Fig. 5.3 Fetal ECG extractions with wavelet Kalman filtering algorithm

$$\begin{cases} \varphi_{k+1} = \varphi_k + \omega\delta \\ Z_{k+1} = -\sum\limits_{i\in\{P,Q,R,S,T\}} \delta \frac{\alpha_i \omega}{\beta_i^2} \Delta\theta_i \exp(-\frac{\Delta\theta_i^2}{2\beta_i^2}) + Z_k + \eta \end{cases} \tag{5.2}$$

According to the differential expression of the maternal ECG signal, the corresponding state-space model can be established as the reference, with which the Wavelet Kalman filtering can be iteratively implemented [7]. A representative example for the fetal ECG detection using Wavelet Kalman filtering is shown in Fig. 5.3.

5.4 Results and Discussion

Experimental results of both simulated and realistic signals indicate that the Cloud computing-based household FECG monitoring system is helpful to the working women and having high-risk pregnancy. Meanwhile this remote surveillance framework is also shown to be useful to the busy labor and delivery unit. However, the performance of the assessment of fetal well-being largely depends upon the FECG detection algorithm. It has been found that the proposed Wavelet Kalman filtering algorithm would fail to extract the FECG when its amplitude is very small during the pregnancy of 28–34 weeks.

In this case, other characteristic waveform from the fetal cardiac activity such as the heart sound signal, could be cooperated into the fetal cardiac electrical signal to improve the reliability of fetal health conditions evaluation, which will be studied in the future.

Acknowledgments This study is supported by the National Natural Science Foundation of China (60971016), the Fundamental Research Funds for the Central Universities of China (CDJXS10 16 11 13, CDJXS11 16 00 01), the Research Project of the Education Committee of Chongqing (KJ112201, KJ110508) and the Natural Science Foundation Project of CQ (cstc2011jjA40047, cstc2012jjB40010).

References

1. Suraj P, William V, Sheng N, Ahsan K, Rajkumar B (2012) An autonomic cloud environment for hosting ECG data analysis services. Future Gener Comput Syst 28:147–154
2. McGregor C, Eklund JM (2010) Next generation remote critical care through service-oriented architectures: challenges and opportunities. SOCA 4:33–43
3. Mischi M, Sluijter RJ (2007) A robust fetal ECG detection method for abdominal recordings. Physiol Meas 28:373–388
4. Zhu Q, Wang M (2005) A wireless PDA-based electrocardiogram transmission system for telemedicine. Proc 2005 IEEE Eng. Med Biol 4:3807–3809
5. Hoang DB, Lingfeng C (2010) Mobile cloud for assistive healthcare (MoCAsH). IEEE Asia Pac Serv Comput 1:325–332
6. Zheng T, Girgis AA, Makram EB (2000) A hybrid wavelet-Kalman filter method for load forecasting. Electr Power Syst Res 54:11–17
7. Sameni R, Shamsollahi M, Jutten C, Clifford G (2007) A nonlinear bayesian filtering framework for ECG denoising. IEEE Trans Biomed Eng 54:2172–2185

Chapter 6
Research of Multi-View Stereo Reconstruction for Internet Photos

Zhiyuan Wang

Abstract This paper develops a multi-view stereo approach to reconstruct the shape of a three-dimensional object from a set of Internet photos. The stereo matching technique adopts region growing approach, starting from a set of sparse three-dimensional points reconstructed from structure-from-motion (Sfm), then propagates to the neighbouring areas by the best-first strategy, and produces dense three-dimensional points. We demonstrate our algorithms that they perform robustly.

Keywords Internet photos · Multi-view · Stereo reconstruction

6.1 Introduction

With the technology development of digital photos and the price decrease of portable cameras, the family photos obtain the popularity that increases the large amount of our door photos. There is established a large number of internet photo sets after sending the pictures on the sharing website. On the one hand, these photos have the wide covering square that nearly include the building, sculptures and other places of interest all over the world. A great number of photos express the same scene in the different view, time (four seasons, one day from day to night), illumination condition (cloudy day, sunny day, etc.) in order to obtain low cost [1]. On the other hand, these photos have not measured the illumination condition; the cameras are not passing the standard. The distinguish ability is

Z. Wang (✉)
Image Processing Center, School of Astronautics, Beihang University, Beijing, China
e-mail: zhiyuanwang1@163.com

W. Du (ed.), *Informatics and Management Science VI*, Lecture Notes in Electrical Engineering 209, DOI: 10.1007/978-1-4471-4805-0_6,

different with mussy dodging problems and brings inconvenience to the usage [2]. In recent years, in order to dig and utilize the information on internet photos, decrease the data selection cost, rich the data source and the fast establishment of three-dimensional scenes, the scene modelling of the internet photo set have been the research hotspot of the computer graphics and computer vision filed that based on the internet photo set.

The multi-view stereo reconstruction uses multi-view to shoot the same scene with many pictures in order to establish the three-dimensional model. The basic thinking is using the dependence information between image and image [3]. To reconstruct the scene three-dimensional model, most methods are using the available images at the same time. With the images numbers' increasing, this method will face great impact. For each seed point selects the suitable image set as the active image set while estimating the neigh boring three-dimensional points, this article provides the new image selection frame. It means each seed points choose one standard image, use the standard image to calculate the neigh boring image in order to replace all the images. This can avoid the large repetitive computation and improve the algorithm effectiveness.

6.2 The Common Method of Multi-View Reconstruction

In Fig. 6.1, the present multi-view reconstruction is commonly through the depth map of each input photo. Then fuse the depth map to obtain the final reconstruction. However, the depth images have the noise and there have lots of calculations for each depth image. Therefore, these methods are usually used for the post processing in order to eliminate the noise and the depth image fusion. Lots of perfect multi-view reconstruction methods directly obtain the object or the scene model through reconstruction in the input images [4]. The global method avoids the redundancy calculation and do not need the post processing. However, it will limited by the scale.

Based on the multi-view reconstruction, lots of writers' successfully find out the various methods of multi-view reconstruction [5]. Aim at the long image sequence, the basic thinking of multi-view reconstruction is calculating the depth map of

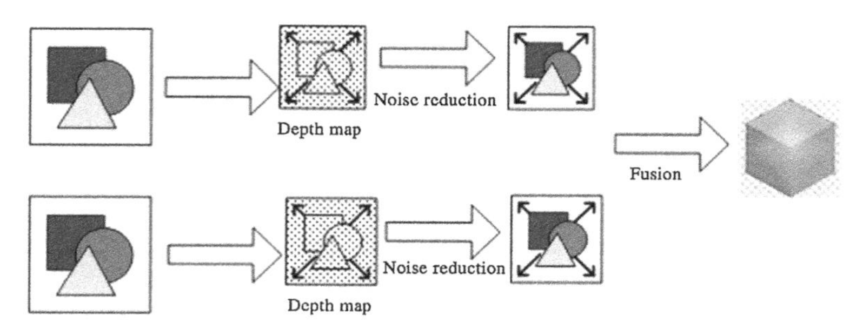

Fig. 6.1 The common method of multi-view reconstruction

each input image. Then, do the noise reduction that based on the neigh boring images. At last, calculate the rid model based on the fused depth map. Aim at the multi-view reconstruction of the internet photo set and aim at the selection of baseline and image resolution ratio, we can control the correctness of depth map.

6.3 The Method in this Article

From Fig. 6.2, we can find out the multi-view reconstruction methods of the internet photo set in this article have two parts. They are pre-processing and multi-view reconstruction. In the part of pre-processing, first the article uses the re-projection error to reject the sparse three-dimensional points in order to obtain the high correctness. Moreover, treat the left sparse three-dimensional points as the seed points. Then, the image layer selection needs each image to select each input image should have the enough characteristic matching point and satisfy the scale and basic requirement of neigh boring images. The multi-view reconstruction will use a region growing method to estimate the three-dimensional points in the neigh boring region of seed points. In order to improve the robustness of stereo matching during the region growing, this article provides the layer image selection of three-dimensional point. Each seed point will select the suitable photo in the input image set as the standard image. Otherwise, for the result correctness, this article also supports the quality test in order to estimate whether the new three-dimensional point can satisfy the setting condition. If it is not satisfied, the new generated three-dimensional will be rejected.

6.3.1 Pre-Processing

This article input the sparse three-dimensional point that produced by SfM for the fast reconstruction result and determine the high correctness. Before the reconstruction, we need to prepare the relative works. Pre-processing has two parts: filter the spare three-dimensional points and image selections. The detailed introduction is in the following.

6.3.1.1 Filter the Spare Three-Dimensional Points

In this article, the input image set is {Ii} and operates SfM. The created camera parameter and spare three-dimensional points are {Ci} and {Pj}. At the same time, the visible image set of each three-dimensional point Pj is {VjK}. SfM uses SIFT method to abstract and match the characteristics. For the repeat texture or texture problems, SIFT can create the error characteristic point or error matching during the characteristic matching. That will lead the inaccurate three-dimensional point creation. During the process of multi-media reconstruction, this article treats spare

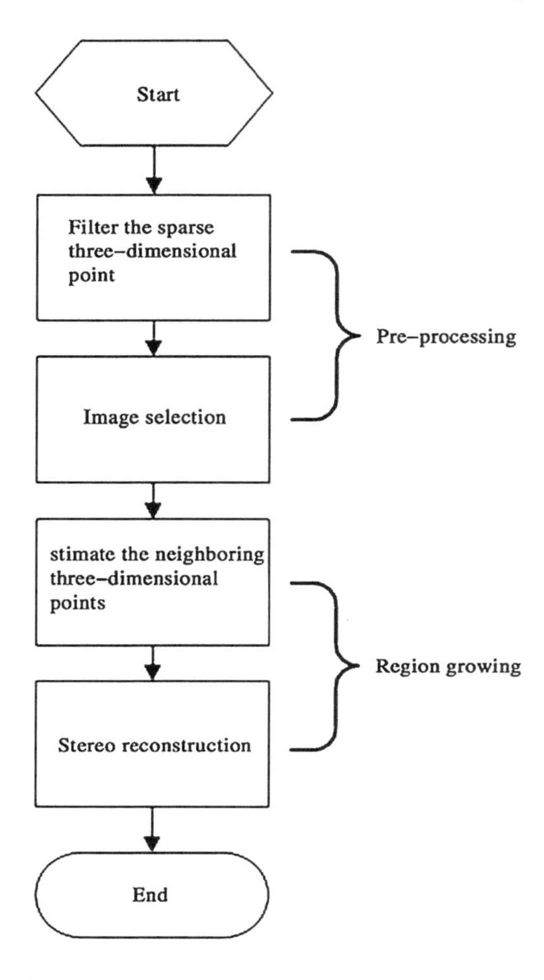

Fig. 6.2 The method of stereo reconstruction in this article

three-dimensional point {Pj} as the input. In order to determine the high correctness of dense reconstruction, we need to filter {Pj} at first, reject the possible noise point.

Judge every three-dimensional point is Pj is the nose point, we need to calculate the re-projection error of each visible image VjK. If k makes $|fjK - CjK(Pj)| > \lambda$ (fjK is the characteristic point of Pj that in the kit image, λ is the appoint threshold). In usually, 1.0 means Pj is the noise point that we need to reject it. In this article, after filtering the noise point, the three-dimensional set is {Pj'}.

6.3.1.2 Image Selection

So many multi-view reconstruction methods need to utilize the entire input image. With the image number increase, the reconstruction speed and the result will obtain the greatest impact. Therefore, in order to improve the method efficiency and robustness, it is very important to select the suitable image set for each three-dimensional point. The image selection in this article will use two parts: the image selection of the image layer and the image selection of the three-dimensional point layer.

The image selection of image layer means each input image selects several neighboring images in the input image set in order to form the set. These neighboring images are finished under the similar sunlight, weather, and exposure. The detailed expression is the similar appearance and resolution ratio, enough characteristic matching, and some basic line requirement. Based on the above condition, this article uses the global image selection to select the image layer which provided by Goesele et al. [3].

The image selection of the three-dimensional point layer has two steps: the first is select one image Pj for each three-dimensional point as the standard image. Then, search the satisfied image subset in the neighboring image set of standard image. This subset is the active image set during the stereo matching process.

The standard image selection is before the optimization of each three-dimensional point. The target is searching the image subset A from the neighboring image set of standard image as the active image subset during this three-dimensional point optimization. We use the following method to select the standard image of each three-dimensional point. First, calculate the camera direction ligature between three-dimensional point and each visible image. Then, calculate the angle between the ligature and the normal of this three-dimensional point. At last, select the image that opposite with the smallest angle as the standard image R.

Image layer selection means select the nearest image set N of each image. While estimating one detailed three-dimensional point information, in order to speed up the depth calculation, here we do not consider about any image in the neighboring image set of standard image. We choose the smaller image set $A \subset N$ as the active image set of this point (in common, $|A| = 4$). The given primary depth and ligature information about the three-dimensional point, during the optimization process to update the opposite image set A of this three-dimensional. The images in this set can satisfy the photometric consistency. The photometric consistency means the same three-dimensional point's projective point is the different views has the similar attribute (such illumination, etc.) distribution. In order to quantitative describe the photometric consistency in the different views of the same three-dimensional point; this article uses the method of mean-removed normalized cross correlation. Compare the reference image R of this point with the projection pixel of candidate image V. They are similar. If the NCC value is larger than the given threshold, the candidate image V will add into the set A.

6.3.2 Regions Growing

This article adapts the multi-view stereo matching that based on the region growing. The basic thinking is optimized expand the most stable matching point and matching region during the matching process. The algorithm first finds out the "seed point". These points will help investigating the seed point neighbor and expand the matching region follow the strategy.

The region growing algorithm is in the following. The primary condition we use optimized sequence to store the spare three-dimensional point cloud. And the three-dimensional point cloud will sequence the confidence key word from high to low. Select the highest confidence from the optimized sequence as the seed point. Its projection pixel of the standard image is the center. Select n*n pixel window, initialize the neighboring pixel that opposite the three-dimensional point of following the seed point information. Optimize the energy functional equation, makes the different photos have the maximum similarity in the opposite pixel windows then to calculate the new three-dimensional point. Different from the traditional region growing, this article provides the restriction test that based on the depth consistency constraint. It means to test the new three-dimensional. If the condition is satisfied, the new three-dimensional point will add into the optimized sequence. Otherwise, reject this three-dimensional point. Repeat the above process until no new three-dimensional point production. When do the iterative optimization estimating about the neighboring three-dimensional point of seed point, this article uses the method that provided by Goesele et al. [3]. It adopts the SSD method to measure the similarity of the projection pixel on the active image set. Moreover, add the illumination factor to measure the same object variation under the different illumination conditions in the model.

From the beginning, this article adds the present three-dimensional point into the optimized sequence in proper order. Each three-dimensional point P includes the following information. P = {X, R, A, n, CF}.

In there, X is the location information f three-dimensional point as the three-dimensional coordinate under the world coordinate system. R means the standard image of the three-dimensional point after image selection and algorithm. Stands for the active image set of the three-dimensional point after image selection and algorithm. N stands for the ligature of the three-dimensional point. CF means the confidence of the three-dimensional point. The higher confidence means the more reliable of the three-dimensional point. The confidence of three-dimensional point is the reliable degree of the three-dimensional point. The confidence can balance the similarity of the rejection pixel in the different views. Because the two matching elements are created by the same unit rejection, they have the very similar attribute or the attribute distribution is close. For example, the color attribute, channel information and distribution of gray values. In this article, we use ZNCC to measure the three-dimensional confidence. This means the ZNCC has great adaptability of the image appearance difference. The detailed expression is the partial light variation has invariance that can suit a certain extent of the illumination variation.

This article uses the region growing to extend the searching of new three-dimensional points. The thinking of this method starts from the least commitment principle of artificial intelligence. It means the unreliable strategy should make the final decision after obtaining the enough information. The advantage of the growing method is the speed. Compared with the global optimization, it has few algorithms. However, the region growing cannot correct the error seed point and these seed points will lead the further matching problems. During the experiment, we find the error of region growing has the following two conditions. First, the processing of energy functional solving cannot constrict. Although the process can successfully constrict, they're opposite the error matching. The second condition is the successful constringency will create the new three-dimensional point. If there exists the error matching in these three-dimensional points, and add them to the present set without test, it will bring the important impact of the further growing process. In order to avoid the error matching expand in the unreliable area, this article uses the filter. Test the result of successful optimization one more time. The filter test based on the parallax continuity constraint, it means the continue depth variation of the neighbor three-dimensional points will have not a great leap.

6.4 Summary

This article provides the method of multi-angle solid reconfiguration that based on the internet picture set. It is based on the region growing. Different from the traditional method of region growing, this article provides restrain test strips. It means do the further test of the new estimated three-dimensional points while growing. Reject the error three-dimensional point; select the traditional method that cannot manage the error three-dimensional points in order to avoid the further fit error. It can greatly improve the correctness of three-dimensional model. In addition to this, we provide the new image selection frame. Each seed three-dimensional point selects one image as the basic image. While estimation the new three-dimensional point, use the basic standard image to determine the suitable image subset as the active image. Replace all the input images can greatly reduce the iterative optimization in order to develop the solid reconstruction effectiveness.

References

1. Furukawa Y, Ponce J (2007) Accurate, dense, and robust multi-view stereopsis. Proc CVPR 20(3):487–493
2. Zhao J, Wang Q, Wang X (2006) The real-time three-dimensional animation of highway that based on the digital terrain model. Comput Eng 31(12):196–199
3. Goesele M, Snavely N, Curless B, Hoppe H, Seitz SM (2007) Multi-view stereo for community photo collections. ICCV 92(4):198–207
4. Liu Y, Li J, Huang J et al (2005) Establish the three-dimensional road model from the plan and transverse section. J Huazhong Univ Sci Technol (Urban Science Edition) 22:156–158
5. Otto GP, Chau TKW (1989) 'Region-Growing' algorithm for matching of terrain images. Image Vision Comput 7(2):83–94

Chapter 7
Information Retrieval Scheme Based on Fuzzy Ontology Framework

Rongbing Wang and Na Ke

Abstract In enterprise management information system, user can obtain information from data resources by semantic querying based on ontology. To achieve fuzzy semantic retrieval, this paper has presented information retrieval system based on fuzzy ontology framework. The framework includes three parts: concepts, properties of concepts and values of properties, in which property's value can be either standard data type or linguistic values of fuzzy concepts. The framework is the extension of RDF data model "object-property-value", which is the current standard for establishing semantic interoperability on the Semantic Web.

Keywords Information retrieval · Ontology · Semantic web

7.1 Introduction

As a standard of the knowledge representation of semantic web is use of anthologies to overcome the limitation of the proposed keyword-based search as a semantic web motivation [1, 2]. However, typical of the ontology concept formalism support does not enough to represent not sure information are common in many fields of application, due to the lack of clear boundaries of the concept of the field. Dealing with uncertain information and knowledge, a potential solution is fuzzy theory in ontology. Then we can generate fuzzy selections include fuzzy concepts and fuzzy membership. Lau [3] put forward a kind of fuzzy domain ontology for business knowledge management. Lee [4] proposed an algorithm, and the application of the fuzzy ontology create news summary. Put forward a kind of

R. Wang (✉) · N. Ke
Liaoning University, Shenyang, China
e-mail: rbwang20111@126.com

W. Du (ed.), *Informatics and Management Science VI*, Lecture Notes in Electrical Engineering 209, DOI: 10.1007/978-1-4471-4805-0_7,

fuzzy ontology total generation framework (FOGA) produce not deterministic fuzzy ontology degree of membership information [5]. Abulaish [6] put forward a framework of fuzzy ontology, a concept descriptor is represented as a fuzzy relations between coding of attribute value degree with fuzzy membership functions.

7.2 Fuzzy Domain Ontology Model

Ontology organizes domain knowledge in terms of concepts, properties, relations and axioms, and the fuzzy ontology is created as an extension to the standard ontology.

Definition 1 (*Fuzzy domain ontology*)—fuzzy domain ontology is a 6-tuple $Q_F(I, C, P^C, R, P^R, A_F)$, where:

1. I is the set of individuals, also called instances of the concepts.
2. C is a set of concepts. A concept is often considered as a class in ontology. Every concept here has some properties whose value is fuzzy concept or fuzzy set. And, every concept can have the degree of membership $c(i) : i \rightarrow [0, 1]$ and the degree of non-membership $v_c(i) : I \rightarrow [0, 1]$ of the $i \in I$ in C.
3. P^C is a set of concepts properties. A property $P^C \in P^C$ is defined as a 5-tuple of the form $p^C(c, v_F, q_F, f, U)$, where $c \in C$ is an ontology concept, v_F represents property values, q_F models linguistic qualifiers, which can control or alter the strength of a property value v_F, f is the restriction facets on v_F, and U is the universe of discourse. Both v_F and q_F are the fuzzy concepts on U, but q_F changes the fuzzy degree of v_F.
4. R is a set of inter-concept relations between concepts. The relation type is not only the ordinary binary relation of $r \in C \times C$, but also is the fuzzy relation and the intuitionist fuzzy relation from C to C.
5. P^R is a set of relations properties. Like concept properties, $P^R \in P^R$ is defined as a 4-tuple of the form $p^R(c_1, c_2, r, S_F)$, where $c_1, c_2 \in C$ are ontology concepts represents relation, and $S_F \in [0, 1]$ models relation strengths and has meaning of fuzzy set or intuitionist fuzzy set on C × C, which can represent the strength of association between concept-pairs $c_1, c_2 >$.
6. A_F is a set of fuzzy rules. In a fuzzy system the set of fuzzy rules is used as knowledge base.

7.3 Fuzzy Linguistic Variable Ontology

Fuzzy language variables of the most basic put forward both-branch fuzzy knowledge and fuzzy system. A language variable is a variable; its value is not a number, but a word. Each language value on fuzzy set, each of which has certain membership functions. A subordinate function values describes these of language of Numbers.

Language variables and their membership functions allow fuzzy logic reasoning as numerical performance does not accurate, performed by humans.

Definition 2 (*Fuzzy linguistic variable ontology*)—A fuzzy linguistic variable ontology is a 6-tuple $Q_F(c_a, C_F, R, F, S, U)$ where:

1. c_a is a concept on the abstract level, e.g. "price", "speed" etc.
2. C_F is the set of fuzzy concepts which describes all values of c_a.
3. $R\{r|r \in C_F \times C_F\}$ is a set of binary relations between concepts in C_F. A kind of relation is set relation R_S {inclusion (i.e. $\in$), intersection, discontinues, complement (i.e. $\in$}, and the other relations are the order relation and equivalence relation $R_O \in \{\leqq, \geqq, \in\}$,$C_F$ and an order relation r compose the ordered structure <C_F, r>. There are other semantic relations between concepts, such as semantic distance relation, semantic proximity relation and semantic association relation etc.
4. F is the set of membership functions on U, which is isomorphic to C_F.
5. $S\{s|s : C_F \times C_F \rightarrow C_F\}$ Is a set of binary operators at C_F. These binary operators form the mechanism of the generating new fuzzy concepts. Basic operators are the "union", "intersection" and "complement" etc., i.e. $S\{\vee, \wedge,\}$. C_F And S compose the algebra Structure C_F, S.
6. U is the universe of discourse.

Modeling the linguistic qualifiers, we extend the fuzzy linguistic variable ontology as follows.

Definition 3 (*Extended fuzzy ontology*)—An extended fuzzy ontology is a 9-tuple $Q_F(c_a, C_F, R, F, S, Q, O, L, U)$, where:

1. c_a, C_F, R, F, S, U Have same interpretations as defined in definition 5.
2. Q is the set of the linguistic qualifiers, e.g. Q = {very, little, close to …}. A qualifier $q \in Q$ and a fuzzy concept $c_F \in C_F$ compose a composition fuzzy concept that can be the value of c_a, e.g. "very cheap".
3. O is the set of fuzzy operators on U, which is isomorphic to Q.
4. $L \in (Q \times C_F) \cup (C_F \times Q)$ is a binary relation from Q to C_F or C_F to Q, <q, c_F> or <c_F, q> $\in$ L main the $q \in Q$, and $c_F \in C_F$ can compose a composition fuzzy concept.

To simplify the transform from fuzzy linguistic variables to fuzzy ontology, we introduce the basic fuzzy ontology model as follows.

Definition 4 (*Basic fuzzy ontology*)—A basic fuzzy ontology is a 4-tuple $O_F(c_a, C_F, F, U)$, where c_a, C_F, F, U have same interpretations as defined in definition 5, which satisfy the following conditions:

1. $C_F\{c_1, c_2, \ldots, c_n\}$ Is a limited set.
2. Only one relation of set, the relation of discontinues, exists in C_F, and C_F is complete on U. In the other words, C_F is a fuzzy partition of U?
3. C_F Has an ordered relation $\leqq$, and $< C_F, \leqq <$ is a complete ordered set, i.e. all concepts in C_F constitute a chain $c_1 \leqq c_2 \leqq \cdots \leqq c_n$.

7.4 Fuzzy Ontology Framework

Combining fuzzy domain ontology with fuzzy linguistic variable ontology, we obtain the three-layered fuzzy ontology framework shown in Fig. 7.1.

The framework comprises the set of concepts and relations, set of properties and set of fuzzy linguistic variable anthologies. The relation between concept and property is "property of", and the relation between property and fuzzy linguistic variable ontology is "value of", in which property's value can be either standard data type or linguistic values of fuzzy concepts. The framework is the extension of Resource Description Framework (RDF) data model "object-property-value", which is the current standards for establishing semantic interoperability on the Web [7]. Since considering the essential semantic relationships between fuzzy concepts, the framework facilitates the information retrieval at semantic level.

7.5 Fuzzy Semantic Information Retrieval

Enterprise management information system, and pay attention to the relations between people, and project and production. Have a lot of from human factor fuzzy phenomena. This will be enough to allow the user to get some information in linguistic value and not accurate numerical. For instance, the linguistic values for production level include "top-ranking", "high-class", "advanced" etc., and

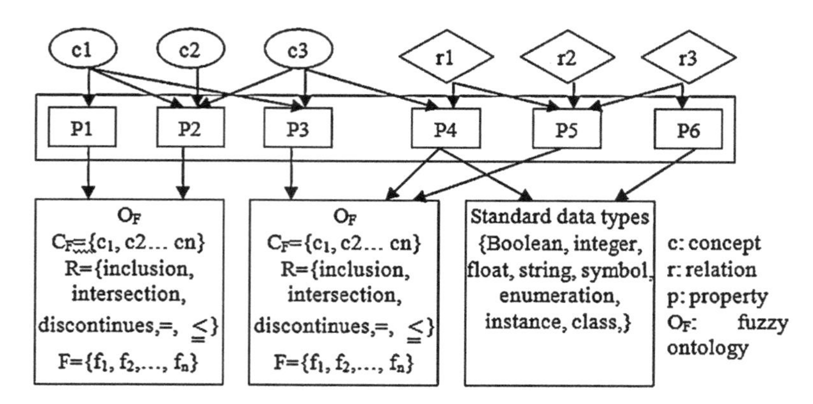

Fig. 7.1 Three-layered fuzzy ontology framework

linguistic values for project size include "small", "medium", "huge" etc. These linguistic values have uncertainty and are fuzzy concepts.

Using the three-layered fuzzy ontology framework, we construct the knowledge ontology structure for person, project and production shown in Fig. 7.2, in which the linguistic values are represented formally through fuzzy linguistic variable ontology. The main fuzzy linguistic variable ontology are as following:

O1 = (age of person, {old, middle-aged, midlife, youth ...});
O2 = (research ability of person, {very weak, weak, neutral, strong, excellent});
O3 = (project size, {very small, small, middle, medium, big, very big, huge});
O4 = (project type, {basic research, application research, development research});
O5 = (production level, {top-ranking, high-class, advanced, average, behindhand});
O6 = (production benefit, {very small, small, middle, medium, big, very big, huge}).

There is a lot of semantic relation between fuzzy concepts. For instance, "top-ranking" = "high-class", "top-ranking" ≦ "advanced", "very weak" ≦ "weak" ≦ "neutral" ≦ "strong" ≦ "excellent", "very small" ≦ ~"small" ≦ "middle (medium)" ≦ "big" ≦ "very big" ≦ "huge", "basic research" = {"application research", "development research"}, etc.

Since the process for information retrieval is based on the knowledge ontology, the semantic and concept research can be achieved. Especially, using linguistic value of fuzzy concept, we can construct the research pattern such as: SELECT instance of concept FROM Data source WHERE (property of concept) <comparison operator> "Linguistic value of fuzzy concept", in which the comparison operators includes: equal comparison (=), unequal (≠), less than or equal (≦) and greater than or equal (≧) etc.

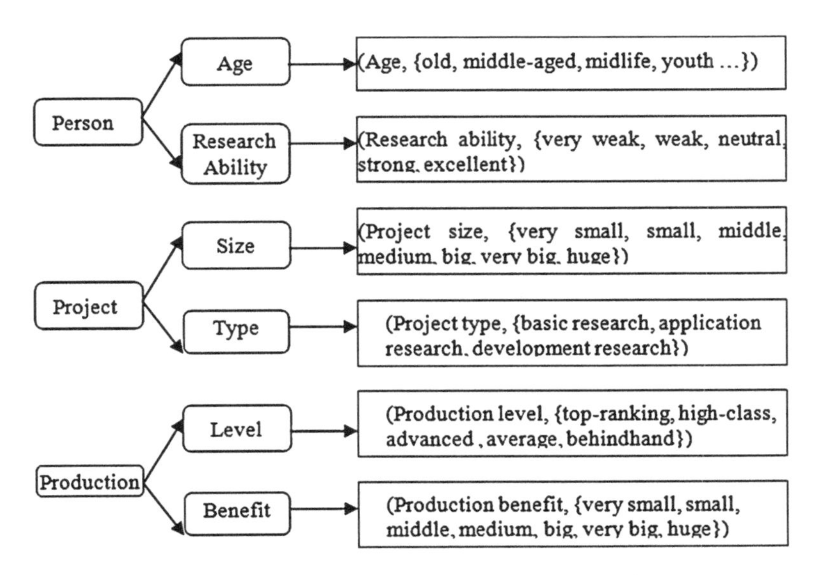

Fig. 7.2 Knowledge ontology structure for person, project and production

For instance, we can retrieve project information through production benefit of property, using the search statement such as: SELECT Project (ID, name, source,…) FROM Data source WHERE Project. Production. Benefit ≦ "big". Using the "order relation" defined in fuzzy linguistic variable ontology: "very small" ≦ "small" ≦ "middle(medium)" ≦ "big" ≦ "very big" ≦ "huge", we can transform the search statement to: SELECT Project (ID, name, source,…) FROM Data source WHERE

Project. Production. Benefit = "very small" or
Project. Production. Benefit = "small" or
Project. Production. Benefit = "middle" or
Project. Production. Benefit = "medium" or
Project. Production. Benefit = "big", in which every sub-condition is ordinary and can be completed easily in SQL engine of DBMS.

When retrieving the project information by the statement: SELECT Project (ID, name, source…) FROM Data source WHERE Project. Production. Level = "top-ranking", using equivalence relation: "top-ranking" = "high-class", we can transform the search statement to: SELECT Project (ID, name, source…) FROM Data source WHERE Project. Production. Level = "top-ranking" or Project. Production. Level = "high-class".

When retrieving the project information by the statement: SELECT Project (ID, name, source…) FROM Data source WHERE Project. Production. Level = "advanced", using inclusion relation: "top-ranking" ⊆ "advanced", we can transform the search statement to: SELECT Project (ID, name, source…) FROM Data source WHERE

Project. Production. Level = "advanced" or
Project. Production. Level = "top-ranking" or
Project. Production. Level = "high-class".

Using the "complement relation" defined in fuzzy linguistic variable ontology: "basic research" = {"application research", "development research"}, when retrieving the information about project by the statement: SELECT Project (ID, name, source…) FROM Data source WHERE Project. Type "basic research", we can transform the search statement to: SELECT Project (ID, name, source…) FROM Data source WHERE Project. Type = "application research" or Project. Type = "development research".

7.6 Conclusions

Information retrieval is an important work for enterprise information management in the WWW. And the semantic information retrieval based on ontology is currently a hot issue. Ontology is the concept of a domain into a human understandable, machine readable format from entity, attributes, relationships, and justice system. Semantic query expansion has set up the concept of fuzzy semantics relations, be helpful for information retrieval in semantic level.

References

1. Fensel D, van Harmelen F, Horrocks I, McGuinness DL, Patel Schneider PF (2001) OIL: an ontology infrastructure for the semantic web. IEEE Intell Syst 16(2):38–45
2. Castells P, Fernandez M, Vallet D (2010) An adaptation of the vector-spacemodel for ontology-based information retrieval. IEEE Trans Knowl Data Eng 19(2):261–272
3. Lau RYK (2010) Fuzzy domain ontology discovery for business knowledge management. IEEE Intell Inf Bull 8(1):29–41
4. Lee CS, Jian ZW, Huang LK (2009) A fuzzy ontology and its application to news summarization. IEEE Tran Syst Man Cybern (Part B) 35(5):859–880
5. Tho QT, Hui SC, Fong ACM, Cao TH (2006) Automatic fuzzy ontology generation for semantic web. IEEE Trans Knowl Data Eng 18(6):842–856
6. Abulaish M, Dey L (2006) Interoperability among distributed overlapping ontologies-A fuzzy ontology framework. In: Proceedings of the 2006 IEEE/WIC/ACM international conference on web Intelligence, vol 26. Hong Kong, pp 397–403
7. Silvia C, Davide C (2007) Fuzzy ontology and fuzzy-OWL in the AON project. In: Proceedings of 2007 IEEE international conference on fuzzy systems conference, vol 32. London, UK, pp 1–6

Chapter 8
Semantic Web Service Discovery Based on FIPA Multi Agents

Wanli Song

Abstract In this paper we propose a framework for semantic Web service discovery that communicates between multi agent system and Web services without changing their existing specifications and implementations by providing a broker. We explained that the ontology management in the broker creates the user ontology and merges it with general ontology (i.e. WordNet, Yago, Wikipedia ...) and recommends the created WSDL based on generalized ontology to selected Web service provider to increase their retrieval probability in the related queries. In the future works, we solve inconsistencies during the merge and will improve matching process and will implement the recommendation component.

Keywords Semantic web services · Web service discovery · Multi agent system

8.1 Introduction

The World Wide Web is the evolution of the sea from service oriented market information and network services (WS) technology is the next wave of Internet computing. Web services are one of the fastest growing areas of information technology in the growing in recent years. Network services of business process exposed to the Internet more business opportunities and commitment by providing

W. Song (✉)
College of Computer and Information Engineering, Hohai University, Nanjing, China
e-mail: wanlisong11@126.com

W. Song
Electronic Information Engineering Department, Nanjing Communications Institute of Technology, Nanjing, China

W. Du (ed.), *Informatics and Management Science VI*, Lecture Notes in Electrical Engineering 209, DOI: 10.1007/978-1-4471-4805-0_8,

a common agreement can use web applications and other nerve cells exchange website [1].

The semantic web service objective was to describe [2] and implementing network service, which makes them more close to the automation of the agent. The method that is the core of represents a network service function clear, the use of the so-called semantic tagging [3]. The semantic web service recently proposed technology integration business process automation. The creation, deployment and call services to meet the needs of individuals and society almost all fields of human endeavor is a kind of culture characteristic of [4].

Combining network services and software agent of sight benefits brought about by application connection held one or another technology: network service should be able to call agency service, and vice versa. However, once the interconnection is established the concept of software agent and technology will help make the new, advanced operation and using web services form [5].

In this context, take the network service is the key component interoperability promote business process and the key technology of software agent as a dynamic found that such process, combining and execute [6].

That is, although the semantic web is committed to providing the content of the program of the meaning of web pages, these entities, only will not be able to make the decision, the interact, and cooperate with other entities. Agent infrastructure, on the other hand, it can provide a lot of problems. Agents have ability to understand and interact with its environment. Because the environment sensitive, autonomy, will help explain the semantic knowledge representation, the ontological agent is a necessary complement, network service, in order to realize the semantic web vision. Several debate is to support the idea of combining the network services and agency infrastructure, including [6] but perhaps no one more moving than the statement, in [7] to make clear the concept, "software agent to realize WS drive running program-two and access them as of computing resources represents a person or organization".

8.2 Proposed Framework

We propose a framework (Fig. 8.1) for semantic Web service discovery based on the technology of agent and semantic Web services. The main part of our framework is a broker which provides semantic interoperability between semantic Web service provider and agents by translating WSDL to DF description for semantic Web services and DF description to WSDL for FIPA multi agents.

Broker comprised of following components:

Control Unit (CU) is responsible for communicating with FIPA multi agent and semantic Web service to provide Web services to them. From agents it will receive queries regarding searching and publishing, in ACL format. Broker executes those queries on DF and sends reply in ACL message. From semantic Web services it will receive queries regarding searching and publishing, in SOAP format. Broker translates SOAP message into ACL format and then execute translated queries on

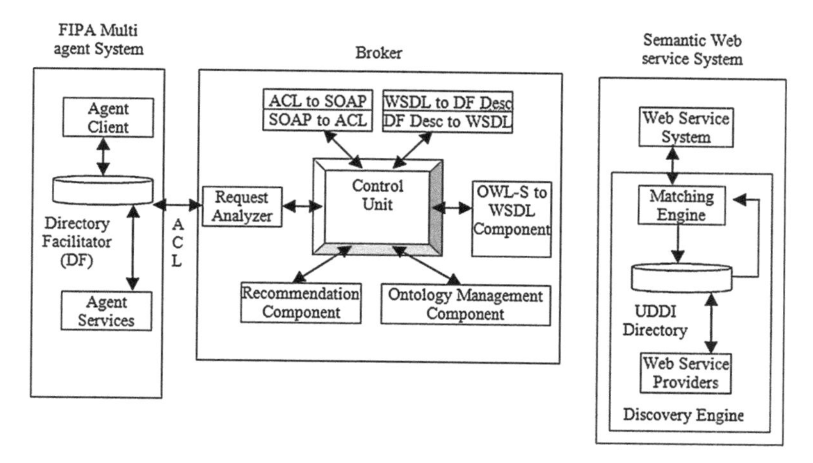

Fig. 8.1 Proposed framework for semantic web service discovery

DF and send reply in SOAP format. WSDL to DF description component is used by CU to translate WSDL to equivalent DF description.

DF description to WSDL component is used by ontology agent to translate DF description to equivalent WSDL.

SOAP to ACL component is responsible for conversion of SOAP message to equivalent FIPA ACL message.

ACL to SOAP component is responsible for translation of FIPA ACL message to equivalent SOAP message.

Ontology management component is responsible for creating the user ontology and merging it with general ontology (i.e. WordNet, Yago, Wikipedia …) and creating new generalized ontology.

OWL-S to WSDL component used to translate new generalized ontology in OWL-S to equivalent WSDL.

Recommendation component is used to recommend generalized WSDL to selected Web service providers.

8.2.1 Assumptions

The following assumptions were made when designing the framework:

All agents are assumed to be FIPA compliant and capable of communicating with FIPA-ACL encoded messages.

All Web services operate using the standard WS stack consisting of WSDL for service descriptions, SOAP for message encoding and UDDI for directory services.

The matching engine and a mechanism of Web service discovery known as discovery engine, ACL to SOAP component and SOAP to ACL component,

WSDL to DF and DF to WSDL component and invocation from agent to semantic Web service infrastructure is assumed to exist.

The framework is registered as an agent service in FIPA Directory Facilitators and as a Web service endpoint in UDDI directories.

8.2.2 Communication of Semantic Web Service with FIPA Multi Agent System

Figure 8.2 describes the communication between semantic Web service and FIPA multi agent system. Whenever SOAP based Web service client needs some service, it performs lookup for the service in UDDI, if the UDDI doesn't have the required service, it redirects its search to the middleware by sending a simple SOAP based UDDI search request. As soon as the broker CU receives message, it passes it to request analyzer. It extracts out three major parameters based on Web services basic functional semantic description:

$$\text{Fb} = < \text{object}, \ \text{action}, \ \text{constraint} >$$

where object expresses the object of a service, Action expresses the behavior which is put on to one object. Constraint expresses the restraints must be obey when service is used. Functional-constraint that is composed of some sub-constraints is as in (1).

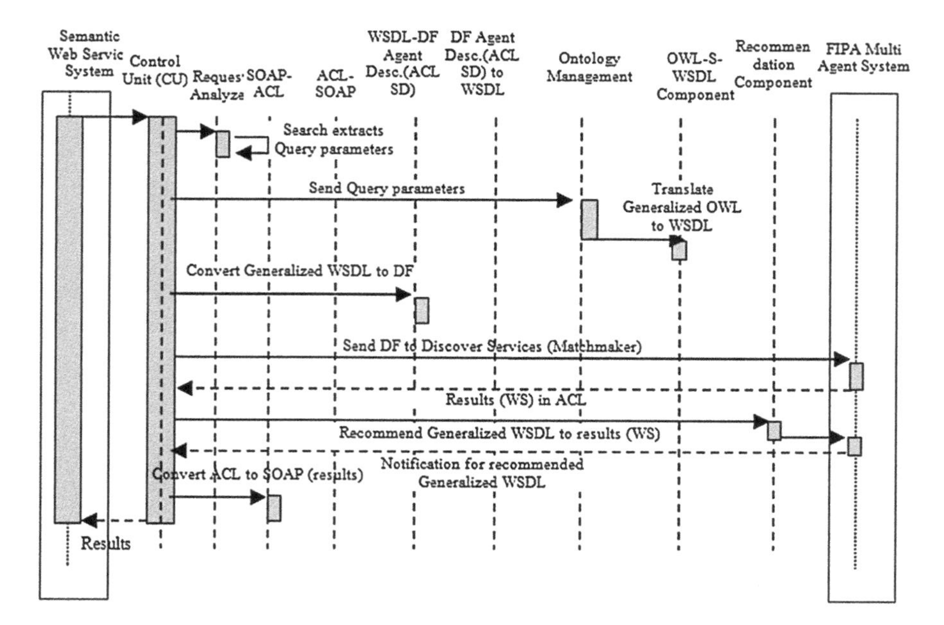

Fig. 8.2 Semantic web service communication with FIPA multi agent

$$\text{Constraint} = R1 \wedge R2 \wedge \ldots \wedge Rm$$

$$R := (\text{Term}, \text{Type}, \text{Value}, \text{PRI}), \ x = 1\ldots m.$$

The sub-constraints are connected by logical operators. Any sub-constraint Rx is a quaternion. $Type_x$ expresses the data type of $Term_x$. It can be numerical, string and some other types. $Value_x$ expresses the range of $Term_x$. PRI^x expresses the priority of the sub-constraint. The request also can be expressed according to the F_b definition [8].

CU sends extracted parameters to ontology management component. Figure 8.3 describes whole scenario of ontology management component. It receives extracted search query parameters and creates a user ontology based on main parameters and then it creates new generalized user ontology to discover appropriate Web services by merging user ontology with general ontology (i.e. WordNet, YAGO, Wikipedia …).

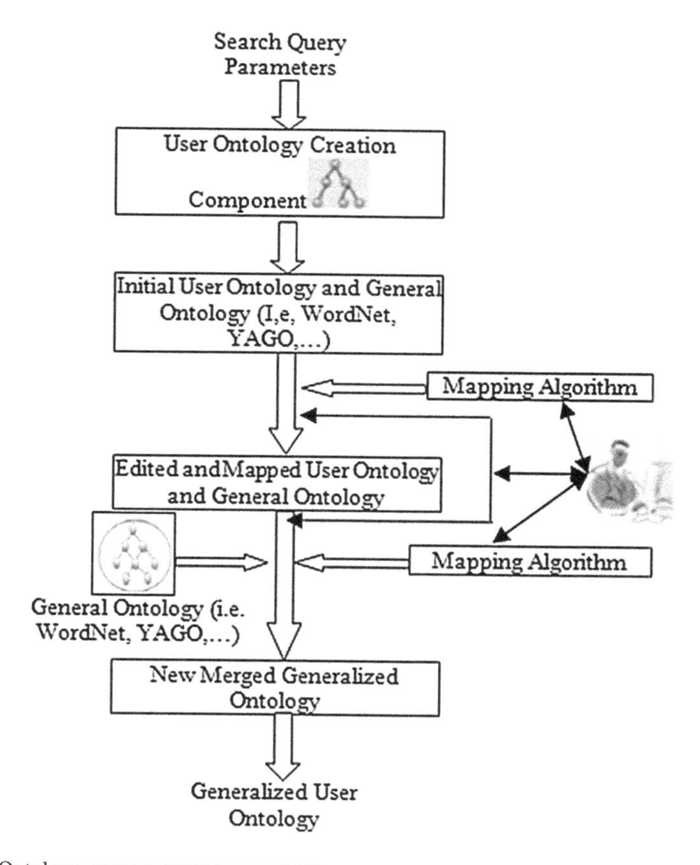

Fig. 8.3 Ontology management components

We add this component because the matching engine matches Web service description based on WSDL of Web services.

The created WSDL is forwarded to WSDL to DF description component and transformation is performed from WSDL of Web services into a form DF agent description that software agents can understand. DF description published and searched in FIPA multi agent system. If required service is found then DF will return the selected and ranked Web services to CU till passed to Web service client and recommendation component.

After this step, the created WSDL sent to recommendation component. Recommendation component recommend the created WSDL to selected Web service providers, because if advertised Web services describe themselves more complete, their retrieval probability in the related queries will increase.

Finally SOAP to ACL component converts ACL result message to SOAP message that is understandable for semantic Web service system.

8.2.3 Communication of FIPA Multi Agent System with Semantic Web Service System

In this section we have explained that how FIPA multi agent system communicates with semantic Web service (Fig. 8.4). Whenever a software agent searches some service, it performs lookup for the agent in Directory Facilitator (DF) of multi

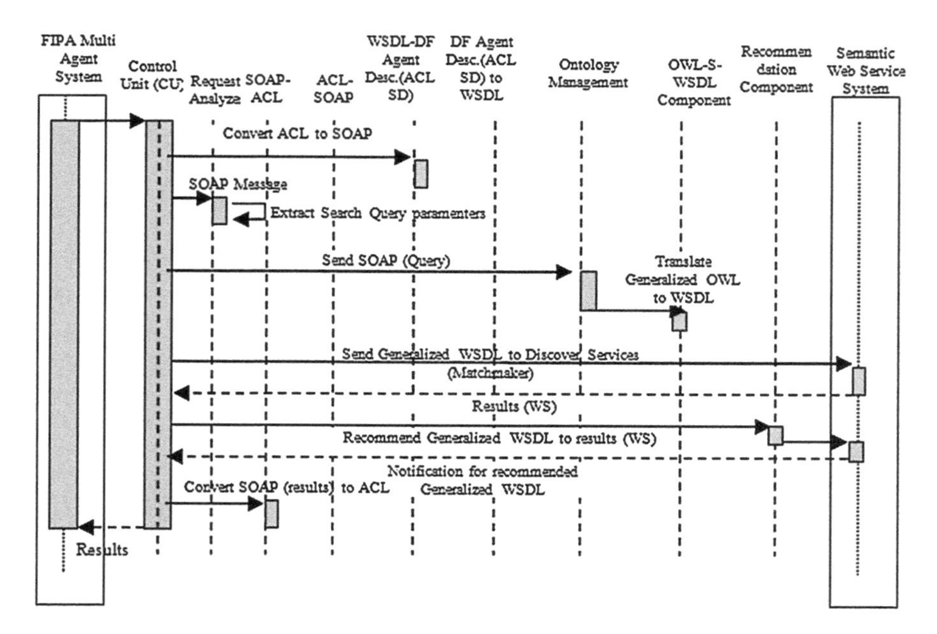

Fig. 8.4 FIPA multi agent communications with semantic web service

agent system by sending required DF-Agent-Description. If DF does not have the required agent registered, it redirects its search to the broker by sending an Agent Communication Language (ACL) message. As soon as CU in broker receives message and passes it to ACL to SOAP component. Afterwards converts the ACL to SOAP sends SOAP message to request analyzer for extracting query parameters.

Extracted query parameters sent to ontology management component to create generalized user ontology and forwarded to OWL-S to WSDL component to create generalized WSDL.

As mentioned before, CU passes created WSDL to discovery engine in semantic Web service system and CU has come to know about the selected Web services. The created WSDL recommends selecting Web service providers. SOAP result message converts to ACL message with help of SOAP to ACL component.

8.3 Conclusions

The multi agent system [7] is a loosely coupled network of problem-solver entities that work together to find answers to problems that are beyond the individual capabilities or knowledge of each entity. Multi agent system relied on semantically rich semantic representation and techniques for communication among agent communities. The importance of agent technology has been realized as one of the important technologies to successfully support the activities like e-business which require autonomous entities to handle dynamism of communication and results in conceptual simplicity and enhance scalability. FIPA [9] is standardization body which governs and promotes agent technology. A lot of work has been done by many other research communities to fulfill this communication gap, which enabled for the first time the major specification governing body of software agents and multi agent systems, IEEE standards committee Foundation of Intelligent Physical Agents (FIPA) to show its interest for adapting semantic Web and Web services standards in its evolving specifications as a sub-group named Agents and Web services Interoperability Working Group (AWSI) [10]. However, there are a number of limitations that exists in context of accuracy and functionality. Major challenges are that both the technologies use different service registries, service description languages and communication protocols.

References

1. Chhabra M, Lu H (2007) Towards agent based web services. In: 6th IEEE/ACIS international conference on computer and information science (ICIS 2007), vol 23, pp 181–186
2. Burstein M, Bussler C, Zaremba M et al (2005) A semantic web services architecture. IEEE Internet Comput 9:72–81

3. Martin D, Burstein M, McDermott D et al (2009) Bringing semantics to web services with Owl-S. World Wide Web-Internet Web Inf Syst 10:243–277
4. Lyell M, Rosen L, Casagni-Simkins L, Norris D (2003) On software agents and web services: usage and design concepts and issues. In Proceedings of the 1st international workshop on web services and agent based engineering, vol 6. Sydney, Australia, pp 230–236
5. Greenwood D, Calisti M (2010) Engineering web services agent integration. IEEE Int Conf Syst Man Cybern 19:235–241
6. Maximilien EM, Singh P (2010) Agent-based architecture for autonomic Web service selection. In: Proceedings of workshop on web services and agent based engineering, vol 6. Sydney, Australia, pp 211–217
7. Foster I, Jennings NR, Kesselman C (2009) Brain meets brawn: why grid and agents need each other. In: Proceedings of the 3rd international joint conference on autonomous agents and multi-agent systems, vol 16. New York, pp 8–15
8. Ye L, Zhang B (2006) web service discovery based on functional semantics. In: Proceedings of the IEEE 2nd international conference on semantics, knowledge, and grid, vol 18, pp 251–256
9. Foundation for Intelligent Physical Agents. http://www.fipa.org
10. IEEE-FIPA Agents and Web Services Interoperability (AWSI) Working Group. Available at http://www.fipa.org/subgroups/AWSI-WG.html

Chapter 9
Research on Applications of Internet of Things in Agriculture

Fujie Zhang

Abstract There is no doubt that efficient fresh agricultural products supply chain management is the key to improve the competitiveness of fresh produce enterprises, and the appearance of things brings a new opportunity for fresh agricultural products supply chain management. At present, there is still a big gap between theoretical research and practical applications, Especially in China, most of the literature of the IOT still remain in the introduction of the IOT itself and its application, but with the development of the relevant technology and the deepening of theoretical research and application, the IOT will play a positive role in the development of the fresh agricultural products supply chain in the future, it will also become an important direction of future research. So, the future of the IOT's application in fresh agricultural products supply chain management is bright, but the road is tortuous.

Keywords Fresh agriculture products · Supply chain management · Internet of things

F. Zhang (✉)
College of Biosystems Engineering and Food Science, Zhejiang University, Hangzhou, China
e-mail: Fujizhang13@126.com

F. Zhang
Faculty of Modern Agricultural Engineering, Kunming University of Science and Technology, Kunming, China

W. Du (ed.), *Informatics and Management Science VI*, Lecture Notes in Electrical Engineering 209, DOI: 10.1007/978-1-4471-4805-0_9,

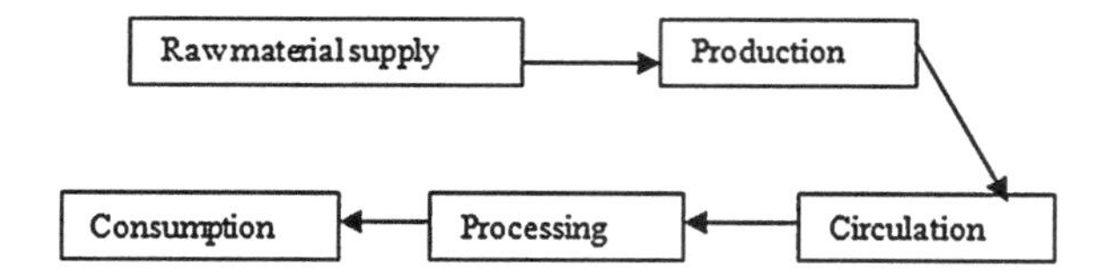

Fig. 9.1 Fresh link in the supply chain of agricultural products

9.1 The Content of Supply Chain Management of Fresh Agricultural Products

Fresh agricultural products include vegetables, fruits, poultry, eggs, aquatic products, meat, dairy products, etc [1, 2]. Supply chain management fresh produce is established on the business process core enterprise, internal and node between enterprise's various businesses as a whole function, in the process of production, and the fresh suppliers, manufacturers, processor, dealers, consumers effectively combined into one production goods. It is for plan, organize, coordinate and control the logistics, distribution and capital flow and information flow in the supply chain, the right number of goods to the at the right time and place, to minimize the cost, to provide the most valuable [3]. To improve the efficiency of the whole supply chain, effectiveness, and added value, so that all the fresh produce trading partner supply chain, greatly increases the economic efficiency of the method [4]. Therefore, supply chain is not only a link to the supplier to the consumer logistics chain, information chain and capital chain, but also is a value the value of agricultural products supply chain because of high additional value to improve the processing, packing, transportation, and the income of about the company. Fresh agricultural products supply chain includes five parts for raw material supply, production; processing, circulation and consumption (see Fig. 9.1).

In addition, consumers are put forward for the change of the taste of the species diversity requirements. Features such as different from other consumer goods, make us more stringent requirements, and put forward the supply chain management fresh produce: first of all, fresh produce is ensured the shortest time from a field or the flow of water table; Second, fresh produce is to make sure to achieve or improve the quality and stable development cycle; Third, is the guarantee to provide consumers with fresh, safety, diversity fresh produce; Finally, reduce the loss of the whole logistics process control production rate and comprehensive preservation of reverse logistics cost.

9.2 The Concept of the Internet of Things

The Internet thing, including all the things in the world of the Internet is the only one thing that code, based on Johnny, cork and computer network, to use RFID, wireless data communication and other technologies [5]. The base of the operation of the Internet thing is electronic product code (EPC) provides a unique identity

Table 9.1 Components of the EPC system

System components	Name	Note
EPC system	EPC standard	Identify the specific code of a single product
RFID system	EPC tag	Be attached to the goods or embedded in articles
	Reader	Read EPC tags
Information network system	EPC savant	The software support system of EPC system
	Object Naming Service (ONS)	Position the corresponding information of goods
	Physical Markup Language (PML)	Describe the information of goods

for physical objects. What the Internet use of computer network, provide based on RFID tag and installation in EPC RFID receive goods delivery transportation device can improve monitoring level for production, distribution, storage and sales of the commodity (see Table 9.1). The appearance of the Internet thing, and it's new idea and monitor each commodity in tracking the whole world will change completely supply chain process and management methods, meanwhile, will give new opportunities for the development of supply chain management of the business enterprise. The unique technical advantages and visual Internet things are China, the United States and other countries pay more and more attention. More and more businessmen and related research and practice have been it. "The Internet".

9.3 The Specific Application of the IOT in Supply Chain Management of Fresh Agricultural Products

9.3.1 Perfect Fresh Agricultural Products Quality Monitoring, and Strict Food Security Sources

Fresh produce is the main source of nutrients, of the Chinese consumers in addition to food, and its consumer occupies a very important position in the daily lives of the consumers [6]. Therefore, fresh produce safety issues are related to personal health people, how to effectively tracking and tracking their problem has become an urgent global problem.

This paper holds that in order to ensure the quality of products and security fresh produce, on the one hand, we need to do a good job of safety monitoring of fresh produce from the source, establish a fresh produce quality tracking, tracking and review mechanism and system, so that, when fresh produce safety problems (especially security threat to the life and health of consumers) appear, we can clearly recall products problems in time, rapidly, and the responsibility of the individuals and departments; On the other hand, we need to set up the comprehensive quality control and tracking system from the source to the market, so that we can be monitoring and management of the key link of fresh produce, such as producing area conditions, the production process, product testing, gift box logo, further strengthen the food safety consciousness and the sense of responsibility of most of the producers.

Application IOT in supply chain management fresh produce can effectively solve the problem of quality and safety of fresh produce. EPC label is embedded in raw material supply chain management from the first step of fresh produce raw material supply, record product information production and circulation of the entire process. This way, when they buy the products, according to the EPC label is embedded products, consumers can easily get the information fresh produce the whole process involving raw material supply, production, processing, distribution and sales. Then based on the information they get, consumers can decide whether or not to buy the product. That is, consumer can be active learning related information through the network product things, not passively accept information goods supplied by seller. It will completely solve the problem of information dissymmetry of the existence of the understanding of the commodity information to consumers and sellers.

9.3.2 Establish Management Information System of Fresh Agricultural Products Based on IOT, and Increase Supply Chain Integration Efforts

At present, the biggest obstacle of the implementation of the supply chain management fresh produces the contradictions between the small-scale production and large market in China. Small-scale production of fresh agricultural production means that a small family farmers out production and management problems, not directly into the market circulation is not effective in its; The big market is refers to the main sales market fresh produce origin, but often geographic span of larger areas, for the sale of fresh produce is an open market.

Therefore, the supply chain integration efforts is inadequate, integration and information sharing between supply chain members is low, so fresh agricultural enterprise can not grasp the changing needs of the customer.

Use IOT the key technologies of the, RFID in supply chain management fresh produce, we can build a fresh produce management information system. Use IOT in supply chain management fresh produce not only make the company the real-time monitoring, each kind of fresh produce, and supervise the whole process fresh produce; Not only the collection, storage, transfer and share information, but also analysis and prediction of the supply chain information every link of fresh produce. In the future development trend forecast, the probability of an accident based on information fresh produce in the current contact, and warning or to take remedial measures, so as to improve the supply chain management level of fresh produce. Supply chain information sharing process to help enterprise between members of each node connected closely, coordinate the members of the behavior, improve the supply chain of the breadth and depth of effective cooperation, so as to improve the supply chain integration efforts. As consumer demand changes, all the supply chain, from the field or water to the table, it can respond quickly to ensure that consumers, can provide fresh, safety, diversity fresh produce.

9.3.3 Reduce Supply Chain Management Cost, and Improve the Efficiency of Supply Chain Management

9.3.3.1 The IOT is Used to Simplify the Workload of the Acceptance and Handling Commodities

When fresh or processed agricultural products labeled EPC enter into storage center (or logistics center), the reader will read the tags at the entrance to finish goods inventory, without unloading and checking off, so as to simplify the goods acceptance link, reduce the goods handling times, save fresh agricultural circulation time, and realize quality stable or improved in rapid circulation.

9.3.3.2 The IOT is Used to Reduce Transportation Costs

By pasting EPC tag in commodities and vehicles, and fitting RFID and accept forwarding device on some checkpoints in transit, suppliers and distributors can timely access to goods position and state information, so as to realize visual management for goods transportation and intelligent tracking for transports, and convenient vehicles scheduling, reduced the number of vehicles and personnel needed to avoid the idle, raised utilization rate. Therefore, the use of IOT reduced the cost of shipping goods.

9.3.3.3 The IOT is Used to Reduce Storage Costs

Based on the real-time inventory and intelligent EPC shelf technology, enterprises realize their inventory efficient management. Through intelligent management, the goods can be placed inside in storage center freely, and it is helpful for improving warehouse space utilization; then fast and accurately understanding storage center inventory levels, reducing the possibility of stock while reducing the number of inventory stock; guaranteeing the accuracy of the in–out warehouse commodity, and reducing needless loss brought by distribution mistakes. In addition, the IOT

Table 9.2 Optimization role of the IOT in links of supply chain of fresh agricultural products

Links of supply chain	The optimization role of the IOT
Supplier	1. Efficient production plan, and reducing inventory
	2. Improve the response to customer demand
	3. Track products, timely recall the products in question, and obtain consumer feedback information
	4. Produce according to the needs, reasonably allocate human resources and production material, and increase productivity
	5. Reduce the costs of distribution and transportation, and improve equipment utilization rate
Transporters	1. Automatic identification, classification and clearance of the goods
	2. Track throughout the entire transport route, so as to improve transport security
	3. Improve the reliability and efficiency of delivery; Improve delivery reliability and delivery efficiency
	4. Improve service levels
Retailers	1. Reduce the stock level,inventory and safety stock
	2. Track products, so as to assure their quality, and reduce the numbers of them which are damaged, misdirected, lost, and stolen
	3. Improve transport efficiency, and reduce operating costs
Consumers	1. Individually purchase, so as to reduce unnecessary waiting and searching time
	2. For further product information and; Obtain information about products and manufacturers
	3. Good after-sales service
	4. Preferential purchase price

can reduce loss caused by such reasons as stolen, damaged, and missing goods, timely discovered these may lead to supply chain of inefficient problem, reduce this part, thereby reducing the whole logistics cost in the process of loss (see Table 9.2).

9.4 Conclusion

The relatively low cost of fresh produce to China has certain comparative advantage in global competition. The twenty first century, and the world economic activity globalization experience unprecedented characteristics, market competition becomes increasingly fierce. In a complicated and volatile market demand, the enterprise of fresh produce our country, if only depend upon their own strength can not stand this risk is brought about by the large investment and long construction period. In order to survive and development, the enterprise must make full use of external enterprise related resources, make full use of the core competitive power of enterprise relationship based on the integration of the core business ability, coordination and participate in the competition, will the traditional association competition between enterprises of the supply chain of the competition. In addition, the supreme achievements gained in some well-known enterprise management practice from supply chain has people more believe, supply chain management of a kind of effective method, for a global competition.

References

1. Gao H (2010) The application of the IOT in the agricultural products supply chain management. Commercial Times 22:40–41
2. Kong HL (2005) EPC and the "Internet of Things"—the fuse causing the supply chain revolution. China Standardization 4:125–128
3. Li J (2010) Thinking of the fresh agricultural products supply chain management. Market 1:17–19
4. Liu XJ (2010) The IOT-the new tools of logistics and supply chain management. Commercial Times 25:40–41
5. Wang H, Shen J, Shi Y (2010) The new trend of supply chain management based on the internet of things. Commercial Times 26:21–22
6. Zhang S (2009) Analysis of agriculture-related supply chain. Logistics Technol 3:36–38

Chapter 10
Improving Flexibility of Supply Chain with Knowledge Management

Ting Liu and Fangsi Zhong

Abstract In the tide of knowledge economy, the supply chain is facing the diversified customs' preference and the uncertain market demand. Core companies should encourage the staff of knowledge innovation and apply scientific means to forecast external conditions reasonably for response market test rapidly. Through the analysis of the problems existing in the flexible management, they would make corresponding measures to solve those problems to improve enterprise performance.

Keywords Supply chain · Flexibility · Knowledge

10.1 Introduction

The competitive advantage of core enterprise in supply chain is not just resources and capabilities of an independent firm, its formation need optimize and recombine the entire process in a comprehensive view. To achieving flexibility of supply chain, members have integrated knowledge of supply chain to response the environmental uncertainty quickly [1]. As knowledge is a key to improve the operational efficiency, the stock and structure of knowledge in supply chain are the most fundamental decisive factors in performance [2]. Enhancing the competitive advantage of supply chain, its importance to formulate an effective knowledge management (KM) mechanism and set up a knowledge sharing platform.

In a dynamic environment, firms have to attach importance to enhance competitivenesses and draw up a reasonable flexible strategy, finding out the flexible

T. Liu (✉) · F. Zhong
Business school, Xiangtan University, Xiangtan, China
e-mail: tingliu211@126.com

W. Du (ed.), *Informatics and Management Science VI*, Lecture Notes in Electrical Engineering 209, DOI: 10.1007/978-1-4471-4805-0_10,

strategic differences in characteristic, scope. From the perspective of the system, supply chains reflect a goal that realizes the whole optimization through constant knowledge innovation [3]. Knowledge determines flexibility and the relationship between the flexibility and the environmental uncertainty demonstrates that KM plays a pivotal role to improve flexibility of supply chain [4]. More specifically, the improvement and completion of KM system is significant in training the flexibility. In addition, organizational learning theory considers that corporate unions have superior to market in the knowledge transfer, knowledge integration and knowledge application. They will maximize the value of inter-enterprise transfer and minimize the cost of inter-organizational learning [5].

It is proved that the supply chain should use KM system effectively to adjust competitive strategy while facing the challenges, and form a function network to provide more opportunities for win–win. Therefore, emphasis of this study is improving flexibility of supply chain with KM, putting on the analysis of current issues, their reasons and relevant solutions in flexibility. By establishing a good KM system, core enterprises can use scientific methods to achieve the optimal performance and long-term development.

10.2 Features of Flexibility of Supply Chain Nowadays

Nowadays complexity, flexibility, timeliness and openness are difficulties on the KM of supply chains. Supply chains should try their best to response market changes and make appropriate adjustments rapidly. And focus on the vital role played by flexible management that can improve predictive power and decision-making efficiency affected by the stock of knowledge and innovation.

10.2.1 The Effectiveness of Knowledge Management

KM in supply chain based on Internet optimize information flow, both supply information and distribution information can be realized agile and fast decision-making process. Information flow is no longer susceptible to time and space, avoiding the information filter effectively. Different facilities and resources linked up and recombined scientifically through the virtual enterprise, so that it's in favor of optimize the service level. This process of absorption and deepen the knowledge, not only improve the quality of the whole chain, but the advantage of enterprise image and products. KM system reflects market dynamic promptly, improving crisis management capability in the supply chain. At the same time, analyzing product manufacturing strategy, sales strategy in KM angle, it will in-depth understanding of supply chain operation and pursue the maximization of profit. Therefore, integrating knowledge effectively is beneficial of increasing influence of core business in the chain network. It depicts the process of KM and the characteristics of the members in a supply chain.

10.2.2 Difficulties on the Flexibility of Supply Chain

System restricted flexibility. Core enterprise grasping opportunity and acclimatizing to the environment will be influenced by the supply chain systems, because it goes across organizational boundaries. As the size of each node is different, the requirements of KM are distinctive. So all kinds of business in dealing with the time, cost, degree of control has a different expression. Enterprises with the highest level in KM are constrained in their original flexibility by other members.

The internal flexible management has low efficiency. The demand of various departments seems inconsistent with outside perception, because of distinctive emphasis on KM of each department. The difficulty that achieves symmetric information affects flexible level and the speed of problem-solving. In addition, the speed of the internal hardware and software updating will reflect the extent of concern for knowledge sharing and innovation [6]. Thus, internal organization should build an flexible culture.

Market changing influence flexible management. Filled with a variety of public information, consumer preferences will be affected by external factors. Those will change customer desires at different times and it is difficult to seize the product preference of consumers in a short term. It involves interaction with multiple enterprises, investigating the ability of KM of the entire supply chain and the speed of information processing.

10.3 Troubles on Improving Supply Chain Flexibility with Knowledge Management

Enterprise's growing depends on the accumulation of knowledge and skills. The flexibility of supply chain is also an important tool, which can optimize the process rationally in a changing environment. Thus, it is meaningful to transfer, share and innovate knowledge of nodes for supply chain flexibility, and make up for "knowledge gap" in the enterprise growth process.

10.3.1 Member Enterprises Pay Different on Flexibility Environment Construction

Enterprises need updating KM system regularly to meet market demands and keeping up with the pace of competitors. Firms undoubtedly invest a large capital. However, the members have different flexible strategies, playing a different proportion for establishing flexible environment in the organizational strategy. The difference on investment affects the using of search engines, and data mining, artificial intelligence, knowledge integration tools. And the business philosophy in the whole chain will restrict the progress of Core Company. It's adverse.

10.3.2 Various Attention of Flexible Makes Flexible Culture Creating Difficult

Core Company must establish a sense of competition and form a flexible corporate culture. Not every company can become a market leader, others may not consciousness to establish a specialized KM team to analyze relevant issues on knowledge acquisition, and staffs are in an inactive attitude. The importance of flexible management in supply chain is not sufficient. Each company has potential to develop new markets with a long-term perspective, based on the original market share. Members need play subjective initiative to solve various problems in the changing market.

10.3.3 Bullwhip Effect is Serious in the Information Networking

In the process of trading, Companies face a high risk phenomenon—the well-known bullwhip effect, that is, information transfer among companies in supply chain are likely to be distorted and shared ineffectively. The arrangement of products production plan and sellers' marketing order will get some influence. Either an ambiguous answer or the lack of constructive ideas makes product development difficulty. In order to weaken the bullwhip effect, companies should use the flexible management way based on KM to reduce the risk of suppliers, managing inventory costs and optimize the allocation of resources.

10.4 Knowledge Management on the Role of the Flexible Supply Chain

Knowledge plays an critical role in the process of enterprises growing. Through the KM platform of exchange knowledge, the function of the flexibility of supply chain with KM can be strengthened by series of scientific measures.

10.4.1 Improving the Supply Chain Flexibility by Establish a KM Platform

It is essential to build an efficient information system. There are inevitably some factors affecting the transmission of knowledge between the nodes in the chain. The KM system platform should in the same level of technology; otherwise it cannot avoid uncertainty and ambiguity in the diversion. Its essence is emphasizing the importance of knowledge in flexible management. Members focus on the overall strategy which sharing knowledge rapidly, are able to response accurately on market opportunities. It is beneficial to control the stock of production, and reduce the costs. Meanwhile, members create a sharing database system which can provide the success or failure of enterprise growing. For the other partners, they are references to deal with conflicts, adjusting the structure and strategy to maintain performance.

10.4.2 Building Knowledge-Based Culture to Promoting the Flexible Management Ability

The behavior of top management and decision-making should reflect a kind of knowledge-based culture. Through stories, rituals and other forms of propaganda, enterprises should implant the culture to the employees. On the other side, some incentives should be set up to encourage outstanding staffs. Taking advantage of this fair system, members obtain win–win cooperation through the advantages complementary and resources sharing. By monitoring activities of the entire chain, companies would master delivery time and product quality. It is able to give a reasonable discounts and flexible contracts to encourage partners sharing feedback information timely. Thus, they can explore the application of knowledge and construct KM platform in common. It is conductive to maintain cooperation relationships, so that the core business accommodates the best quality, most affordable prices of product and service to consumers.

10.4.3 Bullwhip Effect is Serious in the Information Networking

Whether an employee has the explicit knowledge or tacit knowledge is the best human resource to all companies nowadays. Innovative talents are indispensable to promote the sustainable development. Consequently many large companies invite experts and professors to train staffs. That encouraging employees to master new skills through teamwork achieves the best performance. In the process of communication, the collision of thinking brings knowledge innovation; it is the blending process of theory and practice. It is also advantageous to knowledge sharing, and promoting mutual learning among members. Although staff training requires a certain degree of investment, the trainees get a raise of technology level and the utilization rate of knowledge improves significantly, companies receive greater benefits.

10.5 Conclusion

The emphasis of enterprises is on efficiency and accuracy in the fast consumption age. Flexible management with knowledge is the mainstream of modern management. Supply chain should apply it to optimize processes, strengthening cooperation to solve personalized problems. Using modern technology to deliver information in supply chain can strengthen the ability of research and development, improving strain capacity and flexibility and enhancing the internal coordination. Only in this way of establishing the image and product advantage in the market from a strategic point of view can enterprises achieve overall optimization and profit maximization.

Acknowledgments Funded by Humanities and Social Sciences of Ministry of Education (12YJC790122) and Hunan Philosophy and Social Science Fund (11JD67).

References

1. Jiemei Z (2009) A research into the knowledge management of supply chain integration in M&A. Econ surv 4:35–39
2. Bing W, Zhongying L (2007) Review of supply chain flexibility. Sci Technol Prog Policy 24(2):190–195
3. Yili L (2008) Research on supply chain flexibility based on knowledge. Sci Technol Prog Policy 25(11):16–18
4. Nengmin W, Tong Y (2004) Flexibility and knowledge management. Sci Technol Prog Policy 21(11):135–139
5. Jun J, Yuan L (2008) Firm's knowledge acquisition and technological innovation based on alliance network. R&D Manage 20(1):42–46
6. Tao L, Bing W (2003) A research on knowledge sharing of chinese knowledge workers within an organization. Nankai Bus Rev 6(5):16–19

Chapter 11
Research on Dynamic Implementation of Green Human Resources Management

Ting Liu and Pengxin Xie

Abstract Green human resources management is a new management philosophy and pattern in which "green" concept applied to the field of human resource management. Its essential meaning is to take "green" management tools to enhance economic, social and ecological benefits (comprehensive benefits) so as to achieve employees' psychology, human and ecological harmony (Tristate harmony). This paper analyzes the connotation and characteristics of green human resources management, proposing a dynamic model of green human resources management based on the "Tristate harmony". Finally, it explores a way to implement green human resources management dynamically.

Keywords Green human resource management · Tristate harmony · Dynamic model

11.1 Introduction

In recent years, the deteriorating global environment and the continued emergence of green movement jointly form a powerful force which demands enterprises to change their environmental management concepts and establish a green management system [1]. This system includes green production, green technology, green financial, green marketing and green human resources management. At present, there have been many literature researching on green marketing, green production and green financial [2, 3]. However, it appears little discussion has been provided on the green human resource management. In fact, the green management is usually hampered by human resource factors rather than technology [4]. So the

T. Liu (✉) · P. Xie
Business School, Xiangtan University, Xiangtan, China
e-mail: tingliu21112@126.com

W. Du (ed.), *Informatics and Management Science VI*, Lecture Notes in Electrical Engineering 209, DOI: 10.1007/978-1-4471-4805-0_11,

analysis on the green human resource management can not only enrich the green management theory, but also help enterprises to enhance their comprehensive benefits so as to achieve tristate harmony.

Presently, little research exists in the literature examining the impact of such HR factors to the comprehensive benefits and tristate harmony. Guang Yang and Huaqing Huang mainly consider that human resource management (HRM) activities can create "green value" from the perspective of ecological harmony [5]. Jinxiu Wei [6], mainly on the view of "psychology harmony", indicates that green HRM requires caring employees' needs, seeking to create fair and equitable organizational environment. These studies are extensive theoretical foundation, but lack of a comprehensive view on green HRM. This paper provides an overview of the current management literature regarding green human resource factors to achieving comprehensive benefits and Tristate harmony. A dynamic model will be set up to combine green HR factors, benefit goals and harmony values. Based on the literature review, the primary objectives of this study are: (1) describe the connotation and characteristics of green HRM; (2) establish a dynamic model of green HRM based on the "Tristate harmony" view angle; (3) explore a dynamic implementation way to realize green HRM.

11.2 The Connotation and Characteristics of Green HRM

11.2.1 The Connotation of Green HRM

Green management is a comprehensive management activity which guided by the idea of sustainable development. It aims to eliminate and reduce organizational behavior effects on the ecological environment through production, marketing, finance and other activities [7].

Green HRM is one of the most important parts of green management system. The primary mission of green HRM is to take "green" management tools to enhance economic, social and ecological benefits so as to achieve employees' psychology, human and ecological harmony [8]. "Tristate harmony" is not only the essential connotation of Green HRM, but also the basic content of green HRM concept. Ecological harmony means getting along well with nature. Psychology harmony refers to employees' mental health and good adaptability. And human harmony includes interpersonal harmony and the harmony between enterprises and employees.

11.2.2 The Characteristics of Green HRM

Compared with traditional HRM, green HRM has the following characteristics:

11.2.2.1 To be the New Development of the People-Based Management

The traditional human resource management derives from the personnel management. Personnel management is "thing-oriented", regarding people as the tool of completing "things". The traditional human resource management begins to propose the concept of "people-oriented", but people still exist largely as a means of finishing tasks. Green human resources management takes the improvement of the human capacity and self-realization as the ultimate target. It is the new development of the people-based management.

11.2.2.2 Take Enterprise and Staff's Common Sustainable Development as a Goal

Program listings or program commands in the text are normally set in typewriter font, e.g., CMTT10 or Courier.

11.2.2.3 Regard "Tristate Harmony" as a Core Value

There are many "non- green" phenomena in traditional human resources management, such as non-compliance with the principle of integrity, regarded people as a mere means, only paid attention to enterprise development needs, while ignoring the growth of staff, etc. Green human resource management particularly emphasizes that human resources practices should achieve the unifying of employees' psychology, human and ecological harmony.

11.3 A Dynamic Model of Green HRM Based on the "Tristate Harmony" View Angle

11.3.1 The Description of the Dynamic Model

Figure 11.1 presents the green human resource management dynamic model. This model represent the operational interaction of Green HRM factors such as green human resources planning, green recruitment, environmental training, green performance management and green incentives within comprehensive benefits and tristate harmony. The dynamic model consists of three parts, namely green human resources management wheel, benefit wheel and harmony wheel, in which harmony wheel is the core of the model, benefit wheel is a media to connect the harmony wheel and green human resource management wheel, the green HRM wheel is the foundation of this model. The "Tristate harmony" core values determine enterprises to pursue comprehensive benefits, and then indicate green

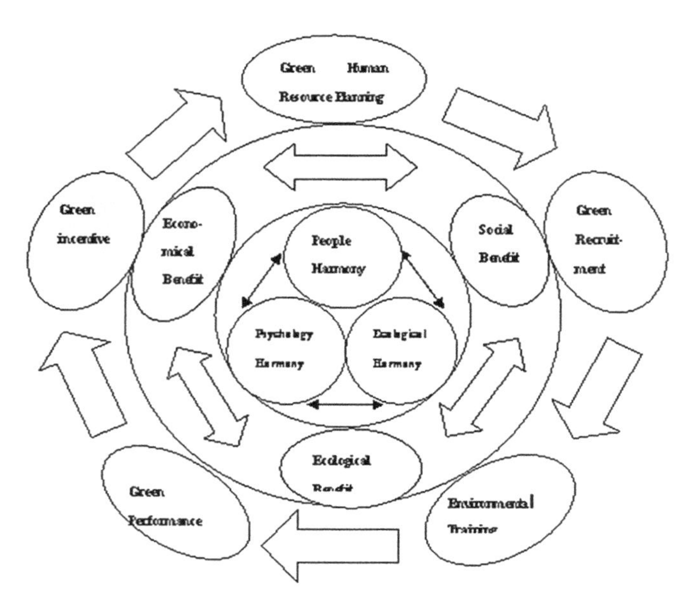

Fig. 11.1 Green human resource management dynamic model

human resource management activities. What's more, enterprises need the power of green human resource management to enhance enterprises' three-dimensional comprehensive benefits, ultimately realizing the "Tristate harmony". These three parts realize dynamic association through continuous operation.

11.3.2 The Operating Mechanism of the Dynamic Model

11.3.2.1 Tristate Harmony is the Core Value Pursue of Green HRM

The ultimate goal of human resources management is to promote people's comprehensive and harmonious development. Zhibin Huang [9] first proposed green harmony management theory, which regards the realization of "Tristate harmony" as the management principle and ideal goal. The theory also emphasizes the unifying of the economic, ecological and social benefits. At the same time, it stresses that the people in enterprise are not only the responsible subject of green harmonious, but also the innovation subject, regarding the "Tristate Harmony" as the root of pursue in the new humanistic management.

"Tristate harmony" is not only the essential connotation of Green HRM, but also the basic content of green HRM concept. The core value pursue of green HRM is to realize the harmony of ecology, human and psychology.

11.3.2.2 The Implementation of Green HRM is the Premise of Comprehensive Benefits and Tristate Harmony

Enterprise must achieve its benefit goal by daily management activities, ultimately realizing its core values. So the pursue of comprehensive benefits and tristate harmony should be based on the effective implementation of green HRM.

Some scholars insist that the main feature of best HRM practices is their universal simulative effect on enterprises' performance [10]. For example, in the selection, Terpstra and Rozell [11] selected five selection activities to discuss the relationship with the organization performance, and the result showed that the selection activities and organizational performance correlated. In training, Russell and Terbory [12] found that staff training programs and financial performance had a significant positive relationship.

Combined with the best HRM practices and green management concept, it can be concluded that the green HRM practices include green human resources planning, green recruitment, environmental training, green performance management and green incentives. The effective implementation of these practices can help to enhance the comprehensive benefits and provide an important guarantee for the realization of tristate harmony.

11.3.2.3 The Enhancement of Comprehensive Benefits is a Media between Green HRM and Tristate Harmony

Delaney and Huselid [13] made a survey of 590 profit and nonprofit organizations, and stated that advanced human resource management activities and organizational performance are related. The practices of green HRM can improve organizational performance. For instance, green selection can choose those talents who agree with green corporate culture and have a green responsibility consciousness, which will create a green culture phenomenon. Environmental training can increase employee awareness of environmental protection and environmental technologies, and provide intellectual foundation for eco-efficiency. Green performance management creates a green indicator system including the economic, social and ecological efficiency indicators, encouraging their employees to improve comprehensive benefits. Therefore, on one hand, enhancement of comprehensive benefits needs the power of green HRM. On the other hand, the comprehensive benefits are essential guarantee to the tristate harmony.

11.4 The Dynamic Implementation Way of Green HRM

Green HRM is the basic power of operating the model. It's significant to explore the way to implement green HRM. The way is based on these green HRM practices and closely contributes to tristate harmony.

11.4.1 Develop Green Human Resources Planning to Instruct the Correct Direction

Green human resources planning can provide strategic guidance for the green HRM practices, and ensure that HRM activities are contributed to the "tristate harmony".

Green human resource planning need to be abided by the "Tristate harmony" value. First of all, green human resource planning should be consistent with the enterprise's total strategy, helping enterprises to economize resources and protect environment. Performance goals should not only focus on cost, market share, profitability and other economic benefits, but to indicators of ecological and social benefits. Secondly, all aspects of green human resource planning should be guided by the green harmony idea, which demands companies to adhere to people-oriented philosophy, promoting the psychology harmony and human harmony. Finally, this planning must be prospective and flexibility, considering the long-term development of the enterprise and adjusting flexibly to adapt to the enterprise internal and external environment changes.

11.4.2 Optimize Green Recruitment and Selection System to Lay the Foundation of Harmonious Culture

Recruitment intends to influence the quantity and the types of candidates for a certain vacancy [14]. The implementation of green recruitment can attract talents who have conservation and environmental awareness. Those talents will have higher satisfaction after entering enterprises, which lay solid foundation for the establishment of harmonious culture.

In the traditional recruitment, enterprises generally pay attention to job seekers' knowledge, skills and attitude. However, the environmental dimension should be emphasized in the green recruitment. So, the human resource department should organize relevant training about the environmental policies and requirements for interviewers so as to increase questions about corporate environmental responsibility and issues in an interview. Meanwhile, enterprises' green image, environmental policy and environmental performance can be published in the recruiting advertisement in order to attract the most suitable talents.

11.4.3 Establish Environment Training Mechanism to Provide Intellectual Support

Establishment of dynamic learning mechanism is the important means to realize enterprise harmony. Environmental training is a systematic process that guides employees' behavior to accomplishing organizational objectives.

In fact, all the employees of a company, and not only the ones linked to certain departments, must receive training in the environment, which is considered one of the main factors for the success in green management. At the beginning, all the employees are necessary to receive basic environmental training, including environmental laws and rules, environmental pollution control and enterprise's environmental policies to enhance their ecological awareness. Then, as senior managers of a company, they need to be familiar with the international and local environmental regulations. Besides, environmental resource managers should master the theoretical knowledge in environmental science, such as environmental economics, management and laws to guide employees' environmental behavior. Finally, it's significant to strengthen technical staff's technology and innovation training. After a series of environmental training, employees will be more receptive to change themselves to maintain organizational harmony and consequently become more proactive.

11.4.4 Strengthen Dynamic Green Performance Management to Provide Performance Guarantee

Green performance management is a whole process of dynamic management system, including the performance plan, implementation, evaluation and feedback. Enterprises should establish green performance indicators and appropriate assessment system to motivate employees to value the environmental responsibility.

First of all, the performance plans should contain quantitative green indicators, such as the fulfillment of environmental responsibility, the energy conservation and emission reduction quantity. Besides, during the performance monitoring stage, it's significant to communicate with all level employees on the green performance plan so as to make them understand their green responsibility. In addition, managers should assess employees' performance based on their green indicators, simultaneously giving feedback to help them understand their achievements and shortcomings in the performance evaluation stage. At last, the use of appraisal results must be combined with the decision-making of human resource management. The effective implementation of dynamic green performance system can guarantee enterprises' comprehensive benefits.

11.4.5 Construct Green Incentives System to Create Harmonious Atmosphere

Green incentives are a new incentive idea which is premised on "green people assumption" to meet employees' green needs, ultimately achieving employees' sustainable and healthy development [15]. Green incentives include both material rewards and immaterial compensation to strengthen staff's environmental awareness and create harmonious atmosphere. Material rewards are one of important incentives to guide staffs to participate in environmental improvement. However, employees are not always motivated by financial incentives. The core of green incentives is to establish an incentive program to help employees to ease their work pressure and maintain physical and mental health because once one is unbalanced in mentality; it is difficult to achieve the balance of ecology.

So it's necessary to improve the communication mechanism and establish new enterprise-interpersonal relationships to construct a relaxing work environment. Meanwhile, a healthy reward system is a new way to encourage employees to concern about their own physical and psychology health.

11.5 Conclusion

Achieving "Tristate harmony" is fallacious without the real involvement of companies in this process. At the core of this process, many companies begin to establish a green management system. Green human resource management is a new philosophy which primary mission is to take "green" management tools to enhance comprehensive benefits so as to achieve "Tristate harmony". However, the present paper generally researches it on the view of psychology harmony or ecological harmony. This model establishes relationships between human resources management and the tristate harmony values. The proposal of this model contributes to a greater recognition of the interaction between human resource management and harmony value in companies based on the tristate harmony view angle.

Scientific progress in this field of knowledge is thought to have the following contributions: (a) organizational leaders recognize the connotation and characteristics of green HRM; (b) a comprehensive perspective is proposed to combine the green human resource management with enterprise's benefit goals and harmony values; (c) give some suggestions about how to implement green HRM to pursuing the harmonious value.

Moreover, further research is required to measure the factors proposed in this model and provide a more scientific way to play the role of green human resource management.

References

1. Polonsky WS (2000) Environmental marketing, vol 3, issue 9. Mechanical Industry Press, Beijing, pp 67–69
2. Jackson T (1993) Cleaner production strategies: developing preventive environmental management in the industrial economy. Lewis Publishers, Boca Raton, pp 122–125

3. Gray R (1993) Accounting for the environment, vol 6, issue 65. Paul Chapman Publishing Ltd., London, pp 88–95
4. Tarricone P (1996) People not products, are the key to pollution prevention, study finds. Facil Des Manag 15(1):18–25
5. Yang G (2003) Green human resources management-the greenization of human resources management. Manag Rev 10:8–15
6. Wei J, Li X (2006) Green human resources management: a new management concept. Gansu Sci Technol 2:113–114
7. Xu J, Wu Y (2004) Green management theory. Bus Res 6:48–50
8. Li J (2009) Discussion on green human resources management. Bus Manag 4:51–53
9. Huang Z (2005) Green harmony management theory-the management philosophy of ecological erea. China social science press, Beijing
10. Zhengtang Z (2004) Theoretical modes of strategic human resource management. Nankai Bus Rev 5:48–54
11. Terpstra DE, Rozell EJ (1993) The relationship of staffing practices to organizational lever measures of performance. Pers Psychol 46(1):27–48
12. Russell JS, Terbory JR, Powers ML (1985) Organizational performance and organizational level training and support. Pers Psychol 38(4):849–863
13. Delaney JE, Huselid MA (1996) The impact of human resource management practices on perceptions of organizational performance. Acad Manag J 39:949–969
14. Ivancevich JM (1995) Human resource management. Irwin, Chicago
15. Xiaolian H, Xiaocong L (2007) Green incentives: the new trend of incentives. Technol Manag 1:135–137

Chapter 12
Analysis of Cloud Computing Applying in Teaching

Feng Xie

Abstract The teaching process is a mutual process of communication, the characteristics of cloud computing make it possible to integrate into the teaching process and play an important role in it. This article discusses the need for universities' cloud computing teaching development and cloud computing to universities' teaching resources, as well as cloud computing teaching platform implementation. It is that at the application level from the digital learning, platform to build cloud-based computing system of university teaching, to develop the above characteristics of cloud computing in the teaching process.

Keywords Data warehouse · OLAP · ODP directory · Network behavior analysis

12.1 Definition of Cloud Computing

Google has come up with a new conception-cloud computing. In 2009, cloud computing has caused a great disturbance, it has been awarded one of the top 10 events in IT in 2009 as well [1]. Many big enterprises such as: Amazon, Google, IBM, Microsoft, Yahoo etc. have started to develop the related products of cloud computing. The big fight in technology of cloud computing has begun.

There is not an agreed standard of cloud computing in academic interests at present. The definition I adopted here is put forward by our herald Liu Peng in cloud computing. He considered cloud computing as a emerging commercial

F. Xie (✉)
Department of Computer Engineering, Guangdong Industry Technical College,
Guanzhou 510300, China
e-mail: fengxie2011@126.com

W. Du (ed.), *Informatics and Management Science VI*, Lecture Notes in Electrical Engineering 209, DOI: 10.1007/978-1-4471-4805-0_12,

computational modeling. It distributes map reduce tasks to resource pool combined by a large amount of computers. It lets sorts of application systems can obtain computation strength, storage space, and sorts of software services according to their necessities. Briefly speaking, cloud computing supplies users computer ability in service way, allow users obtain their necessary service through Internet in the situation of not knowing the technology of the service that supplies, not having the related knowledge and management of the set out.

12.2 Characteristics of Cloud Computing

Cloud computing has become the top technology of IT throughout the world. It has been hot topic of the society. Everyone is talking about how to make better use of cloud computing to serve the society. So, what are the characteristics of cloud computing?

12.2.1 High Reliability

The cloud computing users' data have all stored in the Cloud instead of stored in the hard disk of their own computers in the traditional sense [2]. Cloud is a strong functional server which supplies to users by big companies such as Google, Microsoft, and IBM. This way, data can be promised definitely safe, users won't worry about virus or Hacker's attack; users won't be troubled by losing data any more.

12.2.2 Large Scale

Cloud computing is like cloud in the sky, it has a very large scale. Google's cloud computing has more than 1 million servers, such as Amazon, Microsoft, IBM, Yahoo etc. these big companies' servers added to more than 100 thousand.

12.2.3 Scalability

Cloud computing users can apply to the suppliers of various ranks cloud computing service according to their own realistic necessities. Such as computation strength hard disk's storage, Cloud computing also has enterprise rank and personal rank. If user's necessities vary, they needn't abandon original space to change a new one. They just need to apply more space from the suppliers; meanwhile, their original data won't lose.

12.2.4 Virtualization

Cloud computing supports users to obtain Application Service at anywhere or use any kinds of terminals. Users can log in Cloud to use it through cell phones or PDAs which provides with standard browser. We even can use cell phones to achieve large scale's map reduce tasks.

12.3 Necessity for Universities' Cloud Computing Teaching

12.3.1 Requirements of the Development of Education at Home and Abroad

Cloud computing major has been set up in parts of universities since 2006. One of the initiators of cloud computing, Google, has started Google 101 plan since 2006. This plan aimed to train large quantities of talents in cloud computing. This course system soon became popular throughout the universities. Washington University, California University, Stanford University, MIT, have been added to this plan, set up majors related to cloud computing.

2010.8.14, Beijing University of Aeronautics and Astronautics Software Academy official external release of Majors of Master's Degree of cloud computing major. This major was the first mobile cloud computing, the Master of Software Engineering. The Ministry of education and industry related mechanism drive energetically below, including the Tsinghua University, Peking University, Fudan University, Nankai University, Zhejiang University, Beijing Institute of Technology, Beijing University of Posts and Telecommunications, North China Electric Power University and other domestic colleges and universities are considering cloud computing professional.

12.3.2 Large Scale Computing Market

In July 12th 011, the domestic institutions of research and consultation (CCW Research) study shows, 2010, China cloud computing market dimensions to achieve 55,930,000,000 yuan, grow 29.3 compared to the same period; is expected in 20112015, the cloud computing market compound annual growth rate of more than 50 %. CCW research shows that, in the enterprise users, already has 67.5 % users recognized cloud service mode, and began to use cloud computing services, or in the internal part of the realization of cloud platform sharing.

With the innovation of service mode of application as represented by the landing will be cloud computing and cloud services focused on the development, the SaaS market after years of development, in 2011 development bottleneck, new cloud service providers need to have a strong network resources, broad user base, rich SaaS application software, mature cloud the platform operation mode, comprehensive application terminal support and a complete industrial chain operation system, become the backbone of the cloud computing industry chain.

12.4 Cloud Computing to the University Teaching Resource Effect

According to the teaching resources of the status quo, analysis of the characteristics of cloud computing, it can be found on college teaching resources will have a great effect.

12.4.1 Cloud Computing Can Greatly Save Informatization Investment

At present the university informatization construction cost comes mainly from the hardware and software of the acquisition, maintenance and updating equipment to wait, if these based on cloud computing and service basis, will greatly reduce the capital investment [3]. All services are provided by cloud, without the need to ensure the server running reliability, ensure that the data stored in the server resources security and avoid the network access abnormalities leading to paralysis of the server and the network server response and access number and other restrictions, so the original maintenance, upgrades and other work almost to a minimum, management cost is correspondingly can significantly reduce.

12.4.2 Truly Realize the Integration of Resources, the Establishment of a Unified Resource Platform

The university informatization based on cloud computing and service basis, will be the heavy network information platform construction, curriculum resources, server equipped with storage and management work to the cloud service provider, then the existing fragmented, self-contained, localized network information platform will be transformed into a specific network environment, network the server system, network operating system independent and powerful universal information platform, in this platform to tens of thousands of cloud server as the basis, has a very strong computing function, massive cyber source, the existing network curriculum construction in the presence of soft, hardware resources to repeat investment, virtual teaching operational capacity of equipment support the problem will be smoothly done or easily solved.

12.4.3 The Application of Cloud Computing to Ensure the Information Security of University Teachers and Students

Within the campus network computer virus prevention and control has been a very difficult problem, especially in the multimedia classroom and computer laboratory. A machine poisoning, soon spread to all the machines. Antivirus software is authorized to use cost of college is also not a small expenditure but for virus still cannot effectively control. While in a cloud computing environment, cloud computing provides with advanced technology and professional team in charge of the resource safety maintenance work, teachers and students only through the network, access to their own data. Local no longer any data storage, so don't worry about the destruction caused by virus invasion. Therefore, the application of cloud computing in the colleges and universities not only save the college in the information security costs, and ensure that the university teachers and students of information security.

12.5 Digital Teaching Resources Platform

Digital teaching resource platform in addition to the teaching resources, often in the teaching resources based on network teaching platform, and excellent course platform function, general vocational college teaching resources according to the general professional classification, digital teaching resource construction and management for professional basis, regulate the professional code, various colleges and universities enterprises to use unified professional code the construction of learning resources. Within the library code and curriculum system by the professional, professional is responsible for sorting and construction. Similarly, as a learner can also be in accordance with the professional courses, storage or retrieval and learning. Using the unified professional code, library code and curriculum system construction of learning resources can avoid the two justices, is the fundamental method to eliminate information isolated island. Digital teaching resources platform should also be excellent course network teaching platform, the platform and teaching resources unified planning and integration, to achieve between three data sharing, avoid the teacher repeated upload data, facilitate students to find resources.

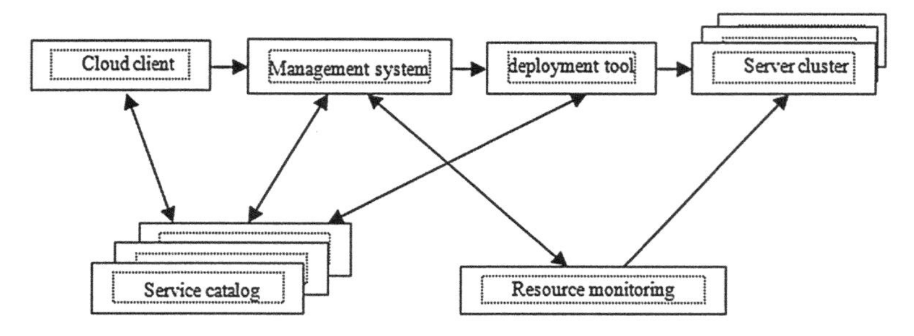

Fig. 12.1 Cloud computing platform architecture

12.6 Cloud Computing Technology of Digital Teaching Resources Platform

12.6.1 Cloud Computing Architecture

Cloud computing is a powerful "cloud" network, connecting a large number of complicated network computing and service, can use virtualization extensions each server capacity, will be their own resources through the combination of cloud computing platform, to provide supercomputing and storage capacity. Universal cloud computing architecture shows in Fig. 12.1.

12.6.2 Cloud Computing Technology Level

Cloud computing technology level from the main system properties and design perspective cloud, is the resource of software and hardware in the cloud computing technology in the role description. From the perspective of cloud computing technology, cloud computing is constituted by the physical resources, virtualized resources, management of middleware and service interface is composed of 4 parts, as shown in Fig. 12.2.

12.6.3 Cloud Computing Operating System

Cloud computing on the Internet is highly integrated computing resources (infrastructure, software, platform) according to user's demand fast custom, cloud computing work to build the system, as shown in Fig. 12.3, user management scheduling "cloud" to provide resources or services, "cloud" is responsible for ensuring that a resource or service availability, safety and quality.

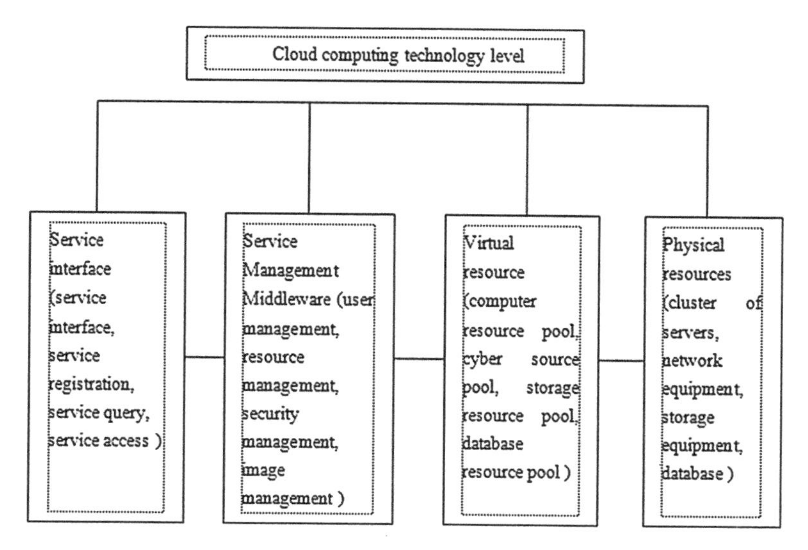

Fig. 12.2 Cloud computing technology level

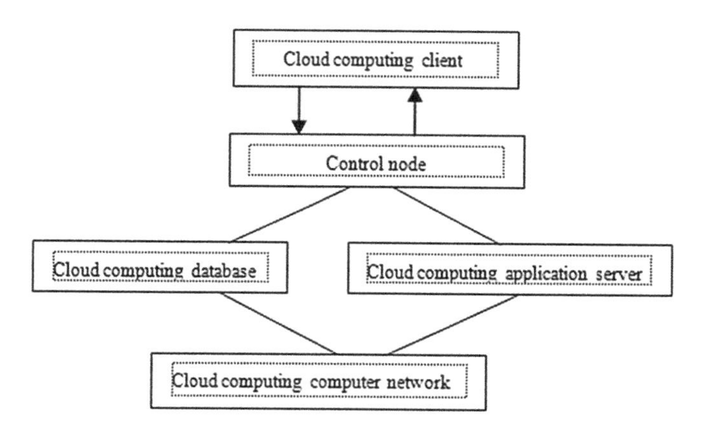

Fig. 12.3 Cloud computing work system

12.7 Cloud Based College Teaching System Framework Structure

12.7.1 The Remote Teaching System Framework Design

Considering the characteristics of cloud computing, according to the actual needs, we design the distance teaching system framework, which has good scalability. Figure 12.3 is a cloud based on distance teaching system structure diagram, which is mainly composed of cloud computing, load balancing equipment, WEB server and database server in four parts. Cloud services in addition to providing computing services, also must provide storage services, cloud services current monopoly in the private sector (enterprise) hands, and they can only provide business credit, for some special users choose cloud services should keep enough vigilance, once the user using large-scale private institutions offer cloud services, regardless of their technical advantages many strong, will inevitably make these enterprises to "data (information)" importance to threaten the user. In the information society, "information" is particularly important. On the other hand, the cloud data for the data owner other than users of cloud computing is confidential, but to provide the enterprise cloud computing is no secret. Based on the above reasons, this article in the architectural design process, using the local database server stored private data, such as student information, course information, professional information and other important data, while the online learning, download, instant communication, interaction and other extensions of all deposited into the cloud server, as a result of these belong to the service data, so that there is no safety problem.

12.7.2 Cloud Computing Platform of Collaborative Teaching Environment

Our bachelor's degree of computer professional courses data structure as the background, using Google Apps Education Suite based on cloud computing platform based on collaborative teaching and learning environment, The use of Google Apps Education Suite, the data structure teaching groups can build a group to prepare lessons platform, the teaching and research group in our teachers' teaching resources in a unified cloud computing platform for integration and optimization, support the teachers teaching seminars, classes, asynchronous collaboration of knowledge communication and sharing, and in order to facilitate data structure of the curriculum reform and development.

Through collaborative teaching environment collective lesson preparation platform construction, in the cloud computing services platform to establish the data structure of the curriculum contents, teaching arrangement, teaching online module, completed the initial teacher resources integration and teaching resource construction.

Digital teaching resource platform in higher vocational colleges, the original shared professional teaching resources platform based on cloud computing, through technology, the establishment of a national vocational colleges digital teaching resources platform, for the higher vocational colleges, enterprises, social learning, sharing, sharing, and inside and outside school, for users to provide professional learning for a variety of resources and online learning services.

12.7.3 For Cloud Computing Experiment Teaching Platform Research and Design

Taking the minimum cost as target, into open source, puts forward and Xiang Yun computing experiment platform architecture, this architecture mainly includes three layers, respectively, the virtualization layer application layer and physical layer. Among them, the application layer to provide users with the mainstream operating system, application software such as Hadoop, Platform as a Service (PaaS), AppScale Google App Engine, at present, mainly by the user to choose the right image file, can obtain and modify the corresponding service. Virtualization layer is mainly through the cloud platform with Infrastructure as a Service (IaaS), virtual dynamic flexibility of the Virtual Machine (VM), the layer is the application layer to provide resources. The existing open source cloud platform with Eucalyptus, Enomalism, abiCloud, Nimbus and so on, this paper uses Eucalyptus as platform. The physical layer is mainly physical machine resources, including CPU, memory, memory, the layer is a virtualization layer, provides basic resources.

In a cloud computing environment, network teaching system are encapsulated into "cloud" of the system depends on the learning resources, such as: teaching courseware, teaching video, electronic resources and other resources can also be packaged as "cloud services". Both for the students of the call, and also can be other education information system to provide services. Using a specific language and framework, using the "loose coupling mode" development of the module, and according to the standard package after the release of cloud computing in the cloud environment, make the resources in the context of the largest share, to avoid duplication of development, resulting in a waste of resources.

12.8 Cloud Computing to Promote Teaching Effect

Indeed, the cloud computing as an emerging technology, it has its shortcomings, but its advantages are main, especially for the lack of funding for the secondary occupation education school, cloud computing will be more and more get everybody's welcome and support it, cloud computing can at least be in following a few respects optimize teaching resources, improve the quality of learning:

Reducing or simplifying the software license charges

Software license is the one that many schools headache, based on cloud application can greatly reduce or exempt from software license cost.

Increase IT resource access channel and scope

With the resource grows in importance, whether teachers or students are required to whenever and wherever possible access to needed resources, instead of receiving the various restrictions. Cloud computing provides a powerful means of searching.

Reduce, eliminate and software version control and update related problems

Based on the application of cloud computing which means that the school will no longer need to worry about the software update, persistent questions, because they occur in automatic cloud.

Learning management system (LMS, SMS) benefits will be more and more obvious

The use of cloud based application virtualization of these services for the school will save a lot of investment. Therefore, the application of cloud computing is bound to the medium occupation schools teaching work is more and more important the process is inevitable and irreversible. We should make full preparations.

References

1. Yu J (2011) Cloud based distance education resource construction mode. China Educ Technol 12:299–301
2. Tang J (2010) Cloud based network teaching system design. J TaiYuan Urban Vocat Coll 3:159–160
3. Xie W (2011) Cloud based network teaching practice exploration. J Guangdong Polytech Norm Univ 1:61–63

Chapter 13
Network Course Graphic Animation Based on Web 2.0 Technology

Weiyan Liang

Abstract Web 2.0 technology has provided new ideas and technical supports for the design of network course; the application of Web 2.0 technology o the design and development of network course has changed into a new research field of educational technology in recent years. In this paper, by applying Web 2.0 technology in the teaching of course Graphic Animation of education technology program, network course Graphic Animation is designed based on Web 2.0 technology. Finally, the author draws up a conclusion on the design experience and shortcomings, expecting to provide a reference for educational technology personnel to develop t network courses.

Keywords Network course · Web 2.0 · Design

13.1 Introduction

At present, Web 2.0 is with great potential for promoting the reform of education; the studies on the application of Web 2.0 ideas and technology are being conducted by people vigorously. In recent years, studies on the application of Web 2.0 in the field of education increase with a gradual step [1, 2].

A variety of online platforms such as Wiki, photos and video sharing, and Google cooperation, which apply Web 2.0 greatly, has been established one after another. The free, cooperative, sharing and collective intelligence ideas are driving Web 2.0 to be applied in education in varying degrees and levels.

W. Liang (✉)
School of Education Science, Jiaying University, Meizhou, 514000 Guangdong, China
e-mail: weiyanliang1d2@126.com

W. Du (ed.), *Informatics and Management Science VI*, Lecture Notes in Electrical Engineering 209, DOI: 10.1007/978-1-4471-4805-0_13,

13.2 Introduction to Web 2.0

Blogger Don, in his article Interpretation on the Concept of Web 2.0, mentioned that Web 2.0 is a new Internet model presented by Flicker, Craigslist, LinkedIn, Tribes, Ryze, Friendster, Delicious, and 43Things.com, cored at Blog, TAG, SNS, RSS, wiki and other social software application, and relying on six degrees of separation, XML, Ajax and other new theories and technologies.

For the definition of Web 2.0, there has been no unified agreement internationally yet. There were experts who gave definition to Web 2.0 from idea and technology, clearly explaining the intension of Web2 [3]: "In terms of idea, Web 2.0 transfers a free, equal and open information exchanging and spreading way to people, and it is a totally utopian information spreading way without restriction from wills of any organizations and social ideologies. From this point of view, there is no a complete Web 2.0 website in real life. However, in terms of core technology, the above point is based on previous ideas and thoughts, and Web 2.0 is a technology application to realize such an idea. At present, these applications include blogger or Blog, instant messaging (IM), social network service (SNS), rich site summary (RSS), encyclopedia (WIKI), and peer-to-peer networking (P2P), etc.". Through the above, it is necessary for us to realize that the more important point of Web 2.0 is its social participation, openness, creativity, individuation, etc.

13.3 Concept and Role of Network Course

The so-called "network course" refers to a kind of experience transferring activities, which is implemented through network for achieving certain expected learning result. In other words, network course is the sum of the teaching contents and activities of a subject presented and implemented through network. Generally, it comprises of two parts: teaching contents and network teaching support environments organized according to certain teaching objective and strategy. The "network teaching support environments" mainly refer to the software tools and teaching resources supporting network teaching, and the teaching activities implemented on the online teaching platforms. From the perspective of education, network course can be divided into higher education network course, training network course, and basic education network course. However, from the perspective of teaching function, network course is divided into tuition network course, auxiliary network course, and learning network course. In this paper, according to learning-oriented network course, the basic characteristics of network course are analyzed, and also the construction principle and model are discussed.

The important role of network course is that it is under the guidance of the learning theory of constructivism. Constructivism is more appropriate for the teaching designs of some complex learning areas and senior learning goals, and is able to make up the limitation (i.e. excessively separating and simplifying teaching contents) of traditional teaching designs by introducing a "new psychological set" to the field of teaching design. Meanwhile, it is helpful for cultivating the team work spirit of students, and expanding way of thinking and creation ability.

13.4 Design of Network Course Graphic Animation Based on Web 2.0 Technology

13.4.1 Course Description and Objective

In this course, introduction is mainly given to the use of Flash drawing tool, graphics editing and bitmap processing, inputting external materials, making of time axis, map layer, components and basic animations, processing of sounds, complex animation production, action script design and application, audio and video animation production, small dub animation production, and testing and publishing of videos. Its purpose is to develop the ability of students to use special software for carrying out multimedia originality, animation design and animation production with animation production.

13.4.2 Analysis of Characteristics of Network Learners

According to the visual appetites of learners, priority is given to a blue interface in this network course, helping students to learn better in a clear and lively interface.

In the mean time, the exchange BBS and blog modules are designed in combination with the advantages of network for learners to have interaction with each other.

13.4.3 Design of Cooperative Learning

Cooperative learning designed in this network course refers to the cooperative learning under the supports of computers. It mainly breaks through the limit of range, and owns the controllability on the computer supported cooperative learning (CSCL) interaction [4].

Through CSCL environment, the problem situations can be displayed comprehensively, and also rich online resources are available. In the network course, cooperative learning is realized through rich resources, Web 2.0 communication BBS, blogs and so on.

13.4.4 Design of Independent Study

In the course website, a knowledge system for course learning is established, and also website navigation menu, site navigation form, resource download, works appreciation, online testing, blogs and related learning network links have been designed clearly.

13.5 Design of the System of Network Course Graphic Animation Based on Web 2.0 Technology

13.5.1 Design of Structural Framework of Network Course Graphic Animation

The network course designed by the author in this paper is mainly divided into front end and back end.

Front end of the website mainly refers to the learning interface of learners.

Back end of the website mainly refers to the renewal and modification that are made by management personnel or teachers on the managements and contents of the whole website.

More specifically, front end structure of the website is mainly divided into eight parts, namely teaching staff, course teaching, teaching conditions, resource download, works appreciation, online examination, communication BBS, and blogs; back end mainly comprises of user management, news management, articles management, conventional settings, and technical support, etc.

13.5.2 Introduction to Related Functional Modules

The learning resources in this website can be downloaded or learnt online without any registration. Users are only required to register if they want to access “online examination”, “exchange BBS”, and “blog” modules.

13.5.2.1 Online Examination System

The function of the online examination system is to allow teachers to set contents for examination in advance and permit each student to access the examination after registration.

In the examination, students can only submit the online paper once. The examination is conducted within 60 min.

When students answer questions overtime in the examination, the system will help them to automatically submit papers.

Although the examinee only has one chance to take part in the examination, he is still permitted to review questions and check his answers. Therefore, he can know his own shortcomings, and will continue to make every effort to the necessary learning aspects.

13.5.2.2 Exchange BBS

The exchange BBS in the website is aiming at providing a space for learners to study. In fact, it is a dynamic web page system, which is supported by a powerful database on back end, and also allows users to make an adjustment to its front-end displaying style and customize special functions.

It owns not only the characteristics of displaying knowledge by categories like static web pages, but also very excellent interactivity and sharing. Besides, it can allow users to send a private message with inside information system if necessary to have a one-to-one exchange.

Therefore, the functional module (exchange BBS) is specially designed, aiming at allowing learners to have one-to-one and one-to-many discussion and exchange, and thus making an enhancement to the interactivity and sharing of website.

13.5.2.3 Blog Diary

The blog diary functional module is designed based on Web 2.0 technology. It owns powerful interactive and sharing features. In such a functional module, learners can write their own learning experiences unconventionally, make discussions on professional technical problems, etc.

In the mean time, teachers and other learners may provide their helps in writing blogs. Such a way is beneficial to the learning of learners and making an improvement to the ability of students in communication and writing. Also, the learners can freely choose the favorite styles and skins according to their actual needs, making their own blog more special online.

Through the exchanges in BSS, the communication between teachers and students can be improved. When there are problems in the self-learning of the learners, they can leave a message in BSS. Then, teachers and others will provide them with a solution when they receive the message.

The exchange BBD of this network course includes three parts mainly: technical exchange module, homework discussion module, and recreation and friends-making module. Through the combination of these three modules, learners can carry out an in-depth discussion on professional Flash technology and assignments from teachers, and also can make friends with people having a common goal, thus laying a solid foundation for them to learn the course.

Besides, the learners can write down their own learning experience through the blog diary function of the network course. In such a way, they can feel progress every day.

Therefore, learners can not only learn knowledge, and also know experience of more learners, promoting them to save twists and turns in learning process and get easier in learning the network course.

At the same time, the emergence of blog diary can also make an increase to the writing ability of students from a different perspective.

In addition, there is a "management" item designed in the blog diary function, allowing users to change skin on their own and make their own space more personalized.

13.6 Conclusion

From above analysis, it can be known that a simple homepage has been designed for the website. Through clear navigation system and convenient searching tool, learning can become much simpler for learners.

Also, the simple and user-friendly interface design can help learners enter learning better and better and improve the learning efficiency.

Contents in the website include works appreciation, exchange BBS, resource download, online examination, teaching team, and teaching conditions, etc. With the website, students can not only learn with courseware online, and also download it to their own computers for learning.

Through online examination, teachers can get a real understanding of what students have learnt, and simultaneously students can test their own level. The management of these contents can be implemented in website back end through classification, aiming at promoting content management and update of network course to become more convenient.

For example, new functional modules and course contents can be added; examination contents or works appreciation contents can be increased or changed; blog diary, BBS contents and membership managements can be implemented; materials can be modified or downloaded.

In addition, the teaching contents can be changed by teachers at any time in accordance with the actual conditions of learners, aiming at promoting learners to accept knowledge better and better.

However, the shortcomings of the website are also in existence. For example, the functional modules such as the blog, BBS and online tests of the website are still independent systems of each other, and have not been integrated in the website further; learners are required to register as a member if necessary to access blog, BBS and online tests, and then are allowed to enter these systems for exchanging, learning and testing.

References

1. Shang Q et al (2009) Web 2.0 technology and its application in education. Primary Middle Sch Educ Technol 1(9):184–189
2. Jiang J (2004) Discussion on characteristics, construction principles and construction model of network courses. E-education Res 03(7):72–76
3. Zhao H (2007) Discussion on Web 2.0-based interactive environment design of network courses. China Educ Info 10:37–43
4. Zhang H, Wang H (2008) Research on the function framework and development technology of course based on Web 2.0 technology. Mod Educ Technol 11:456–461

Chapter 14
Efficient QoS Scheme in Network Congestion

Bencheng Yu and Chao Xu

Abstract When all kinds of unexpected traffic under certain bandwidth resources for an indefinite time will cause severe network congestion, reasonable design and management the bandwidth to avoid network congestion problem is probably the main technology. This essay uses QOS technology, sets the bandwidth allocated bandwidth to do the speed limit enterprise resource management through committed access rate (CAR) and configure traffic shaping (GTS), enabling data transmission flow uniformity, to avoid a large burst of network congestion.

Keywords Network congestion · QOS · Burst · Data flow

14.1 Introduction

The delivery services under best-effort service provided by Network make business transferred as soon as possible, but there is no clear time and reliability [1]. With the rapid development of network multimedia technology, there is the endless stream of multimedia applications on the network, such as distance education, VOD, video conferencing, IP telephony and multimedia real-time business. Internet has moved from the single to the data transmission network of data, voice, video, and other multimedia information integrated transmission network evolution [2]. These different applications require different Qos requirements, and Qos bandwidth, usually measured delay, delay variation and packet loss rate.

It is clear that existing transfer service has been unable to meet the different requirements of various applications on network transmission quality, it needs

B. Yu (✉) · C. Xu
Xuzhou College of Industrial Technology, Xuzhou, Jiangsu, China
e-mail: benchengx284@126.com

W. Du (ed.), *Informatics and Management Science VI*, Lecture Notes in Electrical Engineering 209, DOI: 10.1007/978-1-4471-4805-0_14,

Internet provide a wide range of quality of service type business. And best services will still be available to those who need only connectivity applications.

Qos refers to any combination of performance properties of service. In order to have value, these properties must be available, manageable, scalable authentication and billing, and in use, they must be consistent, predictable, and some properties, and even play a decisive role. In order to meet the needs of various user applications, building optimal and IP that have a variety of network quality of service mechanism is absolutely necessary [3]. Dedicated line services, voice, file transfer, store-and-forward; interactive video broadcasting and video are some examples of existing applications.

Under the limited network resources, network resource robbing situation must exist, then there must be quality of service requirements [4]. Quality of service is a relative of network business, while maintaining quality of service for certain types of business, may be at the expense of quality of service for other business. For example, in the case of fixed total network bandwidth, if certain types of businesses take up more bandwidth, so other businesses will use less bandwidth, may impact other business uses. Therefore, network managers need planning and allocating network resources to be reasonable according to various characteristics of the business, so that network resources are used efficiently.

Many enterprises have access to DDN line, bandwidth is able to meet the needs of enterprises under most cases, but often volatile reaction to the network, especially at a critical moment, network congestion occurs. After analysis and network monitoring, it has been found that, due to individual departments had a larger burst at this time, consuming the most bandwidth, causing the slow speed of the network as a whole [5]. This essay uses Committed Access Rate and Traffic Shaping technology of QOS, through the speed limit or other form to a hard limit to the bandwidth of these sectors to address these issues.

Quality of service (QOS) is a solution to network congestion, the basic idea is to classify data, onto a different queue, and then decide to transfer or guaranteed bandwidth depending on the type of data, which means that the case does not increase the bandwidth, uniform flow of the transmission of data, to avoid congestion.

14.2 Case Simulation

Simulating a QOS application in this article's case, by two technologies: committed access rate (CAR) and GTS (traffic shaping configuration) to address these issues. Topology design as the Fig. 14.1 shows:

From a topology can be seen:

The IP of interface E0/0 of the gateway router R1 is: 200.200.200.1/24, E0/1 IP is: 192.168.1.1/24.

The PC1 was represented our PC that is going to be limited with the speed, its IP is: 192.168.1.2/24, gateway is a router interface E0/1, address of 192.168.1.1.

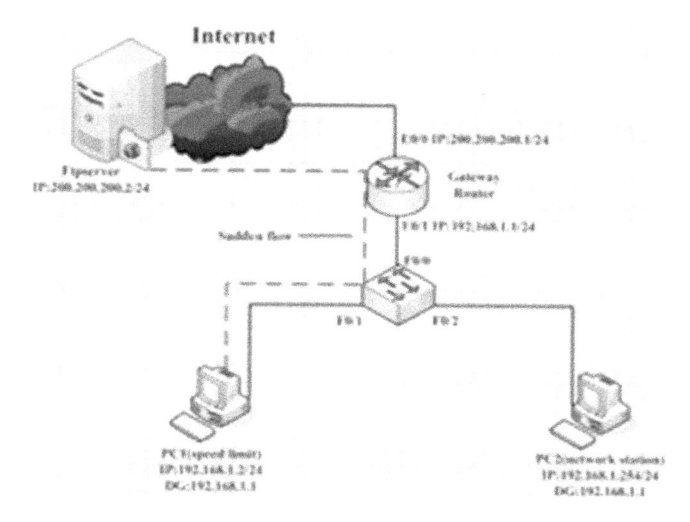

Fig. 14.1 Network topology

The PC2 was represented a network management station and its IP is 192.168.1.254/24, gateway: 192.168.1.1.

IP of FTP server located on the Internet is: 200.200.200.2/24

14.3 Solutions

Imagine that PC1 flow under normal working conditions are: 80 Kb/S, burst: 200 Kb/S, PC1 flow limitation in 100 Kb/S to meet the normal requirements of host PC1, and avoid unexpected traffic congestion on the network. We can achieve this requirement under two kinds of QOS technologies.

14.3.1 Use CAR Technology

Committed Access Rate (CAR) is a QoS feature in your router, they set a standard based on a flexible set and by the IP precedence or QoS group information packet classification limiting the input or output on the interface or interface transfer rates. CAR can be used to rate limiting the traffic based on specific matching criteria, such as the introduction of the interface, IP precedence, or IP access-list standard QoS group. CAR provides configurable behavior, such as when a communication meet or exceed the rate limit, transmission, discards, setting priorities or set QoS group.

1. Configure the ACL, defining needs of plastic flow, as shown in Fig. 14.2.

```
QOS-Router(config)#
QOS-Router(config)#access-list 100 permit ip any host 192.168.1.2
QOS-Router(config)#
QOS-Router(config)#
```

Fig. 14.2 Configuration ACL

Description: traffic will be set to PC1 and plastic

2. Enter the interface pattern, configuring traffic shaping parameters, as shown in Fig. 14.3.

```
QOS-Router(config)#
QOS-Router(config)#int e0/0
QOS-Router(config-if)#rate-limit input access-gr
QOS-Router(config-if)#$input access-group 100 800000 40000 80000 confor
QOS-Router(config-if)#$40000 80000 conform-action continue exceed-actio drop
QOS-Router(config-if)#
QOS-Router(config-if)#
QOS-Router(config-if)#
QOS-Router(config-if)#
QOS-Router(config-if)#
```

Fig. 14.3 Configuration parameters

Note: the above command that matches the access control list of 100 traffic, speed limitation is for 100 Kb/S, the biggest burst of 8,000 byte, meet the traffic being forwarded, excess traffic is discarded; In addition, configuring the CAR at E0/0 or E0/1 on the interface is possible, but note the key configuration parameters are input or output. Note here of units is easy engaged wrong, the first parameter is 800,000 said commitment access rate (CIR) it of range is: 8000–2000 000 000, units for Bits per second, the second parameter is 40,000 said burst volume (BC) it of range is: 1000–512 000 000, units for byte, the third parameter is 80,000 said BE (excessive burst volume or maximum burst volume) it of range is: 2000–1024 000 000, the unit is byte.

3. Check the configuration.

Through the above configuration now to view the configuration information, use the show interface E0/0 rate-limit port speed limit information, as shown in Fig. 14.4.

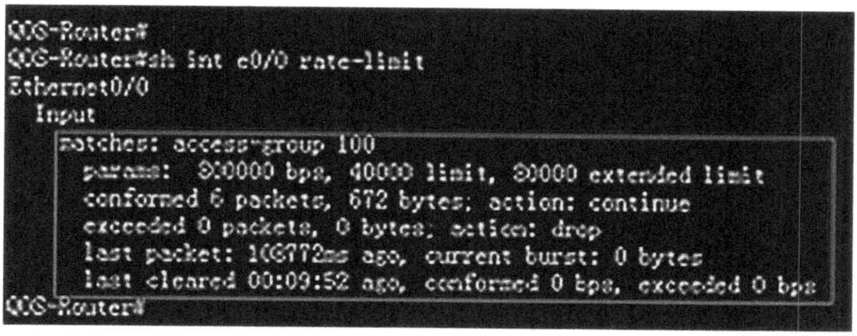

```
QOS-Router#
QOS-Router#sh int e0/0 rate-limit
Ethernet0/0
  Input
    matches: access-group 100
      params:  80000 bps, 40000 limit, 80000 extended limit
      conformed 6 packets, 672 bytes; action: continue
      exceeded 0 packets, 0 bytes; action: drop
      last packet: 106772ms ago, current burst: 0 bytes
      last cleared 00:09:52 ago, conformed 0 bps, exceeded 0 bps
QOS-Router#
```

Fig. 14.4 Configuration inspection

Note: the ACL matched is shown in first row of the box, here are 100, the corresponding is PC1, the second row displays the configured rate limiting, PC1 is limited to 100 Kb/S and burst capacity is 40,000, the maximum burst is 80,000. The third row shows the traffic statistics consistent with it, the action is continuing forward. Fourth line shows the violation of traffic statistics, Fifth line shows the current flow in the sudden volume of information; the last line is the rate of exchange of information.

14.3.2 Use GTS Technology

1. Configure the ACL, defining needs of plastic flow, as shown in Fig. 14.5.

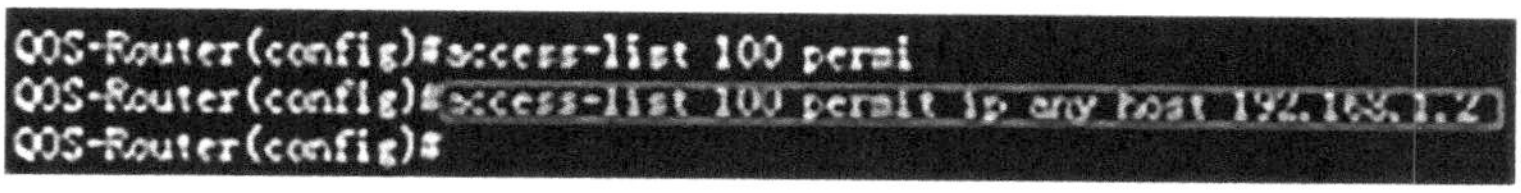

```
QOS-Router(config)#access-list 100 permi
QOS-Router(config)#access-list 100 permit ip any host 192.168.1.2
QOS-Router(config)#
```

Fig. 14.5 Configuration ACL

Description: traffic will be set to PC1 and plastic

2. Class map defined to match a flow or agreement, as shown in Fig. 14.6.

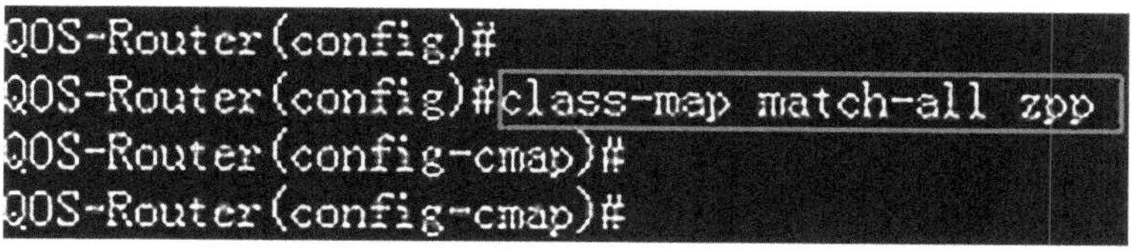

```
QOS-Router(config)#
QOS-Router(config)#class-map match-all zpp
QOS-Router(config-cmap)#
QOS-Router(config-cmap)#
```

Fig. 14.6 Definition class map

Description: ZPP is the class name of the map you created, match-all match class map defined all conditions, is the default configuration.

3. Enter the class map mode, configuration match the criteria, as shown in Fig. 14.7.

```
QOS-Router(config-cmap)#
QOS-Router(config-cmap)#match access-
QOS-Router(config-cmap)#match access-group 100
QOS-Router(config-cmap)#
```

Fig. 14.7 Matching conditions

Description: defined here is to match the configuration of the access control list of 100; Command match Protocol, protocol match Configuration Protocol can also be used. In addition you can also match information such as port, VLAN. Such as ports that match with the command input-interface interface-type interface-number can match incoming traffic, match any matches any packets with the command, use the commands match source-address | Destination-address Mac Mac-address matches either the source or destination MAC address.

4. Define flow control policy, as shown in Fig. 14.8.

```
QOS-Router(config-cmap)#
QOS-Router(config-cmap)#exit
QOS-Router(config)#policy-map zpp
QOS-Router(config-pmap)#class zpp
QOS-Router(config-pmap-c)#
QOS-Router(config-pmap-c)#
```

Fig. 14.8 Flow control strategy

Description: defines the policy map called ZPP, and call the class map ZPP here policy map can be of different names and class names map, in order to facilitate memory can be configured the same. After you configure the policy map and call the class map marked traffic can be configured according to the actual situation, traffic shaping, CAR, and more.

```
QOS-Router(config-pmap-c)#shape av
QOS-Router(config-pmap-c)#shape average 800000
QOS-Router(config-pmap-c)#sha
QOS-Router(config-pmap-c)#shape max-bu
QOS-Router(config-pmap-c)#shape max-buffers 100
QOS-Router(config-pmap-c)#
QOS-Router(config-pmap-c)#
```

Fig. 14.9 GTS configurations

5. Configuration GTS, as shown in Fig. 14.9.

Description: shape average 800,000, indicating that the speed limit is 100Kb/S, average value. Shape Max-buffers 100 defined buffer on line of 100 (the default is 1,000), GTS buffer is used to cache the queue information.

6. Application of policy on the interface, as shown in Fig. 14.10.

```
QOS-Router(config-pmap-c)#exit
QOS-Router(config-pmap)#exit
QOS-Router(config)#int e0/1
QOS-Router(config-if)#service-po
QOS-Router(config-if)#service-policy output zpp
QOS-Router(config-if)#
QOS-Router(config-if)#
```

Fig. 14.10 Application strategy

Description: apply GTS on the interface E0/1, command service-policy output ZPP refers application to the E0/1 interface on the output policy ZPP. We should pay attention to the fact that GTS applies only in output direction, on the E0/1 interface. CAR can be used in both directions.

7. Check the configuration

Now to verify that the configuration, as shown in Fig. 14.11

```
QOS-Router#
QOS-Router#sh policy-map
 Policy Map zpp
   Class zpp
     Traffic Shaping
        Average Rate Traffic Shaping
        CIR 800000 (bps) Max. Buffers Limit 100 (Packets)
QOS-Router#
```

Fig. 14.11 Configuration inspection

Description: we can view that ZPP is the name of the policy (the first line) by command policy map; the name of the class map is also known as ZPP (second row), it is emphasized that the two names can be different; After the third line showing the traffic shaping configuration such as: limit the rate of 100 Kb/S, buffer on line 100 (number of messages), as shown in Fig. 14.12.

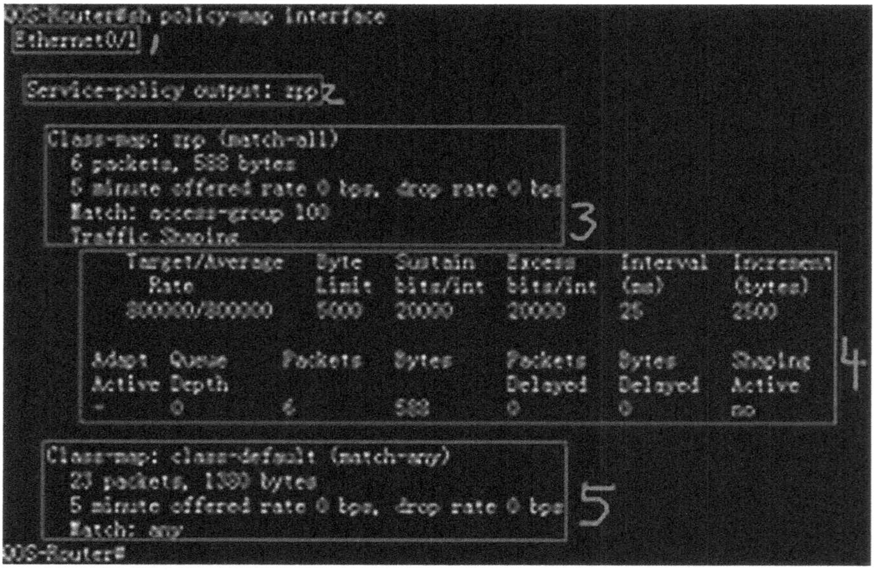

Fig. 14.12 Concrete results

Description: by show policy map interface commands, we can draw the following conclusions:

Box 1: Apply strategies of the definition to E0/1 interface.
Box 2: The application of strategy ZPP and direction of applications is output.
Box 3: The class map name was called ZPP, flow transferred and discarded as well as control lists information.
Box 4: Traffic shaping configuration parameters are displayed, individually explained above, is not perverse in here.
Box 5: The default information

14.4 Conclusions

Now we shall achieve the above requirements, PC1 burst is limited; the network's overall performance has been improved. Through this case, we can see that if enterprises cannot increase the network bandwidth to address network congestion is the method of rational utilization of network bandwidth more effectively. In the network when congestion occurs, according to the nature and needs of your business using the rational allocation of existing bandwidth QOS technology, reduce the impact of network congestion.

References

1. Xi-jiang Z (2012) The application of the port rate limit strategy basing on the Qos ACL. Inf Secur Technol 02:18–22
2. Liang J, Fu Y (2012) Qos guided selection approach for grid resource. J Ankang Univ 11(01):56–59
3. Hua R, Fu Y (2010) Research on QoS-guaranteed resource mapping policy in service grid. Comput Eng 9(15):78–84
4. Weng C, Lu X (2005) Heuristic scheduling for bag-of-tasks applications in combination with QoS in the computational grid. Future Gener Comput Syst 21(2):271–280
5. Foster I, Roy A, Sander V (2000) A quality of service architecture that combines resource reservation and application adaptation. In: Proceedings of the 8th international workshop on quality of service (IWQOS 2000) vol 245, pp 181–188

Chapter 15
Research of Access Control List in Enterprise Network Management

Bencheng Yu and Ran Wang

Abstract With the application of network communications in various enterprise constantly expanded, communications among internal sectors of enterprises and cooperation between different enterprises provide ways to achieve the resources sharing, business more easily carried out and efficiency improved, but while network interconnected also led to confidentiality of data and information of enterprise reduced, security of enterprise information has been affected, at the same time, the fact that staff use network arbitrarily, which caused huge threat on security of network. In order to guarantee the smooth of security network, the enterprises should consider restricting access to websites as well as illegal network access. If trying to manage the entire network through network security device, the overhead of enterprise will be greatly increased, so the factors of enterprise network building should be required to be taken into account. This essay illustrates solutions to demand for enterprise through designing complete possible access control list.

Keywords Network building · Access control list · Network security · Solution

15.1 Introduction

As the Web applications in a variety of business operations is becoming more and more important, the web application provides ways for communications among internal sectors of enterprises and cooperation between different enterprise, as well

---

B. Yu (✉) · R. Wang
Xuzhou College of Industrial Technology, Xuzhou, Jiangsu, China
e-mail: benchengx284@126.com

W. Du (ed.), *Informatics and Management Science VI*, Lecture Notes in Electrical Engineering 209, DOI: 10.1007/978-1-4471-4805-0_15,

as resources sharing [1]. But while network interconnected also led to confidentiality of data and information of enterprise reduced, security of enterprise at the same time, the fact that staff use network arbitrarily, which caused huge threat on security of network. For example, staff feels free to visit the Ministry of Finance's information, download movies, play games, chat, in addition, personnel open Telnet port of the computer in order to facilitate the management, which has a serious threat. In order to protect the sector network, enterprises usually limit their online time and access to the website, try to deny illegal network requests, now there are many ways to implement these requirements, Such as: encryption, administrator permissions enhanced, internal LAN building, But they can only achieve some simple internal function management, lack of flexibility is visible [2]. If you use some of the network security devices to manage the entire enterprise network, so that for larger enterprises, enterprise communications between headquarters and branches, communications throughout the enterprise, the communications between departments, which greatly increases the overhead of enterprise. All these problems can be solved through the access control list, using access control lists to improve the management and maintenance of enterprise network security, while reducing corporate overhead.

15.2 Solution Design

15.2.1 Company Profile

First Division VLAN for 4 departments that is HR, software development, Web site design and the Finance Department [3, 4]. Corresponding to VLAN2, VLAN3, VLAN4 and VLAN5 individually, and then assigns them to the appropriate IP network segment, then according to the needs of enterprises, as shown in Fig. 15.1 uses extended ACL to control restrictions that are based on IP address access between the departments.

Set Int F0/24 of the switch port in trunk mode, and connect the F0/1 port of the router.

Assign Int F0/1 and F0/2 port of the switch to HR vlan2 (alternate one port, departments of their respective expansion switches, following similar).

Assign Int F0/3 and F0/4 port of the switch to the department of software development vlan3. Assign Int F0/5 and F0/6 of the switch ports to the department of website design vlan4.

Assign Int F0/7 and F0/8 of the switch port to the Ministry of finance vlan5

So dividing VLAN in this way staff of each department can take interaction within the same VLAN, joint problem-solving, and visit can not be paid between different departments in different VLAN, but administrative sector has access to various departments.

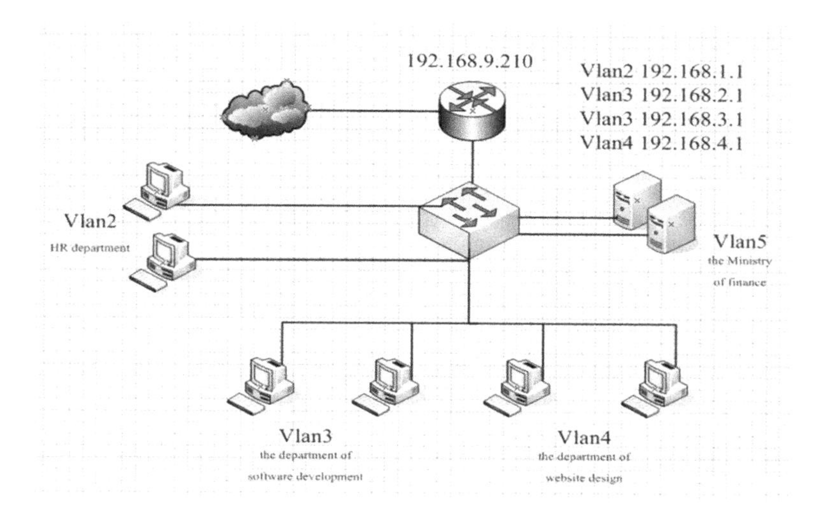

Fig. 15.1 Enterprise department division

15.2.2 The Access Control List Protects the Enterprise Network Security

As shown in Fig. 15.2, using access control lists on the internal servers and clients can protect internal network security business, and when the corporate network connect to the Internet from outside, ACL can prevent from attack by hackers. Internal access control list (ACL) can help to protect the network security from internal damage, for example, employee theft company confidential documents, the competitive activity causes internal damage.

According to the analysis of the security and business requirements analysis, the security policy for the enterprise is determined; the ACL can be used to build a network for the enterprise firewall.

Enterprise network firewall security policies specify the rules, including which external data packet is allowed or prohibited passing by and which network services allowed inside the system. Settings Access control list (ACL) on the router can complete the company's security policy to implement a network firewall feature

15.2.3 Routers and Switches

ACL on Enterprise router can improve security. Limit the type of transmission in the network to improve performance and to reduce network vulnerability, as a consequence, internal attacks, Trojan and worm attacks can be prohibited [5].

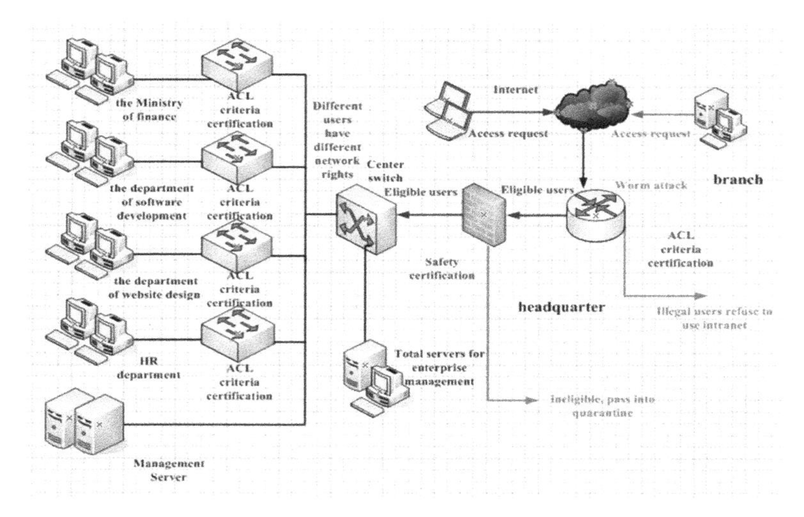

Fig. 15.2 Enterprise ACL function

Through the analysis of virus, we can see the fact that the virus is mainly launching an attack through the TCP 135, 136, 137, 138, 445, port, UDP port 69, 135, 136. The router, as the internal computer, can restrict these ports through, so that it can prevent the virus from the external network as well as prevent viruses within the router go out of it. In the meantime, while VLAN on the switch in the Central Division can also prevent the spread of the virus from one subnet to another.

```
Router(config)#access-list 101 deny tcp host 192.168.4.3 host 192.168.6.2 eq 135
Router(config)#access-list 101 deny tcp host 192.168.4.3 host 192.168.6.2 eq 136
Router(config)#access-list 101 deny tcp host 192.168.4.3 host 192.168.6.2 eq 137
Router(config)#access-list 101 permit ip any any
Router(config)#int e0
Router(config-if)#ip access-group 101 in
```

15.2.4 Ministry of Finance

Because of the confidential financial data of the database server, not anyone can get access to the data above, access control list ACL will be used by now, we can set the list as follows: Which host has the right to get access to financial database servers, and when other hosts outside the list want to visit server of the Finance Department, they will be filtered out.

```
access-list 101 deny ip 192.168.1.0 0.0.0.255 192.168.4.0 0.0.0.255
access-list 101 deny ip 192.168.2.0 0.0.0.255 192.168.4.0 0.0.0.255
access-list 101 deny ip 192.168.3.0 0.0.0.255 192.168.4.0 0.0.0.255
access-list 101 permit ip any any
```

HR department, software development department and Web Design department are prohibited from entering into finance Department server.

15.2.5 Software Development Department

Because software development department will need to write a large number of programs, most of the time is used for programming, but sometimes needs access to the Internet searching for information [6]. We can use time-based ACL to restrict access to the Internet. Software development department employees are only allowed to visit 172.16.3.0 FTP servers at 3–5.p.m. from Monday to Friday. While the FTP resources can not be download in other working hours.

```
Router configuration commands:
Router#show clock
Router#clock set
Router(config)#time-range worktime
Router(config-time-rang)#absolute start 15:00 end 17:00
Router(config-time-rang)# periodic weekdays monday to firday
periodic weekend 15:00 to 17:00
periodic weekdays monday to Friday.
access-list 101 deny tcp any 172.16.5.13 0.0.0.0 eq ftp time-range aaa
access-list 101 permit in any any
int E1
ip access-group 101 out
```

Time-based ACL is more suitable for time management. Users can access the 172.16.3.0 server only in the range from 3 to 5 pm from every Monday to Friday.

15.2.6 Department of Website Design

This department is responsible for designing website work in accordance with requirements. Staff is not allowed to surfing Internet due to the importance of their jobs. On the other hand, it needs constantly communicate with customers so that a specific host may be needed to achieve remote communication, here we can set a specific host to visit outside network and can pass through telnet.

```
access-list 1 permit host 192.168.2.2
access-list 1 deny any
int s0/0
ip access-group 1 out
line vty 0 4
access-list 1 in
```

15.2.7 HR Department

HR department mainly is responsible for recruitment, so it is necessary to go surfing through Internet to seek for the requirements from various enterprises. If during this time someone makes online chatting, plays game and downloads other information that is nothing to do with work, which is bound to cause Enterprise flow wasting, so setting a limitation of some other types of data flow can limit the access to enterprise internal and external network, from example, staff is only allowed to receive external message and visit some fixed websites.

When you use ACL to prevent QQ, QQ landing port (TCP/UDP) 8,000

```
Router(config)#access-list 101 deny udp any eq 8000
Router(config)#access-list 101 permit udp any any
Router(config)#access-list 101 permit tcp any any
Router(config)#access-list 101 permit ip any any
Router(config)#Int e0/0
Router(config)#Ip access-group 101 in
```

We can restrict P2P such as BT entering into network software through router ACL, reducing the waste of corporate network resources, limiting BT and eMule.

```
access-list 101 deny tcp any rang 6881 6890
access-list 101 deny tcp any rang 6881 6890 any
access-liat 101 permit ip any any
```

15.2.8 Server Management:

ACL is mainly used to manage restriction between the various departments, facilitating the Enterprise Manager to know data information about the enterprise easily and timely. By setting the ACL on the management server, so that managers can access all of the departments freely while other departments do not have permission to access the management server.

```
access-list 101 permit ip 192.168.5.0 0.0.0.255 any
access-list 101 deny any
ip access-list extended outfilter
```

15.2.9 Prohibited Network Visit Outside the Sectors for Specific Hosts

As shown in Fig. 15.3, host1 of website design department is prohibited from access extranet

Host1 192. 168. 4. 2
192. 168. 4. 1
192. 168. 5. 1
router0
192. 168. 5. 2
Host2 192. 168. 4. 3
router1

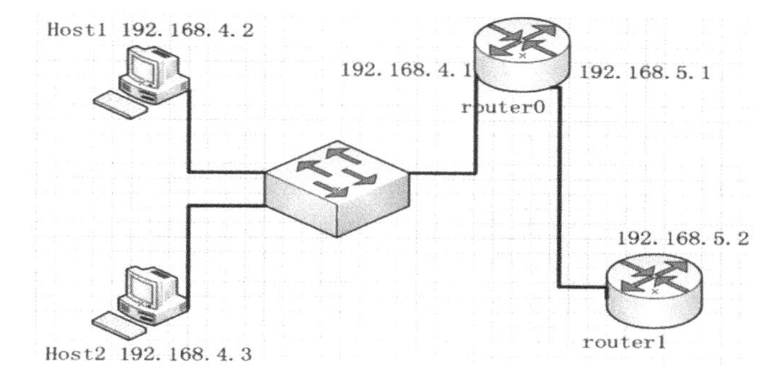

Fig. 15.3 Department of web design

Configuration of the router Router0:

Router0(config)#access-list 1 deny host 192.168.4.2
Router0(config)#access-list 1 permit any
Router0(config)#int s0
Router0(config-if)#ip access-group 1 out

Testing router configuration of Router0:

Show run
Show access-list 1
Show ip int s0

Host1 and Host2 respectively host ping on 192.168.5.2. The results as shown in Fig. 15.4, it showed that destination network does not pass through.

As shown in Fig. 15.5 and Fig. 15.5 host2 ping normally pass, it shows that host1 is prohibited from access to the external network.

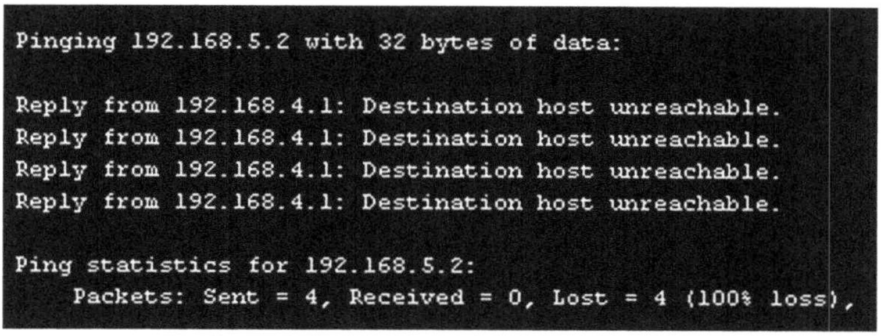

Pinging 192.168.5.2 with 32 bytes of data:
Reply from 192.168.4.1: Destination host unreachable.
Reply from 192.168.4.1: Destination host unreachable.
Reply from 192.168.4.1: Destination host unreachable.
Reply from 192.168.4.1: Destination host unreachable.
Ping statistics for 192.168.5.2:
Packets: Sent = 4, Received = 0, Lost = 4 (100% loss),

Fig. 15.4 Host1 ping

```
Pinging 192.168.5.2 with 32 bytes of data:

Reply from 192.168.5.2: bytes=32 time=62ms TTL=255
Reply from 192.168.5.2: bytes=32 time=31ms TTL=255
Reply from 192.168.5.2: bytes=32 time=32ms TTL=255
Reply from 192.168.5.2: bytes=32 time=31ms TTL=255

Ping statistics for 192.168.5.2:
    Packets: Sent = 4, Received = 4, Lost = 0 (0% loss),
Approximate round trip times in milli-seconds:
    Minimum = 31ms, Maximum = 62ms, Average = 39ms
```

Fig. 15.5 Host2 ping

15.2.10 Limit Speed on Router Port Traffic

Most common service ports are less than 3,000, most of the worms and P2P ports are more than 3,000, so limiting the 3,000 ports and the above can protect the general network application as well as avoid abuse of the network.

```
access-list 101 permit tcp any gt 3000
access-list 101 permit udp any gt 3000
interface f0/0
service-policy input xs
```

15.3 Conclusion

Detailed rules and business requirements are added to the ACL, and finally add the configured ACL to the desired port. Enterprise network administrators can effectively manage employee network of control based on ACL Protocol, various ports, the IP source and destination addresses. For the above solution, if used in an enterprise with many sectors, similar expansion can be made on the basis of this solution, which not only facilitates enterprise network management, but also improve enterprise network security as well as saving additional expenditure on enterprise networks.

References

1. Yuan F, Chen L (2011) Network management and security. Tsinghua University Press, Beijing, pp 23–28
2. Ye Z (2005) Acl security access control systems. Comput Appl Softw 22(3):111–114
3. Fan P, Li H (2004) Network access control technology research based on Acl. J East China Jiaotong Univ 21(4):89–92

4. Kaeo M (2012) Network security design kaeo, Ccie#1287 the People's Post and Telecommunications Press, vol 13, 100–105
5. Tang Z-J, Li H-C (2009) Application of network security management based on Acl. J Sichuan Univ Sci Eng (Nat Sci Ed) 01:79–85
6. Chun JJ, Tai MH (2000) Survey of intrusion detection research on network security. J Softw 11:57–63

Chapter 16
OSPF-Based Network Engineering Design and Implementation

Bencheng Yu

Abstract This essay Designs an enterprise network engineering, which solves the problem of poor device performance bottlenecks, and then improves the scalability of enterprise network engineering. Combining with the project, it discusses the enterprise network requirements, including a division of IP address and a description of network topology design. In addition, this essay focuses on the design and implementation of OSPF protocol-based network, and discusses the use and implementation of the agreement, providing the solution to enterprise network performance for different applications, providing protection for later maintenance and scalability and basic network model for all types of enterprises.

Keywords Network engineering · OSPF protocol · Performance · Safety · Model

16.1 Introduction

According to the enterprise scale and complexity degree of the network system, user demand for networks vary a lot, from simple file sharing, Office Automation to complex e-commerce, ERP and so on [1]. So requirements vary from each other according to Enterprise network performance based on application, network selection should be based on actual requirements [2]. Less network technicians work in non-professional network enterprise, while stronger dependency on the network, therefore network needed should be as simple as possible, reliable, easy to use, and reducing network usage and maintenance costs is very important [3, 4].

B. Yu (✉)
Xuzhou College of Industrial Technology, Xuzhou, 221000 Jiangsu, China
e-mail: benchengx284@126.com

W. Du (ed.), *Informatics and Management Science VI*, Lecture Notes in Electrical Engineering 209, DOI: 10.1007/978-1-4471-4805-0_16,

Open Shortest Path First (OSPF) is an Interior Gateway Protocol (IGP) to a single autonomous system (autonomous system, AS) decision-making routes [5]. Comparing with RIP, OSPF is a link-state routing protocol, RIP is a distance vector routing protocol [6]. The protocol management distance (AD) of OSPF was 110.

16.2 OSPF Dynamic Routing Protocol

16.2.1 Advantages of OSPF Protocol

OSPF is a true LOOP-FREE (no routing loop) routing protocol. The advantages are derived from the algorithm itself (Shortest path tree and link state algorithm).

OSPF convergence speed: it can pass routing changes throughout the autonomous system in the shortest amount of time [7].

The concept of area Division is put forward. The amount of routing information needed to be passed is greatly reduced through the summary of routing information between the areas. After the autonomous system is divided into different region, further avoiding the rapid expansion of routing information following with the network scale.

Control the cost of the agreement itself to the minimum:

Hello packet that does not contain routing information, which is used to discover and maintain neighbor relationships, is very short. Messages that contain routing information is the new mechanism triggered. (Sent only when a route changes). But in order to enhance Protocol robustness, every 1,800 s it will be sent again.

In broadcasting network, using a multicast address (non-broadcast) newspaper delivery can reduce interference to other devices that are not running OSPF network.

In multiple access networks (broadcast, NBMA), through the election of DR, routing and switching between the routers and network segments (synchronization) from o (N*N) time is reduced to o (n) time.

Put forward the concept of STUB areas, making the STUB no longer spread introduction of ASE route in the region. ABR (area border router) support on route aggregation, further reducing the routing information between the area.

In a point-to-point interface types, properties, by configuring the on-demand broadcast (OSPF over On Demand Circuits), OSPF is no longer scheduled sending Hello packets and routing information updates on a regular basis. Only sending updates when real changes occur in network topology information.

Routes are divided through a strict separation of levels (four), providing more reliable routing.

Good security, OSPF supports clear text and MD5 authentication based on the interface.

OSPF adapt to all sizes of networks, up to thousands of units.

16.2.2 OSPF Communication Between the Different Regions and Certification

F0/0 interfaces of R1 and R2 belong to area0, F0/1 interface of R1 and R2 belong to Area1.

F0/0 and F0/1 interface of R3 and R4 belong to arae1.

R1 configuration: the configuration of R1 and R2 is the same.

```
Router(config)#int f 0/0
Router(config-if)#ip add 192.168.10.2 255.255.255.0
Router(config-if)#int f 0/1
Router(config-if)#ip add 192.168.20.2 255.255.255.0
Router(config-if)#router ospf 100
Router(config-router)#net 192.168.10.2 0.0.0.255 area 0
Router(config-router)#net 192.168.20.3 0.0.0.255 area 1
Router(config-router)#no au
Router(config-router)#exit
```

R3 configuration: Configuration is the same for R3 and R4

```
Router(config)#int f 0/0
Router(config-if)#ip add 192.168.20.3 255.255.255.0
Router(config-if)#int f 0/1
Router(config-if)#ip add 192.168.40.2 255.255.255.0
Router(config-if)#no sh
Router(config-if)#int f 0/
*Mar  1 00:16:46.823: %LINK-3-UPDOWN: Interface FastEthernet0/1, changed state to up
*Mar  1 00:16:47.823: %LINEPROTO-5-UPDOWN: Line protocol on Interface FastEthernet0/1, changed state to up
Router(config-if)#int f 0/0
Router(config-if)#no sh
Router(config-if)#int f 1/0
Router(config-if)#
*Mar  1 00:16:52.591: %LINK-3-UPDOWN: Interface FastEthernet0/0, changed state to up
*Mar  1 00:16:53.591: %LINEPROTO-5-UPDOWN: Line protocol on Interface FastEthernet0/0, changed state to up
Router(config-if)#ip add 192.168.50.2 255.255.255.0
Router(config-if)#no sh
Router(config-if)#router ospf 100
Router(config-router)#net 192.168.20
*Mar  1 00:17:16.775: %CDP-4-DUPLEX_MISMATCH: duplex mismatch discovered on FastEthernet0/0 (not full duplex)
net0/1 (full duplex).
Router(config-router)#net 192.168.20.3 0.0.0.255 area 1
Router(config-router)#net 192.17
*Mar  1 00:17:26.319: %OSPF-5-ADJCHG: Process 100, Nbr 192.168.20.2 on FastEthernet0/0 from LOADING to FULL,
Router(config-router)#net 192.168.40.2 0.0.0.255 area 1
Router(config-router)#net 192.168
*Mar  1 00:18:16.771: %CDP-4-DUPLEX_MISMATCH: duplex mismatch discovered on FastEthernet0/0 (not full duplex)
net0/1 (full duplex).
Router(config-router)#no au
Router(config-router)#exit
```

Change the duplex status of the interface R3f0/0 to modified duplex

```
Router(config)#int f 0/0
Router(config-if)#
*Mar  1 00:14:30.999: %CDP-4-DUPLEX_MISMATCH: duplex mismatch discovered on FastEthern
net0/1 (full duplex).sp
Router(config-if)#speed 100
Router(config-if)#du
Router(config-if)#duplex f
*Mar  1 00:14:38.199: %LINK-3-UPDOWN: Interface FastEthernet0/0, changed state to upu
Router(config-if)#duplex full
Router(config-if)#
Router(config-if)#exit
```

Enable OSPF interface redaction MD5 authentication work in Area0

```
Router(config)#int f0/0
Router(config-if)#ip ospf me
Router(config-if)#ip ospf message-digest-key 1 md5 cisco
Router(config-if)#ip ospf au
Router(config-if)#ip ospf authenticationme
Router(config-if)#ip ospf authentication me
Router(config-if)#ip ospf authentication message-digest
Router(config-if)#exit
Router(config)#
*Mar  1 00:19:14.375: %OSPF-5-ADJCHG: Process 100, Nbr 192.168.30.2 on FastEthernet0/0 from FULL to DOWN
```

16.3 Design

16.3.1 Design Principles and Objectives of Engineering

16.3.1.1 Design Follows the Following Principles

Practical and economical: network applications should be implemented in the processes, focus on the practical, building viable network of enterprises by using the principles of economy.

Advance and maturity: when organizing the network, the concept of the nature should be taken into account as well as its maturity, its technology, its equipment, structures, tools, not only shows the advanced properties, but also has the potential to guarantee after several years in a leading position.

Reliability and stability: in the case of advancement and maturity of technologies and products, pay attention to that system, structure, maintenance management, reliability, and stability. Ensure that the network is running properly.

Security and privacy: the purpose of the network is to share resources, a reliable and stable network must possess security and privacy, take different measures in different environments, network security and correct operation can be guaranteed.

Scalability and maintainability: because the network is not the same.

Considering the changes in network, you should design the simplest and most economical network, achieving scalability and maintainability requirements and increasing your network's performance.

16.3.1.2 Purpose of Engineering Design

With the rapid development of enterprise business, current network architecture can no longer meet the needs of enterprises, Enterprise network aimed primarily at the construction of figures in enterprise-wide information management, build a high performance, high security, ease of management and maintenance of enterprise networks to ensure the smooth of the network, easy to apply, will eventually be brought information management to enterprise, improving the quality and productivity of enterprises.

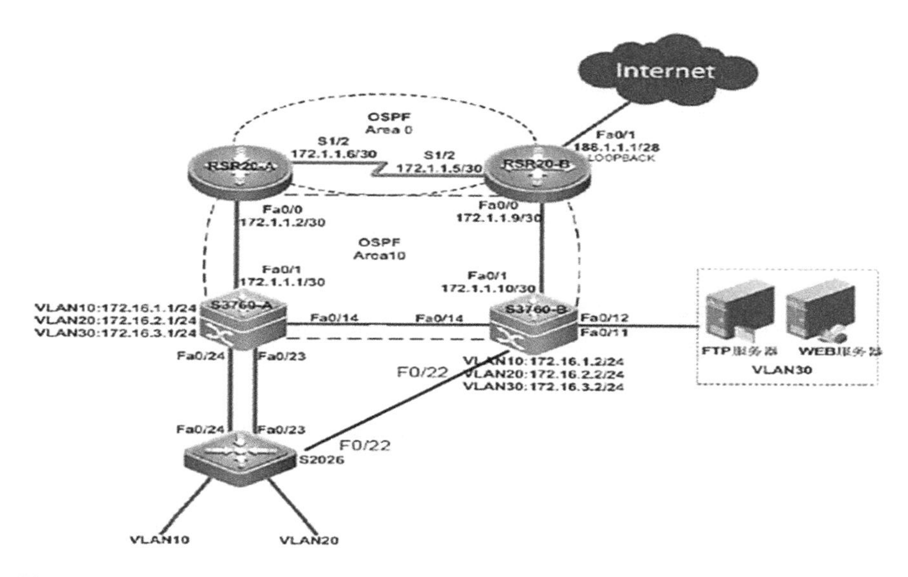

Fig. 16.1 Overall topology

16.3.2 General Topology

Figure 16.1.

16.4 Realization of the Project

16.4.1 Simulators: Stimulate the Entire Topology

Figure 16.2 and Table 16.1.

16.4.2 Connectivity and Technical Support of Full Network

16.4.2.1 In the Core Layer

OSPF routing protocol on the R3 and R4 is necessary by using OSPF in dynamic routing of the enterprise, which not only solved the problem that RIP is only allowed for smaller networks, while also solved that RIP consume too much network bandwidth, processor and memory resources.

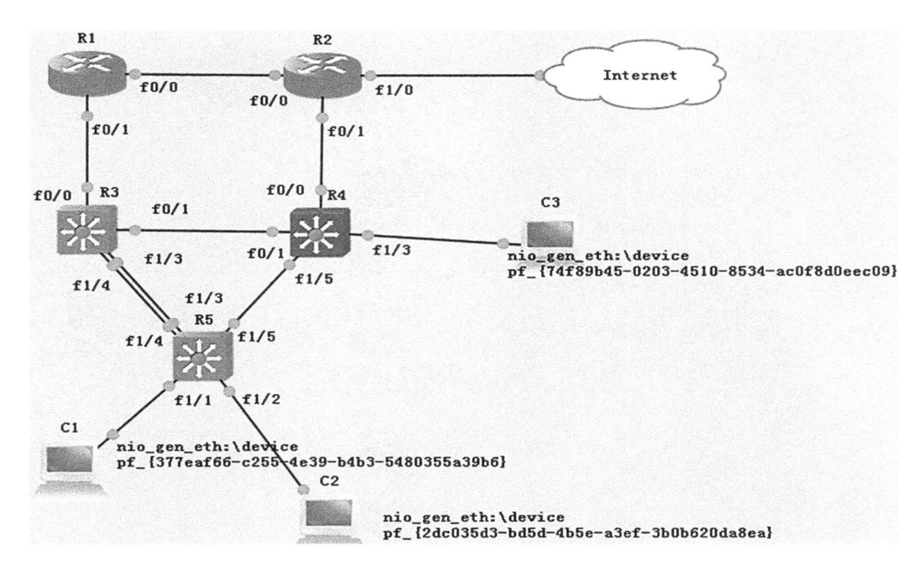

Fig. 16.2 Design with simulator

Table 16.1 IP address distribution

Equipment	Interface	IP	Corresponding equipment	Corresponding interface	Corresponding IP
R1	F0/0	192.168.10.2	R2	F0/0	192.168.10.3
R1	F1/0	192.168.20.2	R3	F0/0	192.168.20.3
R2	F1/0	192.168.30.2	R4	F0/0	192.168.30.3
R3	F0/1	192.168.40.2	R4	F0/1	192.168.40.3
R3	F1/3	192.168.60.2	R5	F1/3	192.168.60.3
	F1/4			F1/4	
R4	F1/5	192.168.50.2	R5	F1/5	192.168.50.3
R4	F1/3	192.168.70.2	C3		192.168.3.2
R5	F1/1		C1		192.168.528.2
R5	F1/2		C2		192.168.2.2

Put F0/0 of the R1 and R2 interfaces into area 0. F0/1 of R1 and R2 belongs to the area of 1. The goal is to realize the communication between different OSPF areas.

Return for testing on R2.

The most important thing is to open the OSPF MD5 authentication in this area on 0. If both interfaces of the router does not have the correct certificate password, an error must occur, the router does not work properly, this router has played very well for privacy and security.

16.4.2.2 In the Aggregation Layer

F0/0, F0/1, F1/0, F2/0 in R3 and F0/0, F0/1, F1/0, F2/0 in R4, F0/1 in R2 also obey OSPF routing protocol, at the same time they belong to the area 1, while using regional certification in area 1.

Because half-duplex is used in interface of Cisco layer three switch. Router interface uses full-duplex, duplex mismatch, so let the duplex mode applied in interface of the switch and router ports connected to the modified three-layer.

Divide two VLAN on the R5, and configure the gateway for C1 and C2 to normal communications.

R3, R4, and R5 can easily cause loops, because the switch does not recognize broadcast, so he will forever be broadcast, causing network congestion. STP spanning tree Protocol must be used to solve this problem. Why? Loop caused much trouble? Because in actual engineering process, something can't be avoided, if we only use a line and someday there is something wrong with this line, it cannot work, entire network will stop, this is immeasurable loss for the enterprise, so the benefits of links are in that if R1 road breaks down, R2 can work instead, during this time we can have adequate time to find the solution to R1 problem, making entire network run security.

Make VTP Protocol in effect on the three sets of three-layer switches, let R5 be client, R3 and R4 do server-side. Achieve unified management for VLAN. At the same time enable STP spanning tree Protocol work, let R3 follow VLAN 10, while V R4 follow LAN 20.

Do link aggregation and load balancing on the R3 and R5. The purpose of connecting the two root line is to accelerate transmission of network, in actual engineering process transmission information of network is large, so it brings much burden on to the device transmission, making its transmission slow and putting effect on efficiency, if connecting so, data package that only transmits in a line can be divided into two separate lines, improving its transmission efficiency greatly, but not for each group of lines because of possibilities of raising loop, so it only can be connected in the main road.

16.4.2.3 In the Access Layer

WEB server and FTP services on the C3, where associated with VM, simulation server. Select the network adapter on the virtual machine, do the FTP and WEB services in the Windows Server 2003 virtual machine.

On the C2 simulation for PC engine, select local loopback port on this computer, the interface is not existed, you need to find the information yourself and add it up, Windows 7 and Windows XP are not the same.

Select the second network adapter on the C1 of the virtual machine, analog PC, test connectivity.

The service can come true in virtual machine, while we can build experimental environment in simulation device, in addition, we can test network of Unicom sexual with this machine, so on to virtual machine and simulation device also be associated together, making it play their roles, it is very practical in actual work, because in actual work it is impossible to be dependent on each other, so it needs to associate with different environments, which is most effective approach feasible.

16.5 Conclusion

Enterprise networks require sub netting, while communication between subnets requires routes. There are two ways of Routes, static routing and dynamic routing. In general there are three dynamic routing in the enterprise network, EIGRP, RIP and OSPF. RIP is just a smaller network, and because of its principle, the larger the network is, the more network bandwidth processor and memory resources Rip consumes. OSPF and EIGRP are considering the need for large networks for larger business network. While EIGRP is Cisco proprietary, some users want to use open protocols. So finally only OSPF can be selected. This engineering can be applied for all types of business, mainly following several features, including the internal division of the business segment, IP allocation, achieving ease of management.

References

1. Du Y (2010) Research on dynamic configuration of OSPF protocol. Xi'an Univ posts telecommun 01:39–46
2. Gao Y (2010) Analysis of working principle of RIP and OSPF Protocol. China's new commun 17:487–493
3. Wang D (2010) Application of OSPF routing protocol in the multi areas. J Chongqing Univ sci technol (natural sci Ed) 02:41–47
4. Wang X, Zou R, Jun P, Li H (2010) Realization of OSPF in campus network. China Sci technol inf 06:401–409
5. Zhang S (2007) Study on OSPF routing protocol. Inf technol educ Fujian 07(03):381–386
6. Gui R, Zhang J, Huang JY (2010) Study on OSPF network types configuration based on different physical link. Micro-comput inf 54(12):1083–1092
7. Guo KH (2003) Design and realization of OSPF campus network routing in multi areas. Comp syst 3(08):1029–1036

Chapter 17
Design and Implementation of Network Fault Location

Chao Deng

Abstract The rapid increase of Internet users and continuous expansion of network scales have made the network maintenance and management highly complex, and the traditional network management model unable to satisfy the needs of large complex network management. In this paper, according to the shortcomings of existed network fault location technology, a fault location method based on topology correlation is proposed. The correlation of network topological structural nodes is applied in this method. According to the infectiousness of network fault, and by constituting input fault data as one-time fault event and using fault correlation algorithm to locate fault event's source, this method can provide users with fault location and improve fault processing time and efficiency via test.

Keywords Networks manager · Topology correlation · Fault location · Event correlation

17.1 Introduction

With the rapid development of Internet technology and network business, the needs of users on network resources grow unprecedentedly fast, and also network becomes more complex. The traditional network management model has unable to satisfy the needs of large complex network management [1]. Therefore, how to ensure the availability, reliability, and service quality of modern network has become an urgent problem to be solved [2].

C. Deng (✉)
Guangdong University of Science and Technology, Dongguan, 523083 Guangdong, China
e-mail: chaodengl259@126.com

W. Du (ed.), *Informatics and Management Science VI*, Lecture Notes in Electrical Engineering 209, DOI: 10.1007/978-1-4471-4805-0_17,

Network fault location is to find the location where fault occurs in network in the shortest possible time, so that fault network can be effectively and quickly removed by network administrators [3]. Because of the infectiousness of network fault, event correlation technology and fault correlation algorithm are applied in fault location [4]. The basic idea of event correlation technology is to filter unnecessary or uncorrelated events for a single concept event through correlating multiple events, and reduce event information provided by network administrators, so as to accurately and fast identify fault source.

17.2 Network Fault and Event Correlation

Accident interrupts of computer network service are common. Therefore, a fault management system is necessary to be available for scientifically managing all faults occurring in network, and also recording the emergence and related information of each fault, so as to ensure the reliability of network [5].

17.2.1 Network Function Model

In network management function model, all network management tasks are divided into five different function domains, which are fault management, configuration management, performance management, charging management, and security management, respectively. The purpose of fault management is to ensure network to provide a continuous and reliable service. The general process of fault management is as follows: discover network fault; search, analyze and separate the reasons for fault; provide network administrators with fault removal helps, or even automatically remove fault. This is shown in Fig. 17.1.

17.2.2 Event Correlation Technology

Event correlation technology is an important fault location tool, and is a hot research focus for a long time. The basic idea is to filter unnecessary or uncorrelated information, separate fault source and reducing faults presented to network administrators by correlating multiple events as a single concept event.

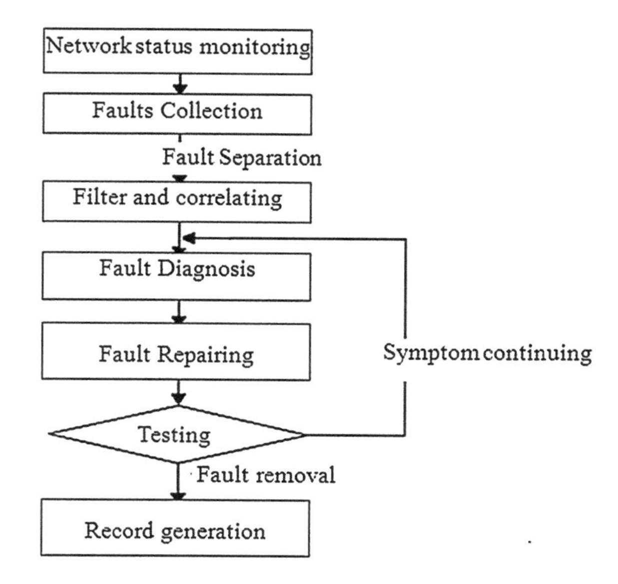

Fig. 17.1 Process of fault management

17.3 Detailed Design

17.3.1 Idea of Fault Location

In real life, no matter home computer or office computer networking, the networking methods are mostly based on a network topological structure. Specifically, a router is connected to many computer terminals or a next router, as shown in Fig. 17.2.

To manage huge network, a distributed management model can be applied. As shown in Fig. 17.2, if a fault occurs in router 5, the data from fault detection may include fault warnings sent from router 5 and computers 10, 11 and 12. This may not be a fault occurring in computers 10, 11 and 12. This is just because of the infectiousness of network fault, making fault management node detect computers 10, 11 and 12 are the devices to have a fault. In this regard, fault source

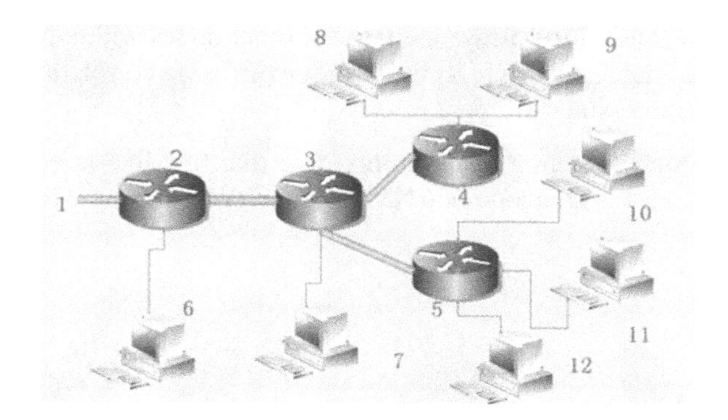

Fig. 17.2 An example of tree network topological structure

location is actually to locate the most likely fault source. Certainly, how about the situation if a fault really occurs in computer 10?

This fault source shows to be router 5 after the fault location of the first time, and then fault detection node will detect there is still a fault. Only computer 10 has a fault after the fault location of the second time. Therefore, is this not meeting the requirement of fault location on locating fault source?

From the tree network topological structure in Fig. 17.2, it can be seen that there is a linkage relationship between network nodes along the direction down from management node. Such a linkage relationship between network nodes is called as correlation. Because of the characteristics of correlation between networks, network fault can quickly spread in network once it occurs.

17.3.2 Creation of Correlation

In the topology structure of Fig. 17.2, management node NMS-1 is correlated with {device 2, 3, 4, 5, 6, 7, 8, 9, 10, 11, 12}; device 2 is correlated with {device 3, 4, 5, 6, 7, 8, 9, 10, 11, 12}; device 5 is correlated with {device 10, 11, 12}, but is not thought to be correlated with {device 10, 11}. These indicate that the condition required by correlation is a whole structure, but not only related devices in theory. In this aspect, it is considered that fault can't be thought to be caused by abnormal conditions of device 5 if warnings are sent from devices 10 and 11. This is because abnormal conditions of device 12 will be caused if device 5 works abnormally. Through network topology discovery module, the link between network nodes can be created. Here, the link between network nodes is saved with a 2D array containing an element of N*N. The N can be understood as "if network topology discovery module sends a network topological structure first, this suggests the structure includes N nodes". To facilitate meeting the requirements of subsequent fault correlation algorithm, N nodes also include network management node NMS-1. If the N*N array is regarded as a link relationship matrix, it can be seen that the matrix is surely symmetrical according to the characteristics of the link relationship. Assume that there is an array temporarily called "link". The definitions of ink[i] [j] == 1 means there is a link relationship between network node i and network node j, and the values of the arrays are 0 as for other nodes without a link.

Link new [N] [N] is used for expressing correlation, and link [N] [N] is used for expressing link.

Step 1 Search from the first line (the first line is corresponding to network management node NMS-1), and order link new [0] [j] == 1 if link [0] [j] == 1 is met.

Step 2 Use j as the line number of starting point of the second searching and search from link[j] [k] (because of the symmetry of link, it can be known k > j can meet requirements and save time), and order link new [j] [k] == 1 and link new [0] [k] == 1 if link [j] [k] == 1 until the same searching with j as line number ends.

Step 3 Turn to step 1 and add 1 into column number, and continue operations of step 1 and step until the first line ends, and then turn to step 4.

Step 4 Return to the state of step 1, and add 1 into line number until it ends.

17.3.3 Implementation of Correlation Algorithm

The core idea of fault correlation is that network nodes correlation and code technology idea are introduced, but fault correlation is a not complete correlation [6]. Because the occurrence of fault is uncertain, as shown in above example and the topological structure of Fig. 17.2, fault can really and simultaneously occur in devices 5, 11, and 10, but device 12 is likely to have no fault and fails only because of the infectiousness of network fault.

If network fault is detected in a detected time, the number of the device with detected fault is used as one-time fault event. In this fault event, because of network topological relationship, some devices actually have no fault, but the way that it needs to pass for acquiring network information fails. Therefore, the device is not detected by fault detection module to be in normal condition.

17.4 Database Design

At present, ODBC technology has been very extensively applied and basically can be used in all relational databases. ODBC is used as data source in this paper. After network topological structure is discovered by topology discovery module, network nodes are necessary to be numbered and saved with a data table for easily unifying fault information, and all field attributes in node table are as shown in Table 17.1: node_id is the number of the node; node_ip is the IP address of the node; node_status is the state of the node (0 means normal, and 1 means fault). The data table is as shown in Table 17.1.

To locate network fault based on topology correlation technology, a network topological correlation structure should be available. Through the above analysis, correlation is created from link relationship. Therefore, link relationship is also saved with a table named as link_table. After the link table of step 2, it is necessary to create a correlation through the above analyzed algorithm in details. The correlation table of topological structure can be saved with correlate_table.

Table 17.1 Node table

Node _id	Node_ip	Node_status
1	NMS	0
2	2.RT_192.168.0.1	0
3	3.RT_192.168.1.1	0
4	4.RT_192.168.2.1	0
5	5.RT_192.168.3.1	0
6	6.PC_192.168.0.11	0
7	7.PC_192.168.1.11	0
8	8.PC_192.168.2.11	0
9	9.PC_192.168.2.12	0
10	10.PC_192.168.3.11	0
11	11.PC_192.168.3.12	0
12	12.PC_192.168.3.13	0

17.5 Software Testing

Before accessing main control module, the network topological structure module and the modules whose link relationship is created by topological structure are found through software simulation. The correlation of the modules relying on link relationship to create correlation is actually created through correlation algorithm, and then software enters main control module. In main control module, analog fault data input module, fault location module, fault source display module and fault event display module. The fault location of software is shown in Fig. 17.3.

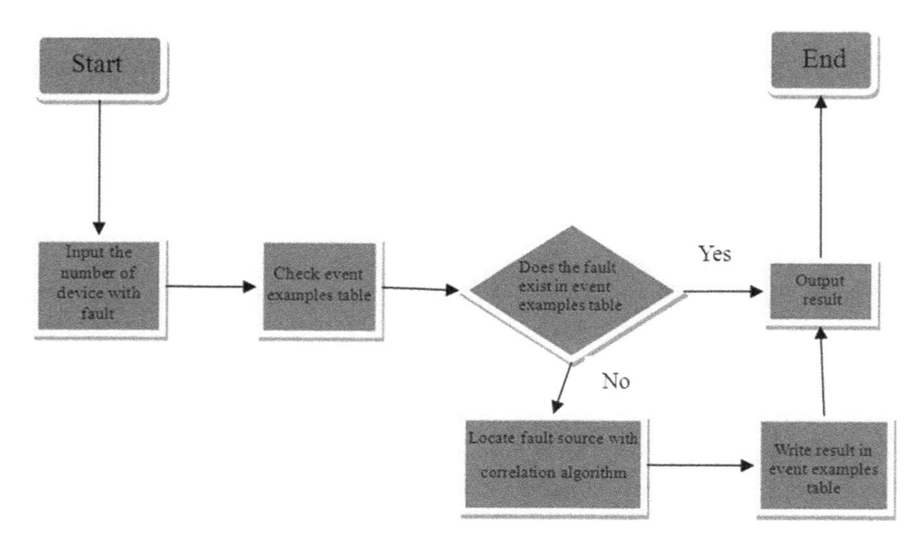

Fig. 17.3 The process of fault location

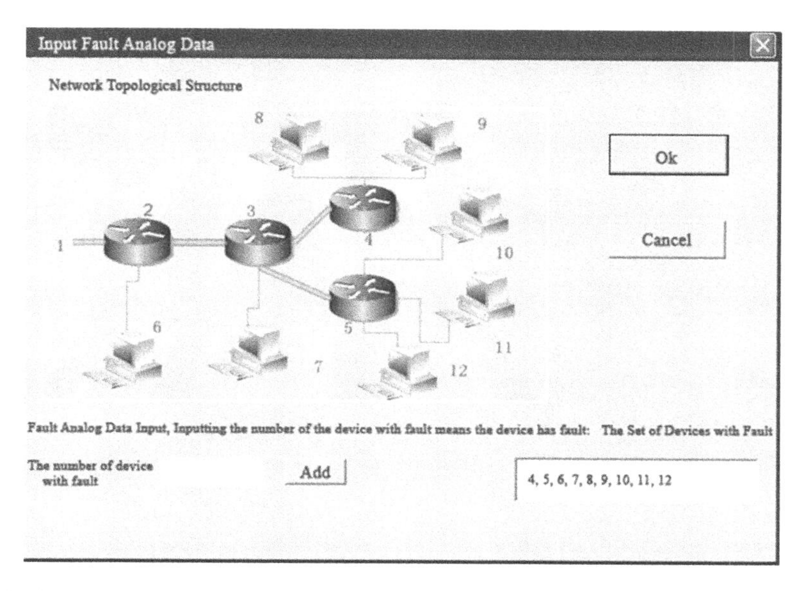

Fig. 17.4 Fault analog data input

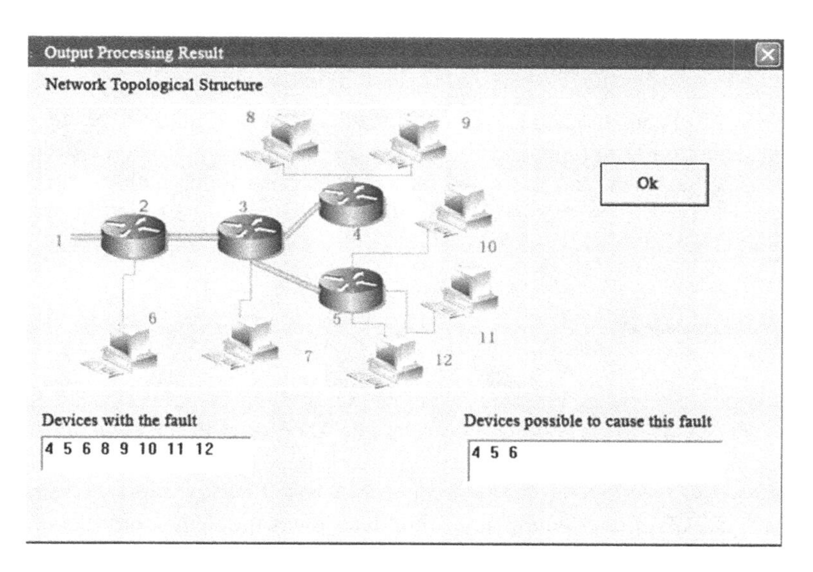

Fig. 17.5 Result output

Test example 1: Devices with fault are devices 4, 5, 6, 8, 9, 10, 11, 12, and the number of the device with fault is input in analog fault data input module, as shown in Fig. 17.4.

Then, processing results can be seen at the fault source output dialog box in which fault source is processed through analog, as shown in Fig. 17.5.

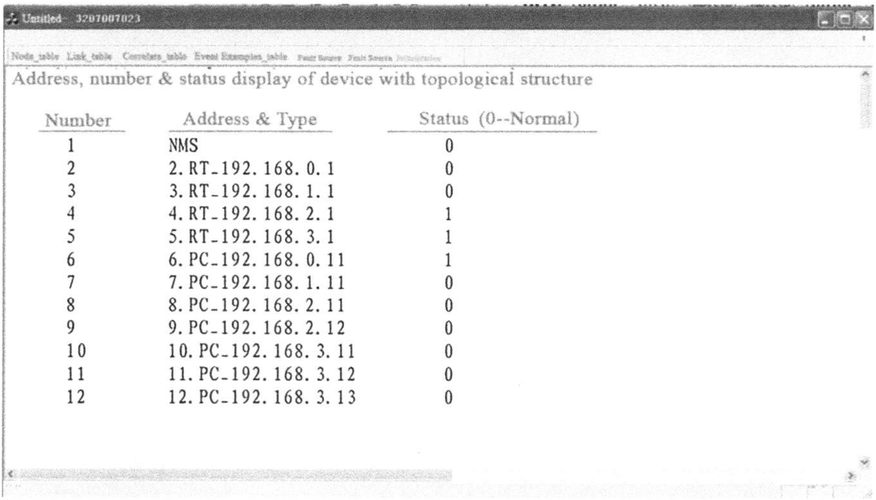

Address, number & status display of device with topological structure

Number	Address & Type	Status (0--Normal)
1	NMS	0
2	2. RT_192. 168. 0. 1	0
3	3. RT_192. 168. 1. 1	0
4	4. RT_192. 168. 2. 1	1
5	5. RT_192. 168. 3. 1	1
6	6. PC_192. 168. 0. 11	1
7	7. PC_192. 168. 1. 11	0
8	8. PC_192. 168. 2. 11	0
9	9. PC_192. 168. 2. 12	0
10	10. PC_192. 168. 3. 11	0
11	11. PC_192. 168. 3. 12	0
12	12. PC_192. 168. 3. 13	0

Fig. 17.6 Detailed results output

To further know the types, IP address and other detailed information of the devices with fault, fault source output text box can be used for viewing and the results are as shown in Fig. 17.6.

The software can only locate the most possible fault source in the first time. Therefore, to ensure the accuracy of the results, warning can still be detected if the first detection is not completely correct although the fault source detected in the first time has been recovered, and then fault can be located with the first device as failure device for the second time. In this way, fault can be removed gradually.

17.6 Conclusion

In this paper, the implementation principle and method of network fault location are introduced based on topology correlation. According to the infectiousness of network fault, the correlation between network nodes should be clearly defined. On this basis, the fault correlation algorithm based on topology correlation is proposed. This fault correlation algorithm determines the fault source according to the correlation between network faults, and therefore, effectively plays a role in fault filtering and location.

Acknowledgments Science and Technology Planning Project 2010 (No. 201010815401) for Scientific Research Institutions of Colleges and Universities in Dongguan.

References

1. Li X (2008) Study on network fault monitoring technology based on SNMP, vol 21(4). Changan University, Shaanxi, pp 158–163
2. Liu Q (2009) Research and implementation of a network fault detection system based on WSDM, vol 11. National University of Defense Technology, Hunan, pp 181–187
3. Liu Z (2008) Study on network fault diagnosis and positioning system based on correlation technology, vol 05(11). North China Electric Power University, Beijing, pp 23–28
4. Wang P (2008) Study on intelligent network fault diagnosis based on network topology and case reasoning, vol 08. Southwest Jiaotong University, Chengdu, pp 52–58
5. Li P (2008) Study on network fault management based on event correlation, vol 06. Central South University, Hunan, pp 178–184

Chapter 18
Security Framework and Correlative Techniques of Next Generation Network

Bigui He

Abstract Along with the development of the information network technology in the modern times, the next generation network (NGN) based on the soft-switch technology has become the mainstream network platform with a gradual step. Especially, its diversified server-side is an innovative development of the network technology. In this paper, a discussion is mainly conducted based on the characteristics including service characteristics of the next generation network, and then the construction of the NGN security framework is introduced, and finally the three core technologies of the NGN are emphatically analyzed.

Keywords Next generation network · Security framework · Soft-switch technology

18.1 Introduction

Along with the rapid development of the telecommunications industry in the modern times, next generation network (NGN) based on the soft-switch technology has changed into a completely new network platform with a gradual step [1].

Multimedia is innovatively added in its functions, and also its server-side tends to be diversified.

In operation, the security of the network system is very important, and especially the secured construction of the hierarchical network architecture is the foundation for the realization of the secured network communication.

B. He (✉)
Chongqing College of Electronic Engineering, 401331 Chongqing, China
e-mail: biguihe34h@126.com

W. Du (ed.), *Informatics and Management Science VI*, Lecture Notes in Electrical Engineering 209, DOI: 10.1007/978-1-4471-4805-0_18,

In the mean time, broadband technology, and other core technologies are required to provide supports for the effective construction of the NGN, mature soft-switch technology.

18.2 Review of Next Generation Network

As an open network platform, NGN has the ability to independently control transmit information, and simultaneously owns a good network service quality [2]. Multimedia service is added in the service form of NGN [3], and also the unified communication of the IP technology is realized by the speech technology based on the soft-switch technology.

As a comprehensive network platform, NGN provides a variety of businesses for the traditional network server-side. In the comprehensive network platform of speech, multimedia, and data transmission, a well-improved network system with high frequency is constructed [4]. What is more, the price and cost of the local calls are reduced by NGN to a certain extent, and also this plays a very important role in the realization of cost-effective and well-improved network service.

As a brand new network platform, NGN owns obvious characteristics in the technical research and development and service functions. Also, the diversified network service is realized by multimedia technology based on the soft-switch technology.

18.2.1 NGN Characteristics

Based on the soft-switch technology, NGN has attained an innovative development in both network construction and network structure, thus making the traditional network platform further enriched and improved.

18.2.1.1 Openness of Network Construction

Based on the soft-switch technology, the network construction of NGN makes the individual control of the functional modules realized, and also the interface protocols of all functional modules are standardized [5]. And then, in the construction of an open network platform, NGN is with a powerful security performance in contrast to the traditional IP.

18.2.1.2 Protocol-Oriented Standards of Packet Based Network

NGN is to make the traditional IP network standardized through a protocol of standards [6]. In NGN, three traditional network platforms are standardized and synchronized for management. Also, based on the standardization of the IP protocol, the facilitation of network is realized.

18.2.1.3 Dynamic Innovation of Network Technology

The technical research and development of the NGN is under a state of dynamic innovation. Especially, the gradually matured soft-switch technology has promoted the service form of the NGN more diversified. In business services, the flexibility of the NGN makes the network forms under an independent individual control more effective. At the same time, the development of the NGN technology owns a broader space [7].

18.2.2 Service Characteristics of NGN

The service form of the NGN tends to be diversified, and especially the construction of a multimedia service platform has enriched the service functions of the Internet.

18.2.2.1 Openness of Service Platform

The service functions of the NGN are mainly based on standardized data interface ends, and provide a secured and effective network service for user side. In an open service platform, the service efficiency of network business becomes much higher and higher.

18.2.2.2 Media-Oriented Service

Based on the soft-switch technology, the multimedia services platforms are realized by the NGN, and subsequently the traditional single network services are richened [8]. At the same time, based on a multimedia services platform, the usages of network for users tend to be more diversified.

18.2.2.3 Virtualization of Service Platform

In the virtual service platform of the NGN, virtualization is implemented based on valid individual identity information, and subsequently it is easier to implement virtualization on number business.

18.2.2.4 Intelligence of Service Flow

Intelligence service model is one of the characteristics of NGN business development. The all server-sides of business are effectively integrated by NGN, and then intelligent business services can be provided for all data interface ends.

18.3 Security Framework of Next Generation Network

NGN is a network platform that is centered at data transmission, and also its secured network framework is a foundation for realizing its good business, especially the security of hierarchical network framework. The hierarchical network security system is mainly based on the soft-switch technology in contrast to intranet and extranet.

Based on the complexity of the hierarchical network environment, the construction of security framework is required to be based on the volume of business on server-side, and also a reasonable service window is required to be set on the server-side of extranet. Moreover, a firewall can be set at port, for the purpose of shielding and intercepting the external interference information. The security and protection of the data interface end of hierarchical network plays a critical role. In data interface end, it is necessary to make recognition on the identity of user-end, and then users are allowed to enter the network service area. In the mean time, for the diversification business services, it is necessary to set reasonable business authorization, and it is especially to set a perfect access restriction for the handling of "sensitive business". In the construction of the security system of hierarchical network, the secured protection on the identity information of users is also a key part of framework architecture. The data processing end in the hierarchical network system, it is necessary to set an effective control layer, and then carry out verification on request data.

In the construction of the security framework system, the construction of the security of the carrying network is the most important, and can play a direct effect on the handling of NGN network business. NGN, in the construction of carrying network system, is still based on the traditional IP technology. Therefore, it features openness for the construction of carrying network. In general, the IP protocol normal form is applied in the construction of carrying network, so as to ensure the running environment of carrying network. Based on IP protocol, the attacks from hackers or viruses can be effectively prevented, and thus the running efficiency of network is improved. Besides, the network business of NGN tends to be diversified. In the construction of the carrying network security system, the control on the security of network nodes is also very important. In the control on the security of network nodes, it is necessary to control the network faults that are caused by a single node. Also, NGN is based on the diversification of business. Therefore, for the setting of nodes, it is necessary to set network nodes at multiple levels.

Besides, the setting of effective basic nodes can help maintain the security of the whole network nodes.

The security construction of the data control layer of NGN is also very important. In the network operation based on the soft-switch technology, it is necessary to control network and especially ensure the security of operation equipments. This is the key to the construction of the data control layer. A non-level network authorization mechanism is applied in the equipments at all nodes, and then data is effectively and independently controlled, and then data is independently processed, ultimately achieving the purpose of business services.

The security construction of the business handling layer of NGN plays a direct influence on the effectiveness of services. Therefore, based on the soft-switch technology, independent control at different level is implemented, making an isolation state formed between network information and lines. With the growing popularity of the Internet, the effectiveness of information communication is the core of NGN. It is necessary to make a confirmation on NGN information and especially the affects of fake or theft information sources on network security. In the soft-switch environment, business information is processed in network, and also the risks of information business are controlled with the form of network segments. That is, the effective separation between NGN and Internet is a foundation for realizing the construction of NGN security framework.

18.4 Analysis on NGN Core Technology

NGN, as a brand new open network platform, is necessary to be supported by core technologies in the construction of network system. And it is especially necessary to be based on the application of the soft-switch technology, making the innovative development of NGN realized.

18.4.1 Soft-Switch Technology

The soft-switch technology, as a brand new technology, is necessary to give consideration to the communication between itself and the traditional network in the process of applying NGN.

Moreover, the soft-switch technology is in the control center of NGN, making the independent processing of the call and media control based on a hierarchical network realized.

The soft-switch technology is still in a developing state. Therefore, the improvement of its functions and system are necessary to be based on standardized protocols and API, and then an open network structure system can be constructed.

Therefore, the soft-switch technology is the core of the next generation network, and this plays a very important role in the construction of the NGN security system.

In the process of constructing the soft-switch system, its core is mainly the construction based on three concepts, and especially the creation based on an open interface is an important aspect of the soft-switch technology.

18.4.1.1 Creation of Open Business Interface

The business model of the soft-switch technology is mainly based on server and AIPI, and is aiming at providing a diversified business service for user-side.

However, the intelligence of service interface based on the soft-switch technology makes the services of business more secured and effective.

18.4.1.2 Completeness of Access Protocol

The soft-switch technology supports the access of multiple protocols. In such a way, it is easy to manage the equipments on the access side, and simultaneously the data process on the access side can become sound, thus ensuring the effectiveness of information.

In this case, the realization of diversified functions of NGN plays a very important role.

18.4.1.3 Construction of Network Interworking

In the construction of the functions of the next generation network, the network interworking plays a critical role.

However, the soft-switch technology, based on the form of hierarchical network, makes network interworking realized.

18.4.2 Optical Fiber Transmission Technology

In the next generation network, importance is attached to the high efficiency and capacity of data transmission.

Therefore, in the realization of the functions of the next generation network, it is necessary to be based on optical fiber transmission technology, so as to realize the high efficiency of data transmission.

In the mean time, importance is also attached to the flexibility of business processing and high-efficient transmission performance.

It is also necessary to be based on optical switching, and then the intelligence of networking can be realized. Furthermore, with the continuous development of the telecommunications industry, the transmission technology of the next generation network is continuously improved.

18.4.3 Broadband Technology

Based on the development of the 3G network, the construction of the functions of the next generation network is required to be based on broadband connection technology, and the effectiveness of the next generation network can be realized.

The broadband technology of the next generation network is mainly based on Ethernet, so as to realize the high-speed digital transmission on user-side.

18.5 Conclusion

The next generation network, as a brand new open platform, makes its business processing diversified based on the soft-switch technology and broadband connection technology, etc. Especially the media-orientation of NGN business makes the forms of network service enriched. This is an innovative development in contrast to the development of network beyond all doubts.

References

1. Luo S, Kong C (2009) Analysis on digital content industry development under the background of next generation network. Spec Zone Econ 12:22–27
2. Li J (2008) Study on routing and switching key technologies of next generation network. Radio Commun Technol 02:181–186
3. Jiang P (2011) Analysis on next generation network technology and application risks. Police Technol 11(05):234–239
4. Liu X (2009) The discussion on next generation network technologies and development prospects. Comput Knowl Technol 12(33):84–88
5. Zhang G, Deng W (2011) Study on transition security problems of next generation network. Sci Technol Inf 21(35):191–195
6. Wang M (2010) The application of the soft exchange technology in electric power communication network. Telecom Technol 18(S3):251–256
7. Chang J, Li J (2011) Next generation network technologies. Inf Technol 22(04):67–73
8. Zhang Y (2010) Analysis on next generation networking technology. Mod Sci 13(08):43–47

Chapter 19
Wireless Autoregulation System of Greenhouse Temperature and Humidity Based on Internet of Things

Jian Liao

Abstract It is one of the most important methods for agriculture automatization in Greenhouse's measurement and control. Wireless Autoregulation system for greenhouse's temperature and humidity based on ZigBee technology was very satisfied with the request of low power consumption, low-cost and scale expanding afterward. We select the transceiver CC2430 as key chip set and designed the hardware interface between CC2430 and MCU Atmega128 and descript software flow for communication, presenting a scheme of wireless sensor network node based on CC2430 chip set, realized the real selfcontrol, it can Monitor every greenhouse's temperature,humidity and control and regulate Sprinkler valve, Experiment proves the design system can remotely and automatically monitor areas online, it has the economical, effective, real-time, convenient character and low energy consumption,which have a important meaning in promoting the Greenhouse to advance.

Keywords Temperature and humidity · Autoregulation · Measurement and control · Zigbee · CC2430 · DHT11

19.1 Introduction

With the development of computer technology and the popularity of the various sectors of business to carry out the dependence on the computer is becoming increasingly serious, the greenhouse has become our source of vegetables is the most important component [1]. Therefore, to strengthen the monitoring of the

J. Liao (✉)
Hunan Mechanical and Electrical Polytechnic, Changsha, China
e-mail: jianli2g11@126.com

W. Du (ed.), *Informatics and Management Science VI*, Lecture Notes in Electrical Engineering 209, DOI: 10.1007/978-1-4471-4805-0_19,

operation of the greenhouses, to maintain the normal operation of the greenhouses and the agricultural sector has become one of the cores. At present, domestic greenhouse environmental monitoring in general is still stuck in the simple model of human surveillance, monitoring mode can not the engine room temperature, humidity, electricity, water spray and other key automatic monitoring and alarm. In this context, the use of the Internet of Things technology, design a greenhouse environment monitoring system, making the computer room to the remote, real-time, automatic environmental monitoring and alarm in a timely manner.

19.2 Technology About the Internet of Things

The Internet of Things is an important part of the new generation of IT. The English name of the Internet of Things "The Internet of things". This has two meanings: First, the core and foundation of the Internet of Things is still the Internet is an extension and expansion of Internet-based network; second, between the extension and expansion of its client to any object and the object information exchange and communication. Therefore, the definition of things: radio frequency identification (RFID), infrared sensors, global positioning systems, laser scanners and other sensing devices, according to the agreed protocol, any object with the Internet connection, the exchange of information and communication in order to achieve the intelligent identify, locate, track, monitor and manage a network of objects.

ZigBee technology is a short-range, low complexity, low power, low-rate, low-cost two-way wireless communications technology. Mainly used for short distance, low power consumption and high transfer rate between the various electronic devices for data transmission as well as typical periodic data, intermittent data, and low-latency data transmission applications.

19.3 The Greenhouse Environment Monitoring System for the Overall Design Based on the Internet of Things

19.3.1 Design of System Architecture

Based on the Internet of Things greenhouse monitoring system structure shown in Fig. 19.1, the various types of wireless sensor data collected by the ZigBee wireless network transmission to the gateway, the gateway device collected a variety of greenhouse monitoring data into RMON database on the server. Remote monitoring server on the service management system for data analysis and statistics, if the data value beyond reasonable limits, notify the administrator by e-mail, SMS, sound and light alarm and the person in charge. The administrator can also remotely through the WEB log in system to view real-time data analysis, master a variety of environmental monitoring of the engine room.

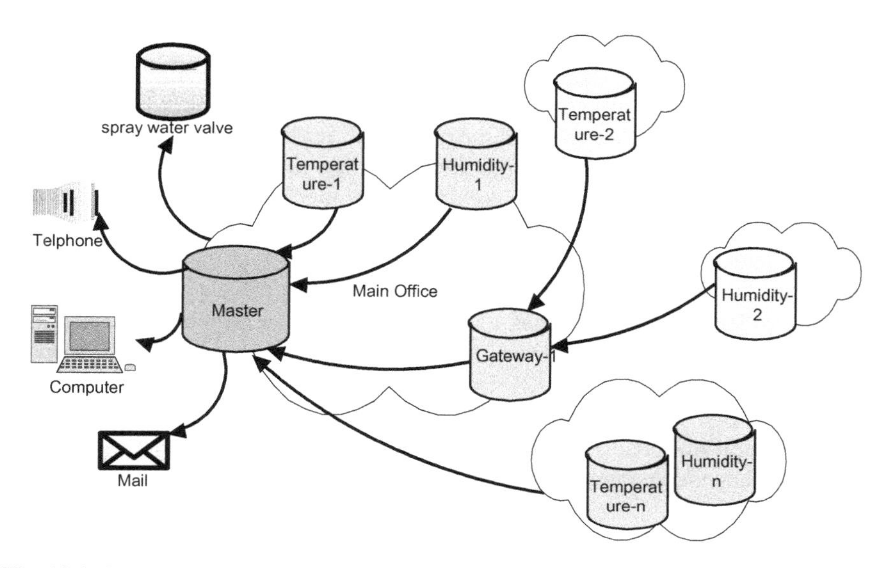

Fig. 19.1 System block diagram

19.3.2 System Features

Using the ZigBee technology, in line with the IEEE802.15.4 standard;

Using TCP/IP communication protocol;

Information management system based on database technology, the WEB page real-time view monitoring information to achieve the integration of management;

On-site real-time monitoring, stable and reliable;

Diversification of alarm;

Less investment and quick, greenhouse management applies to all sizes.

19.4 The Hardware System Design

19.4.1 Data Collection Terminal

The data collection terminal is mainly responsible for the collection room of the humidity, temperature, ventilation window, sprinkler valve signal data and uploaded to the gateway device through ZigBee wireless network gateway device and then data are stored in the database on a remote server. Main hardware for the corresponding sensor, the sensor module from the ZigBee network's point of view, an RFD node, powered by two batteries. The hardware design of the sensor is as follows:

19.4.1.1 The Temperature and Humidity Sensor

Temperature and humidity sensor the DHT11. DHT11 digital temperature and humidity sensor is one that contains the composite sensor calibration familiar with the signal output of the temperature and humidity, and apply it to a dedicated digital modules collection technology and the temperature and humidity sensing technology to ensure that products with high reliability and excellent long-term stability. The sensor includes a resistive sense of wet components and NTC temperature measurement devices, and connected with a high-performance 8-bit microcontroller. The product has excellent quality, fast response, strong anti-jamming capability, and high cost. Each DHT11 sensor calibration is accurate humidity calibration chamber [2]. Small size, low power consumption, signals transmission distance up to 20 m, making it the best choice for even the most demanding applications to the class of applications. Products for the 4-pin single row pin package, convenient connections. Shown in Fig. 19.2.

19.4.1.2 Sprinkler Valve

As shown in Fig. 19.3, Electromagnetic sprinkler valve, control solenoid valve coil power, power to control the solenoid valve sprinkler system.

19.4.2 Communication Module Base on ZigBee Remote Wireless

To reduce costs and power consumption when the engine room of a variety of environmental data collected, data collection terminal of the data processing operation and wireless communications by the microcontroller the CC2430 [3] completed. On the one hand the system connected by ZigBee wireless network with various types of sensor nodes, to read the operating parameters of the wireless sensor nodes and a set time interval, is simultaneously stored in memory, complete the data collection.

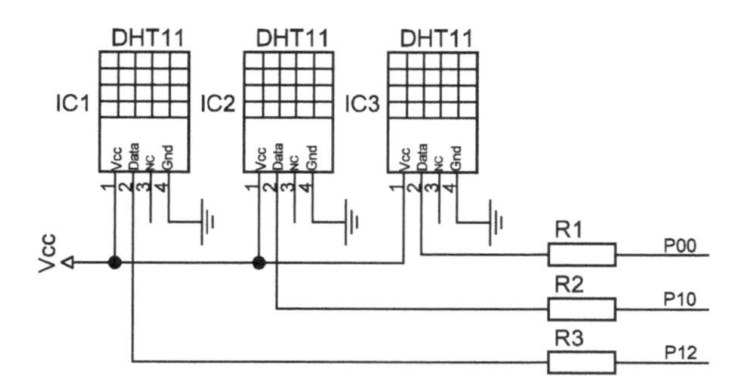

Fig. 19.2 DHT11 schematic diagram

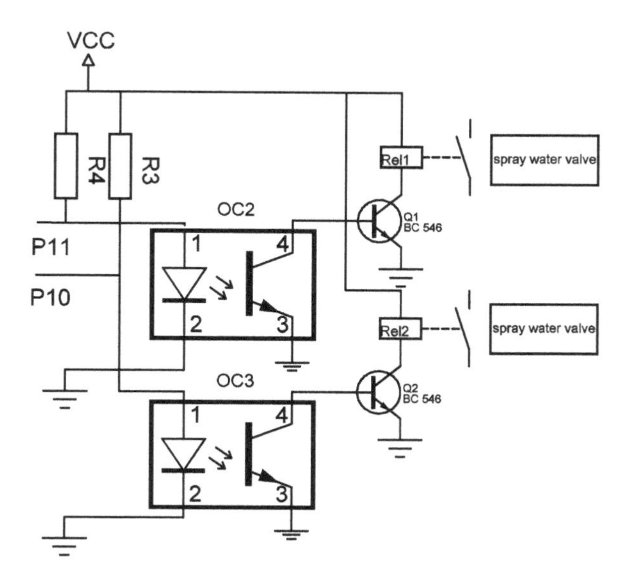

Fig. 19.3 Spray water valve

The CC2430 is a true system-on-chip (SOC) CMOS solution. This solution can improve performance and meet the 2.4 GHz ISM band applications based on the ZigBee low-cost, low power requirements. It combines a high-performance 2.4 GHz direct sequence spread spectrum (DSSS) RF transceiver core and an industrial grade compact and efficient 8,051 controller. Figure 19.4 is a schematic using the CC2430 wireless communication module.

19.4.3 Design of the Zigbee Gateway Equipment

The gateway device is a transfer station of wireless sensor networks and the wired device is connected, the transfer of equipment to complete the two-way data exchange between the Ethernet and ZigBee networks, query commands sent to the subordinate node receiving the request of the lower nodes and data the information collected through the serial port to the PC for processing with data fusion, the request for arbitration and routing function.

The gateway device by the CC2430 + PC machine, the CC2430 ZigBee network, send and receive data and dump the PC machine is responsible for sending and receiving Ethernet data, between the two exchange data via RS-232. An ordinary PC, RS-232 serial port baud rate to a maximum reach 115 Kb, ZigBee, the theoretical bandwidth of up to 250 Kb, between the two rates roughly an order of magnitude, taking into account the transmission control commands and data flow, so they can match use.

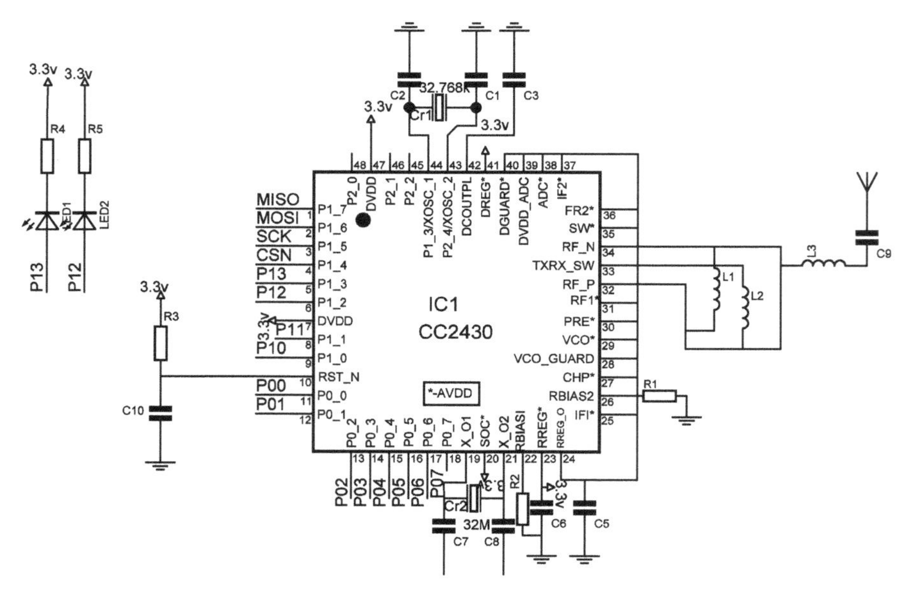

Fig. 19.4 Schematic diagram of the CC2430

19.5 The Software System Design

19.5.1 The CC2430 on-Chip Programming

Selection of the professional development of ZigBee wireless network systems C51 RF-3-PK ZigBee Edition development system as a development platform, it is CC2430/CC2431 professional development system to fully meet the IEEE802.15.4 standard and ZigBee standard for wireless network design and development. Contains all the hardware needed to build a variety of ZigBee network, software development tools, both simple exploration of buttons, and LCD and sensors.

The main program is mainly responsible for the subsequent processing of the wireless data transmitting and data feedback; timer is responsible for timing data transmission sensor to the central node; watchdog timer is responsible for monitoring node situation node reset will be necessary to ensure system reliability and stability. The main program control flow shown in Fig. 19.5.

19.5.2 Remote Service Management System

Dynamic graphical display of remote service management system is responsible for the collection of environmental data to receive and keep the transmission of

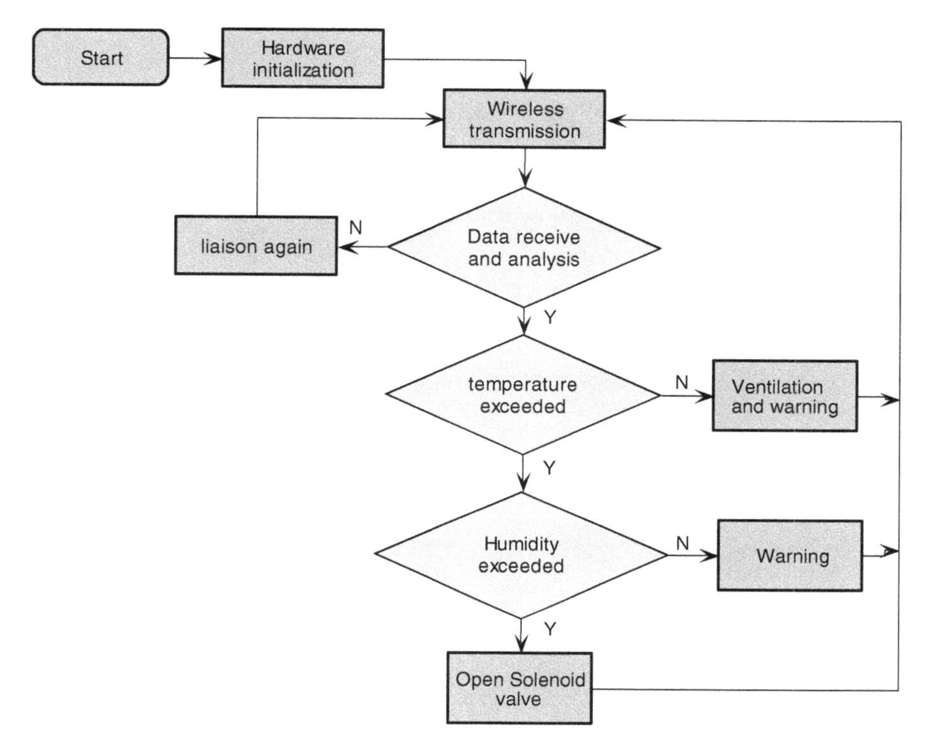

Fig. 19.5 Main flow diagram

wireless sensor network gateway device, released to the wireless sensor network query and command functions, and room temperature, humidity and voltage provide historical sensor data query and trend analysis. When the data exceeds the set threshold range, the system will send SMS to start the alarm, such as notify the administrator. The administrator can interact through wireless networks and mobile devices (laptops, cell phones, etc.), complete real-time data sharing and remote control for wireless sensor networks.

19.6 Summary and Outlook

Internet of Things as an emerging technology and concepts in all walks of life has extremely broad application and the incomparable advantages. And the system is the overall planning, given the system's overall hardware and software design. From the theoretical analysis, the program is effective and feasible.

References

1. Ruan Dian X, Fang T (2008) Application of ZigBee based wireless sensor network in underground coal mine environmental monitoring. Coal Mine Machinery 29(6):163–166
2. Yulan C, Jun N, Li H (2006) The implementation of wireless sensor network node based on CC2430. Comput Knowledge and Technol 06:114–117
3. Liu Y, Tao Z (2008) Design of the environment monitoring network node based on Zigbee. Bus and Net 11(8):22–25

Chapter 20
Remote Control System of Classroom Based on Embedded Web Server

Yongzhe Shi

Abstract A multichannel remote control system for classroom based on the STC89C54 chip was designed with the technique of embedded Web server. The control system can monitor 255 room signals and eight control signals of one classroom at the same time, and can be connected to the internet by the TCP/IP protocol. So the field control information can be shown dynamically in a remote computer by way of web pages. The system has high convenience and friendly monitoring interface, and then especially is fit for the large school, storage and laboratory that need multipoint monitoring and frequent switching door.

Keywords Classroom control · Embedded web server · Multichannel · STC89C54 · Hall sensor CS1013

20.1 Introduction

In recent years, many 8- and 16-bit microcontroller chip to successfully migrate the TCP/IP protocol stack, which was facilitated by using the embedded Web services for embedded applications based on these micro-controller chip. Web server embedded in the device to any access to it is the legitimate users of the network to provide a unified browser way of operating and control interface, the browser as the front-end control panel of the device [1–3]. This changed the situation of the traditional control user interface difficult to master, learning difficulties, and access to the network using TCP/IP protocol, the remote control and a wide range of information sharing made possible.

Y. Shi (✉)
Xi'an Radio and Television University, Xi'an 710002, Shaanxi, China
e-mail: yongzhi45s@126.com

W. Du (ed.), *Informatics and Management Science VI*, Lecture Notes in Electrical Engineering 209, DOI: 10.1007/978-1-4471-4805-0_20,

The embedded Web server application, the Web server in accordance with the actual application requires the introduction to the field testing and control equipment, the appropriate hardware platforms and software systems to support the testing and control equipment was changed to TCP/IP as the underlying communication protocol, Web technology is the core of Internet-based network testing and control equipment [4]. It has the basic function is to give response in accordance with the user's request, the form of a Web page on the remote computer and the user's command interpreter and passed to the implementation of the embedded field devices, thus completing the appropriate action.

Universities and other occasions, the need for monitoring of a large number of multimedia classrooms, currently widely used screen decentralized monitoring system, and the whole system needs a special person to operate, operation and observation are very inconvenient. The STC89C54 microcontroller and CS8900A Ethernet controller chip and devices constitute an embedded hardware platform embedded Web services technology-based multi-channel design of the platform of the classroom door and multimedia switch monitoring system, composed of low-cost, efficient, convenient, but also has networking capabilities of the multi-point monitoring system. The RJ45 interface to connect to TCP/IP protocol and the remote computer, on the remote computer's browser to graphically visualize the classroom doors, windows, multimedia computers and other information, while providing an alarm to the critical point set of historical data extracted interaction control functions.

20.2 System Structure

Figure 20.1 depicts the architecture of the embedded Web multi-channel monitoring system, which is mainly composed by several parts of the front-end switch sensor unit, data acquisition and processing unit, embedded Web servers, network communication unit, and Internet remote access. Solenoid valve sensor will switch physical quantities of information converted into electrical signals and sent to the data acquisition and processing unit. Standardization in the electrical signal of the unit to complete transformation. Network communication unit using the TCP/IP protocol to complete the network data exchange.

20.3 Hardware Design

Because STC89C54 micro-control chip integrates the ISP, so the circuit of the system is simple, high reliability, debugging is very convenient to download. Hall sensor range of options is quite extensive, taking into account the performance and cost, and standardization in the design and versatility, the Hall sensor selected CS3013 relay imported devices the SRD-05VDC-SL-C.

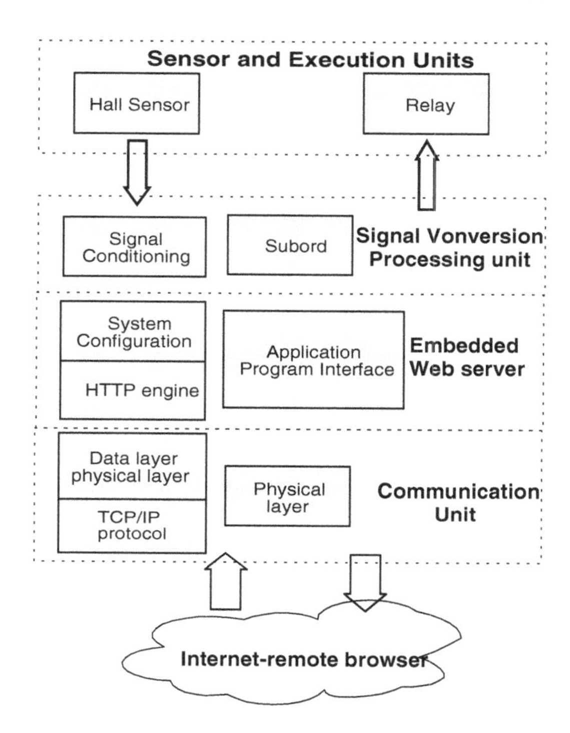

Fig. 20.1 The structure of the embedded monitoring system

20.3.1 The Main System Hardware

STC89C54 eight high-performance RISC architecture microcontroller chip has 16 KB of flash memory and 1 KB of RAM, it has 32 general purpose input and output ports, can not only facilitate the achievement of the connection and network control chip, and can be achieved other features. Network control chip Ethernet LAN controller chip the CS8900A its highly integrated design makes it no longer requires expensive external components necessary for other Ethernet controller, the power supply voltage of 3 V, it is suitable for the use of low voltage power supply of the STC89C54 connected. CS8900A integration of 4 KB on-chip RAM, 10 Base-T transmit and receive filters, as well as direct ISA bus interface with 24 mA drive. The STC89C54 I/O port can be used directly as the CS8900A data bus and address bus. In addition to the highly integrated, of unique PacketPage structure can automatically adapt to changes in network traffic patterns and existing system resources, thereby increasing the efficiency of the system. Hall sensor switching performance is good, therefore, eliminating the need for complex signal amplification and conditioning circuitry. Relay with transistor drive connected to the transistor's base-level I/O port lines.

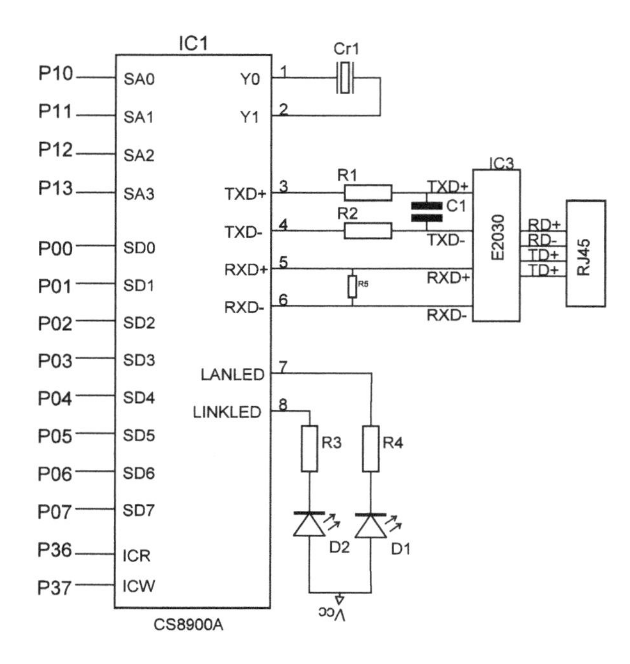

Fig. 20.2 The system's main circuit schematic

20.3.2 The Circuit Structure

The system's main circuit schematic shown in Fig. 20.2. CS8900A I/O space mode 8-bit data bus and STC89C54 P0 port connection, the address bus 4, two control lines. Set the effective I/O status to the address bus and control lines are set low, data can be 8-bit data bus transfer. Data transmission, the CS8900A can directly drive the light-emitting diodes, to alert the network connection or data transfer situation. Between the CS8900A and the network cable isolation transformer E2023, complete isolation of the systems and networks, and are hot swappable. The actual use of the Hall sensor connected with the CPU phase can use the transistor, play a level conversion and impedance matching.

20.4 Embedded Web Server Design

Embedded Web server system compared with traditional Web applications, has its own characteristics. Due to the limited processing power and storage capacity of embedded hardware platforms, embedded Web server must be simplified and only need to meet the most basic remote access and interactive features can be. Interactive methods to select the most suitable for embedded occasions Common Gateway Interface CGI mode [3]. The system software uses a modular design approach, all the code using C language. In accordance with the TCP/IP protocol hierarchy, the whole software is divided into the Ethernet driver module, the TCP/IP protocol stack modules, application-layer HTTP server module.

20.4.1 Ethernet Driver Module

The module implements the CS8900A Ethernet controller configuration and driver. The configuration of the network interface will be implemented in this module inside, and by providing a simple function to the data transmission. The initialization of the pin, software reset to the CS8900A and set its MAC address and other information by calling the module initialization function (Init8900 ()). Initialization is completed; the data transmission can be by calling the other functions.

20.4.2 The TCP/IP Protocol Stack Module

Complete TCP/IP protocol stack to take up storage resources, usually a few hundred KB of ROM and RAM in order to meet the requirements, is clearly not suitable for resource-limited embedded systems, therefore, an appropriate reduction to be based on the specific application. System, a portable version for 8-bit microcontroller, the version with a very small volume and good compatibility. Although only a few dozen KB of transmission speed and limited functionality, but still able to meet the 8-channel temperature and humidity monitor. TCP/IP protocol stack module plays a connecting role is one of the key modules of the entire system. It uses the Ethernet driver module provides functions to complete the data transceiver functions, and provides an API to the application layer HTTP server module. Its main features include ARP, ICMP, IP, and TCP protocols to achieve the basic functions of RFC791, 792 and 793 standard, the API contains a subroutine set for the data to send and receive, and check the flags register.

20.4.3 Application Layer HTTP Server Module

The HTTP server module that contains the HTML Web page embedded in web pages dynamically change the teachers switch status values embedded in the page inside a string variable. When data is sent to the browser before, it will first perform InsertDynamicValues () function, the function in the transmit buffer search string variable, and with the input value to replace the corresponding location. Periodic in the browser to refresh the Web page content (i.e, periodic reload the page each time you reload performs InsertDynamicValues () function, which real-time monitoring), set in the HEAD section of the Web page source

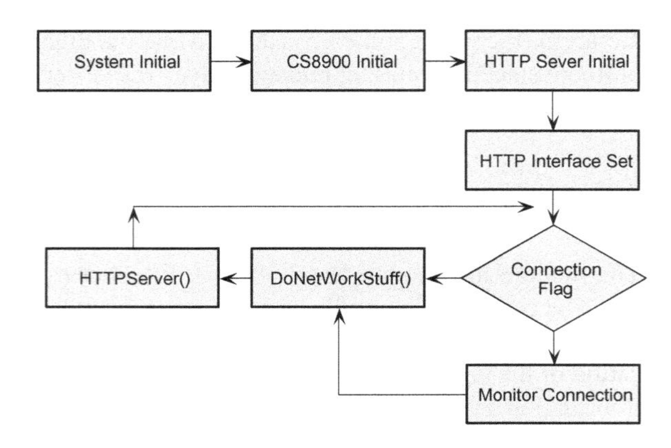

Fig. 20.3 System flow chart

code into the REFRESH. To achieve the following Web page HEAD this C programming code to refresh the browser page every 10 s.

```
<head>
<meta http-equiv="Content-Type" content="text/html; charset=gb2312"/>
<title> remote control system of classroom </title>
<meta http-equiv=refresh content="10">
</head>
<body>
</body>
</html>
```

20.4.4 The Web Server Running Processes

Software system design does not use embedded operating systems, the HTTP server and TCP/IP as a single process, the application layer program through an internal callback function interface and TCP/IP protocol stack communications. Combined with the TCP/IP protocol stack allows only the characteristics of a TCP connection, the main program using a periodic call. DoNetworkStuff () function and HTTPServer () function network connection detection and transmission of information, its flow chart shown in Fig. 20.3. DoNetworkStuff () function is provided by TCP/IP protocol stack. API functions to complete the TCP/IP layer information transmission control function. HTTPServer () function to complete the functionality of the HTTP service, including the HTTP response, the HTML file transfer. In the Web page, the status bar in the form of a graphical display of temperature and humidity values, in fact use only the instructions HTMLtable consider the storage capacity limit, page photographs, and other resource-intensive content, and do also as far as possible concise.

20.5 Conclusion

On many occasions, with the increasingly stringent requirements of environmental indicators, the collection of a variety of different types of information is increasing, less independent application of a single sensor, multi-sensor information fusion and processing has become an inevitable trend. Therefore, multifunctional multi-channel sensor networking capabilities will have more applications. Prove that the actual use of embedded Web-based multi-channel temperature and humidity monitor with good networking and information sharing capabilities, but also has high accuracy, the hardware circuit is simple, easy to use features, to overcome the traditional multi-channel temperature and humidity monitoring system network complex and difficult to use and other shortcomings. Also has a friendly interface, high intelligence, temperature and humidity test points and more as well as versatility.

References

1. Zhu W, He L, Han Ding, Youlun Xiong (2002) Embedded web sensors based on internet. Instrum Tech and Sens 8:1–4
2. Han G, Wang J, Lin T, Zhao H (2002) Design and realization of embedded web server based on web management. J Northeastern Univ (Nat Sci) 23(11):1021–1024
3. Li L, Yang B, Hu W (2003) Design and implementation of embedded web server software. Comput Eng Des 24(10):100–102
4. Han L, Zhao Y, Du Y (2010) Research and design on security of embedded web gateway system. Comput Security 2:55–59

Chapter 21
News Spreading Model Based on Micro-Blogging Platform in Network Era

Yuehui Zhou

Abstract This paper focused on building a mathematical model of news spreading on Micro-blogging. To do so, we have shown that well-known deterministic compartmental epidemic models can be extended to explain dynamics of trend spreading for various types of trends including real-time news as well as social events. As epidemiology has been extensively studied, it is quite useful to be able to express a process as an epidemic model, which opens up an array of analytically rich tools that are known to work in real life situations. We also pointed out this advantage by showing that one such tool can be readily extended for detecting change in trend dynamics on Micro-blogging.

Keywords News spreading mode · Micro-blogging · Network era

21.1 Introduction

Recently, micro-blogging and concrete micro-blogging, has become one of the fastest growing and an index of the user group growth. Because the real-time nature of it and frugal end user customers, Micro-blogging has become the effective method of information transmission-helping it is the official into a big media and reflector is not only a kind of influencer of real-time news.

This paper aims to establish a mathematical model to explain how in Micro-blogging news actually spread. Specifically, in this study, we investigate how to Micro-blogging activities can be used famous deterministic model of infectious disease. It is simple enough to deterministic model of actual useful, also can help

Y. Zhou (✉)
Jilin Business and Technology College, Changchun, China
e-mail: yuehui5zhou@126.com

W. Du (ed.), *Informatics and Management Science VI*, Lecture Notes in Electrical Engineering 209, DOI: 10.1007/978-1-4471-4805-0_21,

us to determine control elements in the most popular insist and stability of a particular news trend. No research results in the known-connection epidemic model and the news spread, and may result from such a study will help us know in information dissemination in Micro-blogging dynamics.

21.2 Epidemic Models

The traditional epidemiological studies are linear model, considering individual disease as a stand-alone unit of observations. The process is based on the Newtonian physics-however complex disease machinery; impact on the relationship is straightforward. After introduced by the French mathematician Henry Poincaré the twentieth century began, a new paradigm of complexity has been introduced into epidemiology. Recently, chaos theory has been developed, that highlights the importance of nonlinear phenomena in affected the course of the disease. Usually two types of model to research on infectious disease: Random certainty and model. The stochastic model change among-individual opportunity to rely on the risk of exposure, disease and other factors-allow population heterogeneity. But it is hard to set up random model and need more data and much simulation to get useful predictions. Deterministic model, also called compartmental models, tried to describe and explain what the average level of the population scale is.

Most models of infectious diseases are deterministic process, because they need to use less data, and relatively easy to set up. In this paper, we focus on deterministic model only we concentrate on the behavior of the population scale. For small populations, should use the stochastic model [1].

Certainty model is classified into different group's individual (room). The model of SEIR, for example, includes four compartments representative sensitive, exposure, infectious diseases and recovery. The model also specifies the transfer rate of between cubicles. Modeling of a kind of disease, it is necessary to have a biological reality of the performance of the disease-continuing invalid period, incubation period, immune status infection and so on.

For example, the SEIR model considers the infected phase accounting for a latent period of disease—when infected individuals (exposed) go through a latent period before being infectious [2]. On the other hand, the SIR model assumes that individuals are infectious as soon as infected-no latent period to be taken care of. Some models assume long lasting immunity after infection (SIR and SEIR) while other models posit recovers become susceptible again (SIRS and SEIRS). To analyze deterministic models, they are usually represented by differential equations describing the transitions between the different disease compartments using continuous time steps. For example, the SIR model can be represented in Fig. 21.1.

Where

$S(t)$ Number of susceptible at time t.

$I(t)$ Number of infectious at time t.

Fig. 21.1 SIR model

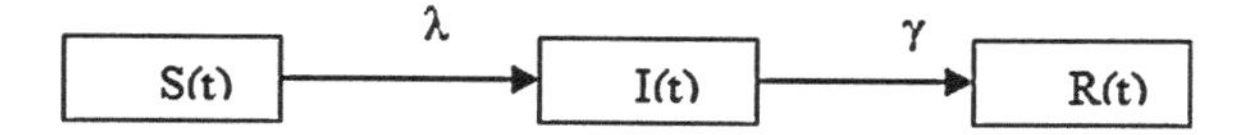

$R(t)$ Number of recovered at time t.
λ The rate of infection per unit time.
γ The rate at which an infectious individual recovers per unit time.

Then, the differential equation system corresponding to the SIR model is:

$$\frac{dS}{dt} = -\lambda g S(t)$$
$$\frac{dI}{dt} = \lambda S(t) - \gamma \bullet I(t)$$
$$\frac{dR}{dt} = \gamma g I(t)$$

Here, $\frac{dS}{dt}$ means change in S per small unit time dt. To be more specific the equation

$$\frac{dS}{dt} = -\lambda g S(t)$$

Means that the compartment of susceptible depletes itself by $\lambda\ S(t)$ as susceptible become infectious by the time interval dt. Similarly, for the infectious compartments, new $\lambda\ S(t)$ individuals are being added and $r \cdot I(t)$ individuals become recovered by the time interval dt, and so on.

As the propagation of a disease depends only on the ability of infectious agents to transmit the disease to the susceptible, the number of the newly infected at each time step depends on the contacts between infectious and susceptible individuals. So, if we know the probability of an effective contact β, then the rate of infection can be effectively expressed as

$$\lambda = \beta g I(t)$$

In that case the SIR model can be re-written as follows:

$$\frac{dS}{dt} = -\beta g S(t) g I(t)$$

$$\frac{dI}{dt} = \beta g S(t) g I(t) - \gamma \bullet I(t)$$

$$\frac{dR}{dt} = \gamma g I(t)$$

Depending on disease biology and available data, we can build a more complex model to have a better understanding of how a particular epidemic sets up in a population. For example, if we want to allow entries of the new susceptible by birth and mortality in the course of time as shown in Fig 21.2, where

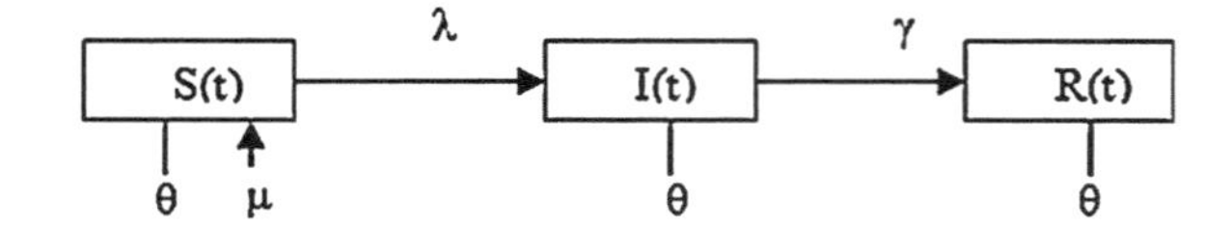

Fig. 21.2 A SIR model with birth and death rate

μ Birth rate per unit time
θ Death rate per unit time

Then the corresponding SIR models can be represented by the following differential equations:

$$\frac{dS}{dt} = -\beta g S(t) g I(t) + N(t) g \mu - \theta g S(t)$$

$$\frac{dI}{dt} = \beta g S(t) g I(t) - \gamma \bullet I(t) - \theta g I(t)$$

$$\frac{dR}{dt} = \gamma g I(t) - \theta g R(t)$$

where $N(t) = S(t) + I(t) + R(t)$, is the size of population at time t.

The potential of infection in a population depends on the basic reproduction number R0 that is defined as the average number of persons directly infected by an infectious individual during his entire infectious period when he enters a totally susceptible population.

The threshold theorem established by Kermack and McK-endric [3] says that if R_0 gets smaller than 1, the disease eventually disappears from the population because, on average, each infectious individual cannot ensure transmission of the disease to one susceptible resulting in lesser amplitude of the disease spreading phase comparing to preceding ones. If R_0 equals to 1 then, the disease remains endemic as one infectious on the average spreads the disease to one susceptible individual. On the other hand, if R0 is greater than 1, an epidemic ensues. This also explains why the introduction of infectious individuals into a community of the susceptible does not automatically give rise to an epidemic outbreak.

21.3 News Spreading on Micro-Blogging

In our approach, there is a basic similarity between the news dissemination on Micro-blogging and the transmission of an infectious disease among the individuals. In other words, each news topic on Micro-blogging spreads like ‘a contagious disease’, where

The infectious are the Micro-bloggingers who have participated in news spreading by tweeting about that topic.

The susceptible are the set of Micro-blogging who follow the infected Micro-blogging as they receive those tweets (infectious contacts) on their stream and as a result they too can tweet about that topic (risk of being infected).

As regency is an important issue in news spreading, to penalize older contents, we assume that infectious individuals lose their ability to spread news after a certain amount of time—becoming the recovered in epidemiological terms.

To develop an epidemiological model for news spreading on Micro-blogging, it is necessary to pick a model and its corresponding parameters that portray a complete and realistic picture. We choose the SIR model because of the following observations.

To highlight the importance of recent tweets, we put more emphasis on newly infectious individuals. In other words, infectious individuals cannot remain infectious forever, which excludes SI-related models from consideration.

When individuals tweet about a topic, it appears on the streams of the susceptible immediately. So, there is no latent period of the spreading, excluding SEIR-related models from consideration.

Usually participating on news spreading is a one-time shot per news cycle, making SIRS models unusable for this case.

We also allow the entry of a new susceptible similar to the birth rate in traditional epidemiology as tweets from infectious individuals reach to their followers' stream causing the population size to grow, As shown on Table 21.1. But, unlike in traditional epidemiology, new susceptible can be introduced only from infectious individuals. So, our proposed model can be represented by the following differential equations:

$$\frac{dS}{dt} = -\beta g S(t) g I(t) + I(t) g \mu$$

Table 21.1 Model parameters in epidemiology versus news spreading

	Epidemiology	Information diffusion on Micro-blogging
S(t)	Set of susceptible individuals at time t	Set of users who have received tweets from infectious individuals at time t
I(t)	Set of infectious individuals at time t	Set of individuals who tweeted about that topic at time t
R(t)	Set of individuals who have recovered at time t	Set of infectious individuals who have been inactive for a predefined period of time by not tweeting about that topic
β	Force of infection: Infection rate	Spreading rate
μ	Birth rate	Number of new followers who receive tweets from infectious individuals per unit time per infectious individual
γ	Recovery rate	1/average duration of infectiousness

$$\frac{dI}{dt} = \beta g S(t) g I(t) - \gamma \bullet I(t)$$

$$\frac{dR}{dt} = \gamma g I(t)$$

21.3.1 Parameter Selection

As different diseases have different dynamics determined by the demographic and biological characteristics (transition rates), the next step after selecting a model is to collect data and find appropriate values of parameters that can explain dynamics of disease spreading.

One important assumption in this model is, all new infectious individuals can arise from the susceptible set. But, as a news item becomes a more mainstream media topic, this assumption may not reasonably hold in our case, as individuals can also get infected from the outside of the Micro-blogging population thus becoming an infectious without ever being in the susceptible set. To measure this effect, we focus on three kinds of trends.

Events internal to Micro-blogging: Events that arise and die away within Micro-blogging without any external interference. For this, we focus on 'Follow Friday' trend, where on every Friday users suggest other user(s) to follow. This has been a recurring event on Micro-blogging after introduced by a user on January 16th, 2009 [4].

Real time news events, in other words, the traditional news: For this, we focus on the games in the world cup soccer between USA and Ghana on June 26th, 2010.

Social events: which are not news in the traditional sense, but, as each important social event usually becomes a trending topic on Micro-blogging because of the sheer number of tweets related to it, we decided to track one such an event. For example, the Memorial Day in the USA.

21.3.2 Data Set: For Each of These Events, We Maintained a Set of Infectious, Susceptible and Recovered Individuals Over Time by Using Micro-Blogging API

We used stream API to track a particular keyword and in every time epoch Δt, we updated the set of infectious individuals $I(t)$ by retrieving the users who tweeted about that topic in the $[t - \Delta t, t]$ interval. Though, unlike traditional epidemiology where the duration of epochs usually ranges from weeks to months, in our case the duration of an epoch is understandably much smaller-ranging from 1 to 4 h.

We then retrieve all the followers F of each infectious individual $i \in I(t)$, and after filtering out any followers who are also in $I(t)$, add them to the susceptible set $S(t)$

We also remove infectious individuals from $I(t)$ who have not tweeted about that topic in the $[t - 2\Delta, t]$ interval to mimic a recovering process and add them to the recovered $R(t)$ set.

For the Memorial Day event [5], we collect 546, 320 tweets starting from 28th of May over 7 days. Similarly, for 'Follow Friday' we collected 115, 300 tweets starting from 20th of May to 22nd of May. And, for the match between USA versus Ghana, we collect 165, 779 tweets starting from 25th of June for 3 days.

21.3.3 Simulation Results

Given the above data set, our objective is to determine appropriate values of parameters that reasonably explain the spreading of news. To do so, we perform a multi parameter least-square fit by using the optimize module provided by SciPy [6].

From Fig. 21.3, we can see that our model does fairly well except in the later region. That's because, Memorial Day—being a social event, does not follow our assumption that infectious individuals can only arrive from the susceptible set. In other words, the number of infectious individuals entered from outside of the population is quite high. But the point is, even though the assumption of mass action principle is arguable here, our model predicts the rise of the trend quite well.

In contrast to the Memorial Day, the event of world cup match being played between USA and Ghana is more of traditional real-time news. As we can see in Fig. 21.4, our model performs much better than the Memorial Day event. We believe that the better performance of our model here is due to the fact that the population (the susceptible, infectious and recovered) in this case is more connected by either spatial or user-preference similarity. In other words, "reciprocity" is high between infectious and susceptible Micro-blog gingers causing the presence of homophile and less sporadic outside interference.

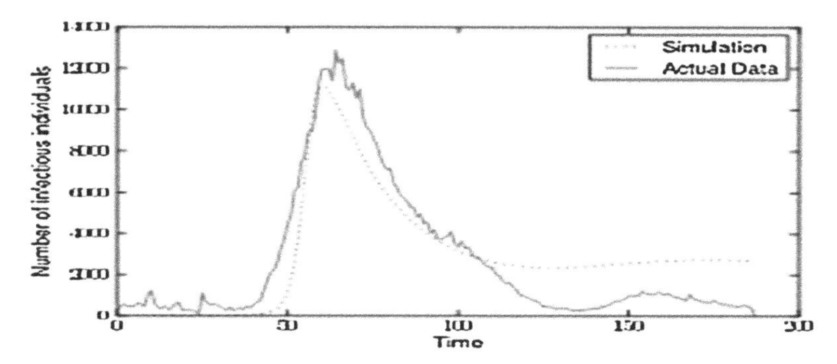

Fig. 21.3 Memorial Day simulation with $\mu = 4.33 \cdot 10^{-05}$, $\beta = 5.38 \cdot 10^{-02}$, $\gamma = 1.02 \cdot 10^{-01}$

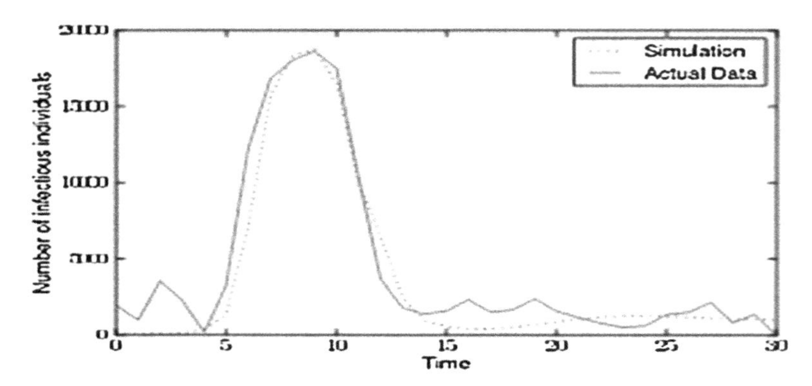

Fig. 21.4 World Cup USA simulation with $\mu = 1.4 \cdot 10^{-04}$, $\beta = 3.43 \cdot 10^{-01}$, $\gamma = 6.635 \cdot 10^{-01}$

21.4 Conclusions

For future work, we plan to focus on two possible extensions. One is to build a real-time on-line detection system of early paid attention to collecting the trend from the stream tweets. Another potential extended to our work will be modeling of information transmission in Micro-blogging as a stochastic epidemic model. These can better control and more reality system as we can consider the effect of user spread the news.

References

1. Baron M (2010) Bayes and asymptotically pointwise optimal stopping rules for the detection of influenza epidemics. Case studies in bayesian statistics, vol 6. Springer, pp 153–163
2. Brauer F, Castillo-Chavez C (2010) Mathematical models in population biology and epidemiology. vol 6. Springer, Verlag, pp 345–349
3. Cha M, Haddadi H, Benevenuto F, Gummadi KP (2010) Measuring user influence in Micro-blogging: the million follower fallacy. In: Proceedings of the 4th International AAAI Conference on Weblogs and Social Media (ICWSM), Washington DC, USA vol 5. pp 1236–1242
4. Cheong M, Lee V (2010) Micro-blogginging for Earth: A study on the impact of microblogging activism on Earth hour 2009 in Australia. Intell inf database syst 34(06):114–123
5. Memo D (2006) Prior distributions for variance parameters in hierarchical models. Bayesian analysis 1(3):515–533
6. Java A, Song X, Finin T, Tseng B (2007) why we Micro-blogging: understanding microblogging usage and communities. In: Proceedings of the 9th WebKDD and 1st SNA-KDD 2007 workshop on Web mining and social network analysis, ACM vol 11. pp 56–65

Chapter 22
Study on Education Management Scheme of Micro-Blogging Network Culture

Wenting Wang

Abstract With the rapid development of IT, the Internet, as a new media, has a profound impact on all aspects of human life. The university is the forefront of IT development, and the network has become an important channel for college teachers and students to get access to knowledge and information. Thus it affects the outlook on the world, life and values of the college students. Network development and application is a two-edged double-edged sword, which not only brings positive development opportunities, but also poses a severe challenge. It is a problem that should be answered by every university administrator that how higher education management should deal with this new situation. In terms of university micro-blogging, it rises perhaps in response to this challenge.

Keywords Network culture · University education and management · University micro-blogging

22.1 Introduction

On January 16th, 2012, "The 29th Statistical Report on Internet Development in China" published by CNNIC in Beijing announced that the scale of Chinese netizens have reached 513 million by the end of December 2011 [1]. Chinese mobile phone users have reached 356 million. Internet culture, as a new cultural form, gradually penetrates into the life of the people, especially the students in colleges and universities.

W. Wang (✉)
Zibo Vocational Institute, Zibo 255314, Shandong, China
e-mail: wentintw459m@126.com

W. Du (ed.), *Informatics and Management Science VI*, Lecture Notes in Electrical Engineering 209, DOI: 10.1007/978-1-4471-4805-0_22,

Internet culture is the information culture based on the Internet. It is with no borders and regardless of the area. In today when science and technology develops rapidly, more updated knowledge can be acquired through using the Internet to be involved in the network culture. It has become a kind of fashion pursuing by more and more people. In this process, network culture acts like a double-edged sword to ideological education in colleges and universities, with both the pros and cons at the same time.

22.2 The Double Impact Analysis of the Current Network Culture on College Students and Counselors

In order to obtain data support, this paper has done the early investigation and in-depth analysis. The survey is divided into the "teacher questionnaire" and the "student questionnaire". It adopts stratified cluster sampling method to select the 173 students including the junior college students and undergraduate in Zibo Vocational Institute. 25 counselors who are relevant to the sample of students are selected to do the research. The survey results have gone through statistical comparative analysis. The survey results accurately reflect the current situation of the communication between the instructors and students as well as the role of the network popular culture. Considering this, we have the results of this survey and make analysis of particular summary. As shown in Table 22.1.

22.2.1 The Influence of the Network Culture on College Students, with both the Pros and Cons

The positive effects: the first performance is that the equality and openness of the Internet culture has brought huge convenience to the college students. The second performance is that the convenience and easiness of the Internet culture has

Table 22.1 Teacher questionnaire and student questionnaire

Students	Instructors
1. Current situation of network application (network objective, frequency, time, etc.)	1. The learning degree of network culture
2. Influences of network culture on students' outlook on values, emotions and behaviors	2. How to review the influence of network culture to university students
3. Recognition degree to network culture	3. Current situation of communicating with students (whether integrating network culture elements into work) and its effect
4. Current situation and effect of communication with instructors	4. Expectations on the self-improvement work utilizing network culture
5. Expectations to counselors on network culture	

brought convenience to the college students. The final performance is that the economy and interactive characteristics of the Internet culture has brought convenience to university students.

The negative effects: the cultural diversity of network information values and outlook on life, alienation of the college students. In addition, the disorder of the network culture, no control, "no government", "liberal" tendency to lead to a lot of false information and spam, it will seriously harm college students' physical and mental health.

The proliferation of network entertainment culture has caused the students' mental deformity. First of all, the performance is the proliferation of online video and film. It is followed by the performance that the proliferation of online dating is quite popular.

22.2.2 The Influence of the Network Culture on Counselors' Work and the Pros and Cons Analysis

22.2.2.1 The Internet has Expanded the Ideological Education Space

The formation of Internet culture has enabled the students' ideological tentacles to stretch out of the campus and the introduction of a wide range of social space to make up for extensive defects of the original campus culture coverage [2]. Internet culture within the wider range of students' ideological education has provided a broad space for the college together. It explores the ideological education in the socialization, which has provided a great amount of favourable conditions.

22.2.2.2 Network Information has Greatly Enriched the Ideological Education Resources

According to reports, the Student Association of Tsinghua University has established the "red site". Since its opening, it has accepted visits from over 10 million people inside and outside the campus. Just imagine, if there is no Internet link, how can a student society create such extensive contacts? Accordingly, it is not difficult to imagine, as long as we are determined to spend pneumatic building your own website, good ideas educators wholeheartedly into the network culture construction, we must be able to attract tens of thousands or even hundreds of thousands of people visit the site, receive health education.

22.2.2.3 Efficient Dissemination of Information has Improved the Efficiency of Ideological Education Information Dissemination

The network has spread fast and wide coverage. The huge speed advantage is the most important feature of the information network. Traditional moral education, most of the lectures, moral educators tend to spend a lot of time and human effort to find information, write the script, students passively accept the "indoctrination" in a closed space. The use of the network has greatly improved the efficiency of information dissemination of ideological education.

22.2.2.4 Network Moral Education has Higher Work Efficiency

It is conducive to promoting the development of the Students' sense of independence.

It promotes the building of the students' spirit of openness.

It helps to promote the cultivation of students' innovative spirit.

The network helps to promote students' awareness of democracy and helps to promote the legal awareness of the sound of students' moral.

It is not difficult to see from the research and interviews that the current manners and contents for university students communicate through the network are various. In order to adapt to this new change, as a college education and guidance leader, they must establish as soon as possible a good channel of communication with students, making them continue to accept new things. The establishment of university micro-blogging is undoubtedly the best choice for the higher education managers currently.

22.3 The Coming of Universities Micro-Blogging Era has Become the Effective Way for Higher Education Managers to Deal with the Impact of the Internet Culture

Micro-Blogging is a platform of information sharing, dissemination and access based on customer relationship. The users can renew the information with about 140 words and realize the instant sharing through the WEB, WAP and a variety of client-side components of individual communities.

Using micro-blogging, a form that is loved to see and hear by university students to manage and perform education guidance. It is inevitable to bring about unexpected results:

Through the micro-blogging, the ideological context of university students can be grasped. Micro-blogging, as a kind of "face-to-back" innovative interaction way, can help the ideological and political workers to keep abreast of the students' change in thinking, and grasp the students' ideological context. It can thus perform targeted ideological and political education. First, the college campus Moral Education website as the basis, the micro blogging platform to create, promote online interaction between the schools and students, students with students, to grasp the students' ideological trends, strengthen ideological education and guidance. Second, college students mobile phone subscriber base as the basis, the establishment of a mobile phone micro blogging platform for teachers and students, students peer-to-peer information interaction for the majority of students, especially the part do not have a computer, Internet handicapped students open up a convenient information delivery channels. Third, college students QQ group as the basis, the establishment of the mass mailing of micro blogging platform to promote the school, college, class, students between the multi-level, flat, equal exchange. Fourth, colleges and universities can be a realistic group of students rely on to micro blogging classification platform to further enhance the relevance and effectiveness of the ideological and political education.

Second, with the micro-blogging, the "Chicken Soup" can be cooked for the students. Ideological and political education should be a choice of colleges and universities, both into micro blogging. Universities should take advantage of micro blogging convenient, fast, and the advantage of a wide area, and improve the methods of ideological and political education. With the help of micro blogging, the "spiritual garden", it can carefully prepare "Chicken Soup" for students in order to expand the coverage and influence of the ideological and political education.

Of course, the opening of micro blogging is a once and for all thing for the college students. Things can be wonderful only by doing with heart and soul. From the university has opened micro blogging, some famous universities such as Tsinghua University, Wuhan University and other colleges and universities "fan" groups popular Chao Wang, while some college micro blogging only a few dozen fans, very lonely. According to the analysis, mainly because some universities micro-blogging updates slow, too narrow positioning, form a single. The lack of expert management, micro blogging non-existent, leading micro blogging updates slow crux of the problem. According to the survey, the college has opened micro blogging, many a month just to update the number of messages and the most "best" micro-blog has more than 6 months without an update, basically in a state of "shock". The micro blogging advantage is the fastest, most refined language for transmitting information is useless if the update rate cannot be guaranteed. As some colleges and universities position the micro blogging in an excessive narrow scope, only one or a few fixed groups can focused on particularly. The lack of information nutrition results in the decline of attention. The concern of the network society, in fact, is in both directions. Therefore, colleges and universities "bib" can provide more for public services, education, and truly become the window of the community to understand colleges and universities. Communication University of China, micro blogging, not issued service announcements, recruitment information, and even funny inspirational video, and gathered a lot of popularity, to forward the high volume. Also, some colleges and universities just simply paste the news on the school website, to fill the micro-Blogging, making the form of expression over simple. It is lack of interaction and assumes a posture of telling their stories. As everyone knows, the charm of the micro blogging lies in interaction.

References

1. Hao S (2010) Ideological and political work versus twitter. J Yangzhou Univ High Educ Study Edition 10(4):146–153
2. Wang X (2011) College microblogging and the ideological and political education. J Yangtze Univ: Soc Sci 18(11):35–39

Chapter 23
Data-Driven Adaptive Control Paradigm of Supply Chain in Pharmaceutical Chain Enterprises

Hua Wei, Wei Xia, Xiao-Dong Wei, Dai-Yin Peng and Chuan-Hua Huang

Abstract There are some problems in researches on supply chain control of pharmaceutical chain enterprises under dynamic environment. The authors establish a new paradigm of computer simulation analysis and adaptive control, the core strategies of which are data-driven and agent's computing. The paper constructs a feedback closed-loop cycle mode that is data collection and processing, data-driven mode discovery and performance identification, data-driven control output. Then an adaptive dynamic control strategy is established based on active induction and emergency intervention control. Contents of the new strategy include: supply chain modeling of pharmaceutical chain enterprises based on CAS, adaptive control paradigm of supply chain in pharmaceutical chain enterprises, program design of control and system behavior assessment based on epsilon machine.

Keywords Pharmaceutical supply chain · Comples system modeling · Computer simulation · Adaptive control

H. Wei (✉) · D.-Y. Peng · C.-H. Huang
Anhui University of Traditional Chinese Medicine, Hefei Anhui 230038, China
e-mail: weihua@cssci.info

W. Xia
College of Electronic and Information Engineering, Anhui University of Architecture, Hefei Anhui 230022, China

X.-D. Wei
Department of Computer Science, The Anhui Zhong-ao Institute of Technology, Hefei Anhui 230000, China

W. Du (ed.), *Informatics and Management Science VI*, Lecture Notes in Electrical Engineering 209, DOI: 10.1007/978-1-4471-4805-0_23,

23.1 Introduction

Pharmaceutical chain operation is a kind of modern mode of pharmaceutical business and organization form. Under this business mode, pharmacies combine together in the form of common stock or franchising to realize service standardization, operation specialization, and management normalization. Pharmaceutical chain operation should maintain uniform purchasing, uniform promotion, uniform service specifications, uniform work process, and uniform operation specifications [1, 2]. With the adjustment of new medicine policies, the pharmaceutical retail operation has entered into the age of chain operation.

There are special requirements for the Chinese pharmaceutical retail operations:

(1) According to the requirements of the relevant laws and regulations, Chinese pharmaceutical retail chain enterprises should operate according to pharmaceutical wholesaling firms. It should be in charge of such things as medicine purchase, acceptance check, storage, maintenance, and delivery and after-sale services and so on. The pharmaceutical retail chain stores should operate according to the social drugstores. It should not purchase drug on their own and independently. The drugs for sale should be supplied only by the delivery centre of chain general headquarters.
(2) The drugs in different stores should not be allocated in a parallel manner. It should meet the delivery requirements and return to the company delivery centre first. And then deliver to the other stores by the delivery centre.
(3) Strict examinations for unqualified drugs.
(4) Strict management for reverse logistics [3].

The existing pharmaceutical retail supply chain faces the following difficulties:

(1) The delivery distance is long and the cost is high. The chain stores are distributed everywhere or some of them even have to be delivered trans-regionally.
(2) Dealing with the lack of products in the store takes a great deal of time.
(3) For drugs that are unqualified and imperfect with damages, pollution, and overdue, it takes a long time to report and ruin. The process is also complex.
(4) It is difficult to do revise logistic management.
(5) It is difficult to adjust drugs in stores.
(6) There is a lack of efficient control over alliance stores.

Focusing on the above problems, this paper tries to adopt the Agent technology [4, 5]. It uses "communication and coordination" to replace the simple "technical economic multiple goals optimization". It takes the "multiple agents supply and demand", subject and environment that is able to carry out "interaction-coordination-adaptation". On this basis, a new normal form is built. It bases scientifically on the complex adaptable system (CAS) theory. The new normal is data driven and uses the Agent calculation as the core strategy. It is the stimulation analysis and adaptable administration of the supply chain computer for the pharmaceutical chain enterprises (Fig. 23.1).

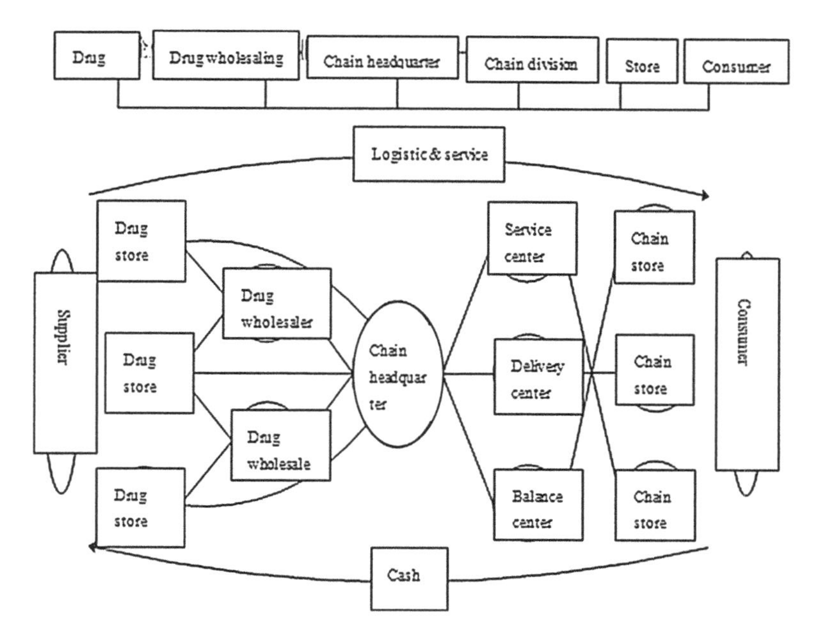

Fig. 23.1 Pharmaceutical retail chain supply structure picture

23.2 Pharmaceutical Retail Chain Supply Chain Modeling Based on CAS

The pharmaceutical retail chain supply chain is compound of several relatively independent entities. It is a cooperative network system. It has such features as adaptability, autonomy, nonlinear and so on.

23.2.1 Agent Modelling

The agent modelling of the pharmaceutical retain chain supply chain is divided into three steps: (1) Build stimulation-reaction behavioural system modelling. (2) Establish credit confirmation system. (3) Offer discovery rule methods. That is exchange and mutation.

23.2.2 Macro Systematic Modelling (CAS System Modelling)

The design subject has three basic parts: (1) Attack symbol—use to get in touch and contact with the other subject. (2) Defensive symbol. (3) Resource base—use to store process resources. On the basis of these, CAS system model can describe:

the whole system includes several subjects. The subjects carry on communications, exchanging resources and information.

23.3 Normal Form Analysis of Adaptability Regulation and Administration of Pharmaceutical Retail Chain Supply Chain

23.3.1 Normal form Analysis of Adaptability Regulation and Administration of Pharmaceutical Retail Chain Supply Chain

This paper has put forward the adaptive regulation and administration normal form for the pharmaceutical retail chain supply chain. The paper constructs a feedback closed-loop cycle mode that is data collection and processing, data-driven mode discovery and performance identification, data-driven control output. Then an adaptive dynamic control strategy is established based on active induction and emergency intervention control. It has formed the dynamical regulation output based on active induction and emergency intervention control [6].

This model focuses on the complexity and adaptive mechanism of the supply chain management for pharmaceutical retail chain. It has divided the environment and level of the system. In addition, it builds the subject modelling of the relationships between the outputs, input, behavioural rules, and learning mechanism on the descriptive micro level. It has founded the system macro modelling constructed by multiple subjects. On the basis, it carries on computer stimulation analysis of the complicated adaptive process for pharmaceutical retail chain supply chain.

The system is led and changed from the parts behavioural situation variations to the whole system evolution process through the design adjustment subject, operation evaluation subject, performance identification subject, induction and invention control subject, and through the evaluation, feedback mechanism towards the visible performance indexes of the system.

23.3.2 Scheme Design Based on ε Machine Regulation and System Behavioral Evaluation

Crutchfield, the American theoretical physician who introduces and reforms the ε machine new method in modelling calculation, discovers the reason and result model in complex system (Fig. 23.2).

Adopt data collection—model calculation and performance indentify—induce and invent control output closed loop model.

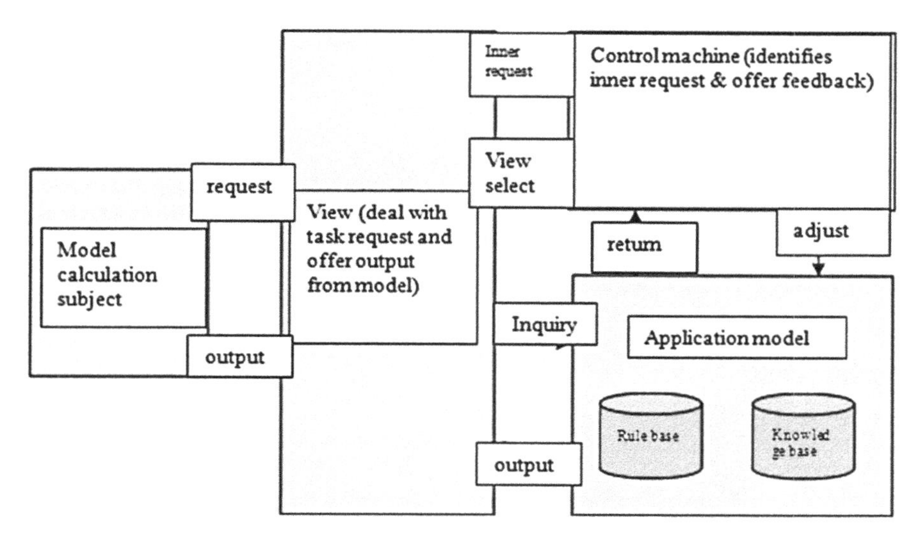

Fig. 23.2 Knowledge system structure in coordinator subject

(1) Collect Agent behavioural information and divide the system into N auto units.
(2) Abstract sample from database.
(3) Divide the signal sequence into several subsequences; reconstruct ε machine.
(4) construct vector set and measurement of system multiple subjects.
(5) Set target set, restrictions set and do adjustments.
(6) Construct decision-making tree. Do calculations according to the restriction tree on possible decision-making schemes.

23.4 Conclusions

Focusing on the problems existed in the research of supply China regulation and administration for pharmaceutical chain enterprise under the changeable environment, the non-traditional complex modeling theoretical method is adopted which is quite different from the traditional precise modeling. On this basis, a new normal form is built. It bases scientifically on the complex adaptable system (CAS) theory. The new normal is data driven and uses the Agent calculation as the core strategy. It is the stimulation analysis and adaptable administration of the supply chain computer for the pharmaceutical chain enterprises. The pharmaceutical retail chain supply chain stimulation model is constructed. Through constructing the closed-loop feedback cycle model of "data collection and treatment-model discovery and performance identification of data drive", design supply chain adaptability dynamic regulation and administration strategy based on active inducement and emergency interruption control.

The further research work is to select the classic pharmaceutical retail enterprises as the demonstration research platform. The theoretical method is technological indentified in the computer stimulation test system.

Acknowledgments This paper is supported by the 2010 key project of Humanities and social science research in universities in Anhui Province (N0:2010sk251zd), provincial research project of natural science of universities in Anhui Province (No. KJ2011Z217), and 2010 general project of humanities and social science fund in Anhui University of Traditional Chinese Medicine (N0:2010rw011B), etc.

References

1. Wei H (2001) Brief introduction on the development of drugstore chain management. Anhui Med Pharm J 5(2):37–55
2. Yang C (2007) The study on the performance evaluation system of chinese medicine supply chain. Harbin Institute of Technology School of Management doctoral dissertation, vol 75(6). pp 99–133
3. Huang GP (2004) Reverse logistic and its characteristics of medical retail chain enterpresizes. Chin Med Store 43(3):54–55
4. Li CM, Chu LZ et al (2008) Multi-agent interaction and cooperation simulation using swarm of inter-basin water transfer. J Syst Simul 20(8):2146–2151
5. Ni JJ, Wang JY et al (2005) An agent hybrid architecture model for simulation of complex adaptive system. J Hohai Univ (Nat Sci) 33(2):207–211
6. Wei XD, Xu LZ (2011) A data-driven adaptability control method on the complex system of water resources. In: International conference on communication software and networks (ICCSN), vol 37(4). pp 658–661

Chapter 24
User-Oriented Subject Service in Network Environment

Qiong He, Wankai Wu and Yuecong Chen

Abstract The current status of subject service in libraries was discussed in this paper, according to the statistical surveys of information users' behavior in North China Electric Power University; it discussed the new module of user oriented subject service in university libraries.

Keywords Subject service · User oriented · Database

Libraries are always considered as an irreplaceable organization for information acquisition, but such a status has been broken or wavered today. The OCLC report "Acknowledgement of Users to Libraries and Information Resources" showed that: 84 % of users used search engine to search information, 90 % of users were satisfied with the information got from search engine. To this it can draw a clear conclusion that: users use search engine as the primary to get information, that is, the relation of user and search engine is closer and closer, but libraries to users are more estranged and strange [1]. This new information environment provides new challenges to the traditional organization module and service module of libraries.

24.1 The Present Status of Subject Service

Subject service is a new service module of libraries that is closed to users to adapt to the new information environment and user requirement centered thinking. Along with the popularization of network and information resources, subject service is the development direction of library services. According to the

Q. He (✉) · W. Wu · Y. Chen
Library of North China Electric Power University, Beijing 102206, China
e-mail: qiong2044he@126.com

W. Du (ed.), *Informatics and Management Science VI*, Lecture Notes in Electrical Engineering 209, DOI: 10.1007/978-1-4471-4805-0_24,

information environment and document resources status of engineering science subject, the American Stanford University established SEQ2 Library, German Max Planck Society (MPS) set up Max Planck Digital Library (MDPL). Through the collaboration with institutes, SEQ2 and MPS provided infrastructure for institutes with scientific information and scientific intercourse. Tsinghua University in China firstly introduced "Subject Librarian System" and started subject service in 1998; in 2006, focused on subject service and information research, National Science Library of CAS promoted document information services for front-line research users actively, put forward to establish service module that put into front-line research users services. And the module was implemented in the following years. Under network environment, domestic and abroad libraries all choose to develop subject service and the construction module changes from meet the document requirements to meet the personalized requirements of users [2].

24.2 Survey of Users' Information Behavior

The service object of university libraries is teachers and students in university, more accurately, is teachers and senior students with more professional knowledge and PhD students and postgraduate students, whose information demand is characterized as high specialization and high crossed, and there are many differences between each other. The survey shows that: despite the great differences of each level student to use the library resources, the junior students that use library resources are only 4.9 % of all, the senior students are about 31.7 %, but the graduate students are almost 75 % of all. And for teachers, more than 700 people applied for VPN services up to 2010, which are over one half of the teachers full-time and on the regular payroll. So focusing on teachers, keeping up with the users' information behavior, understanding their information requirements are the premise to do a good job of subject service.

In recent years, with the improvement of network environments and the building of database resources in universities, the users' information behavior appears to be more and more professional and personal. In the questionnaires of teachers, it was accomplished according to the individual departments. In the investigation forms, in the item of "The main acquisition path of information", close to 50 % of teachers in literal arts such as humanities department and foreign languages department, etc. chose "paper document". And 70 % teachers in science and technology, such as energy, power and mechanical department, electrical and electronic engineering department, economic and management department, mathematical and physical department, etc. chose "professional database in library". But users chose to use "free network resource" were more than 50 % of all teachers investigated, and 93 % were teachers from foreign languages department. This demonstrate that databases of literal arts are absent in the library, and teachers of literal arts are unfamiliar with the use of academic database resources in library.

Table 24.1 The survey of requirement of use training

Department	Introduce of library (%)	Search knowledge of database (%)	Citation and retrieval check of papers (%)	Sci-tech novelty retrieval (%)	The use of chinese database (%)	The use of foreign database (%)	The acquisition of network information resource (%)
Literal arts	10	79	61	24	28	52	24
Science and technology	10	45	35	21	23	33	22

In the survey of the database use condition of teachers, 95 % of users encountered obstacles in the course of using database. 40 % of these are search skill problems, 48 % are obstacles in acquisitions of original resource, 11 % are viewers such as Adobe, CAJ viewer, etc., and 42 % are network problems. And in this survey, more than 50 % of users that encountered obstacles are in departments of literal arts. In the survey of teachers' requirement of use training, Table 24.1 was collected by the humanities and engineering departments. Table 24.1 is as follows.

It follows that users of different departments vary greatly to the training requirements of "Search knowledge of database" and "The use of foreign database".

24.3 Measures of Providing Subject Service

There are two forms of the subject service of university library. The first form is scientific research-oriented service, which includes integrity of user information resource, information push, providing thematic information, subject introduction, individual user training, etc. The second form is teaching-oriented service, which includes resource database of nonesuch courses, the database of graduate thesis, the virtual reference service, the use training of library, etc. [3].

24.3.1 Put Forward the Subject Resource Integration

According to the multiform, multilevel and personalized information requirement to put forward the subject resource integration is not only the inevitable trend in the development of information resource organizations, but also the only way to improve the utilization rate of digital resource and to raise the quality of service. In the purchased databases of North China Electric Power University Library, part of databases has overlapping content, low association degree of information resources and lack of subject navigation system, etc. All the above influence the information choose and acquisition of users. So the most urgent task is to draw up reasonable purchasing plan and integrate the available resources according to subjects.

24.3.2 Extend the Service Content of Subject Information Resource

Aiming at the requirements of individuation and specialization of information from users, exploring the information resource in network, the library should provide valuable network resource time-effective and high pertinent to develop and deepen the traditional information service. For example, to establish important user record for academic leaders, etc. with a clear aim to provide with information research service, information process service, information analysis service, etc. According to the users' information requirements, the libraries push information service periodically, which makes the users unnecessary to access the fixed website many times to get the newest science information and resource of relevant subject field.

24.3.3 Improve the Propaganda and Training Mechanism of Databases

At present, the training forms of university libraries are unitary and most of the trainings are database use trainings at a given time and location, the users can not have a free hand to arrange the time and learning content according to their own subject and interesting. So the library should draw up the training plan combined with the users' training requirements as well and set up the flexible training mechanism that users can choose independently. For example, users can make an appointment of the training about one given database before training; the library can set up course of special subject to one department users with high specialization.

Acknowledgments 2010 Beijing University Library Research Fund Project

References

1. Chu JL, Zhang DR (2008) Subject librarian 2.0 and subject services. Libr Inf Serv 30(2):6–10
2. Chai ML (2009) Study on subject resources construction in subject services. J Mod Inf 55(1):46–48
3. Li WG (2009) Subject management—the strategic selection of university libraries. J Libr Sci 77(3):47–49

Part II
Business Intelligence and Applications II

Chapter 25
Evaluation Index System and Evaluation Method for Sustainable Development of Industrial Park

Hongyan Li

Abstract This article aims at achieving sustainable development of industrial parks, combined with the principles of evaluation design. It builds industrial park sustainable development evaluation system from 34 sub-indicators in four areas such as resource support and utilization, economic development, environmental standards and management level. It is combined with the characteristics of each index to use comprehensive weight, standard value score, maxim-minimum scoring evaluation of quantitative indicators and three-level scoring evaluation of qualitative indicators, sustainable development, to achieve the scoring criteria of industrial park sustainable development.

Keywords Industrial park · Sustainable development · Evaluation indexes · Evaluation methods

25.1 Introduction

As an important carrier of the industrial economy, Industrial Park plays a very important role in promoting economic and social development. It is becoming new highlight of economic development, main battlefield of opening up, large base of job placement, and experimental area of innovative mechanism [1]. It effectively promotes industry and Small and Medium-sized Enterprises (SME) development and local economic growth [2]. At present, the entire China is still in the middle

---

H. Li (✉)
Department of Management Science, Henan Institute of Engineering,
Zhengzhou 451191, China
e-mail: hylihn@126.com

W. Du (ed.), *Informatics and Management Science VI*, Lecture Notes in Electrical Engineering 209, DOI: 10.1007/978-1-4471-4805-0_25,

stage of industrialization. In the global information age, enhancing industrial park development potential to achieve sustainable development has important significance for accelerating the pace of industrialization, rapidly increasing the level of industrialization [3]. Research on sustainable development indicators system of industrial park, on the one hand benefits all government departments' comprehensive and objective assessment and evaluation on the Park, and promoting the development of the park through policy guidance, on the other hand benefits the Parks' better understandings of their state of development, and achieving sustainable development through horizontal comparison, discovery and consolidation of strengths, and finding and filling gaps [4].

25.2 Evaluation Index System

25.2.1 Structure of Evaluation Index System

Evaluation system should be built in combination of impact factors and the actual development of industrial parks, drawing on domestic and foreign research results. It reflects 34 sub-indicators in four subsystems such as resource support and utilization, economic development, environmental standards and management level, constitutes the entire evaluation system group [5]. At the same time, each indicator of the subsystems should be bound together according to certain design idea, with full consideration of its importance, to ensure the system group to form an organic whole. N represents the sustainable development ability level of the textile industry, includes P_1, P_2, P_3, P_4 [6].

Resource support (Y_1) and resource utilization (Y_2) are two secondary indicators under resource support and utilization. Resource support covers 4 indicators such as job rate Y_{11}, power rate Y_{12}, water rate Y_{13}, raw materials Y_{14}. Resource utilization covers 7 indicators such as comprehensive energy consumption per unit GDP Y_{21}, annual industrial added value per unit land area Y_{22}, industrial water recycling rate Y_{23}, raw materials recycling rate Y_{24}, industrial waste gas comprehensive utilization rate Y_{25}, reclaimed water reuse rate Y_{26}, industrial waste heat utilization rate Y_{27}.

Economic development subsystem covers economic development level (Y_3) and economic development potential (Y_4). Economic development level refers to 3 indicators such as annual GDP growth rate Y_{31}, per capita GDP Y_{32}, investment productivity Y_{33}. Economic development potential refers to 3 indicators such as the proportion of the new product sales to total sales Y_{41}, the proportion of professional staff to total staff in the park Y_{42}, the proportion of R&D investment to GDP of the park Y_{43}.

Environment subsystem covers pollution intensity (Y_5), pollution control index (Y_6) and environmental protection index (Y_7). Pollution intensity covers 3 indicators such as gas emission per unit GDP Y_{51}, wastewater emission per unit GDP

Y_{52}, solid waste emission per unit GDP Y_{53}. Pollution control index covers 4 indicators such as COD emission per unit GDP Y_{61}, industrial waste gas treatment compliance rate Y_{62}, industrial wastewater treatment compliance rate Y_{63}, industrial solid waste innocuous treatment rate Y_{64}. Environmental protection index covers 3 indicators such as garbage innocuous treatment rate Y_{71}, park greening rate Y_{72}, the proportion of environmental protection investment to fixed asset investment Y_{73}.

Management subsystem consists of infrastructure (Y_8) and policy and management (Y_9). Infrastructure covers 3 indicators such as the network infrastructure maturity Y_{81}, waste treatment sharing facilities Y_{82}, information platform completion Y_{83}. Policy and management covers 4 indicators such as rules and regulations perfection Y_{91}, park development planning and implementation Y_{92}, contact with the Sector Association Y_{93}, policy support Y_{94}.

25.3 Evaluation Method

25.3.1 Indicators Weight Determination–Analytic Hierarchy Process

(1) Determine the estimation matrix

Experts use 1–9 proportion scale method, to make qualitative description to the relative importance of each level of indicators, and quantify the exact number. The pairwise comparison matrix through experts' scoring as per Table 25.1.

With the hierarchy model based on the indicators established by consultants, to test P level factors on the relative importance of N-layer of factors, and to determine the N–P matrix.

Using yaahp AHP software to list the structural framework of indicators; then construct matrix, select group decision support, add all the expert information, weight, and input the experts' scoring data to matrix. For each matrix, the program will automatically check for consistency.

Table 25.1 Scale order

Scale aji	Definition
1	i equally important to j
3	i somewhat important to j
5	i more important to j
7	i very important to j
9	i absolutely essential to j
2,4,6,8	The intermediate state between the two determination
Reciprocal	If i compared to j, the determination is a_{ij}, then $a_{ji} = 1/a_{ij}$

Table 25.2 Average random consistency index RI

Order	1	2	3	4	5	6	7	8	9	10	11	12
RI	0.00	0.00	0.58	0.90	1.12	1.24	1.32	1.41	1.45	1.49	1.51	1.48

(3) Consistency test

To calculate the consistency index CI and the consistency ratio CR, and to test its consistency.

$$\lambda_{\max} = \sum_{i=1}^{m} \frac{(Nw_i)_i}{nw_i}, i = 1, 2, \ldots, m \tag{25.1}$$

$$CI = \frac{\lambda_{\max} - n}{n - 1} \tag{25.2}$$

where, N is the N–P matrix, n is the order, $\lambda_{\max}$ is the largest eigenvalue.

While the consistency is higher, CI value is smaller. When CI = 0, the matrix is exactly the consistent. However, in the establishment matrix, critical thinking impact of inconsistency is just one of the reasons of consistency, the result of a ratio of 1–9 scale as two two-factor comparison is also the reason of consistency. CI value set based solely on an acceptable standard of inconsistency is clearly inappropriate. In order to get a critical value applicable to different divisor, the impact of the order of the matrix must be eliminated.

In the AHP, the consistency ratio is to solve this problem. To adopt average random consistency index RI, and RI is the correction factor to eliminate inconsistent effects caused by the matrix divisor. Specific values as per Table 25.2.

$$CR = \frac{CI}{RI} \tag{25.3}$$

Under normal circumstances, the order of the matrix $n \geq 3$, when $CR \leq 0.1$, the relative error CI that $\lambda_{\max}$ deviated from n does not exceed one-tenth of the average random consistency index RI, the consistency of the matrix is generally acceptable; otherwise, when $CR > 0.1$, the deviation from the consistency of the matrix is too large, the matrix must be adjusted to make up with a satisfactory consistency.

Industrial park evaluation index and its weights are summarized in Table 25.3, and listed according to the weight of index to target.

25.3.2 Quantitative Indicators Evaluation

(1) Standard value scoring. Standard value scoring is to determine a standard reference value on the basis of the relevant data in recent years. The standard value is mainly to take the world average, the state developed standards and standards in

Table 25.3 Industrial park evaluation index system and weight

Target	Rule	Index	Sub-index	Weight of sub-index to index	Weight of sub-index to target	Ranking
N	P_1 0.2172	Y_1 0.1403	Y_{11}	0.1281	0.018	22
			Y_{12}	0.2009	0.0282	12
			Y_{13}	0.2009	0.0282	12
			Y_{14}	0.4701	0.0659	3
		Y_2 0.077	Y_{21}	0.1346	0.0104	26
			Y_{22}	0.3987	0.0307	10
			Y_{23}	0.1102	0.0085	27
			Y_{24}	0.1102	0.0085	27
			Y_{25}	0.0782	0.006	30
			Y_{26}	0.0828	0.0064	29
			Y_{27}	0.0852	0.0066	28
	P_2 0.3765	Y_3 0.1695	Y_{31}	0.3104	0.0526	5
			Y_{32}	0.1946	0.033	9
			Y_{33}	0.495	0.0839	2
		Y_4 0.207	Y_{41}	0.4762	0.0986	1
			Y_{42}	0.2793	0.0578	4
			Y_{43}	0.2445	0.0506	6

line with China's national conditions. Actual scoring is according to specific indicators. The standard reference values and weights are set into the formula.

If the actual value Ri, the standard value Si, the weights ω_i, the assessment score of i is:

When Si is Positive indicators

$$f_i = \frac{R_i}{S_i} \tag{25.4}$$

When S_i is Negative indicators,

$$f_i = \frac{S_i}{R_i} \tag{25.5}$$

Then, the total score using standard indicators is:

$$\sum_{i=1}^{m} F = \omega_i \times f_i \tag{25.6}$$

In formula, m is the total number of indicators in the standard value scoring.

According to the nature of indicators, the following indicators adopt standard value scoring: $Y_{11}, Y_{12}, Y_{13}, Y_{14}, Y_{21}, Y_{31}, Y_{32}, Y_{33}, Y_{51}, Y_{52}, Y_{61}, Y_{62}, Y_{63}, Y_{64}, Y_{71}, Y_{72}$. If the index exceeds the assessment criteria, 100 points (full points) is the result of the scoring. Specific values as Table 25.4.

Table 25.4 Standard value scoring index

Rule	Index	Sub-index	Standard value	Character
P_1	Y_1	Y_{11}	100	Positive
		Y_{12}	100	Positive
		Y_{13}	100	Positive
		Y_{14}	100	Positive
	Y_2	Y_{21}	0.5	Negative
P_2	Y_3	Y_{31}	15	Positive
		Y_{32}	3	Positive
		Y_{33}	1.6	Positive
P_3	Y_5	Y_{51}	3	Negative
		Y_{52}	4	Negative
	Y_6	Y_{61}	3	Negative
		Y_{62}	100	Positive
		Y_{63}	100	Positive
		Y_{64}	100	Positive
	Y_7	Y_{71}	100	Positive
		Y_{72}	40	Positive

(2) Maxim–minimum scoring

Maxim–minimum scoring is to give a range of values on each indicator. Linear interpolation scoring will be used in this range, i.e. between maxim and minimum. Out of the range, one end is 0, the other end is 100 (full points). It is applicable to the indicators scored in the range. It is a useful complement to score only by specific quantitative criteria.

Specific formula as follows:

If indicator i is set with actual data, weight, maxim and minimum standard value as Si, ω_i, Bs, Bx, when Si is adverse index, its value is better as smaller. The scoring formula is:

$$F = \sum_{i=1}^{4} \omega_i \times F_i \tag{25.7}$$

When Si is positive index, its value is better as bigger. The scoring formula is:

$$f = \begin{Bmatrix} (S_i - B_x)/(B_s - B_x), & B_x \le S_i \le B_s \\ 100, & S_i > B_s \\ 0, & S_i < B_x \end{Bmatrix} \tag{25.8}$$

Then the comprehensive score is $\sum_{i=1}^{m} F = f_i \times \omega_i$.

According to the nature of indicators, the following indicators adopt maxim–minimum scoring: Y_{22},$Y_2$3,Y_{24},Y_{25},Y_{26},Y_{27},Y_{41},Y_{42},Y_{43},Y_{21},Y_{51},Y_{52},Y_{53},Y_{73}. Because these indicators are all based on the ratio of meeting certain conditions with the total gain, so their scores do not fit the single standard value, but are more suitable for the value

Table 25.5 Maxim–minimum scoring index

Rule	Index	Sub-index	Maxim	Minimum	Character
P_1	Y_2	Y_{22}	15,000	2000	Positive
		Y_{23}	85	30	Positive
		Y_{24}	85	25	Positive
		Y_{25}	90	30	Positive
		Y_{26}	90	20	Positive
		Y_{27}	95	30	Positive
P_2	Y_4	Y_{41}	30	5	Positive
		Y_{42}	20	5	Positive
		Y_{43}	10	1	Positive
P_3	Y_5	Y_{53}	25	0	Negative
	Y_7	Y_{73}	8	4	Positive

determined by a range of values determined the interpolation of status indicators between maxim and minimum value.

The maxim of the indicators is the ideal standard of national standards, while the minimum is the lowest limit of specific indicators according to experts' advices, and 0 points in case lower. Specific values as per Table 25.5.

25.3.3 Three-Level Scoring Evaluation of Qualitative Indicators

This evaluation divides the indicators into three levels according to people's view, details as per Table 25.6.

If each indicator's weight is ω_i, and score is f_i, then the score is:

$$\sum_{i=1}^{m} \omega_i \times f_i \tag{25.9}$$

25.4 Comprehensive Scoring

The total score of industrial park sustainable development is:

$$F = \sum_{i=1}^{4} \omega_i \times F_i \tag{25.10}$$

Among them, $\omega_i$is the weight of the i rule; F_i is the score of the i rule. For the industrial park, the total evaluation index of sustainable development can fully reflect the park's sustainable development and capacity. F_i is between 0 and 100.

Table 25.6 Three-level scoring

Level	Bad	Normal	Good
Score	33	67	100

Table 25.7 Index of different levels of sustainable development of industrial parks

Levels of sustainable development of industrial parks	Evaluation index
Strongly sustainable	$F \geq 80$
Intermediately sustainable	$60 \leq F < 80$
Weakly sustainable	$40 \leq F < 60$
Not sustainable	$F < 40$

This evaluation system divides sustainable development of industrial park into four levels. According to the actual situation of the park, the evaluation index of different levels of industrial park are shown in Table 25.7.

The Industrial Park with evaluation index of sustainable development (score) higher than 80 can become a model for sustainable development.

25.5 Conclusion

On the basis of the indicators weight achieved through consulting experts, sustainable development of economy is not isolated. To maintain rapid economic development is the basis of sustainable development of industrial park, while higher growth potential, good environmental and processing capacity and complete development planning and management capacity is the important safeguard to achieve sustainable development of industrial park. In the algorithm, the characteristics of the establishment of indicators should be combined, and the applicable comprehensive evaluation method should be adopted.

Acknowledgments This work is supported by Youth Fund Project of Human Social Science Project of Education Ministry of China (No.10YJC790237), Henan Provincial Government Decision-Making Tender Key Project (No. 2011A04), Henan Province Department of Education project (No. 2011B630003).

References

1. Li F, Li H (2010) The evaluation model of the eco-industrial park and enterprise development capability in the park. J Liaoning Normal Univ 33(2):260–262 (Natural science edition)
2. Sun J (2010) Discussion and suggestions for the sustainable development of shanghai creative-clustering parks. Sci Technol Ind 10(5):22–25

3. Liu Z, Liu H (2010) Study on the campus of economic development and regional economic interaction of the dynamic mechanism. Econ Res Guide 14:40–41
4. Yang H, Chen X (2009) The evaluating index system and the method of evaluation for sustainable developing of regional economy. Value Eng 7:18–21
5. Chen X, Wang Y (2008) Research on evaluation system of eco-industrial parks to sustainable development. Sci Technol Ind 8(7):29–33
6. Tang Y, Shi H (2005) Development index evaluation system for developing industrial parks under the guide of the new industrialization strategy. Syst Eng 23(7):89–93

Chapter 26
Research on Mechanism of Women Participating Construction of Ecological Civilization in Economic Transition

Shujun Li

Abstract For China social labor division, women in social and family for different tasks in, so in the construction of ecological civilization process, gender influence factors should not be ignored. Women of the construction of ecological civilization and the will of the target different, which in turn have an impact on the ecological environment. In combination with the actual situation of the construction of ecological civilization to women in participation mechanism and related problems are analyzed and explained.

Keywords Ecological civilization · Women · Participation mechanism

26.1 Introduction

Women and ecological problems in the world today is in deteriorating environment, economic transformation and upgrading of human society faces under the background of the age problem. In China, the construction of ecological civilization and harmonious society, is communist party of China the comprehensive construction well-off society's important tasks, and is also a "1025" period, carry out the cause of socialism with Chinese characteristics new content, which contains all the women, the people's desire and basic interests. In the process of the construction of ecological civilization, the status and role of women cannot be ignored, establish and perfect the construction of ecological civilization female participation mechanism, is the basic requirement of the construction of

S. Li (✉)
Zhejiang A&F University, 311300 Ling'an, China
e-mail: Shu_junLi@yeah.net

W. Du (ed.), *Informatics and Management Science VI*, Lecture Notes in Electrical Engineering 209, DOI: 10.1007/978-1-4471-4805-0_26,

harmonious society [1–4]. With our current population distribution situation, women make up about just over half the population, has become a socialist economic construction, democratic political important participants, decision makers, and wore socialist culture innovation and spread the important task of modernization, is indispensable to the social construction of strength, will in the construction of ecological civilization occupies a place. Women and ecological problems should be the construction of ecological civilization and in the topic of constructing harmonious society of righteousness [5].

26.2 Ecological Feminism, the Construction of Ecological Civilization and Women Participate in Connotation

Civilization is a national social progress and people's spiritual concentration reflected, representative of a country or region's economic, political, and cultural development level [6–8]. In fact, ecological civilization is the ecological environment civilization, mainly refers to the humanity in the transformation objective in the process of the physical world, need to constantly dealing with a possible to produce a variety of negative influence, straighten out between people and nature, the harmony between the relationship, create good ecological environment and ecological operation system operation and gain spiritual civilization and material civilization of harmony. In the present ecological environment deterioration of the intensity, it is urgent to establish a set of perfect the construction of ecological civilization system, based on the ecological theories, and play women in the important position of the construction of ecological civilization, formation and characteristics that adapt to the civilization in China, master of human thought civilization, and put an end to unilateral pursue short-term benefits of environment and resources and caused damage [9]. The construction of ecological civilization, should be in order to respect the ecological environment and maintain ecological value for essential starting point, follow the "sustainable development" principle, in people use natural process, set up the human and the nature harmonious development of consciousness, all economic activity must be in the overall interests of the natural system as the starting point, in the life, work on the ecological environment and resources on limited force, promote China's economic, social, political and environmental sustainable development, contribute to the building of a harmonious society.

In western society civilization, will reflect the feminist ecological and ethical point of view called "ecological feminism," the French feminist scholars F • Opal in work, put forward a new concept, the feminist movement began tend to "ecological feminism" [10]. In China the construction of ecological civilization in process, sex cannot be ignored, so must change the traditional "gender discrimination" consciousness, but full understanding to the women in the role of ecological construction, including consciousness, influence, consumption behavior affect influence all aspects, etc.

26.3 Women's Role in the Construction of Ecological Civilization

26.3.1 Women's Role in Ecological Consciousness

For a long time by people the influence of the thinking of qualitative, a lot of people don't see women in ecological consciousness formation of positive impact, ignore them in the construction of ecological civilization in the important strength, and will also women in economic development, environmental protection of vulnerable position. Therefore, must change the prejudice consciousness, to realize the power of women. With the development of society, accept higher education, learning women scientific knowledge, promote the moral level, in ecological consciousness constantly updated, in ecological consciousness in form to play a positive role, and constantly develop ideas, can form the main driving force, and continuously drive the whole women's comprehensive quality.

In rural areas, many staying women served as the care for the elderly, the task of raising a child; If not put the powerful force into the ecological agriculture, so the environmental protection consciousness and behind behavior, will restrict the construction of ecological civilization. For example, according to relevant data, according to the survey, in rural areas, about 50 % or more of the rural staying women think combustion on atmospheric pollution caused by the straw, and therefore no straw burning application modes of production and life; Some 23 % of rural staying women think that use straw burning to cook a meal, can save coal; Another 27 % of women are not considered straw burning is appropriate, but the straw stalk carelessly dispose. At present, most women have been known to rural rear ecological the importance of environmental protection, but with a very small part does not have this consciousness; they think it and they have no direct relation, even in the severe pollution near residential construction enterprise, as long as it can bring to the farmers' income, so can accept. Visible, in rural areas, part of the women's ecological consciousness is relatively backward, not for ecological environment protection make positive contribution, still stays in the economy is more important than the environmental protection of the traditional thinking stage.

26.3.2 The Influence on the Ecological Production Behavior of Women's Participation

Recent years, with the rural "staying women" more and more, they gradually become the main force in the construction of rural production. These agricultural women not to suffer the shackles of traditional ideas, also out of the home, and gradually into society, they also participate in the construction of ecological civilization, is a necessary factor in the development of new countryside of power.

Just think, if these women lack basic ecological consciousness, in the farming land destroyed in the process and cause the pollution of soil, in the life destroy the water environment, influence ecological balance, in rural ecological construction can only on paper. Therefore, we must strengthen the ecological construction in rural women of the attention, let them become the construction of ecological civilization practitioners, and continuously learn civilized manners, environmental protection, common sense, develop good culture, scientific farming, positive business, in the process of production improve civilization quality, with host state of mind to join in the construction of ecological civilization.

26.3.3 The Influence on the Ecological Consumption Behavior of Women's Participation

As women in the family of the position and role, decided they are the main body of the social consumption, women's life and individual consumption mode, will have a direct impact on the whole family and even the social spending pattern. Can say, a women's view represents a family, the behavior is not only affect the husband, also can affect children and the elderly. Therefore, women by changing the family consumption idea, can choose the more good for environment protection way of life, this also is further curb environmental degradation of one of the main methods. Consumption is justified or not, whether the way affect the environment and so on, which are an important content of the construction of ecological civilization. If women into buy the products, ignore the influence on ecological protection, will surely promote the environmental protection consumption patterns, in social form.

26.3.4 The Influence on the Ecological Life Behavior of Women's Participation

Out of the family, out of the community, women in social function in the construction of ecological civilization is also very important. For example, in the daily life, the public places, women pay more attention to the life habits and saving, these are the important content of the construction of ecological civilization. Women in the daily life of a good habit, often can give social civilization to bring greater influence and appealing. After a survey, in life and work process, most female is only occasionally use disposable tableware, and most of the women very attention to saving, active driving around the neighbors, friends, colleagues or even strangers to participate in environmental protection construction, to improve the environmental protection consciousness. In the ecological environment health protection, most women more pay attention to environmental protection awareness and raise the consciousness of environmental protection of the environment clean and promote the whole beautiful.

26.3.5 The Women's Influence Power in Ecological Community

Women willing to join in community activities, to get more development space, constructing the harmonious community, neighborhoods, and the experience of the construction of ecological civilization itself back society, promote the development of the harmonious community. According to relevant data show that about 80 % or more of the women will join or occasionally attend community events; and more than 90 % of the women actively participate in community organizations of various activities, such as volunteer activity, ecological construction, etc.

In addition, many women hope community can conduct more some training, education activities, such as job training and women's health, family education BBS knowledge lecture. And women in stop destroying community environment behavior, and beautify the community environment of the play an important role, they expect to construct a harmonious and beautiful community environment. Visible, women in the community family civilization construction, the construction of ecological civilization irreplaceable role play, become the bridge and harmonious communication neighborhood contacts and harmonious development of nature the link, optimization community cultural environment.

26.4 Female Participation Mechanism in Perfecting the Construction of Ecological Civilization

26.4.1 Establishing and Perfecting the Mechanism of Women's Participation

In the new era, the governments at various levels must recognize that women in promoting human development, promote the construction of harmonious society, the importance of women in actively establish political, economic, social, environmental mechanism, through the women's intelligent wisdom, and lay the foundation for the harmonious social development. If want to give full play to the women in the important role of the construction of ecological civilization, must be in the social context of creating good social atmosphere. Therefore, governments at all levels, the personnel department of employment and unit, should strengthen to the women's attention. For example, all enterprises and institutions in hiring female worker, besides missing is not suitable for women is in the position or work outside, cannot refuse to hire or gender to improve the standard of female employment for; Governments at all levels actively implementing the female employment task, ensure the important role of women in employment, develop the productive forces, etc. to full play. In addition, organizations around departments should also increased dynamics, improve the women participating in the consciousness and power, cultivate a group has the potential for the development of women cadres, to improve the ability of participating politics; Also through exchange, discussion meetings, taking the way, strengthen the communication between the post and communication, to ensure that women get more knowledge and experience, better feedback society.

26.4.2 Improving Women's Protection Law of Rights and Interests

Through the legislation, to increase women's legal regulations propaganda, maintain the lawful rights and interests of women. At present, the United Nations economic and social council, has established "the status of women members committee", mainly by the 15 member states that constitute, the main research women in social, economic, political, education, the environment, and other aspects of the rights and status, this paper puts forward the legal system to maintain status of women. In addition, China in further on the basis of the constitutional revision, and constantly improve the population and family planning law, the marriage law and other related laws and regulations, the women's lawful rights and interests of complete basic guarantee system of laws and regulations construction, and make sure that all women's problems can be solved through legal channels. Although the government department heads, staff, such as the theory of the policies and regulations have some knowledge of, but more citizens even to the intellectual knew very little about the law, so to strengthen the publicity of regulations and policies of the education work, and constantly laid the social position of women, give play to the construction of ecological civilization of important guidance.

26.4.3 Strengthening the Foundation Construction of Rural Ecological Culture

In rural areas by hanging posters, billboards and other forms set, enhancing ecological culture propaganda education; Realize garbage classification, set up saving signs and better through the culture means, carry out the ecological culture idea, in various forms of carrying out green culture activities, in the rural areas to create a good ecological environment: first of all, through the "science and technology market" means, improve the women of the science and technology quality, using current XuanChuanChe, extend the flyer, hire a professional technology personnel means such as lecture, teach production experience and environmental protection consciousness; Second, the use of various cultural propaganda positions, such as the Internet and television, to increase women's science and technology culture teaching; Again, still can by setting up agricultural technical consultation hotline, agricultural information means such as text messages, give women more get science technology and information level of channel, imperceptible ascension science and technology quality and ecological environmental protection consciousness.

26.4.4 Advocating Scientific and Civilized Way of Life

Advocating scientific and civilized way of life, and to strengthen the expression civilization construction to have the important meaning. One is through the radio, television, Internet and other news media, ecological civilization consciousness will permeate to the women in all walks of life and production, and then improve the women sense of responsibility; Two it is the publicity of ecological culture, change people's production consumption structure, with the female popular entertainment activities, rich and colorful cultural activities, and constantly improve the quality of life and culture women taste; Three it is China's rural areas for the "staying women" phenomenon, and constantly improve their knowledge, and especially to strengthen science and technology culture training, production skills training, in order to form the scientific way of life, advocating civilization.

26.4.5 Promoting Green Technology

With the rural women's ecological culture construction importance, as well as the thought consciousness to the current situation, must strengthen the training of rural staying women technology, with a variety of ways, incentive rural women in the ecological construction, let them to recognize the importance of science and technology development, ecological culture construction to carry out the production process of life. On the one hand, the content of the training should be consistent with the actual demand of rural women, through the simple and understandable language, with flexible, rich and colorful form strengthen training, transfer of rural women good leaders enthusiasm, looking for volunteers, rural development in a positive role. Farmers in the production of the final goal to become rich, so will the knowledge and improve by green certificates, and the combination of deep into the actual production of the rural field, the full implementation of the green technology, pay attention to the theory and the practice; On the other hand, pay attention to training method to rural staying women have a comprehensive understanding of the full play of the importance of technology in agricultural development, will more professional language, rural staying women can accept, can understand the way, improve training efficiency.

26.5 Conclusion

From the foregoing, in social development today, women and the construction of ecological civilization are closely linked. Especially with the quickening pace of global integration, China must set up the new view of environment, development, with the social problems, and the combination of promoting the development of society. Through the legal regulations, propaganda education and other ways, promote the development of women in society the important position, to their own good habits and moral accomplishment, infection, neighborhoods, social relationship between family members. Therefore, in the construction of ecological civilization, must consider the important role of women factors, and promote the harmonious development of the ecological civilization.

Acknowledgments The research is supposed by Education of Humanities and Social Sciences in general project (project number: 09YJC840046) and the project development fund (2010FR054) of Zhejiang agriculture and forestry university.

References

1. Hui G, Wenhu Y (2011) Discussion on environmental social system and ecological civilization construction of the four kinds of basic relationships. China Popul Res Environ 6:17–21
2. Li S (2008) Female consumer life style social construction and reflection. J China Women's Univ 6:196–198
3. Hongfa F, Xiao Z (2009) The construction of ecological civilization in the consumer "green". Ecol Econ 11:47–52
4. Jin Y (2008) Ecological civilization: from conception to practice in China. Clean Technol Environ Policy 2:312–317
5. Emery KF (2008) A Zoo archaeological test for dietary resource depression at the end of the classic period in the petexbatun, guatemala. Human Ecology 5:191–194
6. Qu Z (2008) Ecological civilization in need of environment ethics and legal protection. J Nanjing Forest Univ 3:46–51 (Social Sciences Edition)
7. Roman MW, Hall JM, Bolton KS (2008) Nurturing natural resources: the ecology of interpersonal relationships in women who have thrived despite childhood maltreatment. Adv Nursing Sci 3:42–46
8. Yaoxian W, Wei L, Mingming Y (2011) The establishment of environmental quality evaluation index system for the enhancement of the public environmental quality. Environ Prot 6:102–108
9. Yu L (2007) On the whole society to develop ecological civilization education. Environ Sci Inst Liaoning Province Acad Ann Meeting 16:422–428
10. Xinshi Z (2010) About ecological restoration and recovery of speculation and scientific meaning and the approach of development. J Zool 1:261–268

Chapter 27
Study of Urban Wetland Park Development in Chongqing

Xiao-Bo Li

Abstract Urban wetland parks have become an important and effective way in the protection, reconstruction and rational utilization of wetland. This paper analyzes the superiority and weakness of urban wetland park development in Chongqing, and proposes some measures and proposals on the protection and development of urban wetland park by taking development requirements of Chongqing ecological construction into consideration.

Keywords Chongqing · Urban wetland park · Development pattern

27.1 Introduction

With the rapid development of urbanization, city has become a high-density mankind gathering place, how to deal with the contradiction between human activities and ecological environment has become an important issue in a livable city. As an emerging and developing park, constructed wetland has the natural resources, landform features and other edge features of both terrestrial environment and water environment. It has good ecological benefit, visual effect and cultural feature, is the development trend of urban ecological park in recent years, and is gradually approved by people. In recent years, the authorities of Chongqing

---

X.-B. Li (✉)
College of Urban Construction and Environmental Engineering,
Chongqing University, Chongqing 400045, China
e-mail: 51fei@126.com

X.-B. Li
Chongqing Education College, Chongqing 400065, China

W. Du (ed.), *Informatics and Management Science VI*, Lecture Notes in Electrical Engineering 209, DOI: 10.1007/978-1-4471-4805-0_27,

has gradually attached importance to the construction of wetland, many large-scale wetland parks is under construction or have already been constructed, such as Caiyun Lake Wetland park, Longtou Temple Park, Kwan-yin Pond Wetland Park, "wetland for migratory birds" of West Garden Expo that is under construction and so on. This paper studies the sustainable development model of Chongqing urban wetland park on the basis of its planning and construction.

27.2 Concept of Urban Wetland Park

Urban wetland refers to the wetlands that are distributed in city, which includes various types of natural and artificial wetlands. At present, the international classification of urban wetlands is not clear. The wetland in this paper refers to the city park that aims at wetland protection. And these city parks are restored from the original reservoirs, ponds, swamps and other natural or artificial waters in the city or suburb.

"Encyclopedia of China • Construction Planning of Garden City" defines park as "Park is a type of public green space, which is built and operated by government or public groups for public rest, appreciating, entertainment and so on" [1]. Compared with other types of parks, the concept of urban wetland park includes three aspects which are "urban nature", "ecological nature" and "park nature". Therefore, I think the concept of urban wetland park can be defined as following. Urban wetland park is a public garden that locates in urban area or near suburbs, it provides sightseeing, leisure, popular science education and other activities which are in harmony with natural ecological process, the main environment is constructed by reserving, imitating and restoring wetland habitat, the sewage is disposed through a constructed wetland system, the main target of this park is to protect and construct the diversity and self-succession ability of the local ecosystem, meanwhile, ecology, art and technology are combined in this park.

27.3 Function and Value of Urban Wetland

The function of urban wetland park is mainly reflected in the ecological environment and social aspects. The function can be specifically listed as following: (1) Improving microclimate. It affects local humidity and the amount of rainfall mainly through evaporation and precipitation patterns, this can ease the urban heat island phenomenon in a certain extent and also promote the sustainable and healthy development in city. (2) Flood prevention and flood storage. Wetlands are generally located in the low-lying land of city, most of which are converted from original middle and small reservoirs. It can store surface runoff and slow down peak during rainy seasons, and at the same time stores natural water resources.

(3) By adopting wetland treatment technology, the sewage that is disposed after treatment is developed and utilized as second water source in city, in order to raise public awareness of water resources utilization. (4) Eliminating and conversing the harmful substances. The speed of water will be slowed down while it is flowing through city wetland, and the toxic substances and impurities in the water will precipitate, or be absorbed and reduced by plants and microbes. (5) The park provides important habitat for plants and animals, and keeps the diversity of city species. (6) Realizing the diversity of landscape. The park provides an important recreational place for urban residents, and meets people's emotion of loving water, near water and viewing water, which riches people's lives. And as an environmental education place, it also raises people's awareness of environmental protection. (7) Affecting the layout of urban planning and the trends of economic and social development. Appropriate water area improves urban developing environment, promotes the development of tourism industry and real estate industry, improves urban quality, creates a favorable investment environment, and thus speeds up the city's sustainable development. It can be said that wetland is one of the basic contents of harmonious and healthy living environment planning.

27.4 Location Analysis

Based on the basic requirements of "The outline of the 12th five-Year (2011–2015) plan for national economic and social development in Chongqing", Chongqing will highlight economical resource utilization and ecological environment protection, build a resource-saving and environment-friendly society, take green development road and enhance sustainable development ability. During the time of the "12th Five-Year", "five Chongqing" will be basically accomplished, the forest coverage rate will reach 45 % and the quantity and quality of urban green space will be improved. With the promoting of innovative city construction, Chongqing will be established into a national central city with element convergence, perfect function, habitability, enterprise-adaptability, landscape garden and specific charming. The natural beauty and historical and cultural heritage of "mountain city" and "River City" will be highlighted, the city landscape with open and beautiful scenery, natural area, mountain architectural features and Chongqing historical context will be elaborately created, and the harmonious co-existence between city and river, mountains, forests, green land will also be promoted. Therefore, more attention will be paid to the environmental protection and urban sustainable development in Chongqing, and the construction of urban wetlands will play a part in this trend which would promote the overall development and the green island function accomplishment in the city.

27.5 Analysis on the Development Advantages and Deficiencies

27.5.1 Advantages

27.5.1.1 Resource Superiority

The main city zone of Chongqing is located beside Yangtze River and Jialing River, so it has relatively rich water resources which can generally meet and adapt to the requirements of urban social and economic development. But as the city zone is located on the back ends of reservoir, with the increase of urban population and economy, wastewater from urban life and production is increased year by year, the capacity of sewage treatment plant is increased, large amounts of water is consumed by irrigation for green land, and the water for waterscape is confronted with deterioration of water quality and decrease of water area because of some design mistakes or economic reasons. Therefore, the development of wetland within city area can ease sewage treatment pressure, reduce the impact of urban non-point source pollution to Three Gorges reservoir area, increase the recycling use of water, and improve urban environmental quality. Meanwhile, it plays an important role in the construction of forest city and sanitary city for Chongqing.

27.5.1.2 Landscape Superiority

Water is the most vibrant element in urban landscape organization because of its active quality and penetrating power. Water environment in the main city zone of Chongqing is formed into stoke, beach, island and some other places, and there places are supplemented by wetlands. These wetlands are evolved from city ponds, reservoirs, artificial lakes, etc. that formed under mountain conditions. This phenomenon creates the charming space landscape of "a jade plate full of pearls of different sizes". The varied water environment patterns provide a rich living environment for organisms, and improve the development space of urban biodiversity.

27.5.1.3 Policy Superiority

In 2004, Chongqing municipal government issued "Circular of further strengthening wetland protection management" in order to implement the spirit of "Circular of further strengthening the wetland protection management by the General Office of the State Council". This circular defines the responsibility of every government functional department, and gives a comprehensive plan for the city wetland protection management. Meanwhile, the city government sets up a coordination and leading group of city wetland protection management, which strengthens the organization and leading of wetland protection management. In recent years, the city zone of Chongqing has made a great progress in wetland construction; many large-scale wetlands are under construction or have already been constructed. The Caiyun Lake Wetland Park which was open to public for free in 2010 had become the first national urban wetland park in the city zone. And the "wetland for migratory birds" of Garden Expo covers an area of 350 acres, some typical wetland plants such as reed, miscanthus floridulus, metasequoia, taxodium ascendens and so on are planted here, landscapes like "Crossing water dam", "Happy fish ponds", "Terraced flowers", "Bird watching platform" and so on are built inside the wetland, which fully represents the harmonious unity of garden and nature.

27.5.2 Deficiencies

There is a land contradiction between wetland park construction and urban construction. Chongqing is in the climax of urbanization development, with the urban construction and outward expansion, wetland in the urbanized area have been developed and utilized in large scale, this inadequate protection is because of insufficient understanding. Urban wetland is confronted with serious threat from ecological degradation, the retained water area almost becomes the place to store city sewage, which is resulted in tremendous change of wetland hydrology, geomorphology, ecology and natural conditions, and the weakening or death of ecological service function.

There is a contradiction between wetland park construction and operation fund raising. Wetland Park covers a large area, has long construction period and heavy investment, and the cost of its development and operation is relatively high. As a city with high-density development and precious and limited land, it will be very difficult to obtain a long-term development without considerable support from government, overall planning, and widespread support from society. Some social investment concerns about the real estate construction in the surrounding land of the completed wetland park, and the wetland is used as a gimmick, once the house has been sold in large scale, the investment in wetland may be stopped.

There are some weaknesses in the study of wetland. Wetland Park has a history of several decades, but the scientific research theory of wetland is relatively weak, technology is relatively backward, research strength and fund are insufficient, and systematic in-depth exploration and research in biological diversity structure, wetland functions, succession laws, protection and exploitation measures, etc. of wetland are scarce at present [2]. Especially, many places in our country attach great importance to construction rather than management, which is resulted in the loss of wetland function in many wetland parks. The subjects of ecology, art and design, management, etc. must be merged together in the wetland park construction; only through this can the wetland be retained in city forever.

The construction of wetland protection laws and regulations system is lag behind. At present, the existing laws and regulations in wetland protection and management are not perfect or sound, the problem of the illegal destruction to wetland resources is either difficult or unable to rely on the law.

Propaganda and education work should be strengthened. Wetland protection as an emerging subject, the breadth and depth of its protection, propaganda, and education are relatively weak at present, and the community also shows indifference to its protection. Because the propaganda that only relies on the department in charge is far from enough, the educational function of wetland ecology has not been well known. Many residents or visitors do not realize the importance of wetland parks and the difference between wetland and ordinary parks, so the whole society's understanding of wetland needs to be further deepened and improved urgently.

The management system is not complete. The protection, management, development and utilization of wetland parks involve many departments like department of forestry, environmental protection and water conservancy. Many departments protect and utilize wetland from their own interests, the division of use rights, protection responsibilities, benefits distribution and so on is not clear, which is resulted in many conflicts like wetland conservation and rational utilization, engineering construction and ecological functions and so on, meanwhile, the responsibility for the wetland destruction still has no one to take.

27.6 Some Aspects in the Development of Wetland Park in Chongqing

There are various protection forms of wetland, such as wetland nature reserve, wetland protection area, wetland parks of various types, wetland area with multipurpose management, wildlife habitat and so on. Therefore, some wetland parks are premeditatedly constructed within the city on the basis of ecological status, which plays an important role in the protection of the severely damaged natural wetlands. The sustainable development of urban wetland park is a difficult and long-term task, it not only needs to meet the modest size and self-succession of wetland, but also need to be coordinated with the urbanization process and the livable environment. Some theories about urban wetland management had been proposed abroad, for example, the 3R management policy, there are also some explorations on the construction of urban wetland parks in China. Since the first national urban wetland park was built in Sanggou Bay of Rongcheng City in Shandong in 2005, more than 40 national urban wetland parks have been built until now, and there are also a large number of ordinary urban wetland parks. The Living Water Garden near the Funan River of Chengdu is especially characterized by its wetland purification function, which sets up a successful model of the combination of river comprehensive treatment and environmental education. This marks the flourishing development of urban wetland reconstruction and engineering restoration in China. The following aspects need to be achieved while the building of urban wetland park in Chongqing.

The connotations of urban wetland park need to be emphasized. Urban wetland parks are open systems rather than closed, isolated systems, the connotations like integrity, diversity, process and so on need to be emphasized. Therefore, urban wetland parks should aim at maintaining the balance and development of regional ecosystem and finding solutions on the basis of overall condition, instead of focusing on building plant communities, improving the park environment, etc. Meanwhile, the succession of biological communities in the park can not be achieved in one day, a long-term and developing perspective need to be applied for the design and evaluation [1].

Scientific planning and sustainable development need to be realized. Detailed survey and evaluation are applied in urban wetland parks in Chongqing, planning, scientific layout and orderly promotion are carried out according to the principle of "ecological priority, minimum intervention" [3, 4]. The wetland should be rationally utilized and protected according to different standards by adopting the sustainable development theory, and the environment ecological construction is the key. The vitality and environmental self-purification capacities of wetland are maintained, in order to make it become a place with natural factor, ecological factor and human culture. Then, repeated demonstration is needed before the implementation of the program, trying to avoid design flaws that would cause irreparable damage. Taking the case of a water park in Chongqing, it was started to construct in 2003 and completed in 2005. The lack of overall consideration of the waters nearby leaded to condition that no living water can be introduced into the park. If the treated water from domestic sewage is used as water source, municipal sewer line need to be re-buried, and no enough land can be used to the pre-treatment. This park had been left unused for a long time, and its land and water facilities were mostly deserted now [5].

The completed management of water needs to be strengthened. The construction of wetland is only the beginning of environmental improvements, and the following management is the key. The scientific study of wetland parks is relatively weak at present, so the department of park management can build a long-term cooperation relationship with universities or some other research institutions. The wetland information and data management system and monitoring system can be built, in order to grasp the dynamic change of wetland and research the way and technical means to wetland restoration and reconstruction. This will provide a scientific basis for the protection, rational use and development of wetland. Meanwhile, the professional training institute will give professional skill training to the screw of the management department regularly, which will enhance the professional quality of the staff, and at the same time ensure the healthy operation of wetland.

The sewage from wetland treatment can be developed and utilized as a second water source. The current demand for landscape water in city is increasing, and treated water from wetland can meet the quality requirements of landscape water. The Government should adopt the mode of overall management, and have the wetland drainage within the city area under unified control and comprehensive utilization, in order to reduce the demand of the disposable water supply.

Management support and mass basis for the wetland parks are provided by completing the existing regulatory system. The construction and management of foreign parks mostly rely on the supervision and assistance of social groups and people [6]. This can be an important reference for wetland park protection in Chongqing. Firstly, the government should make and complete relative laws and regulations according to development and protection need of wetland parks, like "Regulations of urban wetland park protection and management". Secondly, the property right and management right need to be clarified, protection responsibility needs to be defined, and laws should be strictly enforced. Thirdly, the propaganda needs to be strengthened, in order to improve people's environmental awareness and encourage the public participation. Letting the public join environmental protection activities consciously and initiatively, forming a new style, and maintaining the ecological environment. Fourthly, during the wetland construction period and operation period, a intermediary institute should be established, it is used to manage the government and social investments as a whole, in order to ensure adequate fund. The supervision from investors and the public are accepted, to ensure the project quality and the smooth operation.

27.7 Summary

With the increasing demand of urban ecological environment protection in recent years, urban wetland park as a comprehensive park with ecological function, recreational function, science popularization has had an unprecedented development. The distance between people and natural eco-space is closer and people's ecological protection is strengthened through the construction of wetland parks. However, the planning and design of wetland parks is a very complex and comprehensive engineering, there are still many aspects about the construction of urban wetland park need to be improved. Government, development agencies, regulatory agencies and the public need to strengthen their environmental awareness, and the concept of sustainable development need to be applied during the planning, construction and operation period on the basis of actual situation, having overall coordination and improving managing methods, in order to create an ecological landscape space. Only through this can the wetland park develop in a long period of time inside urban area 13399812923.

Acknowledgments This research was supported by the science and technology project of Chongqing Education Committee (No: KJ101503), and thanks for the Chongqing Wetland Protection Management Center.

References

1. Yi D (2007) Planning and design methods of urban ecological park. Architecture and Building Press 22(3):389–396
2. Sun G, Wang H, Yu S (2004) The advance of urban wetland study. Prog Geograph 23(5):94–100
3. Jianyong R (2006) Several considerations on the construction of jinghu country city wetland park. Gardening Technol 2(3):3–5
4. Qingqing Yu (2007) Discussion on some aspects of urban wetland park. Sight Singing Environ 26(5):57–64
5. Gao S, Shao Y, Zhang M et al (2010) Research on construction and management of Wetland Park in Beijing. Wetland Sci 8(4):389–394
6. Qu J, Luo G (2007) Preliminary analysis of community participation in wetland park management. Wetland Sci Manage 3(3):54–57

Chapter 28
Analysis on Local Transfer Pattern for Rural Surplus Labor Forces

Huixin Jin, Wei Li and Guohong Li

Abstract With the development of society and economy, more and more surplus labor forces are coming forth in rural area of China; To guide and foster the rural surplus labors to engage in non-agriculture is necessary for china to solve the "three-agricultures" issue, promote construction of small towns and establish a humorous and well-fare society, as well as a strategy for construction of rural modernization. This paper discussed the necessity of building a double-track transfer pattern under which the rural surplus labor force could realize local transfer and the double-track pattern itself: the local employment track for rural surplus labors in municipal suburbs and prefectural suburbs; the local employment track for rural surplus labors in rural area.

Keywords Rural surplus labor · Local transfer · Travel and tourism industry

28.1 Introduction

As one of the earliest coastal open cities, Qinhuangdao, Hebei province harbors 3 regions and 4 countries, including 30 countryside governments [1], 45 township governments and 2,265 village committees. The rural population in 2009 amounted to 2,020,147 (Statistics Bureau of Qinhuangdao in 2009). The rural surplus labor force which need to be transferred and employment issues emerged in the flood tide of reform. Based on the report offered by Municipal Bureau of Statistics, the number of migrant workers returned home, affected by the financial

H. Jin (✉) · W. Li · G. Li
Hebei Normal University of Science & Technology, Qinhuangdao, China
e-mail: wblxljs@126.com

W. Du (ed.), *Informatics and Management Science VI*, Lecture Notes in Electrical Engineering 209, DOI: 10.1007/978-1-4471-4805-0_28,

crisis in 2009, amounted to 10,848 and the total number of people who returned home added up to 30,000 (early 2009 in Hebei Province, 56.59 million unemployed migrant workers returned home). In addition, the current employment situation in Qinhuangdao is very difficult [2, 3]: there are 110 thousand of surplus rural labor force not employed, 19 thousand of the new growth of the labor force and 3 thousand demobilized soldiers, which have brought Qinhuangdao tremendous employment pressure, especially in rural area. Thus, to research on the local transfer pattern in the rural area of Qinhuangdao city is to find a available way for the surplus labor force, as well an important method to relieve the current pressure.

28.2 Importance to Build a Local Transfer Pattern for Rural Supus Labor Force in Qinhuangdao City

28.2.1 It is Conducive to Social Stability and Harmony

Based on the report offered by Municipal Bureau of Statistics, the number of migrant workers returned home, affected by the financial crisis in 2009, amounted to 10,848 and the total number of people who returned home added up to 30,000 (early 2009 in Hebei Province, 56.59 million unemployed migrant workers returned home). In addition, the current employment situation in Qinhuangdao is very difficult: there are 110 thousand of surplus rural labor force not employed, 19 thousand of the new growth of the labor force and 3 thousand demobilized soldiers, which have brought Qinhuangdao tremendous employment pressure, moreover the rural employment situation is not optimistic. The issue of rural surplus labor force is in relation to the China's stability, development and the vital interests of the overwhelming majority of our population—the farmers. The proper placement of surplus rural labor force will inject a powerful driving force to China's economic development, and it is conducive to turn the population burden into human resources advantages. At the same time it is a significant pushing force in solving the "three-agricultural issues", coordinating the urban and rural social development, narrowing the urban–rural wealth gap and building a stable and harmonious society. Thus, to research on the local transfer pattern in the rural area of Qinhuangdao city is to find an available way for the surplus labor force, as well.

28.2.2 It Provides Theory Support for the Government to Develop Policies for Migrant Workers Training and Employment Service System

Through various means, create conditions of employment, strengthen labor protection, improve working conditions, and on the basis of expanded production, increase remuneration and benefits; the state will give necessary job training for the pre-employment citizens. This provision is also adaptable to citizens of the rural labor force, it is not only the civil right but also state's obligation to improve their work skills and promote their employment [4]. Qinhuangdao City was identified as the first national comprehensive reform pilot of service industry and the first comprehensive reform of the city tour, which is a major strategic opportunity for the local transfer of rural surplus labor. The State Council has published successively some documents which are conducive to the training of migrant workers, urban employment guarantee, but the training of migrant workers and urban employment service system were just initially established and inadequate protected, however, there still exists policy discrimination in local.

28.2.3 It is Helpful to Accelerate the Pace of Urbanization in Hebei Prov and Achieve the Strategic Goals

In 2010, the new transferred rural labor amounted to 1.5 million; 1 million people were trained in Hebei province. By 2015, the sum of transferred rural labor force in Hebei province will increase by 6–18.42 million compared with that by the end of 2009. Qinhuangdao city will work hard to increase 230 thousand township employment posts and realize stable and sufficient transfer employment of rural labor force. Despite ranked first in Hebei in four consecutive years at the national assessment of urban environment comprehensive improvement and the urbanization rate reached 48.7 % (source: 2011 Qinhuangdao city government work report), Qinhuangdao city must accelerate the pace of urbanization to achieve the goal of smoothly transfer 230 thousand people by 2015. Located in the northeast of Hebei province and surrounding by Bohai Sea, near Tianjin and Beijing, Qinhuangdao city, striding the conjunction of two major economic zones, is a coastal open city as well an important port city. So the innovative research on local transfer pattern of rural surplus labor force in Qinhuangdao city is a necessary approach to effectively reach the goal of smoothly transferring 18.42 million surplus labor forces by 2015 in Hebei prov.

28.2.4 Its Favorable to Increase Farmers' Incomes, Reduce Rural–Urban Income Gap and Adjust Social-Economic Structure

The per capita disposable income of urban residents increased from 9,394 Yuan in 2009 to 17,118 Yuan in 2010, while the rural per capita net income ascended from 3,376 Yuan in 2009 to 6,214 Yuan in 2010, there is a larger income gap between urban and rural residents. Statistically, 41 % of farmers' net income came from wage income, 52 % from the family operating income, the remaining 7 % from transfer payments and property income. It is the strategic objective of Hebei Province that rural labor income account for more than 55 % of net farmer income by 2015.

Due to financial crisis, not only the coastal export enterprises in Hebei prov were affected, but also the rural industry, which reduced the transfer employment income of farmers in Qinhuangdao. Only by enhancing the skills of migrant workers and adjusting social-economic structure between the urban and rural, can they find higher-salary job in the economic mechanism which is transition and upgrading, and ultimately achieve the purpose of employment and raising incomes. Therefore, we conduct the innovative research on the local transfer pattern of the rural surplus labor force in Qinhuangdao city to discuss and solve the existing social-economic issues.

28.2.5 It is in Favor of the Rational Allocation of Production Factors and the Optimal Combination

With a large number of workers released from agriculture, man-land relationship greatly eased and the condition of per capita disposable agricultural resources for farmers improved, the marginal productivity of agricultural labor increased, resulting the farmers' average level of output increase [5], correspondingly, the income level of farmers will be improved. They work outside the home and work locally, on the one hand the family income was increased, on the other hand, part of income was consumed in the place where they were working, which promoted the rational allocation of local factors of production and optimization of combination, therefore it promoted local economic prosperity.

28.2.6 It is Helpful to Improve the Quality of Rural Surplus Labor Force and Lower Unemployment Rate of Migrant Workers, as Well to Promote Fair Employment of Urban and Rural

In 2009, there were 31 thousand transferred rural labors and 28 thousand exported labors in Qinhuangdao (2010 Qinhuangdao city government work report). When the financial crisis hit, the migrant workers bore the brunt of impact, many of them lost their jobs in the cities and towns and had to return hometown one after another, the flood of migrant workers returning home during this period resulted in great pressure to the society such as Underemployment and instability. Due to insufficient education and training, the migrant workers presented most vulnerability during the financial crisis, so they were scarce of competitiveness in the process of urban employment and difficult to resist the impact of the economic downturn. So through the innovative research on the local transfer of rural surplus labor force in Qinhuangdao, we discuss the employment approach of migrant workers and protect their right of full employment on the basis of equality.

28.3 Core part of Local Transfer of Rural Surplus Labor Force in Qinhuangdao will take Double-Track Mode

The core part of local transfer of rural surplus labor force in Qinhuangdao will take double-track mode: the local employment track for rural surplus labors in municipal suburbs and prefectural suburbs; the local employment track for rural surplus labors in rural area.

28.3.1 Local Employment Track for Rural Surplus Labors in Municipal Suburbs and Prefectural Suburbs

For those rural area located in certain region of municipal suburbs and prefectural suburbs, we need to speed up the pace of urbanization; when the arable lands were reduced to certain percentage, centralized management should be taken; meanwhile the villagers should move to residential buildings to reduce the occupancy of soil by bungalows and avoid the formation of villages amongst city. The farmers could move to the near city and get job locally to accelerate to pace of transfer of rural surplus labors, which could be done by adjusting the industrial structure to absorb surplus rural labor force and increase employment opportunities.

After 60 years construction and development since the founding of new China, Qinhuangdao city, said "back garden of Tianjin and Beijing", have adjusted it's industrial structure 3 times, a "three-two-one" industrial structure has been formed which is more in line with international trends. The Tertiary sector has become the backbone of the economy of Qinhuangdao and has contributed tremendously to the her rapid economic development, but the internal structure of the tertiary industry is not quite reasonable, ports, warehousing and other transportation in the tertiary industry occupy a prominent position, the tourism, however, due to it's seasonal feature, takes on a the low proportion. The third industrial structure should be reshaped while Qinhuangdao is now ushering in the fourth industrial restructuring. Such as the solgan "city development by tourism" said, promoted by Qinhuangdao government, we can take full advantage of the unique natural and geographical resources, unique cultural and historical resources, industrial tourism resources, tourism resources, agriculture, sports and tourism resources, convenient way to air and sea superiority, the Government's development mechanism, the importance of policy support and the development of a larger space to absorb local employment of rural surplus labor force. Though Qinhuangdao city government promoted solgan to achieve the purpose of expanding the rural surplus labor force, there are some questions to consider:

28.3.1.1 Establishment of Far-Famed Brands of Tourism

Antiquity: Taken Shanhaiguan as the main attractions, we plan to mold more statues of historical figures in the relevant areas and excavate profoundly in the construction, food, culture, entertainment and commodity areas to cultivate antiquity boutique.

Natural beauty: Based on nature landscape such as Changshou mount, Zu mount, Lianfeng mount, Tianma mount and Jieshi mount, we design to create the mountain feature; meanwhile we will build seashore garden based on the 487 km coastline of natural site and the green seashore tourism which featured mainly by Shanhaiguan-Haigang-Beidaihe-Nandaihe.

Ecological agriculture: Taken the Jifa botanical garden, Chile grape valley and Strawberry valley as the main attraction, we aim at building cultivation, harvesting, processing, and catering and other agricultural eco-tourism brand.

Modern city: Mainly based on Haigang district, we intend to create a city brand which combines with business, city tours and specialities.

Industrial Technology Expo: There are a couple of famous companies in Qinhuangdao, such as: the cradle of Chinese glass industry—Yaohua Glass Group Corporation, the cradle of the bridge industry—Zhongtieshanqiao Bridge Group Co. Ltd., China's largest container terminal, unloading technology, the seventh subcompany who owns the world's leading unloading technology—the world-famous of the Qinhuangdao Port, and the largest place of origin and cradle of china's dry red wine and-Changli, it's manufacturing process can be develop to a tourism landscape.

28.3.1.2 Expansion of Tourism Hardware

On the foundation of 40 odd original view spots, we continue to invest on tour infrastructure and integrate tour resource, conducting in-depth development and innovation, to form a scene of multi-business operation. We will change the pattern of seasonal tour and develop it into whole-year tour: climbing mountain in spring, swimming in summer, fishing in autumn and skiing in winter. In this way, combining with local cultures, Qinhuangdao will be developed into a fabulous international tour site which can harbor tourists all year round.

28.3.1.3 Conjunction of tour Software

We need to enhance promotion on national television, broadcast radio, Internet and newspapers, so that can we raise the popularity of Qinhuangdao at home and abroad; we must effectively manage the tour service system by: standardization of tour sightseeing commentary through holding of tour sightseeing commentary contest and annually training accession, which will enhance the quality of tour employ mental staff and further raise the level of tourism services; improving the quality of citizen through molding the awareness that every citizen is the representative of Qinhuangdao; maintaining the Qinhuangdao's image of sightseeing spot through increasing supervision. Visitors with a relaxed frame of mind come from all directions, and they take the new charming image of Qinhuangdao away with content when they come back, what' feedback, will be the overvaluable political and economic prospects.

28.3.2 Local Employment Track for Rural Surplus Labor Force in Farther Rural Area

The policy for the development of farther rural area is establishing rural agricultural industry clusters and speeding up the pace of urbanization, so as to absorb the rural surplus labor force. Based on the definition of agricultural industry clusters, a region can develop it's agricultural industry cluster according to the feature of local agriculture, the benefits is obvious: it not only can foster the deep processing of farm products and extend the agricultural industry chain and increase the additional value of farm products, but also it can enhance the capability to resist the market risk and industrial competition as well it will contribute to development of relevant industries and service industry, which will add more opportunity for rural surplus labors; the farmers leave their soil but still live in hometown, thus the it can avoid the population jamming in cities and can boost urbanization as well as the integration of urban and rural.

28.3.2.1 Development of Agricultural Industry Clusters can Extend Employment Chains

The development of agricultural industry clusters will promote the extension of agricultural industry chains and further the employment chains that have to say it can increase employment categories and expand employment capacity. Qinhuangdao is rich in agricultural resources: abundant apple, peach, pear, apricot, grape, chestnut, walnut, peach, hawthorn, cherry and other fruit, as well some crops including rice, corn, sweet potato and others. It has already developed some distinctive gardens such Jifa botanic garden, Changli grape garden and Wangyu village as the leads with a wealth of experience. We can still further extend the industrial chains, for example: fresh fruits can processed into pickled fruits and canned fruits; dried fruits and grain can also be further processed, so that some intermediaries such as paper manufacturers, processors, technical training institutions, government, farmers and trade agent can be involved in, thus it can absorb a large number of rural surplus labor force to work in local.

28.3.2.2 Development of Agricultural Industry Clusters can Generate Employment Multiplier

The development of agricultural industry chains will generate the domino effect which can lead to a substantial increase in the number of firms within the cluster and rapid increase in production and sales of products, which will promote the formation and expansion of specialized markets. Joint demand of enterprises will drive the development of other industries such as production factors market, transportation industry, postal and telecommunications industry. With the gradual centralization of population, it will also boost the development of the tertiary industry such as the local real estate, commerce, catering, hotels, finance, education and entertainment industry. Multiples of the cluster growth will bring multiple jobs, showing jobs multiplier effect, it will be a great solution to the demand of local employment of rural surplus labors.

28.3.2.3 Development of Agricultural Industry Cluster can Improve the Quality of Rural Surplus Labors

Development of the agricultural industry cluster has greatly promoted rural industrialization and urbanization process, a number of new farmers with high-quality and new ideas, capable of skills and operation, emerged in the process of development. On the one hand, in order to improve economic efficiency, the companies within the cluster lose no time to train the farmers with relevant skills; other hand, since the companies within the cluster have certain requirement on the agricultural technology, in order to meet the employment requirement, the farmers would like to actively ask for improvement of their quality through attending varies training courses to learn the relevant knowledge's, such as all respects of skills from planting to storage. Moreover, due to the local transfer reduced their living cost and education cost, they are more likely to learn techniques actively. In addition, the Sunshine Project launched by the government is free of charge for farmers, so the overall quality of rural surplus labor force will further improve and it will better promote local suburbs transfer of rural surplus labor.

28.3.2.4 Improvement of the Ownership Structure and Spatial Distribution of Township Enterprises During the Developing Process of the Agricultural Industry Clusters

Township enterprises have always been the main forces in absorbing the migrant workers, however, due to their own structural defects of management and the fierce competition with state-owned enterprises and foreign-funded enterprises, township enterprises are generally facing the situation of operational difficulties in recent years. Objectively speaking, the technology of township enterprises is relatively backward, their funds and personnel are lower than state-owned enterprises and foreign-funded enterprises, ownership structure are not reasonable, and there is too much intervention by the township government. Under the situation of that state-owned enterprises have successively accelerated capital deepening and upgrades products, township enterprises should establish a modern and efficient corporate governance structure to strengthen their own advantage of flexible mechanism. At the same time, except for a few strong township and village enterprises, most of the township enterprises should self-select the right location, using the advantage of their location close to rural areas and abundant cheap labor resources to develop labour-intensive industries, especially for those industries which can combine with local resources and those which are closely related to agriculture, they should divide the work horizontally and vertically with state-owned enterprises, foreign-funded enterprises to shun direct confrontation. Basing on the current scattered distribution that almost every village has some small workshops, we should grasp the opportunity of vigorously developing the middle size and small town, promoting the township enterprises to appropriately concentrate to central towns. By this, it will not only reduce the occupancy of the land, but also can play its accumulation effect, formatting scale efficiency. At the same time, we can reduce costs by external economies and raise popular focus by the concentration of enterprises, promoting the development of tertiary industry and enlarging employment, and eventually, we will realize the synchronizing urbanization and local employment of rural surplus force.

References

1. Li Y, Ma J (2010) Lishengju impact on farmer's employment by organized export of labor force under government involvement-based on the case study of Linkou county, Hengshui county, Liaocheng county and Ningyang county. Academic J Shandong Agric Univ 4:102–103
2. Wang Y (2011) Public employment service and transfer employment of rural labor force. Times Finan 24:68–69
3. Xu J (2011) Research on transfer of rural surplus labor force in anhui province 5:36–37
4. Yue W, Zhong W (2011) Research on historical concept evaluation of china's rural surplus labor force transfer over modern times. Finan and Econom 8:80–81
5. Ma J (2011) Research on transfer of rural surplus labor force and farmer's income in Tuoketuo county, inner mongolia province. Inner Mongolia Agric Univ 6:45

Chapter 29
On Role of E-Business in Foreign Trade and Countermeasures

Guiying Zhang

Abstract E-business, a new trade form in information age, has a great impact on foreign trade, bringing commercial opportunities as well as severe challenges. After introducing the importance of E-business in foreign trade, the paper analyzes the promotion and challenges E-business brought to foreign trade enterprises and put forward the countermeasures, to provide guidance and advices for developing E-business in China's foreign trade enterprises.

Keywords E-business · Foreign trade · Countermeasure

29.1 Introduction

With economic globalization and network of computers, E-business, with a huge impact on China's foreign trade enterprises, not only can improve productivity, reduce transaction costs and purchasing costs, but also can improve the international competitiveness of foreign trade enterprises [1, 2]. However, E-business brings commercial opportunities as well as severe challenges to China's foreign trade enterprises.

G. Zhang (✉)
Department of Business Administration, Chongqing College of Electronic Engineering, Chongqing 401331, People's Republic of China
e-mail: zhangyinmg@163.com

W. Du (ed.), *Informatics and Management Science VI*, Lecture Notes in Electrical Engineering 209, DOI: 10.1007/978-1-4471-4805-0_29,

29.2 Important Role of E-Business

As a totally new technological means of economic activities in the 21st century information age, E-business has become one of the most extensive application areas of Internet. Its advanced network technology implants new impetus to enterprises, and thus traditional operation mode of trade is confronted to the challenge of new ways [3]. The application of E-business among enterprises has become an international trend. If foreign trade enterprises to adopt E-business, will be more than the traditional mode of operation, enabling businesses to the business for all of its details, customer information, changes in demand are the most timely understanding, enabling enterprises to firmly grasp the rhythm of the market. In addition [4, 5], E-business back to foreign trade enterprises the following benefits: help to reduce business costs, E-business trading platform through the network business and stakeholders to achieve real-time docking in space to eliminate the traditional trade restrictions on the parties to the transaction, help to reduce business risks, implement real-time information exchange through the Internet and online purchasing, which enables enterprises to understand the latest market conditions, reduce risk, help enterprises to improve operational efficiency, the Internet can reduce transaction both asymmetric information, especially the use of buyers and sellers of electronic contracts, bills of lading, invoices, letters of credit, etc. can achieve instant online delivery, help foreign trade enterprises to improve their international competitiveness, foreign trade enterprises set up via the Internet website or trade-related sites around the world can communicate with product supply and demand information, access to international market competitiveness. Therefore, foreign companies need to develop E-business.

29.3 Promotion of E-Business to Foreign Trade Enterprises

29.3.1 Providing an Effective Trade Platform

The emergence and rapid development of E-business to promote economic integration, countries around the world to promote foreign trade enterprises to accelerate the use of E-business take advantage of the pace of China's foreign trade enterprises are the same. E-business is the information age, especially popular Internet products, information technology, globalization, the Internet become one of the best channels of communication, it has more to interact with a variety of media, subject to the constraints of time and other characteristics, provide a convenient, cost-effective way to facilitate the people to a variety of transactions [6]. The emergence of E-business, providing a favorable international environment and space, to promote the development of China's foreign trade enterprises, but also for China's foreign trade enterprises to provide an effective trade platform for the global economy in such a platform for business contacts and trade transactions will become more frequent, E-business is a global trend in countries' external trade, but also China's foreign trade development.

29.3.2 Reducing Transaction Costs and Purchasing Costs

In traditional international trade and purchase, foreign trade enterprises require a lot of manpower and materials and prone to human errors in operation [7]. The E-business is to rely on computer networks as a medium for information exchange, and it greatly accelerated the speed of information processing and transmission. Foreign trade enterprises through the computer network information processing switch, greatly improving the automation of business transactions, and a significant reduction in a variety of processing costs, reduce procurement costs, while providing a variety of corporate procurement channels, but also to facilitate the required foreign trade enterprises in the online Find information, greatly accelerate the marketing and procurement departments respond to market forecasts, trade both sides of the ordering cycle has also been significantly reduced.

29.3.3 Reducing Dependence on Physical Infrastructure

In the past, foreign trade enterprises in international trade business must have the appropriate infrastructure, and the emergence of E-business can make the operation of virtual enterprise, the construction does not require the construction of shopping malls, store rental cost savings, eliminating the inventory of goods the pressure. Network technology built up through the supply chain system to complete the product design, production to distribution, record storage and after-sales service, formation of closely co-ordinate with the upstream and downstream enterprises of the global market, allow enterprises to rely on the extent of physical infrastructure greatly reduced. Enterprises in business activities can be 24 h for working hours, employees can be self-control, in response to customers in more quickly, and not over the impact of jet lag.

29.3.4 Serving Customers More Effectively

Measuring the success of sales cost of enterprises is rooted in customer satisfaction, the survival foundation of enterprises. Through the use of E-business, online companies can introduce their products to customers and provide technical support, order processing information, and details of cargo query, you can search through the network and understand customer preferences and requirements, so that not only facilitate the, fast, economical, and also allows the enterprise to handle more complex customer service issues, the business relationship with customers to adjust at the same time, through the network, enterprise customers can understand the first time a new product, so allows enterprise customers more satisfied.

29.3.5 Improving International Competitiveness

Through the company's own Web site or sites related to the establishment of foreign trade, foreign trade enterprises will take the initiative to publish information on supply and demand to go out, and not in the establishment of overseas sales network, and the office of the establishment. Through the network, companies can more easily understand the situation of overseas markets, thereby eliminating the business across the region, the gap caused by the space in the world anywhere within the scope of production and business activities can be carried out, and around the world consumers to establish contact, the company's products and services extended to all corners of the world, companies can also network, to publicize their corporate image in order to gradually expand the company's international reputation. In short, the use of this advanced E-business business forms, can help to improve the international competitiveness of foreign trade enterprises.

29.3.6 Establishing Core Competencies

Traditional foreign trade enterprises mainly took intermediary trade as main business model, namely, buying out or purchasing and then adding sale price, to make price difference, just as the role of trade intermediaries. The business model of trade intermediaries often lack adequate control of the supply chain capabilities, from the virtual business more difficult, not core competencies. Especially now, brokers are often subject to the supply chain at the same time squeeze, the interests of the brokers had a great threat. Foreign buyers reduce costs, and gradually began to try to cross-border procurement or procurement centers in the exporting country to establish direct contacts with manufacturers. The domestic production enterprises with the right to export development and bypassing the middleman deal directly with foreign buyers, and even directly to the overseas marketing department set up to carry out marketing activities. In this regard, a large professional foreign trade company resorted to the business model to industrial operations, or engaged in supply chain management based virtual business. However, small and middle enterprises, due to limited funding and manpower, most of the way or stick business brokers in the market a competitive disadvantage. The generation of E-business was in small foreign trade enterprises to change their management methods to create the conditions and opportunities for small and middle enterprises to create a new connection.

Foreign trade enterprises can participate in the global information technology supply chain, enhance its control of the supply chain, foreign trade business model into the middle of trade not only for purely business, but also to manage the supply chain or provide information services, so as to bring more business opportunities for foreign trade enterprises is conducive to foreign trade enterprises to establish their own core competencies.

29.4 Challenges Brought by E-Business to Foreign Trade Enterprises

29.4.1 Security of E-Business

As an emerging trade practices, in order to have a safe trading environment, we must be good laws, regulations, policies and security system for their protection. Its scope includes the legal electronic contracts, the legal validity of digital signatures, online trading disputes, computer crime, how to protect user privacy and other issues. Behind the law to bring online transactions cannot guarantee the legitimate rights and interests of trade caused by the risk, because the legal effect of the number of contracts has not been recognized by law and thus lose their legal protection. Therefore, the corresponding laws and regulations to guide the healthy development of E-business are necessary.

Low-level Electronic Hinders the Development of China's Foreign Trade Enterprises.

Still in the period of using traditional management models and tools, the majority of enterprises in China have not yet achieved electronic. Nowadays, China's large state-owned loss-making enterprises, but also some companies also are engaged in solving the problem turning around losses, due to very little computer and related software, information processing and handling methods are behind the times, not good at tracking dynamic information, acquisition, analysis and sorting, only pure information-based technology products, companies ask for much spontaneous information, which is very low demand for E-business. These reasons are impeding E-business development, resulting in China's foreign trade and economic developments are severely constrained cause.

Network Infrastructure cannot meet the Needs of Expanding E-business.

To achieve real online trading conditions is the need to network with very fast response speed and high bandwidth, and foreign trade but also by the constraints of the development of E-business information, there is no good network infrastructure also directly affects the progress of communications services, hardware must provide support for high-speed network. The network bandwidth is not enough and so has been plagued by all network users. China's use of E-business development of foreign trade is completely limited by advances in network technology, at this stage there are still very difficult.

29.4.2 High Telecommunication Charges Restrict Application of E-Business

China's telecommunications system inefficiencies, high operating costs, administration and management is set in a monopoly system, which makes this system a very high standard tariff. China's information flow costs in absolute monetary value of conversion, then the flow of information than Americans to buy the same price many times higher. While China's per capita income is far lower than the United States. In other words, the situation now, our E-business if you do not change the status quo and want to develop and progress in a short time is very difficult, so the costs in foreign trade has been good development on the use of E-business is still present phase should be promptly addressed the key issues. Foreign trade in order to enhance their competitiveness, they can only develop E-business is the key.

29.5 Countermeasures of Developing Foreign Trade Enterprises

29.5.1 Speeding up Information Infrastructure Construction

The development of electronic commerce depends on a certain scale of the Internet population and the number of Internet companies, which by the Internet access speed and cost impact. Therefore, the Government should develop infrastructure to improve the quality of the Internet. Information network infrastructures are public goods, which mainly rely on government investment, but also to the introduction of private capital and foreign capital participation. These include: the basis of the existing facilities, further improve the network infrastructure to consolidate, encourage strong companies or individuals to develop advanced technologies suited to the condition, in order to fund the tax on to its support, strengthen the foreign trade enterprises of information technology construction training to enhance their awareness of E-business, get rid of the monopoly, the introduction of market competition, reduce costs and increase the level of professional service companies and service quality. Based on the current forecast for China's foreign trade, develop the information needed to develop E-business infrastructure development strategy, a unified planning and development, to reduce unnecessary waste.

29.5.2 Strengthening Legal System

Current virus infections occur frequently, leading to the invasion of hackers leak trade secrets, fraud and other integrity issues do arise, so that people on the network transactions generated fear, which also shows the current standard for

E-business laws and regulations related to the lack of. Although the 10th National People's Congress in 2004 voted the "People's Republic of Electronic Signature Law" to determine the electronic signature with the handwritten signature or seal has the same legal effect, but the current E-business-related laws and regulations should be improved. We know that E-business development is the result of the new technological revolution, the rapid development of productive forces will inevitably lead to the adjustment of the relations of production, so the need for new laws and regulations to accommodate this new relations of production. Therefore, the Government should learn from the experience of advanced countries, based on the combination of the specific development of China's foreign trade enterprises on the one hand continue to improve the existing legal system, other new laws to adapt to new international trade situation.

29.5.3 Emphasizing the Training of E-Business Professionals

Current economic competition is essentially a talent competition. Therefore, China's foreign trade enterprises in order to develop E-business, we must strive to develop not only mastered the trade of E-business knowledge and familiar compound talents. First, in order to solve current needs, you can set up training bases in foreign commerce, or the introduction of such senior personnel from developed countries, especially have practical experience of overseas students, this can quickly make up for market demand. Second, the most important thing is to strengthen basic education in schools at all levels of foreign trade to open E-business programs, reform the traditional teaching methods, the institutions of professional education and the actual needs of business combination, the introduction of foreign advanced education model even consider the universities of experienced foreign cooperation in running schools.

29.5.4 Enhancing the Understanding of E-Business

In the era of knowledge economy, the rapid development of science and technology, especially information technology change is rapid, so we must take the initiative to adapt to these changes. First, we must change the concept, so as to better participate in the global economy running, and will not lag behind. Therefore, the foreign trade enterprises must recognize the importance of E-business, new ideas, market-oriented, and enhance market competition. In particular the corporate management of each individual not only to understand the commercial value of E-business, but also to actively absorb the latest knowledge, to adjust their own ideas, and effective communication of this new concept to the corporate staff, with the latest concept to arm enterprises.

29.5.5 Gradually Implementing Long-Term E-Business

From the current form, E-business will become the mainstream way of commerce and trade, so China's foreign trade enterprises must from now on rational planning of their own E-business development, in order to step ahead in the new situation. Therefore, the foreign trade enterprises should be combined with the actual situation of the enterprise itself to do long-term planning, the development of E-business business development, short and long term planning and goals. First, to achieve the basic management information, through their own site management planning, sales promotion in order to achieve the best results, secondly, to adapt to the requirements of E-business, business process reengineering, by improving the internal ERP system, the internal reform business processes, improve efficiency, third, through the electronic trading platform to build the company's virtual trading floor, to integrate their supply chain management, electronic trading platform that is used to achieve the production, procurement, finance and sales of unified management, in order to reduce inventory, cost control and on-time delivery rate, and real-time query and reporting conditions, final realization of E-business information management and integration of international supply chain, international commerce, so that domestic and foreign suppliers, the business and domestic external customers through the Internet, online and real-time collaboration, the three parties even as a whole, the whole supply chain to improve the level of cooperation and operational efficiency.

29.6 Conclusions

E-business, with a lot of positive impact on foreign trade enterprises, can not only reduce manpower and material costs through improving trade efficiency, but also bring more commercial opportunities to foreign trade enterprises and establish core competitiveness of enterprises. Along with problems like security and low utilization of E-business, foreign trade enterprises need to combine with reality and their own characteristics to develop appropriate developing strategies, make use of E-business, the newly emerging transaction management toll to shorten business cycles, simplify foreign trade transfer process, reduce import and export cost, improve efficiency, totally enhance the international competitiveness of industry and enterprises, and develop in international trade.

References

1. Zhu HB (2011) Strategy of Domestic Sales in Foreign Trade Enterprises via E-commerce. Journal of Wuhan Commercial Service College 3:13–18
2. Wang Y (2008) Small and medium foreign trade enterprise applied electronic business's controlling factor and strategy analysis. Spec Zone Econ 6:34–39

3. YU CY (2005) E-Commerce a new opportunity for chinese foreign trade enterprises. Journal of Ningbo College 4:14–16
4. Huang L (2009) The problem and countermeasure of the small and medium-sized foreing trade enterprises under subprime Crisis. Journal of Nanyang Institute of Technology 5:5–69
5. Wu YL (2009) The impact of E-commerce on the competitive advantage of foreign trade enterprises. Jinan University 14:159–163
6. Cai ZD (2009) Development E-commerce. World market 4:11–13
7. Ming J (2003) Gement Innovation of Foreign Trade Businesses Under E-business. Commercial Research 9:32–35

Chapter 30
Research on Coupling Mechanism of Related Travelers Humanities Psychological Quality and Tourism Development

Xiaoying Lin

Abstract The uncivilized behaviors of Chinese travelers and the growing emphasis on the country's soft power in international competition make it more urgent to improve the quality of the national civilization. Tourists don't have high quality is because they lack contact and cultural knowledge; the overall quality is not high, and the actual impact on the society; all these make a large number of tourist objects lose their attraction to tourists, and cannot make tourists consumption potential into reality. This article is based on the humanity quality of southern minority nationalities to study the fetters of the southern minority nationalities' humanity quality to the tourism industry. First, we analyze the misunderstanding of the past uncivilized behaviors, then give appropriate coping strategies for the status of uncivilized tourists, and finally propose solutions.

Keywords Southern minority nationalities · Uncivilized behaviour · Humanity quality · Coping strategies · Solutions

30.1 Introduction

Ethnic Groups in South China, means areas which lie south of the Huaihe River Qinling Mountain, not including Tibet and Qinghai, and whose minority population accounts for more than 50 % of the mountain area. Traveler is the term temporarily created in this article which refers tourism-related personnel [1]. The quality of humanity or liberal arts refers to people with the humanities and

X. Lin (✉)
Scientific Research Department, Liuzhou City Vocational College, 545002 Liuzhou, China
e-mail: xiaoying_lin21@yeah.net

W. Du (ed.), *Informatics and Management Science VI*, Lecture Notes in Electrical Engineering 209, DOI: 10.1007/978-1-4471-4805-0_30,

social scientific concepts that related to quality thinking of words and deeds, such as the value concept that men are created equal belongs to the humanities field; if a person has this idea, he will be accustomed to treat people equally in both words and deeds. What we mean here is this high-quality humanity to everyone. The relationship between humanity quality of related travelers and tourism is unquestioned. If related travelers and tourists have high quality of liberal arts, then the overall efficiency of the tourism industry is also high. Overall, the liberal arts quality of related travelers in southern minority nationalities areas is far from ideal, as some ethnic minority areas engage in ethnic customs and provide tourists with food and lodging, which was very attractive to tourists, but the quality is not high, because they do not pay attention to give tourists enough science, health and beauty, resulting in many high-level visitors stop looking. For this reason and together with other factors, the south national mountain tourism has a considerable gap with the developed countries [2]. In the World Tourism Organization rankings in 2010, China is the world's third largest inbound overnight travel country, but the competitiveness ranking is only 39. In Europe and the United States in recent years, the shopping proportion of total tourism spends up to 70 %, but China's only close to 35 %. Southern minority nationalities areas are less developed than the developed areas in the whole country.

It started from some uncivilized photos online of China's outbound tourism citizens from 2005. In 2006, the Central Civilization and the National Tourism Administration announced the "top ten tourist uncivilized behavior" and uncivilized traveling behavior in our thoughts has become a lot of people's focus of discussion. However, when it comes to uncivilized behavior, we will immediately think of uncivilized quality, improving the inner quality, so the physical behavior has evolved into the intangible qualities [3]. As a result, everything went back to a long project to improve the quality of outbound tourists uncivilized behavior, which is still be frequently exposed on websites, newspapers and other news media.

30.2 Analysis and Research

Through field investigations, visits and exchanges, and network search, we conducted in-depth research and analysis to find literature on ethnic mountainous areas in the south tourists' uncivilized behavior. Focus on a scenic, after investigation, we found that in the scenic area, there are a lot of tourists uncivilized behavior, and to ensure coordination and long-term development of an industry and in order to eliminate the adverse effects of this uncivilized behavior, we analyzed the status and problems of the scenic spots in the survey through the dissemination of the questionnaire. We issued a total of 200 questionnaires, with 198 recovered and 195 valid questionnaires. Survey theme: the manifestation and causes of tourists' uncivilized behavior.

Table 30.1 Satisfaction on the scenic environment and the frequency of the uncivilized behavior

Objects of the survey and the proportion (%)	Satisfaction on the scenic environment			Frequency of the uncivilized behavior	
	Very satisfied (%)	Satisfied (%)	Not satisfied (%)	Frequent (%)	Unfrequent (%)
Local residents (23)	80	12	8	17	83
Tourists (48)	79	17	4	24	76
The scenic managers (20)	70	20	10	17	83
Other persons (9)	64	24	12	14	86

Table 30.2 Uncivilized behaviors of the tourists

Objects	Random discards (%)	Not abide by rules (%)	Tarnish (%)	Other behaviors
Local residents	39	41	9	11
Tourists	46	39	7	8
The scenic managers	54	32	11	3
Other persons	34	46	6	12

30.2.1 Investigation of Satisfaction on the Scenic Environment and the Frequency of the Uncivilized Behavior

According to field investigations, we can the following data, as is shown in Table 30.1.

According to the data in the table, we can draw the vast majority of respondents to the scenic environment are very satisfied; part of tourists satisfaction to the scenic environment are general; very few respondents are not satisfied with the scenic environment.

In addition, from the questionnaire it can be seen that local residents, tourists, scenic managers, or other relevant personnel involved in the scenes have low frequency of the uncivilized behavior. This means that mostly related travelers follow the principles of civilized travel.

30.2.2 Investigation on the Performance of the Tourists and Uncivilized Behavior

According to the survey, we can get the following data, as is shown in Table 30.2.

The data show that tourists' uncivilized behaviors are mainly in the first two kinds of performance. People who think that the first and second kinds of uncivilized performance cause Guilin affected are local residents that accounted

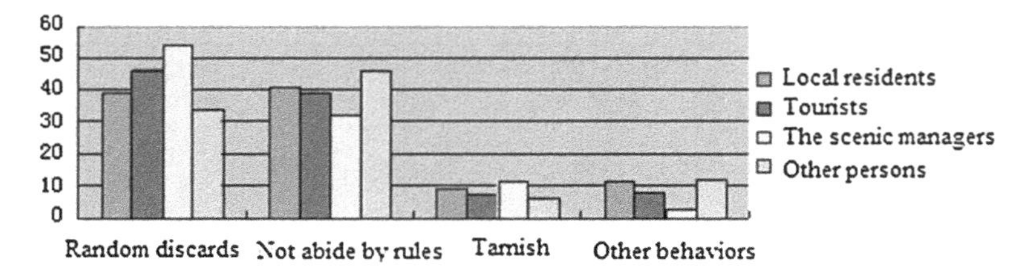

Fig. 30.1 Strip figure of uncivilized behaviors of tourists

Table 30.3 Tourists' uncivilized behavior causes

Objects	Poor quality (%)	Facilities imperfect (%)	Lax management (%)	Inadequate publicity (%)	Other causes (%)
Local residents	71	14	10	3	2
Tourists	65	13	15	4	3
The scenic managers	76	12	9	2	1
Other staff	69	8	19	1	3

for 80 %, the scenic managers accounted for 86 %, tourists accounted for 85 %, and other personnel accounted for 80 %.

Its strip figure is shown in Fig. 30.1.

30.2.3 Investigation on the Cause of Tourists' Uncivilized Behaviors

According to the survey, we can get the following data, as is shown in Table 30.3.

Tourists' uncivilized behaviors have the following reasons: first, tourists are not with high quality and their environmental awareness is not strong; second, scenic infrastructure is inadequate; third, training and management of the employees are not in place, and the scenic spot management mechanism is lax; fourth, the publicity is not enough; and also there are other reasons. Local residents accounted for 71 % consider the first cause, tourists accounted for 65 %, the scenic managers accounted for 76 %, and other personnel accounted for 69 % (Fig. 30.2).

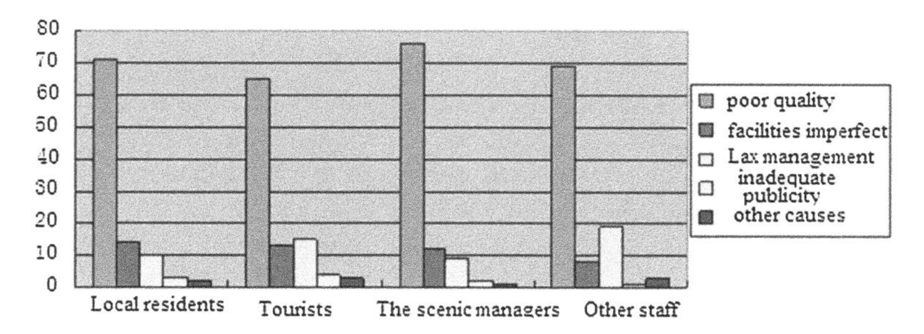

Fig. 30.2 Bar graph of the cause of tourists' uncivilized behaviors

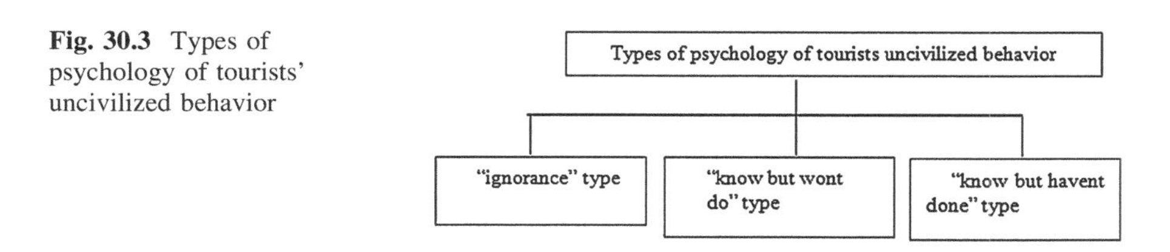

Fig. 30.3 Types of psychology of tourists' uncivilized behavior

30.3 Psychological Laws of the Tourists' Uncivilized Behaviors and Their Changes

30.3.1 Mental Type of Tourists' Uncivilized Behaviors

Specific performance of the tourists and uncivilized behavior is diverse. Central Civilization Office and the National Tourism Administration announced the "Top Ten Travel uncivilized behavior" reflects a more general uncivilized behavior, in addition to other uncivilized behavior. Human behavior is dominated by the psychology, and only learning about the uncivilized behavior on the psychological basis can we find a breakthrough behavioral change fundamentally. However, the currently available research, yet lack of results of research in this area [4]. Therefore, in order to find a fundamental behavior change strategies from psychology that produces uncivilized behavior, the tourists' uncivilized behavior is divided into "ignorance" type, "know but won't do" type, and "know but haven't done" type, as is shown in Fig. 30.3.

30.3.1.1 "Ignorance" Uncivilized Behavior

This refers to the tourists' uncivilized behavior due to totally ignorance of civilization and the nature of the travel.

30.3.1.2 "Know But Won't Do" Uncivilized Behavior

This refers to the tourists' intentional or laissez-faire uncivilized behavior although they ascertain certain travel behaviors are uncivilized. Due to travel off-site and temporary characteristics of outbound tourists in a foreign country only have a short stay, they may have deliberately indulge psychology.

30.3.1.3 "Know But Haven't Done" Uncivilized Behavior

This refers tourists' uncivilized behavior that although they ascertain some of the travel behavior of uncivilized nature, but by the time the behavior causes the environmental impact, and makes the uncivilized behavior inconsistent with their attitude. Such as they have to do this under the leadership of the local tour guide or go with the flow when passing through a red light, or stepping on the lawn.

30.3.2 Characteristics of Stage of Tourists Uncivilized Behavior Change

30.3.2.1 Subject to the Change: The Beginning of Tourists' Uncivilized Behavior Change

This is the first phase of tourists to change their uncivilized behavior. The tourists are forced by external binding to change their uncivilized behavior. This behavioral change is the result of external requirements and is requested behaviour [5].

30.3.2.2 Assimilation Stage: The Transformation Process of Tourists' Uncivilized Attitude

According to Festinger's cognitive dissonance theory, individuals perceived cognitive dissonance will take measures to adjust. Only strengthening the importance of new cognitive order can we ensure the sustainability of behavior change, otherwise the behavior is likely to occur repeatedly. Tourists' uncivilized behaviors are forced to change, but in the course, before and after cognitive conflicts will arise, the tourists themselves will produce civilized behavior knowledge reconstruction of motivation.

30.3.2.3 Internalization Stage: Realization of Tourists' Civilized Behavior and Attitudes Co-ordination

When behavior change is no longer a group requirement, the individual's inner self-impose will demand consciousness acts to change the way. This change might be the final change of the development of the first two acts, and it may also be the behavior change existing independently; it stems from a change in attitudes of the "internalization" stage.

30.4 Strategies of Changing the Tourists' Uncivilized Behavior

According to the previous analysis, real-time behavioral changes are closely linked with attitude change and enhancing quality. For an individual, if we can give him the appropriate stimulus and help rule out the obstacles to his heart, he can enhance the quality of the process of behavioral change, completed by the requirements of civilized behavior, or a leading civilized behavior to conscious evolution of the nature of civilized behaviour [6]. Therefore, the method seems to be with immediate changing uncivilized behavior actually has the potential to play a long-lasting effect, which in China is urgently needed.

30.4.1 Strengthen the Management of Scenic Spots

The management of the corresponding intensity (MAN) can be achieved by the score of the respondents in the various management measures and it reflects the extent of the respondents' recognized visitor management measures, which can be further broken down into MAN1 and MAN2, respectively. They mean the degree of response of the direct management measures and indirect management measures in response level. The formula is:

$$MAN = \sum_{n=1}^{m} \frac{x_n}{m} \tag{30.1}$$

In the formula, x_n is the score for tourists in the nth management measures, and the scores are between 1–5, which means strongly disagree, disagree, general, agree and strongly agree; In this study, m values for MAN1 is 7, and for MAN2 is 4.

30.4.2 Strengthening the Reconstruction of Tourists Communication Startup Knowledge

At present, the uncivilized travel behavior criterion is not clear. Moreover, even with clear behavioral requirements, with social development, the new uncivilized behavior will continue to appear, and a clear code of conduct may be able to balance those uncivilized behavior, but could not estimate a variety of other uncivilized emergence of behavior. A lot of people on the judgment of the uncivilized behavior are confusing, because with this lack of systematic education, their minds are not in the correct way of thinking. Therefore, society needs to help people build civilization judgment; we should not only report what is uncivilized behavior; more importantly, we should explain why it is uncivilized to let people know about the connotation of civilization and civilization's benefits for ourselves and others. To this end, government departments should introduce and publicize more specific and scientific tourism civilized standards of behavior and associated implementation details.

30.4.3 Improve the Quality of the Liberal Arts of the Related Tourists

The reasons for ethnic Groups in South mountain areas related travelers' low humanities quality [7]:

1. First of all, we attach importance to quality education, compulsory education, and the adults of tomorrow are today's students. According to the fifth national census of 2,000 data, years of education of the southern ethnic minorities in mountain regions are between more than 2.68 and 6 years. In self-study, student extra-curricular books are few and to a considerable amount of adults have a lower amount of reading. China's per capita reading amount in 2010 is only 4.2. Beijing's per capita textbook volume is 19–29.
2. Social guidance. Many southern minority nationalities areas are remote from the socio-political and economic centre in history; with transportation difficulties, they easily fall into separate groups, and they often face harsh natural environment, as well as disputes with other ethnic groups. In order to live we must try to have high efficiency so there is the natural choice of man.
3. Improve the quality of liberal arts of adults. Many southern minority nationalities areas government and society have a misunderstanding of adult education for farmers, which is, we should not do the same as in many developed countries like Israel, to encourage them to study to get high schools and universities diploma, but only to teach them real technical knowledge. We should know there is no systematic knowledge-based understanding of the practical and technical knowledge, and it is basically impossible to raise the level.

30.5 Conclusion

Whether to face the requirements of the system that requires behavior, or cramped in the environmental impact of knowledge reconstruction under the guidance of behavior, in the process of changing uncivilized behavior, people will face some kind of mental tension and behavior, which often obstruct civilized behavior change. But if every time they hesitate with the power of a civilization, their civilized travel behavior will not be hindered; instead, it will be prompted. Because when the tourists arrive in unfamiliar environment, they often fall into a loss state, in which case, those tourists who are considered to be cordial and familiar with the destination situation will affect the behavior of tourists and play the guiding role. Staff providing professional travel services for tourists and those tourists travel companion is most likely to become such influential people.

References

1. Festinger L, Quan Z (1999) Cognitive dissonance theory (tran: Festinger L, Quan Z) vol 3(5). Zhejiang Education Press, Hangzhou, pp 75–80
2. Taifa Z (1999) The social ills and governance of the tourism industry. Gansu Soc Sci 4(05):55–58
3. Meng L, Chunping H (2002) Researches on uncivilized travel behavior vol 2(1). Beijing International Studies University, Beijing, pp 22–25
4. Fengxian S (2011) Soap opera to promote Brazilian "family planning". World Expo 31(24):61–66
5. Beijing Tourism Administration Bureau et al (2000) Beijing tourist attractions overview vol 3(4). China Tourism Press, Beijing, pp 445–447
6. Guihua Y, Yuehua W, Zhaoxia X (2004) Eco-tourism vol 1(2). Nankai University Press, Tianjin, pp 92–99
7. Qin Y (2005) Identification and segmentation of Ecotourists—example: baihuashan nature reserve in Beijing vol 4(6). Peking University, Beijing, pp 906–917

Chapter 31
Analysis of Ecological Environment Construction and Sustainable Economic Development Path

Ming Hu

Abstract Yulin City in Shaanxi Province north, is located in pastoral areas, economically underdeveloped, desertification and soil erosion is serious. In recent years, after construction of the local soil and water conservation, ecological environment has improved rapid economic growth. Based on the 2000–2004 environment control measures in Yulin City, and cost analysis showed that soil and water conservation as the center through the comprehensive management of the ecological environment can be fundamentally changed, to promote sustainable economic development.

Keywords Yulin · Economically · Development

31.1 Regional Situation

Yulin city in the north of Shaanxi province, is located in Jinshan have five provinces (areas) at the junction of the heartland, longitude 107°28′–11°15′, north latitude 36°57′–39°34′, population is 3.335 million, the area of 43,578 km^2, the total area of the 58 %. Basic characteristics as the underground resources are rich, is the national new lines of large energy sources heavy chemical base; Less developed, and is the famous poor areas; Desertification serious, and the ecological environment is bad, is the soil and water conservation and the key area of preventing sand.

M. Hu (✉)
Weinan Teachers University, Weinan, China
e-mail: hm5109@163.com

W. Du (ed.), *Informatics and Management Science VI*, Lecture Notes in Electrical Engineering 209, DOI: 10.1007/978-1-4471-4805-0_31,

31.2 Ecological Environment Construction Progress

31.2.1 Soil Erosion and Land Desertification Get Effective Governance, and the Ecological Environment is Obviously Improved

Yulin is sandstorm areas, land desertification and soil and water loss is very important, and the ecological environment is very fragile. After decades of afforestation, water conservancy, carry on the comprehensive governance and development, vegetation recovers gradually. At present, the area of afforestation save area are 973,000 hm^2, set up 117,000 hm^2 windbreak and sand-fixation large woodland, 129,300 hm^2 hazards of sand of farmland is protection. Since the 1950s vegetation coverage from 1.8 to 38.9 % [1]. In the global desertification land from 50,000 to 70,000 hm^2 speed expansion every year, China land desertification annually to 2,460 hm^2 spread, but desert the reversal of speed is 1.62 % in the narrow.

Hilly-gully wide application in the soil and water loss comprehensive control project and small watershed comprehensive control project, make water and soil loss got preliminary control. Yulin area water and soil loss area of 36,900 km^2, after years of comprehensive management, to 2004 soil and water loss control area of 21,543.3 km^2 accumulated, accounting for loss area ratio 58.4 %. Soil and water conservation in Yulin city, the area is always completed, and a growing trend. The list below: Table 31.1.

31.2.2 Agricultural Infrastructure Obviously Improved

The basic farmland construction is the comprehensive control of soil and water conservation and the important component. Through the construction of basic farmland, which can improve the environment at the same time grain production growth, for the steep slope farmland, grow grass planting laid solid foundation.

Table 31.1 2000–2004 soil and water conservation in Yulin area in the completion of form (km^2)

Years	Base last year	Plan completion			Complete plan	Year reduce	Accumulative total	
		Plan	Completion					
			Total	Small watershed			Total	Small watershed
2000	20,437	1,100	1,232	421.5	111.9	2,051.1	19,617	10,118
2001	19,617	1,650	1,707	712.7	103.5	1,533.2	19,791	10,377
2002	19,791	1,770	1,816	591.5	102.6	1,158.6	20,448	10,606
2003	20,448	1,720	1,749	405.2	101.7	1,170.4	21,027	10,711
2004	21,027	1,200	1,280	377.7	106.7	763.9	21,543	10,918

The "ninth five-year plan" as the country in Yulin city, preventing sand key areas, organization and implementation of the small watershed comprehensive management, returning cultivated land to forestry and natural forest protection, shelter forest construction, a number of key projects, the ecological environment get primary control. All kinds of reservoir built 77 seats, with a total capacity of 994 million m^3. Soil erosion in the area of 26,400 hm^2, green coverage rate is greatly increased. At present in Yulin are terraced fields, such as closing, the irrigable land basic farmland 391,000 hm^2, per capita 0.13 hm^2, about 400 kg of grain per person. Years into the Yellow River silt at the beginning of 530 million ton by founding reduced to 290 million ton [2]. Preliminary improving the agricultural production in Yulin city infrastructure and ecological environment, improve the comprehensive agricultural production ability.

31.2.3 Comprehensive Control of Soil and Water Conservation and Promote the Development of Rural Economy

In Yulin city, is located in the loess plateau and the junction of Maowusu desert, the natural environment is bad, and agricultural development lags behind. However, in the process of comprehensive management, implement a variety of business, the comprehensive development in the rural economy face get change. Yulin region in pay special attention to the food production at the same time, they are focus on development of the sheep, fruit, potatoes and beans, oil and other industry, the output value of 2.6 billion yuan, the per capita net income of farmers more than 1,000 yuan [3]. The energy industry development, industrial expansion, although agricultural development atrophy phenomenon, but the second industry gradually increase proportion, per capita income has improved. 2003 years of agricultural output value of GDP to 18 %, 2005 reduced to 16 %. According to the current agricultural output value in Yulin city GDP ratio annual decrease 1 %, according to the 2011 GDP in Yulin city, the agricultural output value is expected to reach 10 %, basic to the national urban modern quantitative index [4].

31.3 Ecological Environment Comprehensive Management Measures and Cost Analysis

31.3.1 Management Measures

Since 1978, our country started the "three north shelter forest" construction engineering, has greatly improved the ecological environment of the Yulin city. In addition, for a long time in Yulin city, soil and water conservation and ecological construction work in engineering construction, the ecological restoration and supervision and law enforcement for key, woods and facilitate afforestation, economic projects measures [5]. To ecological, economic, and social three big benefit for the target, insist on scientific development outlook, strengthen prevent supervision, the integrated use of the engineering measures, biological measures and cultivation measures to control soil erosion, made outstanding achievements (Table 31.2).

31.3.1.1 Four Field Construction

Four field construction including level terrace, Nian level, and have great closing. From 2000 to 2004 four field construction plan finished 38,500 hm^2, actual finish a total of 40,190 hm^2, including up to 15,010 hm^2 watersheds. By the end of 2004, four fields came to 305,420 hm^2, including small watershed accounts for 165,840 hm^2 (Table 31.3).

31.3.1.2 Water and Soil Conservation Forests, Economic Forest, Planting of Soil and Water Conservation

In 2004, the original foundation in Yulin city, finished on water and soil conservation forests, economic wood fruit, water and soil conservation, respectively planting 53,210 hm^2 (including 3,420 hm^2 timber forest, shrub 49,790 hm^2), 5,770 hm^2,

Table 31.2 The main measures in Yulin city water treatment completion (2000–2004) (10^3 hm^2)

Years	project	Plan completion			The proportion of complete plan (%)	Accumulative total
		Plan	Completion			
			Total	Small watershed		
2000	Four field	8.68	9.69	3.39	111.6	291.14
	Water and soil conservation forests	60.2	66.84	24.26	111	1,084.19
	Economic forest	10.68	12.3	5.15	115.1	168.63
	Planting of soil and water conservation	24.67	28.42	7.2	115.2	254.86
	Facilitate afforestation	3.63	3.58	1.53	98.6	42.74
2001	Four field	7	7.17	2.91	102.3	298.92
	Water and soil conservation forests	102.04	100.43	34.9	98.4	1,105.57
	Economic forest	10.67	11.65	4.43	109.2	172.65
	Planting of soil and water conservation	52.42	54.21	8.75	103.4	10.49
	Facilitate afforestation	14.75	15.71	3.48	106.5	52.32

Table 31.3 The main measures in Yulin city water treatment completion (2000–2004) (10^3 hm^2)

Years	Project	Plan completion			The proportion of complete plan(%)	Reduce
		Plan	Completion			
			Total	Small watershed		
2002	Four field	7	7.17	2.91	102.3	3.67
	Water and soil conservation forests	102.04	100.43	34.9	98.4	66.77
	Economic forest	7.84	9.08	4.5	115.8	6.06
	Planting of soil and water conservation	57.78	59.72	15.41	103.4	37.28
	Facilitate afforestation		2.81	0.61		0.95
2003	Four field	7.55	7.76	2.26	105.6	4.37
	Water and soil conservation forests	98.76	100.47	24.86	101.7	68.57
	Economic forest		6.57	2.51	107.4	4.14
	Planting of soil and water conservation	6.12	57.95	10.35	100.3	38.82
	Facilitate afforestation	57.77	0.02	0.02		
2004	Four field	7.34	7.46	3.17	101.6	4.36
	Water and soil conservation forests	46.66	53.21	16.76	114	27.82
	Economic forest	6	5.77	1.92	96.2	4.07
	Planting of soil and water conservation	58.73	14.56	101.3	38.74	326.91
	Facilitate afforestation	0.83	0.83		0.3	54.73

58,730 hm^2. By the end of 2004, soil and water conservation forest, economy in Yulin city, woods, water and soil conservation, grow grass came to 1.16,286 $\times$ $10^3 hm^2$ (304,190 hm^2 timber forest, shrub 858,670 hm^2), 179,800 hm^2, 326,910 hm^2. These projects for the construction of the regional economic development, to adjust the industrial structure, and increasing the farmers income to lay the foundation.

31.3.1.3 Check Dam Construction

Check dam can block sand ground, the comprehensive function. 2002 years to build check dam in Yulin city, the 403 seats, including a large dam 25 seats, medium dam 53 seats, small dam 325 seats. Key projects of 15 seats and strengthening the dam seepage path 2,516 seats. Total area of 651.7 km^2 to control soil and water erosion 1,132.3 hm^2. 2003 years to build 399 a check dam in Yulin city, including large dam and seat, medium, small dam 29, 71, 299. Key projects of 14 seats, in addition, there are 2,659 dam seepage path reinforcement. Total area of 539.5 km^2 to control soil and water erosion and sediment blocking capacity of 42.4 $\times$ $10^4 m^3$, can silt place 662 hm^2. 2004 years to build warping DAMS in

Table 31.4 Soil and water conservation Yulin completion of the investment classification table (10^4 yuan)

Years	Total	Small watershed	Reservoir management	Demonstrated	Grass	Mechanical construction	Four field	Research funds
2000	4,257	1,725	19	0.5	494.2	465	1,498	
2001	3,585	1,374.5	47.8	147.8	551.3	402	873	
2002	4,821	1,514	103	67	1216	372	1,329	27
2003	3,833	2,100.6	88	20	705.5	160	676	
2004	3,527	1,060.2	55	154	753.3	620	636	20

Yulin city, the 326 seats, including 15 key projects seat can silt place 169 hm^2. The construction of the check dam in controlling soil and water loss, improvement of agricultural production conditions of plays a very important role.

31.3.2 Management Cost

In 2000–2004, total investment 200.2301 million yuan in Yulin city ecological construction. Among them the central financial investment of 97.6868 million yuan RMB, the provincial financial investment is 20.1593 million yuan, and the earth (city) finance finished 1.378 million yuan, the county (city) financial investment of 35.487 million yuan, the Yangtze river and Yellow River ShuiBaoJu middle innings total investment is 56.22 million yuan. In addition, there is a part of the masses to. Specific to Tables 31.4 and 31.5:

From the table can see, in 2000–2004, used for soil and water conservation in Yulin city of the investment amount and no is the trend of the growing, financing source is still very limited. At present, many are still in Yulin city, county, where self accumulation and reproduction input ability is poor, and can be used for the ecological environment is limited to the management of funds. To control water and soil loss at the lowest standard 150,000 yuan/km^2 calculation investment, want to rely on local power to control soil erosion is extremely difficult. Therefore, from the state to the province, city, district want to notice Yulin's ecological

Table 31.5 Yulin city, the classification of sources of soil and water conservation investments

National investment units: Ten thousand

Years	Total		Provincial finance		Changjiang SCS	
	Arrangement	Complete	Arrangement	Complete	Arrangement	Complete
2000	4,256.9	4,257	1,021	1,021		
2001	4,239.9	3,585	538	537.5	914	594
2002	4,916.8	4,821	240	240	1,031	
2003	3,929.1	3,929	50	50	895	799
2004	3,622.5	3,623	168	167.7	989	893

environment construction, increasing investment, support of Yulin ecological environment construction. Generally speaking, the current environment level of management and protection ability is low, environmental management in the region is still a long-term and arduous task. Environment construction must be in control of soil erosion, prevent desertification, improve the environment for the premise, pay attention to the basic farmland construction, realizes the sustainable development of ecological and economic security.

31.4 Conclusion

1. Ecological environment and regional economic development has a direct relationship. Ecological environment change can promote the development of economy, the economic development and promoting the construction of ecological environment. Therefore, should pay attention to in Yulin city ecology, economy coordinated development.
2. Water and soil loss in Yulin city, area of 36,900 km^2, since the founding, the ecological environment in Yulin city, much attention country great effort in ecological improvement. By the end of 2004, soil and water loss control area of 21,543.3 km^2 accumulated. Control and protection of ecological environment is still Yulin economic and social development of the eternal theme is the premise of sustainable development of northwest and guarantee.

References

1. Zongfan LM (1997) Ecological agriculture in the Chinese Loess plateau, vol 21. Shannxi Science Press, China, pp 19–23
2. Zhimao G, Guangshu S (2002) Disscussion on the cultivated land reduction and forest and grass on loess plateau. J J Arid Land Resour Environ 12:62–65
3. Xuetian G, Fenli Z (2004) Eco-Environment construction and sustainable agriculture development in the Loess plateau of northern part of Shaanxi province. J Res Soil and Water Conservation 4:47–49
4. Fang F (2007) Initial analysis city of Yulin in the modernization process. J J Yulin College 23:27–29
5. Baiping Z, Dadao L, Xiaoding M (2006) ConceDtion and scientific foundation of state special. J Eco-region Prog Geography 2:8–16

Chapter 32
Study on Clothing Marketing Strategies Based on Consuming Behaviors of White-Collar Females

Ying Liang

Abstract As the important segment market of the clothing products, the white-collar females' market is of great marketing value. So small and medium-sized clothing enterprises must pay more attention to this group and make innovative marketing strategies of brand according to a series of sales behaviors, for instance, the white-collar females prefer the famous brands. This article mainly talks the characteristics of the white-collar females' consuming behavior in buying clothes and the marketing strategies of clothing brand in the small and medium-sized clothing enterprises based on these characteristics.

Keywords White-collar females' consuming behaviour · The small and medium-sized enterprises · Clothing brand · Marketing strategy

32.1 Introduction

With more and more enterprises entering into the clothing industry, small and medium-sized clothing enterprise is facing much more pressure of competition. for the group of women white-collar, enterprises are sure to know the consuming behavior characteristics of the white-collar females well firstly, aiming at getting rid of disadvantages and keeping an invincible position in the environment of the increasingly fierce competition. In addition, they work out some effective marketing strategies of brand based on the consuming behavior characteristics of the white-collar females [1].

Y. Liang (✉)
Guangxi Economic Management Cadre College, Nanning Guangxi 530007, China
e-mail: 78366580@qq.com

W. Du (ed.), *Informatics and Management Science VI*, Lecture Notes in Electrical Engineering 209, DOI: 10.1007/978-1-4471-4805-0_32,

32.2 The Consuming Behavior Characteristics of the White-Collar Female on Purchasing Clothes

32.2.1 Tendency of Obvious Perception

White-collar females are full of various kinds of pressure at work, therefore, they are eager to release them. Since white-collar females possess of sufficient and relatively independent disposable income, these characteristics make them a consuming tendency of perception prominently. According to the investigation of Zero Company, people, age at 18 to 35 years old, accounting for 93.5 % proportion have a variety of irrational consuming behaviors, which account for 20 % proportion of the total consumption of women. Having been got paid, 52.8 % of women with their bulgy purse do some sudden kinds of consuming behaviors, and 46.1 % of women,happy or not, do some kinds of emotional consumer behaviors; In addition, 79 % of women who hold the indifferent attitude don't regret for spending the extravagant money driven by the irrational emotions, instead, they all think that shopping can bring a good mood [2].

32.2.2 Emphasis on the External Beauty

Generally, women pay more attention to their appearance than than men does, therefore, women are easier to be attracted by good-looking appearance of clothing. Most of women are tempted with beautiful appearance of the clothing firstly when they purchase clothes. This consuming characteristic is inseparable from the perceptual thinking mode that women judge product, for instance, most of female consumers are willing to choose the products according to the perceptual awareness whether a hair accessory, a car or a house, especially in aspects of purchasing the clothes. To some extent, external factors such as Packing, appearance, color and its affiliated ornaments have influences on the purchasing behaviors of the white-collar females [3].

32.2.3 Emphasis on the Details of Clothing

Compared with men, women pay more attention to the details of products, for most of the white-collar females agreed that delicate exquisite is the symbol of the taste. They usually require the garments whether a coat or small ornaments and shoes, underwears or pajamas should be made excellent and perfect, sometimes even exposed thruma of the cloth will affect the mood and desire of the consumers [4].

32.2.4 The Desire of Purchasing Greatly Influenced by Reputation

As the most influential and appealing way of transmission in light of the consuming behaviors of white-collar females and the desire of purchasing, the spread of word-of-mouth power mainly spreads information from mouth to mouth. Most of the white-collar females received higher education and they become more and more accustom to using the familiar way of purchasing, especially the recommend of friends. They are easy to produce the group psychology and group interaction, thus leading to some kinds of infectious consumer behaviors. According to the analysis, 55.5 % of young women who do not want to buy are urged to buy the products which are not necessary, for they are influenced by their friends when they are shopping. In addition, as the popularity of cell phones and Internet provide the spread of word-of-mouth power with the extended platform. Most of the white-collar females are willing to take an active part in the spread of word-of-mouth power, and to share the suggestions and experiences with their friends as well as to put forward their own feelings and sentiments through MSN, QQ, text message, BBS communities and so on.

32.2.5 Price Sensitive to the Product

Most white-collar females who are good at thrift have strong brand awareness. What kinds of brands they prefer to be so expensive, some of which are in the range of luxury. To the great extent, it stimulates their purchasing desire when their income is not enough to make them frequently and freely to buy the some brands and when the some brand began to be discounted. For example, they even can buy several sets of seasonal clothes and pairs of shoes when products are in discounts and promotions, after purchasing they will feel the sense of achievement.

32.2.6 The Pursuit of the Quality of Life

On one hand, as the social status of the white-collar females and independence of economy are gradually improved, they are more and more sure of the reflection of their own value and they work hard and also do not forget to pursue the quality of life simultaneously; On the other hand, the dual role of the traditional family and professional women let them often feel tired and stressful and also they are eager to release pressure as well as to relax themselves. Even if they are much too tired, they will not forget to buy themselves some expensive cosmetics and beautiful clothes; longing for a beautiful family life and love, they will take off the professional clothes and then put up the casual wear after they come home from work.

32.2.7 *Special Care of Consuming Experience*

Most of white-collar females specially care about the joy of shopping and consuming experience, sometimes even more than the value of goods. Shopping for white-collar females is not only to buy good food and clothing that make their own beautiful style, but also to relax mood and release the stress, thus the white-collar females regard shopping as a relaxed, happy and enjoyable activities. Without planing, they often go shopping and buy what they like casually with some good friends, walking around and chatting. Besides, as for shopping environment, they have higher requirements and they usually like to go to a large-scale mall with the comfortable environments, good supporting facilities, the convenient transportation and diversified entertainment. Moreover, they are fond of the market with the humanized design, high-grade adornment, considerate service as well as the market which can bring them comfort, joy, respect and achievement from the aspects of smell, touch, hearing and vision.

32.3 The Marketing Strategies of Clothing Brand in the Small and Medium-Sized Enterprises Based on the White-Collar Females' Consuming Behavior

With more and more enterprises joining in the clothing industry, the small and medium-sized enterprises are under great pressure, some of which don't attach importance to the marketing strategies of brand. They believe that they will get good sales with attracting the mature consumers. However, the spread of word-of-mouth marketing which gets the brand spread from person to person, has got out of date under the informationization times and under the direct sales giants' surging. Besides, conference marketing is less effective while the consumers become wiser and wiser. And the two marketing strategies are limited by the population, the region and the time. Thus, apart from the high-quality products and the perfect operating system, the small and medium-sized enterprises need to create the innovative brand strategies.

With high economic independence, the white-collar females have a high sense of the clothing brand. In fact, the marketing strategy of brand is more popular for the clothing enterprises to compete with each other and get the market share. Most white-collar females prefer one brand which indicates the trust of it. So the small and medium-sized clothing enterprises have to strengthen the clothing brand because the foundation of a brand is the quality of products. And they should improve the quality in the environmental protection, craft, color, design, materials in accordance with the professional characteristics of the white-collar females. Besides, the clothing is of culture and it reflects the culture. It can show the women's beauty of individuality and the vivid image. Thus, the small and medium-sized clothing enterprises should connect the brand with rich culture, position the bright brand and make best of the internal and the external spreading channels to be appreciated by the women in spirit. Then the enterprises will create the brand belief and enjoy high brand loyalty. Next, I will analyze the marketing strategy of clothing brand in the small and medium-sized enterprises based on the white-collar females' consuming behavior.

32.3.1 Shaping the Brand Culture

Clothing is not only a carrier of culture, but also shows the personalized beauty and vivid image of the white-collar females. Therefore, the small and medium-sized enterprises should give the brand a profound and rich cultural connotation first, re-refine clothing unique selling points, paste campus cultural label to its brand and give a new soul to the brand with the stories behind it. Meanwhile, the background of the brand just meets the need of the corresponding consumers in the market, namely the 18–45 year-old white-collar females. They have strong personalities and are in pursuit of cultural connotation and high-quality petty life. Therefore, brand story can provide the dealers and salespeople with a good entry point to the introduction of clothing. The way to introduce the functions and values of clothing by stories is more appealing to the consumers in comparison to straightly introduces the clothing. Besides, it is an easier access to the consumer's sentimental recognition and a stepping stone of the sales.

32.3.2 The Employment of Cultural Ambassador of Dissemination Brand

Staff of the marketing department can select and cultivate team leaders and Internet network sales with influence, affinity as well as a wide range of interpersonal relationship and appoint them ambassadors of the enterprise's brand culture. Moreover, they should be awarded with the fine letter of appointment with the signature of the most senior leaders in the enterprise as well as their photos with the high-level group. When showcasing the clothing to the consumers, they could present them with these items and put them up on the walls or place them on the office desks. This may play a significant role on appealing and convincing the customers and spreading the cultural brand of our enterprise.

32.3.3 *Making Best of Large Shopping Malls*

As the main way of selling the clothing goods, the large shopping malls are widely stored by the brand clothing and the high-end clothing. The white-collar females have a high requirement for the shopping environment so that they prefer the large stores. So the enterprises need place to put the products in the malls to attract the consumers and stimulate their brand consciousness.

32.3.4 *Opening Chain Stores for Certain Brand*

Chain stores are popular for many brands. They are easily acceptable and highly appreciated by the white-collar females for their unified façade design, pleasant environment and the new clothing styles. In this way the enterprises can not only expand the influence of the brand, but also increase the sales volume. Generally the women have their favorite brands. So they often choose the chain stores for brands where they will be greatly satisfied.

32.3.5 *Network Marketing*

In the information age, all the entertainments go networking. Therefore, the enterprises shouldn't abandon the network while making use of the entertainments for brand marketing. In the network age, entertainment marketing stands out by its fast spread and high interaction. As a new way and concept, network marketing can effectively make the dealwith groups and individuals. It is a new way for the women to buy clothes on the internet. It can provide the free choices whether on time or the space, fashion experience and low prices to the women, at the same time it can realize the product brand and make the enterprises receive the brand effect. In addition, while selling the clothes on-line, the enterprises can directly give the statistical analysis, get the consumers' feedback through the BBS, and promptly adjust the combination pattern to increase the efficiency of the clothing the small and medium-sized enterprises.

32.3.6 *Make Use of Public Relations Activities*

In the end, the enterprises should make use of the public relations activities. Through the activities, they can shape the image of the company and the brand, improve the social reputation of the company and the brand, and attract more women consumers.

Above all, the enterprises must catch the characteristics of women's consuming behavior, including that they have a tendency of obvious perception, they pay more attention to the external beauty and the details of clothes, they are influenced by the reputation of a brand, they are sensitive to the prices, they pursue the quality of life and they care about the consuming experiences and any other consuming behaviors; then the enterprises work out the marketing strategies of brand based on these characteristics.

References

1. Xuefei H (2011) Research of the white-collar females' consuming psychology and behavior based on the professional characteristics. Mod Commercial Ind 10(3):92–97
2. Qianqian Y (2011) Kaltendin–the end of one marketing pattern of clothing brand. Foreign business 03(3):582–588
3. Yitan M (2010) Analysis of one new marketing pattern of clothing brand–network marketing. Mark Modernization 233(09):23–32
4. Ling W (2009) Discussion on the marketing strategies of clothing brand. Liaoning Tussah Silk 32(03):30–37

Chapter 33
Research on Rights Protection of Consumer and Interests in E-Commerce-Taking Functional Department and Industry Association

Qinghua Zhang

Abstract commerce puts forward a new challenge on the protection of consumers' rights and interests. The paper has chosen functional department and industry association as research objects, presented the current situation and existed problems concerning the protection of consumers' rights and interests in E-commerce, then followed by raising relevant proposals to perfect consumers' rights and interests protection policy under China's practice via borrowing foreign experiences. It is hoped that all this will furnish some necessary reference in perfecting the system to protect consumers' rights and interests in E-commerce, making functional departments and industry associations to protect consumers' rights and interests of work more efficiently so that the legal rights and interests of consumers will not be violated.

Keywords E-commerce · Protection of consumers' rights and interests · Functional department · Industry association

33.1 Introduction

E-commerce puts forward a new challenge on the protection of consumers' rights and interests and plenty of studies have been done concerning this area, focusing on the legal level as well as on analyzing the cause and form when the consumers' rights

---

Q. Zhang (✉)
School of Information Management and Engineering, Shanghai University of Finance and Economics, 200433 Shanghai, China
e-mail: zhangqh@mail.shufe.edu.cn

W. Du (ed.), *Informatics and Management Science VI*, Lecture Notes in Electrical Engineering 209, DOI: 10.1007/978-1-4471-4805-0_33,

and interests are violated in E-commerce. This paper, when conducting the research, has chosen functional department and industry association as research objects.

33.2 Demands for Rights and Interests of Special Protection in E-Commerce from Consumers

33.2.1 Right to Know

Consumer Rights Protection Law regulates that consumers shall enjoy the right to obtain true information of the commodities they purchase and use or the services they receive. In E-commerce, consumers trade with sellers under a simulated network condition. Orders are placed through remote network, settlement through E-banking, commodity delivery to the door through distribution companies and for many of the digital products, such as digital video, MP3, software, etc., dealings could even be concluded by means of instant network transfers. Undesirable businessmen often give incomplete or false information via network, violating consumers' right to know.

33.2.2 Right of Privacy

The right of privacy is a personality right which any natural person enjoys, by which, the person can dominate or control his/her personal information, private activities and fields which have nothing to do with public interests. In E-commerce, content of right of privacy mainly refers to: basic information of consumers, such as name, age, identification card, home address, telephone number, etc.; property status of consumers, including credit card number and balance, etc. of consumers when concluding a deal; E-mail address, which may bring sufferings to consumers from the possible advertisement mails and spam if sellers disclose the registered E-mail addresses of consumers to other people.

33.2.3 Right of Fair Deal

Consumer Rights Protection Law regulates that consumers shall enjoy the right of fair deal. A consumer shall have the right to fair terms of trade and conditions such as quality guarantee, reasonable price, accurate measures, etc., when purchasing a commodity or receiving a service, and shall have the right to reject coercive transactions by business operators. In E-commerce, consumers' right of fair deal

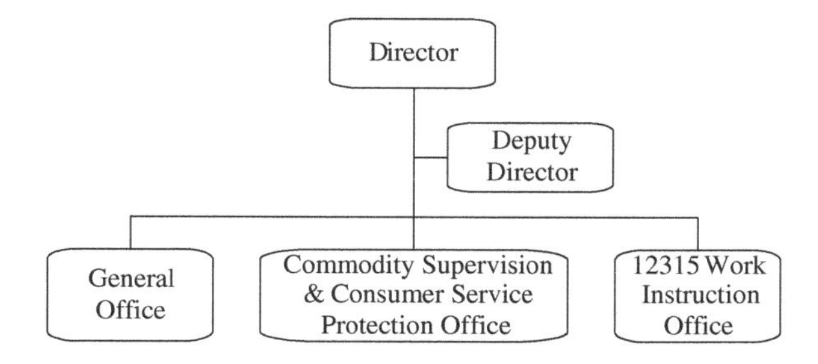

Fig. 33.1 Three functional departments in consumer protection bureau

often is being violated. Main forms when fair trade rights are invaded: Descriptions fail to meet actual commodities to be bought, lack of guarantees for after-sales service and difficulty in changing or refunding, etc.

33.3 Current Situation Concerning Related Functional Department and Industry Association

33.3.1 Consumer Protection Bureau

Consumer Protection Bureau is the subordinator of the State Administration for Industry and Commerce of the People's Republic of China whose main function is: to formulate specific measures in protecting consumers' rights and interests, to undertake quality related supervision and management work for products in circulation area, to conduct work of safeguarding consumers' rights in the related service area, to investigate and to penalize behaviors of selling fake and/or substandard goods as well as to take responsibilities of instructing consumers to consult and to complain, to accept and to handle consumers' reports, including construction of network system. Three functional departments are arranged in regard to the above mentioned functions [1] (Fig. 33.1).

33.3.2 Consumers' Association

China Consumers' Association, established in December 1984 under approval of the State Council, is a national social organization to protect consumers' legal rights and interests concerning social supervision of commodities and service, whose purpose is to conduct social supervision for commodities and service. Also, the association aims at conducting social supervision for commodities and service, protecting consumers' legal rights and interests, guiding popularity to make reasonable and scientific consumption and promoting a healthy development of socialist market economy [2].

33.4 Existed Problems Concerning Related Functional Department and Industry Association

33.4.1 Chaos Function Setting in Consumer Protection Bureau

General Office shall be responsible for both administrative affairs and drafting comprehensive regulations, specific measures or approaches concerning the protection of consumers' rights and interests. Bedsides organizing and guiding commodity quality supervision and inspection, including rights protection, Commodity Supervision and Consumer Service Protection Office shall draft rules and regulations, specific measures or approaches for commodity quality supervision management in circulation area as well as for protection of consumers' rights and interests. Meanwhile, 12,315 Work Instruction Office shall guide to centralize consultancy, complaints and reports of consumers, including data management, analysis and summary concerned.

33.4.2 Inadequate Function of Consumer Protection Bureau

It is difficult for the above mentioned five functions of Consumer Protection Bureau to meet new requirements to protecting consumers in E-commerce. Due to the fact that Consumer Protection Bureau only works to guide consumers' complaints, to help consumers in furnishing evidences to the related departments instead of to directly proceed against illegal sellers, it becomes more difficult for online consumers to provide evidence and the cost for rights protection comes higher, leaving consumers at a weaker position.

33.4.3 Lacks of Complete Industry Related Self-Discipline Guide for E-Commerce

For the time being, there are more than 100 laws concerning protection of consumers' rights and interests featuring in complexity and mixture. The excessive intervention of government not only results in that the E-commerce operators lose their ways, making them feel hard to follow the fast-changing network, feeling psychologically tied up, but also results in poor effect to develop China's information network industry and to protect consumers' rights and interests. Besides the supervision of legal regulations, it is quite necessary to formulate a full set of self-discipline guides for self-management and self-restraint. From the point China Consumers' Association, it is only a national organization standardizing sellers' behavior in sense of morality and protecting consumers' rights and interests, which lacks of a complete industry related self-discipline guide for E-commerce.

33.5 Relevant Proposals to Perfect Consumers' Rights and Interests Protection Policy in E-Commerce

33.5.1 Delineating Function of Consumer Protection Bureau

In regard of the complexity of Consumer Protection Bureau in function arrangement, we can follow example of Federal Trade Commission (FTC) to establish annexed departments of Bureau of Consumer Protection [3–10]. For FTC, it is the only organ at US federal level to own extensive executing power to protect consumers. The subsidiary of Bureau of Consumer Protection is dedicated in protecting consumers' legal rights and interests which should not be violated [11–14].

1. Considering the complicated function, General Office could be divided into four sub-branches: Administration Branch for formulating plans, documents and general documents, including routine administration affairs, e.g. file management. Information Brach for publicizing government affairs as well as informatization, including website construction of Consumer Protection Bureau. Legal Affairs Branch for drafting laws, rules and regulations concerning protection of consumers' rights and interests, including rules, regulations and specific measures which should be prepared by Commodity Supervision and Consumer Service Protection Office for supervision management of commodity quality in circulation field and for protection of consumers' rights and interests in service field. In the meanwhile, the Legal Affairs Branch shall work for formulating rules, regulations and specific measures in accepting and handling consumer' consulting, complaints and reports as well as investigating and penalizing illegal cases violating consumers' rights, which should all be done by 12,315 Work Instruction Office. Research Institute for social investigation, seminars to protect consumers' rights and interests, including in assisting Legal Affairs Branch to formulate and revise each related laws and regulations.
2. Commodity Supervision and Consumer Service Protection Office could be further divided into Quality Inspection Branch and Rights Safeguarding Branch. For the former one, it works to organize, to guide supervision management of commodity quality in circulation field, to conduct quality monitoring and supervising inspection in circulation field, including organizing and guiding activities to penalize selling fake and/or substandard goods which violating consumers' rights. For the latter, it works to directly proceed against illegal sellers on the basis of its original rights safeguarding work in service field.
3. 12,315 Work Instruction Office could possibly be divided into three centers. Consulting Center to provide advice for consumer. Complaint Center to accept

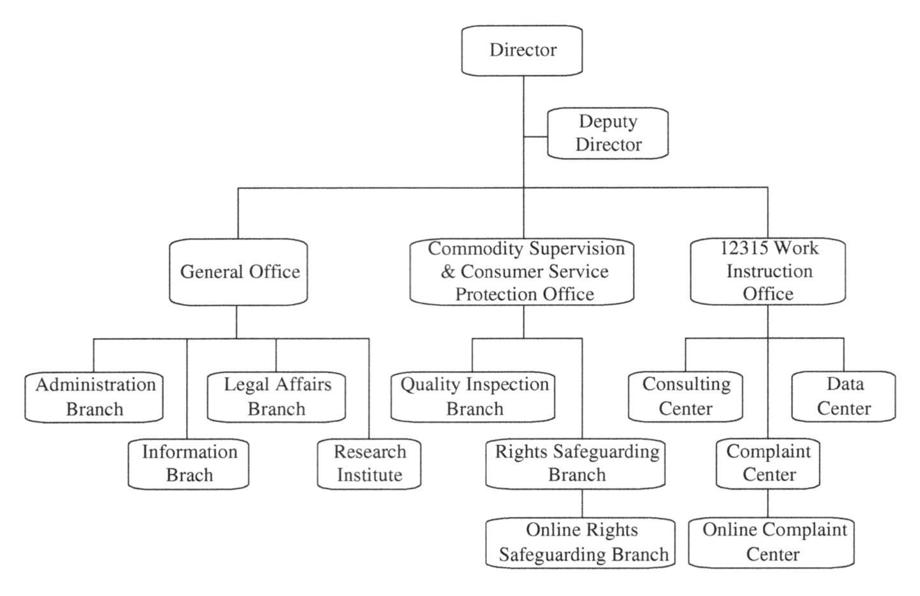

Fig. 33.2 Relevant proposals to function setting of consumer protection bureau

consumers' complaint or report and submit to Rights Safeguarding Branch in case of instituting a proceeding. Data Center to manage and analyze data of consumer complaints from nationwide (Fig. 33.2).

33.5.2 Expand Function Scope of Consumer Protection Bureau

1. To increase legislation scope. To unify and to standardize legal system to protect consumers' rights and interests in E-commerce field, it is suggested to add legislative function in Legal Affairs Branch for special purpose of formulating and modifying laws and regulations, rules and regulations and policy measures in the areas concerned. And, also, legislation scope will be increased for sake of promoting the unification of legislation and law enforcement, further guaranteeing the steady development of E-commerce.
2. To increase appeal scope. To avoid placing online consumers at a weaker position, to enhance consumer confidence as well as to reduce cost of safeguarding rights and interests, it is suggested to add appeal function in Rights Safeguarding Branch, taking legal actions in a direct way against illegal businessmen and warning sellers at the same time.

Some added comments:

1. For Information Brach, a special channel could be established to publicize blacklist of online businessmen, criticizing illegal businessmen or even publicizing punishment measures.
2. Legal Affairs Branch shall make corresponding laws and regulations according to the new demands in E-commerce field.
3. Research institutes shall conduct plenty of seminars and social surveys for sake of protecting consumers' rights and interests in E-commerce, enriching theoretical knowledge and collecting public opinions for further purpose of helping Legal Affairs Branch to make reasonable laws and regulations.
4. Quality Inspection Branch may undertake regular checks for online commodities, ensuring excellent commodity quality in E-commerce deal.
5. Online Rights Safeguarding Branch under Rights Safeguarding Branch may form a set of online dispute settlement mechanism, effectively mediating disputes arising from online deals.
6. Online Complaint Center under Complaint Center will, first of all, accept online trade complaints and then submit to Online Right Defense Division for resolution (Fig. 33.2).

33.5.3 Self-Discipline Guideline Made by China Consumers' Association

It is impossible to fully rely on laws and regulations set forth by government if we hope to perfect protection of consumers' rights and interests in E-commerce, it also needs cooperation from the self-discipline of the industry itself. Consumers' Association may, together with E-commerce industry association, make out "Protection Principles of Consumers' Rights and Interests in E-commerce", stating as following:

1. Advertising publishing self-discipline principle

It is hard for the existing Advertising Law regulating the obligation and relationship of each party in the traditional advertisement to be applied onto network advertisement, which is the main cause of false and fraud advertisement in the net deal. Compared with the nature of passivity, limitation and fixity from advertisement regulating laws, the advertisement ethics comes more active, broader and more flexible. The guideline should clarify the due norm of online advertisement that should be the substantial fact instead of any descriptions of fraudulence or misleading.

2. Information disclosure self-discipline principle

In regard to the related legislation concerning online information disclosure, the laws and regulations in force include Consumer Protection Law, Product Quality Law and the Regulation on the Administration of Company Registration. Information disclosure in these legal documents is targeted at protecting consumers' rights to know under traditional trading mode, in which many regulations appear ambiguous, poor in operation an inapplicable for online trading. The guideline should make clear requirements for scope, method, person in charge and responsibility for the disclosure of E-commerce information, guaranteeing that the reading conditions provided to consumers are definite, clear and complete.

3. Privacy protection self-discipline principle

For personal information provided by consumers when shopping online, sellers are responsible for non-disclosure or not transferring to other people of such information. From the legal point, corresponding duties of sellers should be established in details, defining the related responsibility of sellers. If service is provided, the sellers are liable to protect consumers' information through certain relevant technical support. Sellers are liable for compensation due to damages from using consumers' information. Sellers shall take joint liability for compensation any losses the consumers suffer from sellers' transferring consumers' personal information to the third party. The guideline should clearly regulate that without users' permission, it is prohibited for sellers not to store or collect consumers' personal data without reasonable purpose. Sellers shall not improperly disclose or shall deliberately spread consumers' personal information. Sellers shall not, without authorization, alter users' personal information or shall not disclose wrong messages. Sellers are not allowed to access to email by using consumers' information or to access into any personal online information field.

33.6 Conclusions

When perfecting protection system for consumers' rights and interests in China's E-commerce field, we shall consider the factor of adopting international practices without copying foreign experience. And we shall form our unique model with consideration of our specific national conditions, offering both excellent development environment for E-commerce industry and the best protection in terms of consumers' rights and interests.

Many of the scholars have put forward proposals to perfect protection of consumers' rights and interests from the point of the unified legal system, effective work by justice and administration as well as technical innovation, etc. The precondition to implement such proposals requires that all the functional departments work simultaneously, i.e. to add more functional departments from horizontal point and expand functional scope of all related departments from the longitudinal point. If the functional departments and industry associations are concentrated developed, the protection system for consumers' rights and interests could be fundamentally perfected.

Acknowledgments This work was supported by Shanghai Philosophy and Social Science Foundation from Shanghai Planning Office of Philosophy and Social Science (2009EZH001), and Leading Academic Discipline Program, 211 Project for Shanghai University of Finance and Economics (the 4th phase).

References

1. State Administration For Industry and Commerce of the People's Republic of China, Consumer Protection Bureau, http://www.315.gov.cn
2. China Consumers' Association, http://www.cca.org.cn/
3. The FTC's Bureau of Consumer Protection, http://www.ftc.gov/bcp/about.shtm
4. Starek RB, Rozelle L (1997) The federal trade commission's commitment to on-line consumer protection. John Marchsall J Comput Inf Law XV 1(4):679–684
5. McKnight DH, Chervany NL (2002) What trust means in e-commerce customer relationships: an interdisciplinary conceptual typology. Int J Electron Commer 6(2):35–59
6. Li Q (2007) Research on Chinese C2C e-business institutional trust mechanism: case study on Taobao and Ebay. Wireless Commun Networking Mob Comput 12(4):3787–3790
7. Ma J (2010) The protection of consumers' rights in the framework of the C2C transaction model. J Ningbo Univ 23(4):114–117
8. Turban E, King D, Lee J, Viehland D (2007) Electronic commerce—a managerial perspective (4th edn). China Mach Press 23:382–387
9. Murray BH (2004) Defending the brand: aggressive strategies for protecting your brand in the online arena. Am Manage Assoc 4:229–237
10. Roussos G (2006) Ubiquitous and pervasive commerce, new frontiers for electronic business, 44th edn. Springer, Heidelberg, pp 74–79 (Suppl 1)
11. Pavlou PA, Gefen D (2004) Building effective online marketplaces with institution-based trust. Inf Syst Res 15(1):37–59
12. Spindler G, Börner F (2007) E-commerce law in Europe and the USA, 12th edn. Springer, New York, pp 38–44 (Suppl 3)
13. Gu J (2007) Consumer rights protection on the online auction website—situations and solutions: a case study of Ebay. British and Irish low, education and technology association 2007 annual conference, Hertfordshire, vol. 4, pp. 16–17 (Suppl 1)
14. Hecker M, Dillon TS, Chang E (2008) Privacy ontology support for e-commerce. Internet Comput 12(2):54–61

Chapter 34
Research on Customer Satisfaction in B2C E-Commerce Market

Qinghua Zhang

Abstract B2C has become the trend in future Internet consumption and it is of the highest importance in terms of business operation strategy on how to increase consumer satisfaction. The paper, considered the operational flow of B2C trading, put forward the consumer satisfaction model. Consumer satisfaction is mostly decided by perception of quality from consumers, which will more or less be affected by the four factors of goods quality, website interface, distribution and after-sales service. Specific measuring index has been developed on which questionnaires had complied and distributed to obtain data influencing consumer satisfaction in B2C activities. The consumer satisfaction model was then verified in the way of conducting factor analysis in respect of such data by applying SPSS17.0. B2C e-commerce enterprises may borrow the consumer satisfaction model and measuring indexes to evaluate the weak points of their owns for sake of adopting appropriate improving measures.

Keywords B2C e-commerce · Consumer satisfaction · Measuring index · Factor analysis

34.1 Mental Characteristics of B2C Internet Consumer

Internet consumption usually refers to a procedure when consumers buy goods or service via internet, featuring in the two ways of B2C and C2C.

Q. Zhang (✉)
School of Information Management and Engineering, Shanghai University of Finance and Economics, 200433 Shanghai, China
e-mail: zhangqh@mail.shufe.edu.cn

W. Du (ed.), *Informatics and Management Science VI*, Lecture Notes in Electrical Engineering 209, DOI: 10.1007/978-1-4471-4805-0_34,

For B2C, it focuses on network retails by applying the natures of globality, interaction and individualization of Internet for purpose of providing customers with more convenient, more individualized and more competitive products and service, simplifying buying procedure of customers, lowing trading cost and increasing customer satisfaction and customer loyalty at the same time [1].

The mental characteristics of B2C Internet consumer are as follows: (1) Growth of consumption initiatives. Consumers tend rational and influence onto web consumption from the advertisement sees limited. Consumers often search Internet for products when they feel it necessary to buy. They obtain information related with the products, analyzing and comparing, then placing orders to buy at the last moment. (2) Lower stability of consumer mentality. The network times change so rapidly and consumers are facing a market with extremely rich commodities, which is also a market with tremendously updated speed, the stability and loyalty of consumers are seen at a lower level. (3) Pursuit of convenience and efficiency. The main character of web consumption lies in un-limitation from time and space, allowing consumers to get what they want within doors. Due to the speeding-up life pace in modern society, people are pursuing a clear and simple website, a fast and convenient purchasing procedure as well as a timely and rapid distribution while they go web shopping. (4) Pursuit of low price and high quality. Internet helps to save enterprises plenty of exhibition and circulation cost, making goods price in web market slightly lower than those in the traditional market, which is of great attraction to consumers in sense of e-commerce. Cost for consumers in web market to compare prices can almost be neglected; hence the sensitivity to price from consumers appears relatively higher. (5) Emphasis of shopping experience. The particularity of web consumption, e.g. consumers can only get to know commodities from images and descriptions, makes sellers to design their websites and to provide service from the angle and standpoint of consumers. For consumers, from the moment to open an E-commerce website to successfully buy something, it requires series of behaviors, during which, any inconvenience or dissatisfaction will directly lead to the loss of consumers.

34.2 Customer Satisfaction Model in B2C E-Commerce Market

The so-called satisfaction refers to a pleasant or disappointed state of feeling when someone compares the effect of what he/she senses about a certain product with what he/she expects. If perceivable effect is below expectations, the consumer will dissatisfy. If perceivable effect is equal to or above expectations, the consumer will accepts.

With development of the Internet, researchers have taken up studying questions concerning consumer satisfaction under web environment. The adopted studying

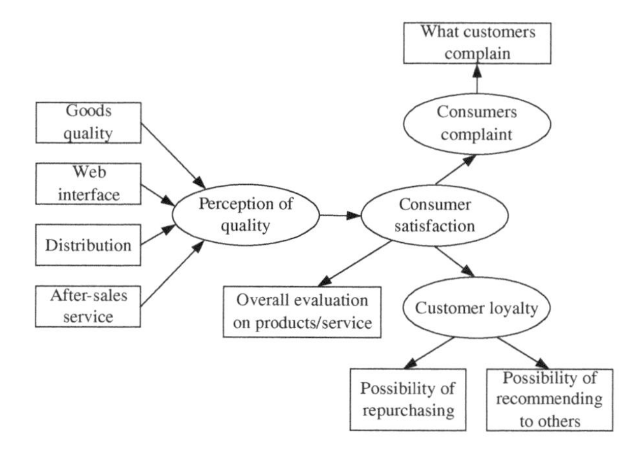

Fig. 34.1 Customer satisfaction model in B2C market

methods and models are almost similar to those for traditional consumer satisfaction, which only varies slightly in the component.

Starting from the psychological characteristics of B2C web consumers, we analyze process when consumers buy and use products or service, founding that there exist many factors to form satisfaction and there also exist certain cause-effect relationship. By consulting the previous research result [2–5], we put forward customer satisfaction model in B2C e-commerce market (Fig. 34.1).

Four variables (perception of quality, consumer satisfaction, consumers complaint, and customer loyalty) are included in the model, in which, the perception of quality is a reason variable of consumer satisfaction while consumers complaint and customer loyalty are the result variable in terms of consumer satisfaction. Ellipse boxes refer to latent variables (variable which could not be directly observed) and arrows between latent variables direct the cause-effect relationship between latent variables. Rectangular boxes refer to observable variables and arrows between rectangular boxes and ellipse boxes direct the reflecting relationship between observable variables and latent variables.

34.3 Consumer Satisfaction Measuring Index System in B2C E-Commerce Market

34.3.1 Design Principle for Measuring Index

Measuring index is the key element in measuring customer satisfaction. Views, opinions and requirements of consumers are all feed backed through these indexes, which could be used as important basis in exactly grasping consumers' requirements and maintaining a steady improvement. When designing the indexes, besides the need for accurate reflecting latent variables, the three principles of consumer orientation, completeness and operability should be followed.

(1) Consumer orientation. "Consumers are oriented to decide measuring indexes" is essential requirement for customer satisfaction measuring index design. To exactly grasp need of consumers, we should choose consumer orientation, choosing those measuring indexes which could be sensed and focused by consumers. (2) Completeness. Customer satisfaction measuring index system should reflect comprehensiveness and evaluating indexes should make analysis on the nature of B2C market from the many layers of different ways, with complete and systematic contents. (3) Operability. Customer satisfaction measuring index system should reflect operability, quantification of measuring indexes and obtaining of data should be operable, especially should be convenient for operation.

34.3.2 Structure of Consumer Satisfaction Measuring Index System

Latent variables in the B2C consumer satisfaction models could not be directly observed, which is required to be expanded for making series of directly observed indexes. Among those, the latent variables form two layers for the expansion of "Perception of Quality". The first layer includes four indexes (goods quality, web interface, distribution, and after-sales service) which could be further expanded into more specific indexes according to actual shopping situation 6–8.

(1) Goods quality: Reflecting consumer requirements for price and quality, etc. of goods, which mainly include quality of commodities, price of commodities, and enrichment of commodities. (2) Web interface: Reflecting consumer requirements for service of enterprise websites, which mainly include payment method, shopping safety, website accessing speed, design feature of website, personalized service, and operation convenience. (3) Distribution: Reflecting consumer requirements for distribution, which mainly include distribution method, accuracy of distribution, timeliness of distribution, and distribution fee. (4) After-sales service: Reflecting consumer requirements for after-sales service from enterprises, which mainly include convenience for goods changing and refunding, convenience of making complaints, and outcome of complaint settling.

34.4 Survey Design and Result Analysis

A survey questionnaire will be made on the basis of the above mentioned 16 indexes in Table 34.1. Research object of this questionnaire lies in the consumers of B2C E-commerce market. According to the 27th Statistical Report on Internet Development in China released by CNNIC, netizens of China for the time being are mainly young people under the age of 35 with relatively higher education

Table 34.1 Basic information of the survey sample

Category		Data	Frequency (%)
Gender	Male	193	52.02
	Female	178	47.98
Grade	Freshman	32	8.63
	Sophomore	56	15.09
	Junior	117	31.54
	Senior	153	41.24
	Graduate	13	3.50
Average expenses per month (RMB)	Under 800	5	1.35
	800–1,200	135	36.39
	1,200–1,500	126	33.96
	1,500–2,000	71	19.14
	Above 2,000	34	9.16

background and a monthly income of below RMB2000. Majority of netizens who do web consumption have desire of web shopping or experience, and university students meet these features 9. This is the main reason why we have chosen university students in Shanghai as our questionnaire object. 400 questionnaires are delivered and collected by means of e-mail, instant-messaging software and telephone interview, etc. with 371 effective response received, reaching a 92.75 % collection rate.

34.4.1 Test of Questionnaire Reliability

Reliability, namely dependability, refers to degree of consistency for result when we repeatedly measure the same object with the same method. Test of reliability could verify stability and consistency of surveyed result, i.e. whether the questionnaire could stably measure the targeted matters or variables. In practice, we adopt Cronbach α in measuring the reliability of questionnaires. In Eq. (34.1), K is the quantity of topics in the questionnaire; σ_i^2 is the variance of score for No. i Measured index; σ_{Total}^2 is the variance of total score for the measured index.

$$\alpha = \frac{K}{K-1}\left[1 - \frac{\sum_{i=1}^{k} \sigma_i^2}{\sigma_{Total}^2}\right] \tag{34.1}$$

Generally speaking, for Cronbach α of over 0.7, the reliability is relatively higher for questionnaires. Taking Eq. (34.1), we test the reliability concerning this time's survey sample, obtaining the result of 0.854, which shows that this questionnaire is of a higher reliability.

34.4.2 Basic Information of the Survey Sample

Basic information of the sample is characteristic information of consumers, specific distribution as shown in Table 34.1.

34.4.3 Factor Analysis

Factor analysis is a kind of statistic approach in dealing with multivariate data, revealing relationship among multivariable data for purpose of summarizing and integrating few factors from among the plenty of observable variables. Then, few factor variables are to be used in summarizing and explaining the originally measured information to the maximum, including establishing pretty simple concept system to reveal the essential relationship between matters.

Consumer satisfaction in B2C e-commerce market relates the four aspects (goods quality, web interface, distribution, and after-sales service) and each aspect is called a factor [6, 7]. Relationship between the 16 specific indexes and the four factors could be described as the following factor models [Eq. (34.2)].

$$X_i = \mu_i + a_{i1}f_1 + a_{i2}f_2 + a_{i3}f_3 + a_{i4}f_4 + \varepsilon_i,\ i = 1, 2, \ldots, 16 \tag{34.2}$$

In Eq. (34.2), f_1, f_2, f_3, f_4 are the common factors,respectively representing the four factors of goods quality, web interface, distribution, and after-sales service; a_{ij} is the load on factor f_j from X_i, μ_i is the mean value of X_i, ε_i is the content which could not be released by the four common factors from X_i and is also called special factor.

Characteristic root and variance contribution have been obtained on the basis of the surveyed data, getting a 80.713 % variance contribution, which is over the ideal limit of 80 %, after the common factors abstracted being rotated (Table 34.2). Refer to Table 34.3 for the rotated factor loading matrix.

34.4.4 Result Analysis

From Table 34.3, we find that there is a relatively bigger load on the measuring index of 1–3, from the common factor 1, indicating that all these three indexes are of bigger correlation, which could be considered as a single category. From the point of category they belong to, we can also see that questions reflected by these three measuring indexes should go into the same category, defining common factor 1 as quality factor. The common factor 2 all relate with shopping websites, which are defined as website factor. Similarly, the common factor 3 gives a good interpretation about logistics factor, which is defined as logistics factor. Common factor 4 is defined as service factor.

Table 34.2 Characteristic root and variance contribution

Topic No.	Initial characteristic value			Rotated square and load		
	Total	Variance (%)	Accumulate (%)	Total	Variance (%)	Accumulate (%)
1	6.422	34.450	34.450	4.422	24.740	24.740
2	3.776	20.820	55.270	2.946	21.820	46.560
3	2.764	14.276	69.546	2.783	17.276	63.837
4	2.962	11.167	80.713	2.663	16.876	80.713
5	1.971	4.903	85.616			
6	1.704	2.948	88.564			
7	1.446	2.820	91.384			
8	1.384	2.348	38.462			
9	1.339	1.470	95.202			
10	1.231	1.464	96.666			
11	1.041	1.105	97.771			
12	0.947	0.802	98.573			
13	0.773	0.654	99.227			
14	0.685	0.475	99.702			
15	0.571	0.283	99.985			
16	0.526	0.015	100.000			

Table 34.3 Rotated factor loading matrix

Measuring Index	Topic no.	Common factor 1	Common factor 2	Common factor 3	Common factor 4
		The rotated factor loading matrix			
Quality of commodities	1	0.873	0.130	0.107	0.130
Price of commodities	2	0.764	0.101	0.301	0.166
Enrichment of commodities	3	0.689	0.235	0.064	0.201
Payment method	4	0.172	0.802	0.157	0.091
Shopping safety	5	0.211	0.925	0.231	0.187
Website accessing speed	6	0.238	0.764	0.176	0.176
Design feature of website	7	0.156	0.890	0.198	0.043
Personalized service	8	0.089	0.767	0.208	0.063
Operation convenience	9	0.101	0.914	0.041	0.127
Distribution method	10	0.104	0.203	0.781	0.134
Accuracy of distribution	11	0.230	0.119	0.918	0.202
Timeliness of distribution	12	0.189	0.097	0.909	0.098
Distribution fee	13	0.090	0.276	0.870	0.289
Convenience for goods changing and refunding	14	0.259	0.214	0.257	0.843
Convenience of making complaints	15	0.172	0.134	0.134	0.789
Outcome of complaint settling	16	0.180	0.018	0.167	0.912

In B2C e-commerce market, consumer satisfaction model and measuring index system are verified and perception of the quality is the cause of consumer satisfaction which is decided by the four aspects of goods quality, web interface, distribution, and after-sales service.

34.5 Conclusions

This paper, after having analyzed the mental characteristics of consumers in B2C e-commerce market, has built up the corresponding model of consumer satisfaction which have been analyzed and verified.

Measurement of consumer satisfaction in B2C market is in favor of that B2C enterprises ensure the quality of goods, including the perfection of web site interface, distribution service and after-sales service. Enterprises could conduct consumer survey by using the relevant questionnaires, finding out the weak points, taking improvement measures and enhancing consumer satisfaction and loyalty. Scientific measurement of consumer satisfaction will also help consumers to true and dependable products or service information so that they have more satisfying products or service quality, especially the more excellent after-sales service.

We have to admit that when choosing samples, we have only taken university students in Shanghai as the survey object, without having more extensive sampling information. Our follow-up research work will focus on obtaining survey samples in a wider area for verifying the consumer satisfaction model in B2C e-commerce market, which will be made more reliable. In another hand, the model we have proposed here is only for the satisfaction influence from the "Perception of Quality" of consumers instead of studying the relationship among consumer satisfaction, consumer loyalty and consumer complaints which shall be further discussed at a later time in a separate way.

References

1. He M (2002) Internet consumption: theoretical model and behavior analysis. Heilongjiang People's Publishing House, Haerbin
2. Ranganathan C, Ganapathy S (2002) Key dimension of business-to- consumer websites. J Inf Manage 38:457–465
3. Barnes SJ, Vdgen RT (2002) An integrative approach to the assessment of e-commerce quality. J Electron Commer Res 3(3):114–127
4. vander Merwe R, Bekker J (2003) A framework and methodology for evaluating e-commerce websites. Internet Res Electron Networking Appl Policy 13(5):330–341

5. Liu J, Yu T (2010) Research on the e-commerce customer satisfaction model. Softw Eng 12:43–45
6. Torkzadeh G, Dhillon G (2002) Measuring factors that influence the success of internet commerce. Inf Syst Res 13(2):187–204
7. Li J, Sun J (2009) Empirical research on quality dimensions of e-commerce websites based on factor analysis. Sci Tech Manage Res 10:264–266

Chapter 35
Empirical Studies on Collaborative Relationship Between Enterprise Scale and IT Level

Hongwen Zhu, Qianqian Li and Chunxiao Liu

Abstract To solve the problem of how to make better the matching of enterprise scale and its IT level, we have chosen 94 enterprises with different sizes in Heilongjiang province as sample, analyzed the correlation and the synergy between enterprise scale and IT level. Our research has found out that in the case of large sample, it shows the significant positive correlation between enterprise scale and IT level, but with the decreasing of sample size, the correlation is no longer significant. At last, we have determined that the optimal order parameter between enterprise scale and IT level is 1.49.

Keywords Enterprise scale · IT level · Collaborative relationship

35.1 Introduction

The impact of information technology (IT) on enterprise's performance has gradually depend in the context of rapid development of the information society. On one hand, a large number of researches have shown that IT is fundamental to an enterprise's survival and growth and IT is a key factor differentiating successful firms from their less successful counterparts; on the other hand, enterprises have

H. Zhu (✉) · Q. Li · C. Liu
Management School of Harbin Institute of Technology, 150001 Harbin, China
e-mail: hitzhw@126.com

Q. Li
e-mail: liqianqian.hit@gmail.com

C. Liu
e-mail: liuchunxiao.hit@163.com

W. Du (ed.), *Informatics and Management Science VI*, Lecture Notes in Electrical Engineering 209, DOI: 10.1007/978-1-4471-4805-0_35,

recognized the necessity and importance of IT building. But for their own scale, enterprises don't clear how to develop IT to make better the matching of enterprise scale and IT level, so as to enhance the enterprise competitiveness. There are little studies on this issue, and lacking theoretical and empirical researches in this area.

Based on the China Enterprise Assessment Indicator System of Informatization published by National Information Evaluation Center (NIEC) and the newest enterprise scale's classification plan formulated by National Bureau of Statistics of China, we conducted a questionnaire survey for 145 enterprises of different scales from five cities (Suihua, Jixi, Yichun, Qitaihe and Mudanjiang) of Heilongjiang Province, finally selected 94 enterprises as the sample. In order to gain a clear understanding of the association between enterprise scale and IT level and to determine the optimal order parameter, we carried on an empirical analysis, described the data sources and the methodology used to address the research question. Finally, the results and the implications of the study are presented and some concluding comments offered.

35.2 Literature Review

Despite the widely existed results national and international about enterprise scale and IT level, researches on this two are rarely combined. Both of them are closely linked to the enterprise competitiveness, so we can associate them quite well through indicator system of enterprise competitiveness, and lots of studies are focused on the relationship between them and enterprise competitiveness.

Although some scholars argue that there is no discernible relationship between enterprise scale and enterprise competitiveness, most of them are agree with Professor Porter's view: enterprise scale is closely related to enterprise competitiveness, and a firm's competitive advantages are dependent on the firm's cost control (including information cost) and differentiation [1]. In Porter's research, conclusions about scale economy are: enterprises with scale economy can reduce overall cost, thus establish the cost advantage, and set up the foundation of differentiation. Although there are many factors which can affect enterprise competitiveness, scale economy is the most important one, that's the reason why so many enterprises are seeking on achieving scale economy.

Bolton and Dewatripont [2] take enterprise as an information network, so as to minimize information transfer costs and control losses. Marschak and Radner [3] argue that enterprise's advantage is making itself more specialized though the process of gathering and processing information, thereby reducing overall cost. So it can clearly be seen that information cost is a very important factor influencing enterprise scale.

Gregory L. Parsons is the first scholar who conducted a preliminary exploration on enterprise competitiveness and IT level [4]; after that, Porter [5] summarized enterprise competitive advantages which contain cost, differentiation and strategic target; then Wiseman [6] developed Porter's theory and focused on the theory of cost competitive advantage, but the difference is that Wiseman argued that differentiation is the most critical element to reflect competitive advantages. Since then, this theory became the basis for investigating enterprise IT level and its competitiveness. Some scholars carried on their theoretical researches combined enterprise IT level, product value chain with competitiveness, which proved that information technology can raise enterprise competitiveness [7]. Although lots of researches have shown that enterprise competitiveness is not completely depended on IT level, information technology is a powerful role to improve enterprise competitiveness.

35.3 Collaborative Theory, Enterprise Scale and IT Level

35.3.1 Collaborative Theory

Synergy refers to the coherent relationship between the different components of system, it's the ability of a group to outperform even its best individual member. In a technical context, its meaning is a construct or collection of different elements working together to produce results not obtainable by any of the elements alone. The value added by the system as a whole, beyond that contributed independently by the parts, is created primarily by the relationship among the parts, that is, how they are interconnected [8]. In collaborative theory, the most important concept is the order parameter: the order parameter is normally a quantity which is zero in one phase (usually above the critical point), and non-zero in the other. It characterizes the onset of order at the phase transition. The order parameter susceptibility will usually diverge approaching the critical point, when the order parameter reaches its maximum value; the system will display a macroscopic ordered and organized structure [9].

When the components within the system collaborate each other quite well, the system will be macroscopic ordered, as to enterprise, it means all kinds of resources working together smoothly, ordered, efficiently and collaboratively, and enterprise being more competitive. In the external environment, whether enterprise running orderly or not is closely connected with the degree of freedom of the entire market. When the market is highly competitive, that means a high degree of freedom of market, enterprise possessing sufficient assets and management capabilities, can better coordinate resources and operate orderly. Thereby, our research focuses on finding out a numerical relationship between enterprise scale and IT level which can lead the market freer.

Table 35.1 Classification plan of enterprise scale

Enterprise classification	Sales revenue per year (0.1 billion RMB)	Total assets (0.1 billion RMB)
Mega enterprise	>50	>50
Large enterprise	>5 and ≤50	>5 and ≤50
Medium Enterprise	>0.5 and ≤5	>0.5 and ≤5
Small enterprise	≤0.5	≤0.5

Table 35.2 Investigated enterprise scale classifications

Enterprise scale	Large enterprise	Medium enterprise	Small enterprise
Account	6	38	50
Proportion	6.38 %	40.43 %	53.19 %

35.3.2 Enterprise Scale

National Bureau of Statistics of China formulated the newest enterprise scale's classification plan in October 2010, which is summarized in Table 35.1, the new classification plan classifies enterprise scale according to two indicators: sales revenue and total assets. For example, only if an enterprise could meet the requirement of both the sales revenue per year above 5 billion and the total assets above 5 billion, it belongs to mega enterprise.

According to the new classification plan, we classified 94 enterprises in the sample; the results are shown in Table 35.2:

Among the 94 enterprises, 93.62 % are medium and small enterprises, therefore this article is concentrated on the medium and small enterprises, both of their sales revenue and total assets are under 0.5 billion RMB. In Heilongjiang Province, there are more than 1,547.8 million medium and small enterprises, accounting for over 90 % of the total enterprises and medium and small enterprises contribute a lot to the development of economy of Heilongjiang Province. In last several years, the number of the medium and small enterprises soars sharply, in addition, with the rapid development of Internet and E-Business, which makes enterprise's IT building extremely imperative.

35.3.3 Enterprise's IT Level

Enterprise's IT level is one of the key factors affecting national IT level. Since the late 1990s, National Information Evaluation Center (NIEC) began to investigate and establish the index system which is appropriate for evaluating national enterprise's IT level. The index system comprises 5 primary indexes which are strategic position, infrastructure, human resources, utility and application, information security and 21 secondary indexes, and it's a more comprehensive and systematic system.

In order to establish the index system, NIEC investigated the characters of various sectors, adopted lots of suggestions from authoritative experts and senior managers of enterprises, and combined with data analyzed results; for the purpose of setting adapt weights for each index. In addition, enterprises from different sectors, with different scales and different stages of development have been selected as the standards to ensure that the evaluation of different enterprise can be comparable [10].

35.4 Methodology

According to the index system of informatization and the data gathering, we optimized and selected 15 secondary indexes as the index system of our research. And then we divided these 15 secondary indexes to 5 primary indexes: strategic position, infrastructure, human resources, utility and application, information security, their weights are 10, 20, 15, 50, and 5 %.

According to the data collecting and formula (35.1), we marked every index, and finally calculated the scores of every enterprise's IT level.

$$V = \sum (P_i \times W_i) \tag{35.1}$$

In formula (35.1), V represents the finally score of the index with 100 as the maximum grade, represents the score of the index i, represents the weight of the index I, the descriptive statistics of 94 enterprises' IT level are given in Table 35.3:

To get the weights of the two indicators of enterprise's scale classification, we use the theory of entropy and entropy weight. The entropy of index can be defined as:

$$H_j = -\frac{1}{ln2}\sum_{i=1}^{n} f_{ij} ln f_{ij} \tag{35.2}$$

About $fij = \frac{rij}{\sum_{i=1}^{n} rij}$, when $f_{ij} = 0$. In order to make all of the indexes become positive indexes, we standardized the indexes in two situations: if the value of an index is the higher the better, we call it benefit index, we standardized it according to the formula $r_{ij} = \frac{r'_{ij} - \min(r'_{ij})}{max(r'_{ij}) - \min(r'_{ij})}$; in contrast, according to the formula $r_{ij} = \frac{max(r'_{ij}) - r'_{ij}}{max(r'_{ij}) - min(r'_{ij})}$; and then the weight of index j is:

Table 35.3 Descriptive statistics of IT level

	Num	Min	Max	Mean	SD
Variable	94	12.45	60.3	38.2274	8.56817
Valid value	94				

Table 35.4 Distribution of the 94 enterprises scale

Score	Number	Proportion (%)	Total (%)
0–200	73	77.66	77.66
200–400	12	12.77	90.43
Over 400	9	9.57	100

$$w_j = \frac{(1 - H_j)}{\sum_{j=1}^{m} (1 - H_j)} \tag{35.3}$$

Thus we can get the weight vector of the higher level index W:

$$W = (w_1, w_2, w_3, \ldots\ldots w_m) \tag{35.4}$$

The process of calculating the weights of the two indicators is as follows:

Take one million RMB as the unit to unify the original data, and then standardized them and got the feature matrix;

According to the feature matrix and the formula $H_j = -\frac{1}{ln2}\sum_{i=1}^{n} f_{ij} ln f_{ij}$, we can get the entropy of the two indicators of enterprise scale: 4.903914, 4.856856;

According to the entropy, multiplied the matrix of the original data, we finally got the weights of the sales revenue and the assets, and they are: 0.503032, 0.496968;

The distributions of the 94 enterprises scale are presented in Table 35.4:

35.5 Empirical Analyses

This section will be extended by the hypothesis raising, data analysis and the hypothesis testing, and then, will go to the conclusion.

35.5.1 Putting Forward the Hypothesis

According to the previous studies, on one hand, the enterprises should think about its profit pattern, make a finer management strategy, the present scale and development orientation should also be taken under consideration; On the other hand, the enterprise scale is also an important element to consider, which can give a rapid capital accumulation for the enterprises in growth, otherwise, neither the company could be developed in long term, with a sufficient application of IT facilities and a renewed and developed talents, nor the scale economy for enterprise can be realized. Therefore, the enterprise scale should be connected with the IT level in the enterprise development. In the collaborative development of the enterprise scale and the IT level, we imagine that it exists a certain number, an order parameter. As long as this number isn't beyond the interval of 0–2, the investment of IT construction can be guaranteed proper to the enterprise scale development, so that the enterprise can grow steadily.

Based on the above analysis, we propose the hypothesis as follows:

Hypothesis 1: there is the correlation between the enterprise scale and its IT level;
Hypothesis 2: there is the collaborative relationship between the enterprise scale and its IT level.

35.5.2 Hypothesis Testing

Hypothesis 1: we carried on a correlation analysis between enterprise scale and its IT level, coming some results: the degree of correlation between the two is 0.241, the significant is Sig.(2-tailed) = 0.019, less than 0.05, so the 94 enterprises have a positive correlation between enterprise scale and IT level. However, when we do a division of sample according to their scales, we get a much poorer significant correlation between the enterprise scale and its IT level, as the number of each sample has decreased. For example, when there are only 50 enterprises, the Pearson Correlation is 0.173, Sig. (2-tailed) is 0.23; while when the number of sample becomes 38, the Person Correlation is 0.185, and the Sig. (2-tailed) is 0.266; when the sample has only 6, the Person Correlation is 0.421, and the Sig. (2-tailed) is 0.406.

Also, when we analyze the groups of big and medium, big and small enterprises, we find out that the correlation of the enterprise scale and the IT level is much lower, that is insignificant. When the sample is totally random, the enterprise scale has a significant correlation to the IT level when the number of the sample is big enough, while the correlation is no more significant when the sample is less than 50. There will be some causal factors in the little sample, and the results of the correlation analysis will be affected, and the correlation will be much lower. The case of big sample is better and much optimistic, and the correlation will be significant. In conclusion, the enterprise scale and the IT level have a significant correlation between them.

Hypothesis 2: this section focus on the analysis of the collaborative relationship between the enterprise scale and its IT level, and the most important step is to find the order parameter of the system. In this article, the order parameter only depends on the enterprise scale and the IT level, thus, on considered only these two factors. Because of a wide range of the ratio between the enterprise scale and the IT investment in reality, we chose the reciprocal of these two factors as the order parameter. This order parameter distributes between 0 and 2, by the analysis of the Q–Q graph, we see that the scatters of the graph is well in order, but obviously not along the straight line, that means the order parameter is not a normal distribution;

when we try to apply the natural logarithmic form of the ratio, we find the data is well along the straight line, so that is nearly a normal distribution, finally this form after transformation which obeys the normal distribution is the ideal state we need.

We can confirm that, in the whole market environment, the data analysis from all the enterprises can lead to a straight line of X = Y on the enterprise scale and IT level Q–Q graph. Consequently, $x = \mu = 0$ could be the axis of symmetry of the group of natural logarithmic of the ratio of the enterprise scale and the IT level, that is the group of numbers distribute on both two sides of the axis of symmetry. According the analysis above, on can define the function of the order parameter Y: L is the IT level, S is the enterprise scale, so the Y in the market system should be represented by: $Y = ln\frac{L}{S}$, and each of the enterprise's order parameter in system obeys the normal distribution. According to the theory of the normal distribution, the theoretical maximum value of the order parameter is: $Y_{\max} = \frac{1}{\sqrt{2\pi}}$, we get $\frac{L}{S} = e^{\frac{1}{\sqrt{2\pi}}}$, approximately equal to 1.49, that means when the ratio between the IT level and the enterprise scale reaches 1.49, the order parameter becomes the maximum.

The collaborative relationship between the IT level and the enterprise scale can be explained as the follows: in the whole market environment, if we only consider the factors of enterprise scale and the IT level, that is much bigger the order parameter, much freer the macro-market, and much more fiercely the competition among the enterprises.

For example, we imagine there is an enterprise, which has firstly reached the ratio of 1.49, and other companies are not, we say that this enterprise possesses an advantage in this market; however, when the others have been well adjusted in the IT investment and the enterprise scale, these enterprises in the market will face to the fierce competition. So the objective for enterprise is to reach the value 1.49, because this number is just an ideal state, the strategy of enterprises shall try their best to approach this number.

35.6 Conclusions

The quantitative analysis of the collaborative relationship between the enterprise scale and the IT level is a project with high theoretical value and the practical significance. In our country, this kind of quantitative analysis combining these two factors almost has not appeared, because there are lots of difficulties in the collection of the data. In this article, all the data are from the government of Heilongjiang Province, the 94 enterprises are all the provincial scales. This article applies the related model and the collaborative analysis method, studies the relationships between the two variables, and gets the following conclusion:

1. The size of the sample affects the correlation degree of the enterprise scale and the IT level. When the sample is big enough, the correlation of the two factors

is very significant, which explains that it exist a correlation between the enterprise scale and the IT level; while when the sample is less than 50, this correlation is not significant any more.

2. This optimal order parameter gotten in this article is 1.49, according to which the enterprise should keep a collaborative development of its IT level and its scale. Known by this, we get a basic standard to measure the relationship of enterprise scale and its IT investment.

Although we got some empirical conclusions, there are still some problems we can't ignore. Firstly, the data in this article are all from five regions of Heilongjiang and only three sectors, so the external validity is relatively low, which probably affect the guiding significance for the whole sectors. Secondly, because of the empirical studies nature of this article, the sample of 94 enterprises are a little far from enough to do studies in this direction, although we have gotten all the data from IT level to the finance of these companies. Additionally, we have only one year's data of the IT level and the enterprise scale, for the collaborative studies, there will be a little limit. Well, the collaborative relationship between two factors is a gradual process study; a few years' data can be more trusted.

At last, we suggested that the questionnaire shall be designed comprehensively so that more detailed information can be collected; it is better get more than 3 years' data of enterprises from government so we can do time series analysis; the size of the sample should be bigger; to the enterprises from different sectors, they should be analyzed separately so that the suggestions we supplied could be more pertinent.

References

1. Porter ME (1997) Advantage competitiveness. Huaxia Press, Beijing, vol 12, pp 34–35
2. Bolton P, Dewatripont M (1994) The firm as a communication network. Q J Econ 115:809–839
3. Marschak J, Radner R (1972) The theory of teams, vol 23. Yale University Press, NY, pp 137–146
4. Hitt LM, Brynjolfsson E (1996) Productivity, profit and consumer welfare: three different measures of information technology's value. MIS Q 20(2):121–142
5. Michael E (2001) Porter Strategy and the Internet. Harvard Bus Rev 3:63–76
6. Wiseman C (1998) Strategic Information Systems. Homewood, Illinois: Irwin 8:35–41
7. Jichen J, Tongbin Z (2008) Mechanism analyze of IT promoting enterprise's competitiveness. J Dongbei Univ Finance Econ 4:62–63
8. NIEC (2010) Basic theory of enterprise informatization 5:6–12
9. Synergy (2005) Shanghai. Shanghai Translation Publishing House 23:5–13
10. NIEC (2010) Explanation about the index system of enterprise informatization. http://www.niec.org.cn/qyxxh/zbtx01.htm.2-1

Chapter 36
Study on the Effect of Government Spending on GDP Growth

Yaliu Pan

Abstract Based on the above analytical results, it was concluded that, from the auto-regression model analysis for the relationship of governmental non-tax revenue expenditure and GDP growth, the original hypotheses is supported by this empirical data analysis. The results show that governmental purchasing expenditure has a positive effect with private capital formation, and this is represented by the positive impact on long-run economic growth. However, it cannot be substantiated that government's fundamental construction funding contributes significantly to the long-run economic growth. Nonetheless, it was found that non-tax revenue expenditure has some strong positive impact on short-run economic growth, though not significantly for the long-run GDP growth. Therefore, the impact of this part of government spending on GDP growth should not be ignored.

Keywords Gross domestic product · Decomposition technique · Government expenditure

36.1 Introduction

Non-tax income is government expenses including tax expenditures, the government is an important part of finance expenditure. Together with tax revenue, expenditure non-tax income is a policy tool for government macroeconomic adjustments of a country's internal performance. In most countries, however, only a small non-tax income by spending part of their whole government spending, so

Y. Pan (✉)
Economics and Management School of Wuhan University, Wuhan, Hubei, China
e-mail: yaliupan@126.com

W. Du (ed.), *Informatics and Management Science VI*, Lecture Notes in Electrical Engineering 209, DOI: 10.1007/978-1-4471-4805-0_36,

scholars don't pay attention to this problem. Most of the research has influence on economic growth of government spending is tax expenditure. For example, Ram [1] with a cross panel data made a national measurement of regression analysis, to make sure that there is a positive impact on economic growth. At the same time, Barro [2] to study the influence of the Angle of government expenditure for production and consumption of productive public expenditure and concluded that there is a positive long-term effect on economic growth, and not only short-term effect productive public expenditure. Their research suggests that, pure public private investment products have complementary effect, can be used as a whole economic benefit, and provided a positive influence on long-term growth. Mixed public goods, on the other hand, tend to replace part of private investment, through the short-term effect on economic growth substitution effect.

They follow the thought of the two trends. A thought in the study followed Barro government public spending are divided into productive spending and unproductive spending lead to different effects on the economic growth [3]. Another followed a completely different thinking, looking at the Ram size, the relationship between the government spending and GDP [4], however, these Chinese scholars still not including non-tax revenue part of the analysis [5], when they are spending and non-tax revenue should have only a positive influence on the short-term economic growth [6]. Another followed a completely different thinking, looking at the Ram size, the relationship between the government spending and GDP. However, these Chinese scholars still not including non-tax revenue part of the analysis, when they do their research work. This is a problem, because the Chinese government income structure is quite different, the western countries. The Chinese government tax system is made of a number of non-tax revenue about a third of the entire government tax revenue. Not including non-tax income spending, they survey is only part of the problem.

36.2 The Structure of Chinese Government's Revenue and Expenditure

The Chinese government is by five levels: the central government, provincial (autonomous regions), city and county, township. There were five financial levels, to collect and distribute funds. From planning and management institution, the fund management into two categories: tax and non-tax revenue. Although there are some change and income in the economic system collection system over the years, the basic structure is very stable consistent.

Tax income distribution is to point to by the government income received from the manager and the owner of the policy. This is mainly budget income controlled by the government. Infrastructure income, which has been called extra financial capital, it is to point to not be financial capital management within the budget. However, since 2005, the income was budget control. The government non-tax revenue comes from regular policy implementation, rental income from the government owned properties, sales of capital goods, etc. On average, the fund is about 30–40 % of the total government revenue according to the national bureau of statistics (annual data).

Accordingly, the government's expenditure system can be divided into government spending and tax government non-tax revenue expenditure. The former is the central government's main source of income, can be divided into government spending and investment infrastructure. The latter is one of the main sources of local fiscal revenue and expenditure, special use at the local government spending. Because of the different function that two financial variables, their effects in the growth of GDP should is quite different.

According to the public finance theory, its main function is to provide the pure public products of the central government, local government's main function is to provide mixed public goods, so tax expenditures should have only a positive influence on the long-term economic growth, and non-tax revenue expenditures should have only a positive influence on the short-term economic growth.

36.3 Methodological Issues

The basic idea is to observe decomposition technique based on the analysis of the vibration of the innovation. That is, when the system balance, this will be for the future of the interference and breaks the balance for some reason. If the system is stable, it will restore balance in a certain time ever seen deflection. If we anticipate Y_t according to $\{y_{t-1}, y_{t\equiv 2}, \ldots\}$, we can prove that the conditional expectation is the best linear anticipation which can be devoted by Yt*, that is, Yt* = E (Yt/Yt−1, Yt−2...), well, if the information included in $u_t = Y_t - Y_t^* = Y_t - E(Y_t/Y_{t-1}, Y_{t-2}\ldots)$ didn't refer to Y_t in $\{y_{t-1}, y_{t\equiv 2}, \ldots\}$, we call it innovation. To all its random process $\{u_t\}$ meets: $E\{u_t\} = 0$, $E\{u_t \quad u_t'\} = \sigma^2$, be limited and $E(u_t/X_{t-1}) = 0$. Here (X_{t-1}) is the information (or data) before the t moment; and $\{u_t\}$ is the innovation corresponding to (X_{t-1}). We can express this process like this:

$$Y_t = C + \Theta Y_{t-1} + e_t \tag{36.1}$$

$$(I - \Theta L)Yt = C + et \tag{36.2}$$

$$\begin{aligned} Y_t &= (I - \Theta L) - 1C + (I - \Theta L) - 1et \\ Y_t &= \mu + \sum_{i=0}^{\infty} \Theta^i L^i e_t = \mu + \sum_{i=0}^{\infty} \Theta^i e_{t-1} \end{aligned} \tag{36.3}$$

If we use a 2-dimension-process-hypothesis, to express what we mentioned above, the thought can be shown as:

$$\begin{bmatrix} y_{1t} \\ y_{2t} \end{bmatrix} = \begin{bmatrix} \mu_1 \\ \mu_2 \end{bmatrix} + \sum_{i=0}^{\infty} \begin{bmatrix} \theta_{11}(i) & \theta_{12}(i) \\ \theta_{21}(i) & \theta_{22}(i) \end{bmatrix} \begin{bmatrix} e_{1t-1} \\ e_{2t-1} \end{bmatrix} \tag{36.4}$$

Here $\Theta = \begin{bmatrix} \theta_{11}(i) & \theta_{12}(i) \\ \theta_{21}(i) & \theta_{22}(i) \end{bmatrix}$, and component $\theta_{jk}(i)$ is the multiplier. $\theta_{11}(0)$ Explains how many units y_t will change at t moment after one unit time if it can change one unit at t moment. $\theta_{11}(1)$ Denotes how many units y will change after t + 1 moment while e_t can change one unit at t moment. Here long term multiplier: $\sum_{i=0}^{\infty} \theta_{jk}(i)$ shows the total effect to the kth component which influenced by the jth interference. The impulse functions are:

$$\theta_{11}(i), \theta_{12}(i), \theta_{21}(i), \theta_{22}(i), i = 0, 1, \ldots K$$

Using VAR model to predicts h steps; the variance matrix of h steps equals

$$\begin{aligned} &\sum(h) = \Sigma_u + M_1 \sum\nolimits_u M_1 + \cdots + M_{h-1} \sum\nolimits_u M_{h-1} \\ &M_0 = I, \quad M_1 = \sum_{i=1}^{\min(p,i)} \Theta_j M_{i-j}, \quad i = 1, 2, \ldots \end{aligned} \tag{36.5}$$

For

$$\Sigma_u = ADA' = AD^{1/2}D^{1/2}A' = P^{-1}P'^{-1} \tag{36.6}$$

Using

$$\begin{aligned} &AD^{1/2} = P^1 = \Psi_0, \Psi_t = M_t AD^{1/2} = M_t P^1 \\ &\sum(h) = P^{-1}P'^{-1} + M_t P^{-1}P'^{-1}M_t' + \cdots + M_{h-1}P^{-1}P'^{-1}M_{h-1}' \\ &\sum(h) = \Psi_0\Psi_0' + \cdots + \Psi_{h-1}\Psi_{h-1}' \\ &\mathrm{Var(yi,\ r(h))} = \sum_{j=1}^{n} \Psi_{ij,0}^2 + \cdots + \sum_{j=1}^{n} \Psi_{ij,h-1}^2 \end{aligned} \tag{36.7}$$

Reorganized we can get

$$\mathrm{Var(yi,\ r(h))} = \sum_{k=0}^{h-1} \Psi_{i1,k}^2 + \cdots + \sum_{k=0}^{h-1} \Psi_{in,k}^2 \tag{36.8}$$

$\sum_{k=0}^{h-1} \Psi_{i1,k}^2$ is the effect of Jth innovation the accumulation of n innovations is the whole predict variance, and the proportion between these two parts is decomposition

$$\frac{\sum_{k=0}^{h-1} \Psi_{i1,k}^2}{Var(yi, r(h))}, \quad j = 1, 2, \ldots, n \tag{36.9}$$

36.4 Variables and Model

36.4.1 Definition of Variables and Determination of Variable Units

Definitions of Variables: Due to the emphasis of analyzing the effect of the non-tax government spending (GNR), this study chooses the data of extra budgetary fund expenditure in National Bureau of Statistics of China (Yearly Data) as proxy variable of governmental non-tax, the data of government consumption in the same yearly data as proxy variable of government purchase expenses (GP) and the data of other internal budgetary expenditure as proxy variable of budgeted infrastructure investment (GFI). And, we define that, within the government budget, GFI is for infrastructure, large public water works, large electric power plants, highway systems, and etc. These investments have the characteristics of little or no returns in short run and they do not attract private funds. However, as we mentioned above, these funds should have a long run effect for GDP growth. Moreover, non-tax revenue expenditure is the major finance tool for the local government and its main function is to provide local public goods. This public spending will crowd out private funding because non-tax revenue expenditure mainly is used for city's public maintenance, local road construction, county bridges construction, of which could have been done by private enterprises. So these funds should only have a short run effect for GDP growth.

Data used and their unit root test.: The time period used in this study is 1978–2006. All data are obtained from the Annual of China Statistics and are deflated by constant price of 1990. Most of the time series were not stabilized, so the Augmented Dickey-Fuller (ADF) method to test and adjust the unit root data was used. The test results showed that the four (4) time series logarithm were not stable. However, they were stable after taking one lagged difference. This indicates that these variables in logarithm belong to integration I (1) process (see Table 36.1).

This study uses the Vector Auto Regression (VAR) model to test the long run correlation effect among variables. The assumption was made that there was a co-integration among selected variables. This allows us, under a non-structural condition, to evaluate and examine the mutual impact among all variables on their combined effects dynamically. Due to the small sample (1978–2006) used in this study, there is limited degree of freedom in choosing variables and time lagged periods variables. The time lag is two (2) periods (See Table 36.2). The model is:

Table 36.1 Unit root text (ADF)

Variable	Lagged	Test statistics	Critical value	P (%)
LGDP	2	−3.98	−2.99	5
LGP	2	−4.84	−3.73	1
LGFI	3	−3.47	−2.99	5
LGNR	2	−4897	−3.75	1

Table 36.2 Johansson co integration text

Counteract Series LGDP LGFI LGP LGNR Lag: 1–2					
Eigenvalue	Likelihood	5 %	1 %	Hypothesized	
0.80453	87.4698	62.99	70.05	None **	
0.54592	33.4436	42.44	48.45	At most 1	
0.31262	12.7732	25.32	30.45	At most 2	
0.20212	5.38348	12.25	16.26	At most 3	
Vector error correction					
LGDP	LGP	LGFI	LGNR		C
548.000000	−1.31721	−0.04338	−2.25996	0.30633	10.44
	(0.3743)	(0.0217)	(0.5816)	(0.1096)	
L-Likehood	158.332				

$$LY_t = C + \sum_{i=1}^{p} B_p LY_{t-i} + U_t \tag{36.10}$$

$$LY_t = \begin{bmatrix} LGDP_t \\ LGNR_t \\ LGP_t \\ LGFI_t \end{bmatrix} \quad C = \begin{bmatrix} C_1 \\ C_2 \\ C_3 \\ C_4 \end{bmatrix} \quad B = \begin{bmatrix} B_{11} & \cdots & B_{14} \\ \vdots & \ddots & \vdots \\ B_{41} & \cdots & B_{44} \end{bmatrix} \tag{36.11}$$

36.4.2 Variance Decomposition Analysis

The long-term growth problem is actually a growth ratio problem. So, to analyze the effect of governmental non-tax revenue spending on GDP's long-run growth is to test the variable's ability in explaining the residual errors in various models of estimating GDP's long-run growth. The stronger of the variable's ability to explain the long-run residual error, the more important this variable will be. Based on this scenario, we use decomposition analysis to test government non-tax revenue spending effect on GDP growth.

From the decomposition analysis results of estimating the GDP growth, in the short term, variables GP, GNR, GFI, were not able to consistently explain the major portion of residual errors in estimating the GDP growth together it could explain only 7 % of residual errors. In the long run, from the third (3) period, the ability of variable GC had its ability in explaining the effect of residual error increased, to 18 % (See Table 36.3). However, the variables GNR and GFI did not have strong ability in explaining the residual effect even after the third period. They accounted for only 3.3 and 1.5 % respectively. On the other hand, GDP explained 75 % of its own variations.

Table 36.3 Test results

R^2	0.997698	0.995643	0.956791	0.992147
Adj R^2	0.996324	0.992367	0.946260	0.984568
F	1542.38	457.34	55.3565	206.66

The result indicates that government purchase expenditure, among other government expenditure components, has the largest impact on long-run GDP growth. Although the non-tax revenue spending has less explanatory effect except in the long-run, its ability in explaining the residual effect of GDP growth is larger than other budgetary internal spending. Therefore, it is more important than other budgetary internal spending variables. In addition, the result points out the fact that if we ignore the impact of non-tax revenue expenditure in relation to Chinese governmental public spending on economic growth, it would be both a theoretical and practical oversight. The reason is that our analysis indicates that non-tax revenue expenditure does have powerful pulling power on short-run economic growth and has stronger explanatory power on China's GDP long-run growth than other internal budget expenditure.

36.5 Empirical Analysis

36.5.1 Estimated

The auto-regression model test shows that there is a better effect in the internal structural among variables and conforms to the basic expected results. The effect of government non-tax revenue expenditure on GDP growth is larger than the effect of government investment on basic constructions. The coefficient of GNR spending is 0.059, that of GP spending is 0.053, and that of GFI spending is –0.01. Outlined below are the estimated regression coefficients of each variable in logarithm in the auto regression equation:

$$Y_t = \begin{bmatrix} 1.78 & 1.97 & -0.21 & 2.29 \\ -0.01 & 0.28 & 0.06 & -0.17 \\ 0.053 & -0.45 & 1.11 & -1.02 \\ 0.059 & -0.35 & -0.01 & 0.11 \end{bmatrix} Y_{t-1} + \begin{bmatrix} -0.42 & -0.95 & -0.01 & 0.54 \\ 0.05 & 0.32 & 0.03 & -0.15 \\ 0.16 & 0.98 & -0.02 & 0.99 \\ -0.01 & -3.34 & 0.14 & -5.6 \end{bmatrix} + \begin{bmatrix} 0.78 \\ 0.12 \\ -3.46 \\ -5.64 \end{bmatrix} \quad (36.12)$$

Table 36.4 Variance decomposition for GDP1

	S.E.	LGDP1	LGP1	LGFI1	LGNR1
1	0.016079	100.0000	0.000000	0.000000	0.000000
2	0.020214	93.05691	3.271848	0.254268	3.416970
3	0.021024	79.01879	15.10386	1.530006	4.347337
4	0.023042	75.92285	18.23348	1.559060	4.284612
5	0.024091	74.31188	17.92206	1.428658	4.337403
6	0.024500	76.76674	17.51748	1.427980	4.287802
7	0.024754	75.34338	18.49312	1.719534	4.443967
8	0.024902	74.73762	19.27331	1.776082	4.212982
9	0.025068	74.64577	19.40726	1.789716	4.187253
10	0.025176	74.81805	19.24952	1.749219	4.183207

Ordering LGDP1 LGP1 LGFI1 LGNR1

36.5.2 Variance Decomposition Analysis

From the decomposition analysis results of estimating the GDP growth, in the short term, variables GP, GNR, GFI, were not able to consistently explain the major portion of residual errors in estimating the GDP growth together it could explain only 7 % of residual errors. In the long run, from the third (3) period, the ability of variable GC had its ability in explaining the effect of residual error increased, to 18 % (See Table 36.4). However, the variables GNR and GFI did not have strong ability in explaining the residual effect even after the third period. They accounted for only 3.3 and 1.5 % respectively. On the other hand, GDP explained 75 % of its own variations.

The results show that the government buy spending, government spending in other components, has the largest affect long-term to GDP growth. Although non-tax revenue less consumption long-term effect in addition to explain, it can explain the remnants of GDP growth in the greater influence than other financial internal consumption. Therefore, more important is more than any other financial internal consumption variables. In addition, the result pointed out that if we ignore the fact that the influence of income and infrastructure spending public expenditure on economic growth, the government may are both a theoretical and practical supervision. Reason is that our analysis shows that the non-tax revenue also has strong traction the short-term economic growth and spending with strong extent in China's gross domestic product (GDP) contribution to the growth of the long-term than other internal budget expenditure.

36.6 Conclusions

As long as the government non-tax income for spending, a person should be how to analyze the effect, this part of the revenue to GDP growth is really a huge challenge. Is mainly due to this reason, this paper will try to correct the problem through analyzing the government non-tax revenue growth of spending in effect the growth of national economy use variance decomposition technique.

References

1. Ram R (1987) Wagner's hypothesis in time-series and cross section perspectives. Public Finance 41(3):393–414
2. Barro RJ (2010) Government spending in a simple model of endogenous growth. J Polit Econ 98:103–125
3. Davoodi H, Zou H (2009) Fiscal decentralization and economic growth: a cross-country study. J Urban Econ 43:244–257
4. Acemoglu D, Johnson S, Robinson J, Yared P (2008) Income and democracy. Am Econ Rev 98:808–842
5. Torsten P, Tabellini G (2007) Democracy and development: the devil in the details. Am Econ Rev 96:319–324
6. Xiao-Li W (2010) The dynamic analysis of effects of public expenditure on short-term GDP growth—the experimental analysis based on VAR model. Stat Res 5:26–32

Chapter 37
Income of Employee Model Based on Hierarchical Theory

Guoqing Tao

Abstract On the basis of the traditional human capital pricing theory, this article believes that the employee's income should depend on their internal and external value. Therefore, this article takes EVA as the measurement basis of income of employee in the enterprise and designs a model of total income of the whole employees based on human capital and physical capital. Furthermore, this article calculates the contribution rate of every employee according to Rosen's hierarchical theory, and then design the income of employee model based on hierarchical theory.

Keywords Hierarchy · Human capital · EVA

37.1 Introduction

Theodore W. Schultz not only promulgated the concept of human capital but also was a twentieth century pioneer in human capital theory. Schultz defined human capital by example: "Much of what we call consumption constitutes investment in human capital. Direct expenditures on education, health, and internal migration to take advantage of better job opportunities are clear examples. Earnings foregone by mature students attending school and by workers acquiring on-the-job training are equally clear examples [1]. Yet, nowhere do these enter our national accounts. The use of leisure time to improve skills and knowledge is widespread and it too is unrecorded. In these and similar ways the quality of human effort can be greatly

G. Tao (✉)
Economics and Management School of Wuhan University, Wuhan, China
e-mail: guoqingtao2012@126.com

W. Du (ed.), *Informatics and Management Science VI*, Lecture Notes in Electrical Engineering 209, DOI: 10.1007/978-1-4471-4805-0_37,

improved and its productivity enhanced. I shall contend that such investments in human capital accounts for most of the impressive rise in real earnings per worker." Modem economic theory believes that capital can be divided by physical capital and human capital and the corporation cannot survival and develop without anyone. With the development of economics, human capital has played more and more important role and replaced the leading position of physical capital. Despite that thousands of researches have been done on human capital evaluation by economists, it is still challenging job for people to identify the value of human capital and the income of employees because it is more complicated, uncertain, and dynamic than physical capital. This article wills analysis the evaluation of human capital and point out income of employee model based on hierarchical theory.

37.2 Total Income Model for the Whole Employees

37.2.1 An Overview of Human Capital Evaluation Theory

All current researches on human capital evaluation focus on two methods: one is cost method which evaluates human capital based on investment in it, including Theodore W. Schultz and Garys Becker's historical cost method and replacement cost method. The advantage of cost method is simple and practicable and all costs can be estimated exactly based on enterprise's statistical and historical date; Another is income method which evaluates human capital from the perspective of its foreseeable income, including Baruch Lev and Aba Schwarts's future income discount method, Eric G. Flamholtz's random income method and economic value method and Hekemian, Jones and R.H. Hermanson also contributed a lot in this field. The strong point of income method is that it can evaluate human capital dynamically and overcome cost method's shortcomings that it evaluates human capital grounded in statical historical data. Current human capital evaluation models' main idea: working out the reasonable distribution rate for income between shareholders and the employees according to the value of human capital which is evaluated by cost method or income method. While we find there are some problems in practice: the cost method is too simple and does not consider differences between employees and impact of uncertain factors on employees and enterprise in the future; due to lacking some necessary data, the income method can not evaluate future income of employees exactly, and it is unreasonable and unscientific to identify value of human capital only based on its future income. It is manifest that scientific evaluation of human capital is a prerequisite for health development of enterprise. If we cannot evaluate human capital correctly, it will cause irrational distribution between employees and shareholders and inequality of income among employees.

37.2.2 Total Income Model for the Whole Employees

This article holds the opinion that income of employees should comprises two components: one is internal value of employees, which includes both innate abilities (physical, intellectual and psychological capacities that individuals possess at the time of birth) and acquired abilities (knowledge, skills and experience); the other is external value of employees, which is the employee's contribution to enterprise. In other words, it is enterprise's income which is created by employees with the help of physical capital of shareholders. As is known to all, the enterprise is the profit organization which is made up of physical capital (shareholders) and human capital (employees). Consequently, enterprise is a prerequisite for employee to create income. It is impossible for the employees to make profit without enterprise, and vice versa. Furthermore, the same employee may create different income for the shareholders in different enterprises. For instance, employees are likely to create more income in a leading promising enterprise than in one with difficulties. All in all, the external value of human capital depends on both employees and enterprise.

In sum, the total income for the whole employees in enterprise is:

$$I = W + V \tag{37.1}$$

Function (37.1): I is total income of the whole employees; W is fixed wage, which depends on their internal value; V is floating wage, which depends on their external value.

This article posits that the job market is efficient. Therefore, the number of employees and enterprises is big enough so that everyone must be a price taker. If employee's income is too low, they will try to find better paid jobs; on the other hand, if the enterprise pay employee more than they should deserve, it is easy for shareholders to find others replacing them. Consequently, I is the real value of employees (including internal and external value). The fixed wage W should be the necessary cost of maintenance of employees. We can estimate it according to their internal value (talent, knowledge, skill, experience, and so on). The floating wage V is their external value (the enterprise's income which should belong to the employees). Apparently, the key of function (37.1) is how to identify the floating wage V.

37.2.3 Identifying Floating Wage V of the Whole Employees

37.2.3.1 Value Measurement Basis for Floating Wage V

Due to the fact that floating wage is the income which belongs to employees and is created by them with the help of physical capital of shareholders, this article takes EVA (abbreviation for Economic Value Added) as the value measurement basis for it. We believe that it is scientific and practicable way to take EVA as measurement basis. Firstly, we cannot evaluate employees only according to its historical cost or future income. We should evaluate them from the perspective of shareholders and then take their contribution to the enterprise (shareholders) as measurement basis. From the perspective of shareholders, net income (after deduction of all costs), or EVA is the contribution created by employees. Therefore, EVA is the increased value created by employees with the help of physical capital or the whole employees' contribution to the enterprise. Secondly, EVA is an objective indicator of enterprise profitability and it is easy to get from enterprise's financial statements in the past years. It is the Stern Stewart consulting organization that registered EVA as a trade name in the 1990s and defined EVA: the enterprise's net operating profit after taxes, minus a cost of capital charge for the investment or capital employed in the enterprise. The function is:

$$EVA = NOPAT - WACC \times TC \tag{37.2}$$

Function (37.2): *NOPAT* is profit after taxation; *WACC* is weighted-average cost of capital; *TC* is total cost (equity capital and debt capital).

37.2.3.2 Identifying the Proportion of EVA for the Whole Employees

As has been discussed above, it is the human capital and physical capital to create EVA and it is impossible to do without anyone. Therefore, EVA belongs to both. Then, the proportion for employees is the key of the model. The shareholders should identify the proportion according to scientific and reasonable basis. And then, we can calculate the floating wage V for the whole employees. The function is:

$$I = W + V = W + EVA \times H = W + (NOPAT - WACC \times TC) \times H \tag{37.3}$$

Function (37.3): H is the proportion of EVA for the whole employees.

This article utilizes the famous Cobb–Douglas functional form of production functions to identify the proportion H. The function is:

$$Q = AL^{\alpha}K^{\beta} \tag{37.4}$$

Function (37.4): Q is total productivity (the monetary value of all goods produced in a year); L is labor input; K is capital input; A is total factor productivity; α and β are the output elasticities of labor and capital, respectively. Cobb–Douglas production function indicates that there is a relationship between total productivity, human capital and physical capital. Meanwhile, the bigger α is, the more contribution human capital (employees) makes; the bigger β is, the more contribution the physical capital (shareholders) does. Taking knowledge-intensive industry as an example α is bigger than its counterpart in manufacturing industry and employees make greater contribution to total productivity, vice versa.

Both side of the function are transformed to logarithms:

$$InQ = InA + \alpha InL + \beta InK \tag{37.5}$$

The function (37.5) provides a log-linear form which is convenient and commonly used in econometric analyses. We assume that the scale of the enterprise increase or decrease slightly in the short run. So the total factor productivity A is a constant and the function (37.5) is:

$$InQ = \alpha InL + \beta InK \tag{37.6}$$

If we let $Q = InQ, L = InL, K = InK$, then function (37.6) is:

$$Q = \alpha L + \beta K \tag{37.7}$$

The function (37.7) provides a linear relationship between total productivity Q, labor input L and capital input K. Then proportion H is:

$$H = \frac{\alpha L}{\alpha L + \beta K} \tag{37.8}$$

We can get the historical data of labor input and capital input from enterprise's financial statements and put them into function (37.8):

$$\begin{aligned} Y_t &= \alpha L_t + \beta K_t \\ Y_{t+1} &= \alpha L_{t+1} + \beta K_{t+1} \\ &\ldots\ldots \\ Y_{t+n} &= \alpha L_{t+n} + \beta K_{t+n} \end{aligned}$$

If we solve the equations we will get α and β in different years. Then we can predict current α, β and proportion H by means of mathematical model. Thus the floating wage of the whole employees is:

$$V = (NOPAT - WACC \times TC) \times \frac{\alpha L}{\alpha L + \beta K} \tag{37.9}$$

Then the total wage model for the whole employees:

$$I = W + (NOPAT - WACC \times TC) \times \frac{\alpha L}{\alpha L + \beta K} \tag{37.10}$$

According to function (37.10), we can identify the total wage for the whole employees I. Then we should divide I between the whole employees to determine individual's wage. As has been discussed above, the fixed wage for everyone depends on employee's internal value and the floating wage depends on their external value. The idea of this article is try to calculate the contribution which individuals make to the enterprise and determine the floating wage of every employee according to everyone's contribution, and then work out the total income for every employee. As enterprise's total employees are composed of different levels and kinds of person, this article utilize hierarchical theory proposed by Sherwin Rosen to identify individual's contribution to the enterprise.

37.3 The Income of Employee Model Based on Hierarchical Theory

37.3.1 Hierarchical Model

Firstly, Rosen assumes that the enterprise is a pyramid organization: the CEO is at the top of the pyramid and the production workers are at the bottom. Rosen believes that an organization can either exist as an independent square and circles or merge with a higher organization. For instance, a two-level organization can either be an independent company or merge into the enterprise with three ranks or more. Secondly, Rosen argues that managers are intermediate between higher managers and production works. Consequently, the activities of CEO will have indirect effects that successively filter through all lower levels and be magnified by them [2]. Figure 37.1 is a classic three-level organization of hierarchical theory. R_j is the rank; R_1 is production work; R_2 is middle manager; R_3 is CEO.

Let q_{nm} denotes the productivity of worker nm in the production activity. It is worker's endowed skill and varies from worker to worker. Let r_m represents the skill of managers and also varies from person to person. Let t_{nm} denotes the amount of monitoring or supervisory time that manager r_m allocates to workers nm. Then the product attributable to r_m controlling nm is:

$$x_{nm} = g(r)f(r_n t_{nm}, q_{nm}) \tag{37.11}$$

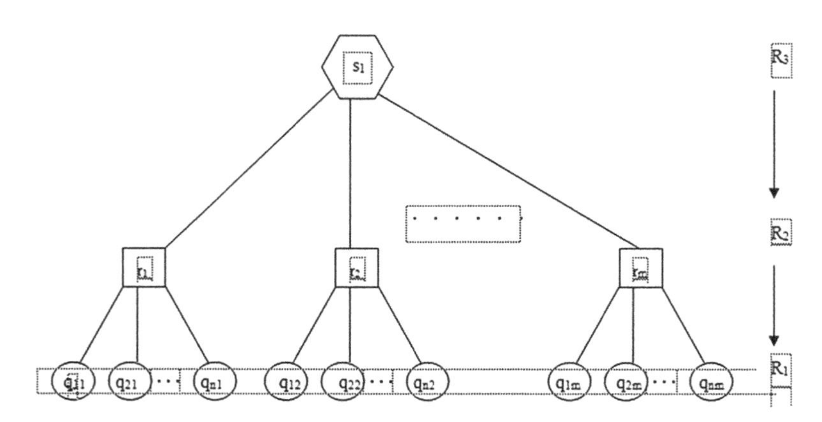

Fig. 37.1 A three-level organization of hierarchical theory

The total output of an R_2 activity is the sum of outputs of all production workers controlled:

$$X_m = \sum_n x_{nm} = g(r) \sum_n f(r_n t_{nm}, q_{nm}) \tag{37.12}$$

Then the total output of the organization in Fig. 37.1 is:

$$Y = \sum_m y_m = \sum_m G(s) F(sv_m, X_m) \tag{37.13}$$

y_m is the output of a person whose talent is s in R_3 and manages output X_m of manager in $R_2, v_m, G(s), sv$ and $F(sv_m, X_m)$ have the same interpretation for function (37.11) and (37.12).

37.3.2 The Income of Employee Model Based on Hierarchical Theory

To simplify this model, let z denotes the skill of employees. Assuming that the output which is cooperated by manager in R_j and his subordinates is $(kz_j + mz_{j-1})$, the output of R_2 and R_3 is $s(kz_1 + mz_0)$ and $s^2[k(kz_2 + mz_1) + mz_0]$ respectively. Then the total productivity of an n-level enterprise is [3]:

$$Y_n = A^n z_n + \sum_{j=1}^{n-1} s^{n-j} A^j m z_j + s^n m z_0 \tag{37.14}$$

Function (37.14): m and k are positive dependent variables and $m + k < 1$; s is management range; Y_n is the total productivity; n is the number of enterprise level and $n \geq 3, A = sk > 1$.

As we can see from the function above: the output of CEO is $A^n z$; the productivity of employee in rank j is $A^j mz$; the output of production worker is mz. It is manifest that the total productivity Y_n is a linear function of the total employees' skills in the enterprise. Furthermore, the skill of CEO is enlarged $(sk)^n z$ times and the manager's skill is enlarged $(sk)^j mz$ times. Therefore, the contribution rate of the whole employees in different rank is:

$$h_j = \begin{cases} \dfrac{A^n z_n}{Y_n} & (j = n) \\ \dfrac{s^{n-j} A^j m z_j}{Y_n} & (0 < j < n) \\ \dfrac{s^n m z_0}{Y_n} & (j = 0) \end{cases} \tag{37.15}$$

There is a positive correlation between output of the enterprise and EVA, so we can take h_j as distribution rate of total floating wage for employees in different ranks in the enterprise. Then the total floating wage for the whole employees in the same rank is:

$$I_j = h_j \times EVA \times H, \quad (0 \leq j \leq n) \tag{37.16}$$

Function (37.16): I_j is the total floating wage for the whole employees in rank j. Therefore, the total wage for the whole employees in the same rank:

$$V_j = W_j + \begin{cases} \frac{A^n z_n}{Y_n} \times EVA \times H & (j = n) \\ \frac{s^{n-j} A^j m z_j}{Y_n} \times EVA \times H & (0 < j < n) \\ \frac{s^n m z_0}{Y_n} \times EVA \times H & (j = 0) \end{cases} \tag{37.17}$$

Function (37.17): W_j is total fixed wage for the whole employees in rank j; V_j is total wage for the whole employees in rank j.

We can divide the total floating wage for the employees in the same rank between every employee according to personal capacity, and management range, assessment and other indicators. The function is:

$$V_{ji} = W_{ji} + V_j \times I_{ji} \tag{37.18}$$

Function (37.18), V_{ji} represent the total wage for any employee; W_{ji} is employee's fixed wage; I_{ji} is distribution rate of V_j for everyone and varies from person to person.

It is manifest that the employee has to increase EVA of the enterprise and distribution rate I_{ji} so that they can earn more income. For the production workers or managers in the low rank, they can try to get promotion so that their effects on EVA can be magnified by their subordinates. To the managers in the same rank, they should try to allocate monitoring or supervisory time scientifically to maximize their contribution to EVA, and then they will maximize their floating wage.

37.4 Conclusion

This article attempts to study the employee's income from the perspective of their contribution to the enterprise. We divide the total wage of employee into fixed wage and floating wage: the fixed wage W_{ji} depends on internal value of employee (physical, intellectual and psychological capacities that individuals possess at the time of birth); the floating wage I_{ji} depends on external value of employee (employee's contribution to enterprise). However, we should make it clear that this model is only applicable to enterprises which have simple hierarchy in nature and where there were no distinct differences of contribution among employees in the same rank. For complex hierarchical enterprises, we will research it in the further study.

References

1. Schultz TW (1961) Investment in human capital, the American economic review. Am Econ Assoc 51(1):1–17
2. Rosen S (1982) Authority control, and the distribution of earnings, the RAND corporation. Bell J Econ 13(2):311–323
3. Coase RH, Hart O, Joseph E (1999) Stiglitz, Lifengsheng (interpreter), contract economics. The Publishing House of Economic Science, Peking, vol 12, pp 223–230

Chapter 38
On Tourism Development Modes of Intangible Cultural Heritage

Jishu Shao and Wei Zhang

Abstract How to conserve and develop the intangible cultural heritage is one of the hot topics in both tourism industry and academic research area. In the past, argument of conflict relationship between tourism and intangible cultural heritage has prevailed, however, more and more researches now question it. Based on the analysis of values and characteristics of intangible cultural heritage, this paper considers that the relationship can be beneficial to both with the appropriate and modest development, meanwhile, this paper summarizes seven tourism development modes, and at last it gives some advice on integrative adoption of those development modes after the thorough comparison of the seven modes.

Keywords Tourism · Intangible cultural heritage · Development mode

38.1 Introduction

At first sight, intangible cultural heritage (ICH) is naturally against, at least irrelevant to, tourism, because ICH usually focuses on the provision and conservation of resources, while tourism is primarily related to exploitation and development of those in order to get more commercial gain. A number of people have suggested that tourism and cultural heritage are incompatible [1] (Boniface 1998; Jacobs and Gale 1994; Jansen-Verbeke 1998), and that because of this

J. Shao (✉)
Economics and Management School of Wuhan University, Wuhan, China
e-mail: jishushao20@126.com

W. Zhang
Business School of Sias International University, Zhengzhou, China

W. Du (ed.), *Informatics and Management Science VI*, Lecture Notes in Electrical Engineering 209, DOI: 10.1007/978-1-4471-4805-0_38,

incompatibility, a conflict relationship is inevitable. The cultural heritage sector argues that cultural values are compromised for commercial gain [2].

According to Bob McKercher (2005), Tourism and Cultural heritage are neither natural allies nor natural enemies [3] the type of relationship that emerges between these sectors at an asset specific level depends on the level of maturity, knowledge and good will each brings to the relationship (Bob, Pamela and Hilary 2005). In other words, if the process is proper and appropriate, the relationship between tourism and ICH would go well. This paper explores the appropriate development modes of ICH without any harm to the relationship between tourism and ICH.

38.2 Literature Review

38.2.1 Definition and Classification of ICH

In October 2003, at its 32nd session, UNESCO declares an announcement—Convention for the safeguarding of the Intangible Cultural Heritage, in which the term intangible cultural heritage was officially given a definition for the first time. The "intangible cultural heritage" means the practices, representations, expressions, knowledge, skills—as well as the instruments, objects, artifacts and cultural spaces associated therewith—that communities, groups and, in some cases, individuals recognize as part of their cultural heritage.

At the same time the announcement classified the intangible cultural heritage into five categories:

1. Oral traditions and expressions, including language as a vehicle of the intangible cultural heritage;
2. Performing arts;
3. Social practices, rituals and festive events;
4. Knowledge and practices concerning nature and the universe;
5. Traditional craftsmanship.

38.2.2 The Values and Characteristics of ICH

The value and characteristics of ICH represent the features, the core and basis of ICH resources; six categories can be summarized as follows:

38.2.2.1 Historical Value

Intangible cultural heritage has specific historical features, reflects the production development level, social structure and way of life, the relationship among people, moral practices. This information can help us to understand the social and economic relations, so as to restore the true purpose of history.

38.2.2.2 Cultural Value

Intangible cultural heritage contains the rich cultural resources, living cultural fossil. The cultural value has three characteristics: the first is its uniqueness, intangible cultural heritage is a reflection of the nation's unique cultural gene, which cannot be repeated; the second is the cumulative, it refers the accumulation of culture of the nation in the long historical development process; the third is the diversity, which is the rich content and the innovation and development of ICH.

38.2.2.3 Spiritual Value

It refers to the national consciousness and national spirit which is deeply hidden in the cultural heritage itself, accumulated in the long-term production and life practice, the positive upward and cohesive. It contains the development experience, survival wisdom, ethical values, group consciousness and group spirit, which is the essence and core of intangible cultural heritage.

38.2.2.4 Scientific Value

Intangible cultural heritage itself contains a lot of scientific factors and components, with high scientific research value, providing us with precious historical materials, such as the Chinese Han and other minority traditional medicine with important scientific value.

38.2.2.5 Aesthetic Value

Intangible cultural heritage shows a nation's life style, aesthetic taste and artistic creativity; like many traditional arts bring us spiritual or emotional aesthetic appeal in image, color, style and art, philosophy and religion, even today it is still worth appreciation and research.

38.2.2.6 Zeitgeist Value

It includes educational value and economic value. Because the intangible cultural heritage contains a large number of various disciplines of knowledge and skills, which constitute the important content of educational activities; economic value refers to the reasonable development of ICH, turning into the realistic economic benefits on the premise of the protection of ICH.

38.3 The Connection Between Resources of ICH and Tourism

From the values and characteristics of ICH mentioned above we can find that the tourism industry and ICH have very close immanent connection. Some of the ICH resources can be transformed into tourism products, which can be used as one kind of unique tourism attraction, promoting the development of tourism, as well as the important means of protecting and inheriting the ICH.

First of all, the historical value, cultural value, spiritual value and aesthetic value of ICH provide rich high quality tourism resources for tourism development. Actually many excellent tourism routes and products are extracted from the value of ICH.

Secondly, the uniqueness of ICH may become rare tourism resources. These rare tourism resources greatly satisfy the tourists' curiosity, so we can take full advantage of the uniqueness of ICH, which can be transformed into high quality tourism products.

Finally, the diversity of ICH provides the foundation for the diversification of tourism products. At present the ICH is divided into five categories, different types of ICH, due to its different values and characteristics can be transformed into different tourism products, which greatly enrich the contents and forms of tourism products, thus meet various demands of tourists with different cultural backgrounds and different age.

In short, the ICH is an important part of cultural heritage, is the witness of our history and the important carrier of national culture, and contains the unique spiritual values of Chinese nation, ways of thinking, imagination and cultural consciousness, which reflect the vitality and creativity of Chinese nation. Tourism industry has to balance the relationship between ICH and tourism, actively takes advantage of the positive effect of tourism on ICH, and avoids the negative effect, so that the development of tourism can get on well with the conservation of ICH.

38.4 The Tourism Development Modes of ICH

38.4.1 Based on the Value and Characteristics of ICH, the Following Tourism Development Modes are Strongly Suggested

The first is the museum mode of ICH. The museum mode of ICH is the collection and display of intangible cultural heritage of a country, region or nation, is also an important method to preserve the ICH via records or videos. Its main characteristic is to display the ICH in solid form. From a practical point of view, this mode can display various ICH at the same time and same place, but this form looks some boring to tourists.

The second is the theme park mode of ICH. It is a kind of theme scenic spot centering on the ICH along with other entertainment in order to satisfy various needs of tourist, with strong participation and knowledge within this area. The folk culture theme park holds the very great proportion among this kind theme park; folk village belongs to the concrete form with condensed native nature. This theme park mode's main features include a prominent theme with ICH, comprehensive entertainment projects, high degree of participation, but the investment period will last long, which would take more risk.

The third is the tourism festival mode of ICH. The festival is the important characteristics of national culture, is the cultural symbol with high local characteristics. This festival mode may display the ICH through the Folklore or Temple Fairs. One is the folk festival tourism, which would transform the traditional folk festivals into tourism activities with the combination of sightseeing and participation. Another kind is the folklore activity tourism, which would hold special cultural tourism activities centering on the traditional folklore activities. While whatever the festive mode, both are seasonal activities.

The fourth is the staged performance mode of ICH. The mode brings the ICH into the stage and exhibited to the tourists in the form of performance. This type of performance not only increases the efficiency, but also aligns with the features of short retention time of tourists. Meanwhile, the performance eases the tourists and satisfies the care-free mood because of its aestheticism. But the shortcoming of the performance is the form is single, simple and more commercial sometimes.

The fifth is the ecological preservation zones mode of ICH. This kind of the model compensates the shortage of static museum from the perspective of participation experience and usually in the form of ethnic culture village, known as ecological museum. This mode is mainly designed for the village which has rich ICH resources and relatively intact ancient styles. The villages rely on the village unique architecture, folk customs, folk arts and crafts to develop the cultural heritage tourism.

The sixth mode is called handicraft production mode. Folk arts and craft is the one that the people make manually or semi-automatically in order to meet the needs for the life and aesthetic requirements. Due to the differences among the custom, geography, and tasty, the folk arts and craft has different characteristics. Those handcrafts and folk arts are a complex collection of items made with various materials and intended for utilitarian, decorative or other purposes, the whole process can be shown in front of tourists, that can easily stimulate the tourist interest and desire to buy those tourism products, some tourists even imitate the process to DIY the products with strong feeling of traditional arts.

The seventh is the tourism merchandise development mode of ICH. This mode may transform some ICH resources into different tourism merchandises in various forms like tourist souvenirs, travel supplies, food, books and videos. Those tourism merchandises can be sold in tourism scenic areas, which would increase the added value of tourism as show in Table 38.1.

38.4.2 Assessment and Advice Out the Seven Tourism Development Modes

For the above seven tourism development modes of ICH, considering the features of tourism, through comprehensive analysis, this paper gives the following suggestions:

First, Integrated development including the combination of several development modes of ICH, or the comprehensive use of ICH tourism and traditional tourism should be adopted during the tourism development. Each of the seven kinds of development modes has its own advantages and disadvantages, one mode as a tourist route appears single, and so the combination of several development modes in one tourist route can maximize the advantages of those modes; Meanwhile the comprehensive use of ICH tourism and traditional tourism routes can enrich the contents and the forms of tourism and improve the satisfaction of tourists.

Table 38.1 Comparison of seven modes

Mode	Advantages	Disadvantages
Museum mode	Well protection of ICH	The form looks boring to tourists
Theme park mode	Prominent theme; comprehensive entertainment	The investment period will last long
Festival mode	The atmosphere is strongly traditional and warm	Are seasonal activities
Staged performance mode	Efficiency; satisfies the care-free mood of tourists	The form is single, simple and more commercial sometimes
Ecological preservation zones mode	Concentration of ICH	Only suitable for some villages with rich ICH
Handicraft production mode	High participation and involvement	Last long time for tourists sometimes
Tourism merchandise development mode	Mass production and sales	Some ICH can't be transformed to merchandise

Second, all the tourism development modes of ICH except the museum mode can put in more tourist experience. Survey from market shows that current tourism market demands more tourist experience; more tourist experience can make tourists have the actual perception about the quality of tourism product or service, satisfy the willingness of tourists participation, fully arouse the enthusiasm of tourists.

Third, during the development of ICH tourism, some of the latest technologies can be adopted, which would make the tourism products show both the massive historical sense of ICH and contemporary sense of the times. Continuous development and innovation in tourism products would make ICH maintain enduring life and satisfy the constant and unceasing needs of tourists, meanwhile the those tourism products, with the help of latest technologies, can not only meet the needs of tourists, but also can improve the added value of heritage tourism.

38.5 Conclusion

"Priority of protection, rescue first, reasonable use, inheritance and development" is now the basic policy for the protection of ICH in China. On the premise of the conservation of ICH, reasonable exploitation of the intangible culture heritage resources and adoption of appropriate development mode will effectively not only enrich the tourist items and stimulate the tourists' interest, but also inherit and develop the intangible cultural heritage, thus the tourism industry will have a broad development space.

References

1. Berry S (1994) Conservation, capacity and cash flows-tourism and historic building management. In: Seaton AV (ed) Tourism: State of the art, Wiley, Chichester, vol 12. pp 712–718
2. Urry J (1990) The tourist gaze. SAGE Publications, London
3. McKercher B, Ho PSY, Du H (2005) Relationship between tourism and cultural heritage management: evidence from Hong Kong. Tourism Manage 26:539–548

Chapter 39
Five-Force Analysis and Market Strategies of Budget Hotels in Small and Medium-Sized Cities

Li Pan

Abstract Budget hotels, as an emerging hotel model, were started and attained a rapid development in China in the last ten years, and have proven to be important part of China's hotel industry. As competitions in the hotel industry become increasingly fiercer, budget hotel brands launch a fight for seizing more market share. Therefore, it is necessary for Xuchang Jinjiang INN as a famous hotel brand to be distinguished from the hotel product strategies of the developed economic areas in the development environment of the small and medium-sized tourism cities, so as to adapt to the characteristics of regional market.

Keywords Jinjiang INN · Budget hotels · Xuchang · Small and medium-sized cities · Hotel operating management

39.1 Introduction

Budget hotels are also called as limited-service hotels, which are targeted at medium and low end consumers, and also can meet the lodging needs of leisure travel and business markets relying on clean hygienic conditions, secured facilities, high-quality services and economic prices.

L. Pan (✉)
Cultural Tourism Resource Planning and Development Institute, Xuchang University, Xuchang, Henan 461000, China
e-mail: lipan189@126.com

W. Du (ed.), *Informatics and Management Science VI*, Lecture Notes in Electrical Engineering 209, DOI: 10.1007/978-1-4471-4805-0_39,

39.2 Industrial Backgrounds for Budget Hotels Started in Small and Medium-Sized Cities

39.2.1 Fast Development of Budget Hotels at Domestic Market

By the end of 2009, there were 303 budget hotels chain brands totally in China. In the first half of 2011, the turnover of China's budget hotels market exceeded 10 billion RMB, attaining an increase of 20 % in contrast to the same period of last year. In the third quarter of 2011, 1,523 new budget hotels started their business totally in China, and 144,095 new guest rooms were increased totally, attaining an increase of 29.75 and 26.48 % respectively in contrast to the end of 2010 [1].

39.2.2 Accelerated Chain Stores and Market Share Increase of Well-known Budget Hotels

Budget hotels feature huge market potential, low investment, high return, short cycle and so on. Jinjiang INN, home INN, 7-day INN, and Hanting INN, as the main power of the domestic budget hotels industry, make an expansion with an amazing speed all over the country. The turnover of the major four budget hotels brands has reached 8.12 billion RMB, accounting for 45 % of the market share At the same time, American "Super 8" and IBIS of French Accor group have successively entered the domestic market for seizing market share [1].

39.2.3 Risk Aversion and Steady Growth of Hotel Groups

Under the situation that market and resources at large cities are divided up, the impact from the financial crisis made the hotel industry worse and worse. The impacts in large and medium cities are especially more serious at that time. The number of traveling people was sharply reduced, and simultaneously the hotel occupancy rate declined overall. Therefore, the strategy of paying attention to the secondary/third class cities, entering small and medium-sized cities and following an overall balance and stable expansion of group proved to be the new ideas for the development of budget hotels.

39.3 Market Selection of Jinjiang INN to Enter Xuchang

39.3.1 Unique Local Cultural Tourism Resources

Xuchang is a city that owns a profound historical culture. It was the place where the Xia Dynasty was established and the culture of the "Three Kingdoms" was originated. The number of the independent tourism resources in Xuchang has reached 1,125, including two 4A-class scenic spots and five 3A-class scenic spots [2]. In addition, the "Three Kingdoms" culture, Jun-porcelain culture, and flowers culture, which are well known all over the country, are of great value for researches.

39.3.2 Great Efforts Made by the Government of Xuchang to Promoting Tourism Development

The tourism plan of Xuchang during the period of the "12th five-year plan" is creating three major tourism brands, which are Ancient Cao-wei Capital Wisdom Traveling, livable flower city leisure traveling, and Shenhou Old Town experience traveling [3]. In November 2011, on the signature ceremony for "about establishing a memorandum for close cooperation mechanism of departments and city" in Xuchang, the development goal of creating Xuchang to be one of the important tourism destinations and distribution center in Henan province and even central China was proposed.

39.3.3 Rapidly Developed Tourism Economy

Only from January to February 2011, there were 47 newly-contracted projects in Xuchang, reaching a total investment of 11.79 billion RMB, including a foreign investment of 8.752 billion RMB. The major projects signed last year are being implemented smoothly; 42 major investment projects, each of which is valued more than 500 million RMB, have been performed completely according to the terms of contract Only in 2011, there was one new 3A-class scenic spot, and ten A-class scenic spots established completely in Xuchang. Among these A-class scenic spots, two are two 4A-class scenic spots, five 3A-class scenic spots, and five national industrial and agricultural scenic spots. In the mean time, the planning scheme for seven tourism towns with Henan's tourism characteristics has been complete as well. At present, there are 83 tourism enterprises in Xuchang, and the numbers of the direct and indirect employees in the tourism industry reached 10,000 and 50,000, respectively [4]. In recent years, the rapid development of tourism has been attained in Xuchang, and simultaneously the tourist reception population and tourism income are substantially increased (see Table 39.1).

Table 39.1 Tourist reception (person-time) and tourism income in Xuchang

Years	Number of accepted tourists (unit: 10,000 person-time)	Tourism income (unit: Billion RMB)
2007	257.53	13.2
2008	365.0	18.0
2009	419.0	23.0
2010	606.0	31.7
2011	696.9	38.04

Data source: statistical material of Xuchang government

39.3.4 Good Location Advantage

Xuchang is located in the center of Henan province, and thus is very superior in the geographical position. It is adjacent to Zhengzhou (capital of Henan province), surrounded by Kaifeng, Pingdingshan and Luohe, and forms the "four-wing and one-core" development pattern of urban cluster development with Xinxiang, Luoyang and Kaifeng in central China. In the mean time, the convenient traffic conditions of Xuchang can make it easy to share tourism resources with Luoyang, Kaifeng, Zhengzhou, etc.

39.4 Five-Force Analysis on Industrial Environment of Xuchang Jinjiang INN

Five-force analysis was proposed by Michael Porter in the early 1980s. In Michael Porter's five forces analysis, a great number of different factors are gathered in a simple model, so as to make an analysis on the most fundamental competition situation of an industry and effectively analyze the competition environment of customers. The different combinations of five forces will ultimately play an effect on the changes of profit-making potentials of an industry.

39.4.1 Existing Competitors from the Same Industry

The prices of high-star hotels such as Fugang Hotel, Hongbao Hotel and Sandinghua Hotel give a reflection to the grades and qualities of hotel. For this reason, it is impossible that great price reduction is put into implementation. Social hotels and guest houses, although their service qualities are not high, will not impose challenges to Xujiang Jinjiang INN from the perspectives of the customers' fixed consumption level and the hotel consumer groups.

39.4.2 Threats from Potential Entrants

Both entry and exit barriers for budget hotels are low, and this feature will attract a great number of investors to enter the market segment. On the one hand, along with the rapid economic development of Xuchang and the continuous deepening of the political system reform, increasingly more hotels and guest houses will be further improved, so as to enter the budget hotel industry. On the other hand, the hotel chain groups from foreign countries have began to race to control market share in China.

39.4.3 Threats from Alternative products

The threats from alternative products on budget hotel enterprises mainly source from three aspects: (1) the cost of alternative products is lower under the premise of meeting customer demands; (2) the cost for customers to turn to alternative products is lower; (3) alternative products are cheaper than budget hotel products. At present, the threats from alternative products of Jinjiang INN are still in existence, such as the emergence of bath center "Daliangtaosha". In addition, some small social hotels and family hotels attract most consumers by relying on their low prices.

39.4.4 Bargaining Power of Consumers

Today's consumers unceasingly put forward higher and requirements on the prices, environment, and traffic, security, sanitation and comfort level of hotels. They desire to receive high-quality products or services with a lower price, making their bargaining power enhanced. Therefore, what hotel is face up with are increasingly more professional customer sources and intermediaries. In the mean time, the bargaining power of consumer market will get stronger and stronger. All these will impose a very important effect on the competition structure and profit-making of Xuchang Jinjiang INN.

39.4.5 Bargaining Power of Suppliers

The bargaining power of the suppliers as very important providers of enterprise's raw materials, energy, equipments and other service can directly play an effect on the profitability of the budget hotels. Because a chain management model is applied in Xuchang Jinjiang Inn, the raw materials, equipments and energy are purchased uniformly. Therefore, the threats from suppliers are weak.

39.5 Market Strategies for Xuchang Jinjiang Inn Hotel Products

39.5.1 Selecting Advantageous Media as Marketing Tool in Combination of Xuchang Media Audiences Habits

The marketing of Xuchang Jinjiang INN should be basically oriented at market demands, and also the best hotel business model should be created and designed with the most acute, the most advanced, and the most unique senses. A variety of ways can be used for the marketing of Xuchang Jinjiang INN. For example, in the early business days of Xuchang Jinjiang INN, advertising was made on Xuchang morning paper, so as to attract the mass people to have dinners in Heart-to-Heart Restaurant of the hotel. In the mean time, it can hold thanks meetings for new and old customers, kindly treat local government and residents, establish a good relationship with mass media, participate in some public benefit activities, and set up a good image.

39.5.2 Innovating the Way of Thinking, Generating a Differentiation Competition, and Experiencing Xuchang Regional Culture

The key to realizing the differentiation lies in creating the unique hotel product sales points in combination with Xuchang regional culture. The contents of the differentiation are diversified. For example, as long as the overall tone, product contents, brand features, enterprise culture, marketing channels, and service models of hotel can promote consumers to produce a different feeling, such a differentiation can become the unique sales points.

39.5.3 Strengthening Restaurant Functions According to the Characteristics of Xuchang Market

Although the tourism industry of Xuchang has attained a very rapid development under the supports of municipal government, it is still in a disadvantage position in terms of the potential market demands and the people's consumption level. There are two major reasons for this situation: (1) the development of economy is relatively lagging-behind in Xuchang at present; (2) the appeal of tourism resources in Xuchang is not so strong although the tourism industry has attained an outstanding achievement. Therefore, if restaurant business can be strengthened, the incomes of Xuchang Jinjiang Inn can be greatly increased. For this reason, it is necessary for Xuchang Jinjiang Inn to make an enhancement to its restaurant functions and comprehensive strength.

39.5.4 Paying More Attention to Family Love, Sincerity and Convenience

To win in the competition, it is necessary for Xuchang Jinjiang Inn to pay more attention to family love, sincerity and convenience in the process of serving customers. Specifically, the professional trainings can be provided for the hotel service personnel so as to make them focus on the customer core demand (i.e. have a good sleep every night). Also, it is necessary to make a difference between customers' articles for everyday use, release shopping and weather forecast information to customers' mobile phones for the sake of businessmen. Also, special columns can be provided on the home page of hotel's website for releasing all kinds of information about the places customers will check in.

References

1. Cheng D, Wu Y (2010) Study on the tourism development of Xuchang. J Xuchang Univ 12(2):131–132
2. Pan L (2011) Discussion about the development of the tourism integration of Xuchang. Three kingdoms culture. J Heibei Tourism Vocational College 11(6):141–143
3. Cheng Y (2007) Study on the development strategies of budget hotels in China. Shandong Univ 29:487–495
4. Jinjing GAO (2011) Analysis on the budget hotels management and development in cities. J Nanning Polytech 16(2):77–78

Chapter 40
Study on Development of Lushan Hot Spring Health Tourism Products

Fenglian Zou and Zhibin Hu

Abstract Through the development of more than 60 years, the hot spring tourism industry in China will confront with a situation of "innovative competition, standard development, and secondary upgrade". In this paper, from the perspective of product integrated marketing theory, by taking Lushan (Xingzi) hot spring as an example, the authors carry out an analysis on some problems in the development of Xingzi hot spring tourism products, propose the framework and model of the integrated development of hot spring health tourism products, and also provide some suggestions for the design of Xingzi hot spring health tourism routes.

Keywords Product integration · Health preservation by hot spring · Route design

40.1 Introduction

Since the arrival of the twentyfirst century, the vacation tour age that is oriented at health and leisure has soundlessly arrived in the world [1]. As a famous Chinese saying goes, hot spring water, like slippery jade, is very proper for you to enjoy for a while.

Lushan hot spring is located in the Hot Spring Town of Xingzi County of JiuJiang that is in the south of Lushan. Hot spring water, which is colorless, odorless, and clear to transparent, features a low degree of mineralization, weak alkali, and high fluorine, and is medical mineral water including silicon, fluorine, hydrogen sulfide, and radon and can be comparable with the hot spring of Pyrenees

F. Zou (✉) · Z. Hu
Xinyu University, Xinyu 338004, Jiangxi, China
e-mail: fenglianz791@126.com

W. Du (ed.), *Informatics and Management Science VI*, Lecture Notes in Electrical Engineering 209, DOI: 10.1007/978-1-4471-4805-0_40,

Mountains that are very well-known in Europe [2]. As the national strategy of vigorously developing Poyang Lake Ecological Economy Development Zone is advocated, it is necessary for Xingzi Hot Spring to seize the opportunity of the state to provide supports to ecological health tourism, so as to make an enhancement to Xingzi hot spring tourism industry.

40.2 Current Situation and Problems of Xingzi Hot Spring Tourism Development

40.2.1 Current Situation of Xingzi Hot Spring Tourism Development

In 2003, a tourism development strategy was established again in Xingzi County. That is, the tourism development idea of making every effort to constructing the "leisure tourism corridor centering on hundreds of meters from Lushan" based on the development of hot spring. In 2010, Hot spring Resort was named as one of the provincial resort areas. The tourist reception volume reached 5.32 million person-times in 2010 from 450,000 person-times in 2002; the comprehensive income from tourism increased to 2.009 billion RMB in 2010 from 17.4 million RMB in 2002; the proportion of the hot spring tourism had accounted for more than 60 % of the tourism industry [3].

40.2.2 Main Problems of Hot Spring Tourism Products Development

40.2.2.1 Identical Product Development Projects and Increasing Competitions

After the development of eight years, a large number of hot spring enterprises have been concentrated in Xingzi Hot Spring, and therefore the industrial clustering effect has been obvious to all [4]. However, the majority of these enterprises are based on the development model of leisure and vacation oriented hot spring resorts, and their hot spring tourism products mainly include three types of projects: (1) health care and preservation project, in which relevant items such as massage, manipulation and beauty care are provided for consumers on the basis of the traditional hot spring baths; (2) leisure and entertainment project, in which health and fitness entertainment sports facilities are constructed on the basis of large outdoor hot spring, including golf, tennis court, badminton courts, and water entertainment facilities, etc.; (3) meeting-based training project. Because of the homogeneity competitions, the price for an entrance ticket of Xingzi Hot Spring is stable in 120 RMB at present and the price for a standard room is specified at 400 RMB. Therefore, hot spring enterprises can prevent falling in the traps of price competition only if product transformation and upgrading of products can be realized.

40.2.2.2 Backward Product Development Idea and Insufficient Market Segments

At present, the development idea of Xingzi Hot Spring tourism products is still in a backward state. On the one hand, high importance is only attached to the mass market; ignorance is given to the market segments; the understanding of the market demands of tourists and the consumption characteristics of different groups of tourists is in an insufficient state. On the other hand, Xingzi Hot Spring tourism products tend to the "noble group" and are oriented at high-end market in the sources of tourists. For example, reception facilities of four-star standards or above and golf course are used as supporting projects, which need great investment and may meet with a great risk.

40.2.2.3 Insufficiently-Exploited Local Cultural Intension in Product Development and Absence of Brand Personality

The developers of hot spring only stay in the concept of hot spring, but give a cold shoulder to the exploration on the intensions of local culture and hot spring. There are plagiarisms in the aspects of landscapes, architecture styles and hot spring products projects designs, which are in shortage of the characteristics of local culture, thus giving rise to the absence of local cultural intension in tourism products and the indistinct personalities of the local tourism enterprises.

40.3 Development Framework of Hot Spring Health Tourism Products from the Perspective of Product Integrated Marketing

The integration of tourism products refer to a process, in which new types of tourism products and benefits will be produced through resource allocation, optimized combination and product transformation and upgrading on the basis of the existing tourism products. In accordance with the overall concept of tourism products and the characteristics of dividing the functions of hot spring tourism destinations, hot spring tourism products can be classified into five levels for development as shown in Fig. 40.1 below, and a different integrated marketing strategies can be put into implementation in a different product cycle. At present,

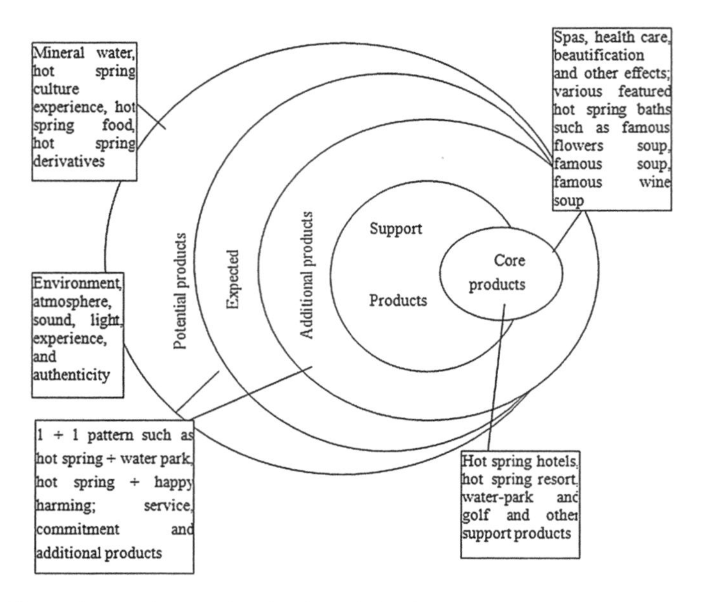

Fig. 40.1 Hot spring tourism product levels and development framework

the integration development of Xingzi hot spring tourism products is still in the stage o growth [5]. According to above theoretical analysis, the main integration types of Xingzi hot spring tourism can be concluded from the aspects in the following.

40.3.1 Brand Integration

The proposal of the brand concept "Great Lushan" is beneficial not only to breaking the shortcomings of the division in administrative regions and straightening out the management system problems accumulated in a long term, but also to integrating all sorts of tourism resources of the entire Lushan, for the purposes of realizing an unified management and also making an enhancement to the overall strength of Lushan in an all-round way. A great number of scenic spots such as Hot Spring, Guanyin Bridge and Xiufeng of Xingzi County are located in the south of Lushan. Xingzi Hot Spring has broken through the fixed development model of the traditional hot spring tourism products by relying on the integration of the local tourism resources, making hot spring bath culture and modern life integrated as a dynamic whole. Therefore, a new brand can be packaged and promoted with the name of "Lushan Hot Spring Resort".

40.3.2 Tourist Routes Integration

Xingzi, as a powerful tourism county in Jiangxi province, is benefited from the great development of hot spring project. Lushan tourist attractions are mainly distributed in two places (south Lanshan and top Lanshan) of Xingzi County. Under the concept of "Great Lushan", the tourism products of the two places can be recommended to the mutual tourists, and simultaneously tourist routes can be designed again, so as to open up a one-route tour combining both south Lanshan and top Lanshan attractions. In addition, there are ten tourist attractions with their own unique characteristics, such as Xiufeng, Guanyin Bridge, Taiyi resort, Bailu Academy, Ancient Nankang City, hot spring, the Peach Garden, and Poyang Lake. All these tourist attractions can be integrated with tour routes of other tourist attractions, for the purpose of creating a series of high-quality tourist routes that are not only complementary in advantages and also own special charming.

40.3.3 Industrial Integration

As part of the tourism resources, it is necessary for hot spring to coexist with other types of resources. In the process of developing Xingzi hot spring tourism, it is necessary to give consideration to the resources situation of the local places in an all-round way, and also develop the hot spring resources with other surrounding tourism resources, industrial resources, forest resources and fishery resources together by centering at hot spring tourism industry, so as to promote the traditional industrial developments of the local places and also make integrated marketing realized. Thus, supported food accommodation, transportation, traveling, shopping and entertainment can be available for improving the chains of hot spring tourism industry, and also a comprehensive hot spring tourism industry system that integrates sightseeing, shopping, vacationing, business, conference, exhibition, and culture, sports, entertainment, food and other functions together can be created for tourists.

40.4 Suggestions for the Integrated Development of Xingzi Hot Spring Tourism Health Products

In the following, some suggestions for the integrated development of Xingzi hot spring tourism health products are proposed by the authors respectively from two aspects (i.e. business development model and product route design).

40.4.1 Development Models of Xingzi Hot Spring Tourism Health Products

40.4.1.1 Integrating Hot Spring and Convention Tour Resorts

This tourism product is primarily oriented at business conference groups with high-end consumption demands, and features the health recuperation of hot spring. More specifically, relevant groups are attracted to hold different types of conference tourism activities through the construction of the indoor and outdoor bath rooms of various modern bathing facilities, different scales of conference facilities, and supporting entertainment and expanding training projects, etc. This kind of development model needs to have sufficient tourist sources through the relationship marketing so as to maintain the business management of project and make a profit, and also puts forward a relatively higher requirement on the marketing ability of travel agencies, etc.

40.4.1.2 Integrating Hot spring and Sightseeing Entertainment Resorts

This tourism product is primarily oriented at urban residents with mass consumption demands. Therefore, it is necessary to design various combinations of bathing sites under the guidance of sightseeing entertainment of the mass people, such as honeymoon site, wedding anniversary site, family union site, friend union site, and private date site and so on.

40.4.1.3 Integrating Hot spring and Ecological Agriculture

The development model of integrating ecological agriculture and hot spring is actually to integrate hot spring, ecology, and agriculture together. In such model, not only the sustainability of ecology can be realized, but also a new rural hot spring tourism business model can be created. In the mean time, geothermal resources are brought into full utilization relying on this model. Therefore, through the applications of modern facilities and high-tech means, an agricultural tourist attraction area, which is oriented at hot spring and high-tech agriculture sightseeing, can be established, so as to attract the city white-collars who are eager to temporarily get away from urban areas and return to the nature.

40.4.2 Route Design of Xingzi Hot Spring Tourism Health Products

40.4.2.1 Religious Bath and Hot Spring Health Tour: Lushan–Xiufeng–Xingzi County–Taiyi Village–Hot Spring Town

The religious activities in Lushan have always been flourishing for a long time. In history, temples were constructed in almost every peak of Lushan. For example, "Five Major Temples" (Guizong, Xiufeng, Wanshan, Xixian, and And Haihui) were well-known in history. In addition, there are many famous Taoist temples such as Jianji, Taiping, Qingxu, Guangfu, Taiyi and Xianren Hole. In the modern society, people's mental burdens are very heavy. For this reason, the tourism products integrating religious bath and hot spring featuring morality cultivations can be developed, so as to allow the urban residences to alleviate physical and mental fatigues in famous mountains by temporarily getting away from busy urban life.

40.4.2.2 Self-Driving Hot Spring Health Tour: Ancient Nankang City–Luoxing Island–Xiufeng–The Peach Garden–Hot Spring Town

Xingzi county is located between the world famous Lushan and the Poyang Lake that is the largest freshwater lake in China. The route integrates famous mountains, great lake, ancient city, hot springs and pastoral scenery together. It can not only give a vivid reflection to the diversity of tourism resources of Xingzi County, but also help richen the sightseeing experience and process of self-driving tour.

40.4.2.3 Visiting-oriented Soup Bathing Health Tour: Xingzi County–Bailu Academy–Xiufeng–The Peach Garden–Hot Spring Town

South Lushan was not only the hometown of famous ancient idyllic poet Tao Yuanming, but also a place where a large number of literary giants, great thinkers, politicians and strategists, medical giants, and Buddhism and Taoism hermits served for people as pure and upright officer, gave lectures, had a travel, wrote articles and lived in seclusion, and thus created many famous poems and interesting anecdotes for latter generations.

40.4.2.4 Happy-farmer Oriented Soup Bathing Health Tour: Poyang Lake–Hot Spring Town and Donggushan Forestry Centre–Hot Spring Town and Happy Farmhouse

Hot springs town is located in the mid-west of Xingzi County, with Poyang Lake in the east and Lushan in the north. In surrounding areas, there are planting, fishing and other agricultural production forms. This route mainly gives expression to the feature that allows tourists to collect agricultural products on maintains, catch fish at lake and cultivate lands at fields. In such a way, tourists can select bathing for relaxing and recuperating after the end of a day's farming experience life, for the purpose of keeping health.

Acknowledgments This paper is supported by Humanities and Social Science Research Project for Colleges of Jiangxi Province (No. GL1167).

References

1. Yanping W, Qiaoyun S (2007) Introduction to studies of hot spring tourism, vol 3(5). China's Tourism Publishing House, Beijing, pp 59–64
2. Shan L, Jie Z (2009) Exploration on the development model of hot spring tourism products. J Chongqing Coll Educ 11(6):45–48
3. Yanping D (2011) Study on the strategies for the hot spring tourism industry of Jiangxi Province. J Yichun Univ 21(6):78–82
4. Kuanrong Y, ChunMei F, Fei C (2009) Discussion on the development of hot spring tourism under the background of products transformation. Jiangsu Commercial Forum 3:74–78
5. Yu Y, Yanping W (2009) The Progress on hot springtourism of China in 30 years. Tourism Forum 16(5):24–29

Chapter 41
Study on Ability of Independent Innovation of Medium-Sized Industrial Enterprises

Shihui Jiang and Keyan Jiao

Abstract The statistical analysis of large and medium industrial enterprises in Henan Province, ability to support innovation environment, awareness of independent innovation, independent innovation capability, R&D capability, manufacturing capacity and effectiveness of six aspects of the innovation dynamic analysis, to fully reveal the past four years, Henan province and medium-sized industrial enterprises in independent innovation and development that exist in different strengths and weaknesses. The results showed that: Henan Province medium-sized industrial enterprises the level of each index of the total increase in varying degrees, ranking basically upstream position in the country, however, most of the relatively Liangjun below the national level, especially in the manufacturing capability and innovation the relative amount of benefits also declined.

Keywords Medium-sized industrial enterprises · Innovation · Development

41.1 Introduction

At present, the general lack of innovation in our province's business products, therefore, built on the basis of technological superiority of the market rules, although many companies expanding in size, but the lack of core technology, lost a lot of accrued benefits. Therefore, how to enhance the capability of independent innovation of enterprises has become the province's business community and

S. Jiang (✉) · K. Jiao
Department of Information Engineering, Henan College of Finance and Taxation, Zhengzhou, China
e-mail: shihuij23o@126.com

W. Du (ed.), *Informatics and Management Science VI*, Lecture Notes in Electrical Engineering 209, DOI: 10.1007/978-1-4471-4805-0_41,

academia to discuss a hot issue, but also caused the government at all levels of particular concern. This research will be fully based on the medium-sized industrial enterprises in Henan Province capability of independent innovation in recent years the development of analysis to identify the development of independent innovation capacity strengths and weaknesses, and ultimately be able to effectively promote a proposed Henan medium-sized industrial enterprises in independent innovation capability of the countermeasures for the relevant government departments to develop and improve the cultivation of independent innovation capacity of policy to provide a reference, and enterprise development to improve independent innovation capability development plan to provide ideas.

41.2 Medium-Sized Industrial Enterprises in the Development of Independent Innovation Capability

Business innovation is the company through their own efforts and explore ways of generating breakthroughs in core technologies and concepts, break technical difficulties, and on this basis, the ability to promote and rely on their own innovative follow-up session to complete, to market new products or the first to use new technology, to complete the commercialization of scientific and technological achievements in order to gain competitive advantage and monopoly profits to achieve the desired objectives [1]. It is developed and finally the market began a systematic process, any link in the chain of independent innovation capability of enterprises has a certain impact.

In this paper, based on the study of international experts, along Market started in research and development and finally achieved, the idea of corporate profits from the environment to support innovation ability, awareness of independent innovation, independent innovation capability, R&D capability, manufacturing capabilities and benefits of innovation in terms of Henan Province, six medium-sized industrial enterprises in the past four years, the development trend of independent innovation capability for detailed analysis [2, 3].

41.2.1 The Ability to Support Innovation Environment

Which, due to changes in indicators Statistical Yearbook, 2006–2008, the social aspects of the three years funded mainly refers to loans from financial institutions, while in 2009 addition to the social aspects of funding [4, 5], including loans from financial institutions, but also including to since in the non-profit organization funded and individual donations. According to the data collected, we established four years from 2006 to 2009, to medium-sized industrial enterprises in Henan

Table 41.1 Innovation environment and medium-sized industrial enterprises in Henan Province, the ability to support dynamic changes

Index			2006	2007	2008	2009
Ability to support innovation environment	Average access to government innovation fund (million)	Nationwide	32.28	39.79	47.81	332.97
		Henan	16.76	28.71	30.38	122.94
		Ranking	25	20	23	26
	Average access to social innovation capital (million)	Nationwide	77.71	73.82	64.76	49.77
		Henan	52.19	72.95	32.16	18.67
		Ranking	16	11	23	27

Note Data from the "Statistical Yearbook of China Science and Technology"

Province and the national innovation environment in support of capacity development momentum for the table, as shown in Table 41.1.

From Table 41.1, we see: Over the past four years, and medium-sized industrial enterprises in Henan Province, the Government innovation funding although the average has increased, but lower than the national average, basically in the country's downstream position. In addition, companies receive an average funding of the social aspects of certain fluctuations, four years, the highest in 2007, ranking rose to 11th in the country, while the 2008 and 2009, two years, an average of the social aspects of business funding decreased, the country ranks only lower level, which may be affected by the financial crisis of 2008 caused. Overall, the removal of the 2008 financial crisis, medium-sized industrial enterprises in Henan Province government-funded innovation efforts is not, by contrast, comes from the social aspects of innovation support is relatively better, which is ease the financial pressure on businesses to play a certain role.

41.2.2 The Awareness of Independent Innovation

Awareness of independent innovation of enterprises refers to a culture of innovation, including not only the authenticity of the aspirations of entrepreneurs to innovation and innovation intensity, but also with the role of business is closely related to each individual. Here, we use the R&D personnel R&D expenditure per capita and per capita R&D personnel R&D projects to reflect [6].

41.2.3 The Ability of Independent Innovation Investment

Independent innovation capability of independent innovation of the premise and guarantee, but also enterprises in independent innovation an important driving force is the reaction of technological change, a sign, from R&D personnel intensity of investment, R&D funding intensity and business R&D facilities in three respects to the measure.

2006–2009, four years, and medium-sized industrial enterprises in Henan Province who invest in innovation to dominate the total annual number of steady growth in R&D personnel, the total number remained in fifth in the country. However, medium-sized industrial enterprises in Henan Province in the proportion of investment in R&D is relatively backward, are below the national average, basically in the middle level of the country. This shows that medium-sized industrial enterprises in Henan Province relative to employees in terms of R&D personnel, or inadequate.

41.2.4 R&D Capability

Innovative R&D capability is the effect of the accumulation of resources, but the ability to substitute for investment in innovation R&D capabilities. Evaluation of R&D capability of independent innovation capability is the traditional indicator of the ability of independent innovation is currently the most representative measure of one of the indicators. Since patents are usually some major technological innovations, test methods and instruments of the great inventions, all with independent intellectual property rights, some of these techniques are not available in other companies, it is more in line with the characteristics of innovation. Therefore, the innovation measure, we use patent indicators to represent the company's R&D capabilities. To this end, we have established nearly four years, and medium-sized industrial enterprises in Henan Province, R&D capacity development trend of the table, as shown in Table 41.2.

Comprehensive analysis of Table 41.2, we see that: in terms of innovation and technological achievements, from 2006 to 2008, three years, and medium-sized industrial enterprises in Henan Province, the number of invention patent applications in the steady rise in ranking from 10th in 2006 up to present the first 8-bit. From the relative indicators, the number of invention patent applications accounted for R&D staff proportion of full-time equivalent also continued to rise, the country ranked first in 2006 from 21 up to 2008's first 16-bit. This shows that

Table 41.2 Medium-sized industrial enterprises in Henan Province, the ability of R&D dynamics

Index			2006	2007	2008	2009
Innovative technological achievements	Patent application (item)	Nationwide	25,685	36,074	43,773	63,011
		Henan	487	761	1,200	1,427
		Ranking	10	8	8	10
	Number of invention patent applications/R&D personnel (entry/thousand years)	Nationwide	36.92	42.06	43	54.37
		Henan	13.48	18.95	26	23.79
		Ranking	21	17	16	23

Note Data from the "Statistical Yearbook of China Science and Technology"

medium-sized industrial enterprises in Henan Province, or effective R&D work, and the efficiency of R&D personnel also increased.

41.2.5 Manufacturing Capacity

As the "China Statistical Yearbook of Science and Technology," indicators of change, there is no 2009 data for this indicator. Therefore, we have established a medium-sized industrial enterprises in Henan Province and the country from 2006 to 2008, three years of changes in manufacturing capacity (as shown in Table 41.3), in order to analyze medium-sized enterprises in Henan Province in recent years, production capacity advantages and disadvantages.

Table 41.3 medium-sized industrial enterprises in Henan Province, production and operation of equipment in the total level of the original price has gradually increased, ranking increased slightly, from 13th in 2006 rose to 12th place in 2008. From the relative indicators, the average production and operation enterprises and original equipment manufacturing microelectronic devices account for the proportion of business equipment, in general, has declined three years, and has been below the national average, ranking is also falling, respectively, from 2006 to No. 22 and No. 17 in 2008, fell to No. 27 and No. 20. This shows that medium-sized industrial enterprises in Henan Province's investment in production equipment is concerned, although increased, but not much intensity, and advanced input devices also need to be further improved.

Table 41.3 Medium-sized industrial enterprises in Henan Province, the dynamic changes in production capacity

Index			2006	2007	2008
Production equipment and advanced level	Production and operation of equipment Price (million)	Nationwide	77,93,66,664	83,10,65,276	1,00,39,94,905
		Henan	2,81,89,881	3,38,80,790	3,74,79,828
		Ranking	13	10	12
	Average production and management company with original equipment (million)	Nationwide	23,873	22,925	24,904
		Henan	18,919	17,974	17,679
		Ranking	22	24	27
	Original micro-controlled production equipment/ production and operation of the original price (%)	Nationwide	12.14	13.08	12.07
		Henan	9.74	9.48	9.33
		Ranking	17	20	20

Table 41.4 Medium-sized industrial enterprises in Henan Province, the dynamic changes of effective innovation

Index			2006	2007	2008	2009
Product market	New product sales revenue (million)	Nationwide	31,23,28,084	40,97,61,681	51,29,19,771	57,97,80,546
		Henan	8,37,71,36	1,11,26,601	1,35,67,047	1,63,12,978
		Ranking	10	10	13	14
	New product sales revenue in the proportion (%)	Nationwide	14.80	15.69	16.02	17.34
		Henan	9.76	9.29	9.01	9.59
		Ranking	20	22	18	19

Note Data from the "Statistical Yearbook of China Science and Technology"

41.2.6 Innovation Efficiency

Therefore, the medium-sized industrial enterprises in independent innovation, efficiency can be reflected out of new product production. Table 41.4 that reflects the 2006–2009 medium-sized industrial enterprises in Henan Province, four years of new product output situations.

From Table 41.4, we can find: medium-sized industrial enterprises in Henan Province, the new product sales revenue increased year by year, to 2009, sales of new products rose to 163,119,870,000 yuan. However, its ranking has dropped from 10 in 2006, the country dropped from 14th in 2009. Similarly, medium-sized industrial enterprises in Henan Province, the new product sales revenue in the main business revenue is lower, four years below the national level, and was basically a declining trend, there is a certain ranking fluctuations and unstable.

41.3 Conclusion

Through the above medium-sized industrial enterprises in Henan Province, the ability of independent innovation and development of a detailed analysis, we see that: horizontal comparisons, medium-sized industrial enterprises in Henan Province, the ability of independent innovation indicators in the evaluation of the total amount of both an advantage basic among the nation's upstream position, mainly because of Henan Province is a province, a large business base, which will inevitably affect the total level of each index. However, the ability of independent innovation of enterprises greater impact factor is the relative index, and in the relative indicators, and medium-sized industrial enterprises in Henan Province are basically lower than the national average, with the national level there is still a gap. longitudinal comparison, the past four years, and medium-sized industrial enterprises in Henan Province in the evaluation of the aggregate increase in varying degrees, however, the relative amount of development is not optimistic. Specifically, medium-sized industrial enterprises in Henan Province, production capacity and effectiveness of these two indicators of innovation in the relative amount of the decline in varying degrees, therefore, raise the level in the country's situation, and medium-sized industrial enterprises in Henan Province in both cases the gap with the national level was greater.

Acknowledgments Henan soft science research projects in 2011 (ID: 112 400 440 071)

References

1. Yulin Z (2006) Innovation economics, vol 12. China Economic Press, Beijing, pp 356–360
2. Liu L, Zhang A (2007) Innovation, non-technological innovation and industrial innovation system. Innovation and Technology 6:14-19
3. Chen J (1994) from the introduction of technology into the learning model innovation. Res Manage 2:32–34
4. Zhang W, Yang Q (2006) Discussion of the concept of independent innovation and define. Sci Res 6:956–961
5. Xie X (1995) Scientific and technological progress, innovation and economic growth. Chin Eng 5:6–9
6. Lin Y, Chen Chun B (1997) Imitate the innovation, innovation and high-tech business growth. China Soft Sci 8:107–112

Chapter 42
Study on Non-Contact Yangtze River Flow Measurement Algorithm Based on Surface Velocity

Zili Li and Caijun Wang

Abstract At present, in China, the flow velocity of river is detected mainly with flow meter and ADCP, and also the flow of cross section of river is estimated through a certain algorithm. However, these flow velocity detection methods are restricted by the navigations of ships and the hydrologic conditions of river. In this paper, based on the detection of the experimental data of Yangtze River Wuhan with UHF radar, the flow velocity of this river reach is extracted, and simultaneously the flow of the cross section of the tested river reach is estimated with a certain algorithm according to the extracted surface velocity, and finally the estimated result and the measured data are compared for verifying the feasibility of using the shore-based UHF radar detection to detect river hydrologic information.

Keywords UHF radar · Flow velocity extraction · Cross-section velocity distribution · Flow estimate

42.1 Introduction

Direct measurement is the primary way that is applied in the measurement of river flow at present [1]. Non-contact river flow measurement is the development direction of river flow measurement, and owns the advantages in measurement

Z. Li (✉)
College of Electronic Engineering, Guangxi Normal University,
Guilin 541004, Guangxi, China
e-mail: zili45li@126.com

C. Wang
College of Electronic Information, Wuhan University, Wuhan 430072, Hubei, China

W. Du (ed.), *Informatics and Management Science VI*, Lecture Notes in Electrical Engineering 209, DOI: 10.1007/978-1-4471-4805-0_42,

efficiency and security in comparison with direct measurement. For example, the non-contact measurement can be implemented without any restrictions during flood period, at upstream and downstream of weir and dam, or in the near-shore flow field. The development of the non-contact measurement is of great practical significance for the measurement of river hydrologic environment.

At the present stage, one of the most important developing directions of the non-contact measurement technology lies in the surface-wave radar remote sensing technology. In the Radio Wave Propagation Laboratory of Wuhan University, a three-channel ultra-high frequency (UHF) river flow detection radar has been researched and developed, and also has been tested on Yangtze river Wuhan [2, 3]. In this paper, by extracting surface velocity information from the river remote sensing data of UHF river flow detection radar, a river cross-section velocity distribution model is established, and then the cross-section flow is estimated, and finally the estimated result and the measured data are compared for verifying the feasibility of the verification method.

42.2 Establishment of the Cross-Section Flow Velocity Distribution Model

According to the principles of the fluid mechanics and the fluvial hydraulics, the inside of river water body is layered based on water depth, and all flow layers are mutually associated and influenced by using open channel flow to control state equation [4]. Therefore, surface flow has a correlation with the distribution of under-surface flow velocity [5]. At the same time, surface flow also has a relationship with the water depth of cross-section. Besides, studies on these correlations are helpful for the establishment of the cross-section flow velocity distribution model, and therefore the water flow information at all in-depths can be acquired.

42.2.1 Relationship Between Surface Flow and Cross-Section Flow Velocity Distribution

The distribution of flow velocity $\bar{u}(z')$ along water depth cannot be acquired from the solving of open channel control equation. Therefore, to acquire $\bar{u}(z')$, it is necessary to make use of some semi-empirical formula methods. According to the different distribution of shear stress of cross-section water flow, the cross section is divided into two parts (i.e. inner zone and outer zone).

In inner zone, shear stress always keeps constant along a vertical line and also is equal to side-wall shear stress. Therefore, according to the semi-empirical formula method of the mixing length theory $l = \kappa z'$, the inner-zone empirical equation can be gained as follows.

$$\frac{\bar{u}}{u^*} = \frac{1}{\kappa} \ln z' + C \tag{42.1}$$

In Eq. (42.1), $u^* = \sqrt{\tau_0/\rho}$ is frictional velocity; κ is Karman constant; C is integration constant and determined according to experimental data. In the range of inner zone, the value of flow velocity $\bar{u}$ depends on side-wall shear stress τ_0, fluid properties ρ, and the distance z' from river bed.

In outer zone, because shear stress not equal to constant and the laws of the inner zone is unlikely to actually compound with those of the outer zone, it is necessary to introduce a function for a modification. Here, by introducing Coles [5] wake stream function, the logarithmic formula is modified, and thus the outer-zone empirical equation can be gained as follows.

$$\frac{U_c - \bar{u}}{u^*} = \frac{1}{\kappa} \ln\left(\frac{\delta}{z'}\right) + \frac{\Pi}{\kappa}(2 - \varpi) \tag{42.2}$$

In Eq. (42.2), U_C is the maximum flow velocity in the cross-section; ϖ is wake stream function; water depth δ is the place where the maximum flow velocity U_C is measured in the cross-section; Π is wake flow parameter. As for balanced water flow, it is necessary for wake flow parameter to keep constant. In open channel flow, there is wake flow parameter $\Pi = 0.2$, which is approximately a constant.

42.2.2 Relationship Between Surface Flow and Water Depth

After the equation of flow velocity along water depth is determined, the numerical value of water depth can be further deduced. It can be expressed with the form of power according to the flow velocity empirical distribution equation [6].

$$\frac{\bar{u}}{u^*} = a\left(\frac{z}{z_0}\right)^m \tag{42.3}$$

In Eq. (42.3), z_0 is the bed-surface roughness feature length, which is used for expressing the distance between zero and river-bed bottom, but also the position of virtual-bed bottom; coefficients a and m are parameters related to river-bed form and flow condition, and also their values are controlled by the empirical approach and different parameters can be selected for describing the flow velocity distribution according to different flow conditions.

In natural rivers, the suggestive values $m = 1/8$ and $A = 9.45$ of Engelund are applicable [7] if the objects are rivers with big water depth and small-size bed.

42.2.3 Relationship Between Surface Flow and Cross-Section Mean Flow Velocity

After the flow velocity distribution along water depth is discussed, the relationship between surface flow and cross-section mean flow velocity will be studied as follows.

First, according to the previous experience [8], the cross-section mean flow velocity U can be determined with the following approximate relationship.

$$U \cong (0.8 \oplus 0.9)\bar{u}_s, U \cong 0.5(\bar{u}_{0.2} + \bar{u}_{0.8}), U \cong \bar{u}_{0.4} \tag{42.4}$$

In Eq. (42.4), $\bar{u}_s, \bar{u}_{0.2}, \bar{u}_{0.8}, \bar{u}_{0.4}$ are the velocities of the given position points, respectively; numerical value is water depth; S is water-surface. In natural rivers, the above calculations are easy to give rise to errors if the lateral bed-surface cross-section distribution is uneven.

Second, according to the cross-section flow velocity Eq. (42.1) and (42.2), the integrals of cross-section flow velocity are calculated based on inner zone and outer zone, and $0.2\,h$ are used as the boundary point of inner zone and outer zone.

Thus, the outer-zone mean flow velocity U_o equation can be expressed as follows.

$$U_o = \bar{u}_s - \frac{u^*}{\kappa}(0.5976 + 0.766\,\Pi) \tag{42.5}$$

The inner-zone mean flow velocity equation can be expressed as follows.

$$U_I = \frac{u^*}{\kappa}\left[\left(1 + \frac{k_s}{6h}\right)\ln\left(1 + \frac{6h}{k_s}\right) - 1\right] \tag{42.6}$$

The cross-section mean flow velocity U can be expressed with the equation $U = 0.2U_I + 0.8U_o$, and thus the ratio between the mean flow velocity and the surface flow velocity can be solved by using the equation $U = 0.2U_I + 0.8U_o$ to divide the flow velocity distribution equation.

$$\frac{U}{\bar{u}_s} = 0.8 - \left[\frac{-(0.2 + \frac{k_s}{30h})\ln\left(1 + \frac{6h}{k_s}\right) + 0.612\Pi + 0.678}{\ln\frac{30h}{k_s} + 1.81\Pi}\right] \tag{42.7}$$

In the logarithm equation, the factors that play an influence on the above ratio include wake flow parameters, river-bed roughness degree, and water depth and so on.

Third, under the condition that the power-times flow velocity distribution Eq. (42.3) is applied, mean water depth integral is selected and thus the cross-section water depth mean flow velocity U can be expressed with the equation below.

$$\frac{U}{u^*} = \frac{a}{(1+m)}\left[\frac{z_0}{h}\left(\frac{h + z_0}{z_0}\right)^m - \frac{z_0}{h}\right] \tag{42.8}$$

By using the mean flow velocity equation to divide the power-times flow velocity distribution equation, the ratio between the mean flow velocity and the surface flow velocity can be solved as follows.

$$\frac{U}{\bar{u}_s} = \frac{1}{m+1}\left(\frac{z_0}{h}\right)^{m+1}\left[\left(\frac{h}{z_0}+1\right)^{m+1}-1\right] \tag{42.9}$$

In the power-times relational expression, the factors that play an influence on the above ratio include not only the water depth, but also the parameter m. However, the main river flow condition has already been included in m.

42.3 Cross-Section Flow Estimate Model

A distance element, which can be detected by radar, is used as cross-section position; in each cross-section, river width B is divided into several segments. For each segment B_i, the mean surface flow velocity $\bar{u}_{si}$ of each segment is acquired according to the surface flow velocity distribution data detected by radar, and subsequently the mean flow velocity U_i and water depth h_i of this segment can be gained according to relevant equations, and the total flow Q of the distance-element cross-section can be obtained with mean section method.

$$Q = \int_0^B (Uh)dx \approx \sum_{i=0}^{n}\left[\frac{1}{4}(U_i + U_{i+1})(h_i + h_{i+1})B_i\right] \tag{42.10}$$

In Eq. (42.10), B is river width; n is the number of segments; B_i is segment river width; U_0, U_n, h_0, h_n are the mean flow velocities and water depths at the boundary respectively and can be regarded to be 0 in calculation.

42.4 Analysis on Experimental Calculation Results

42.4.1 Determination of Calculation Parameters

Yangtze River Wuhan is in the middle of the whole Yangtze River; the value of its water width is approximately 1800 m during the period of experiment; water flow is seen to be two-dimension. Related parameters are selected according to water flow properties and river channel features [9]. First, the river-section energy gradient S_e is 0.0024 % according to the hydrologic data of Hankou Hydrometric Station; the river-bed roughness length k_s is usually about 0.20 mm according to the hydrologic characteristics of Hankou Hydrometric Station [10]. Next, the shape of the cross-section is approximately estimated; according to the river cross-

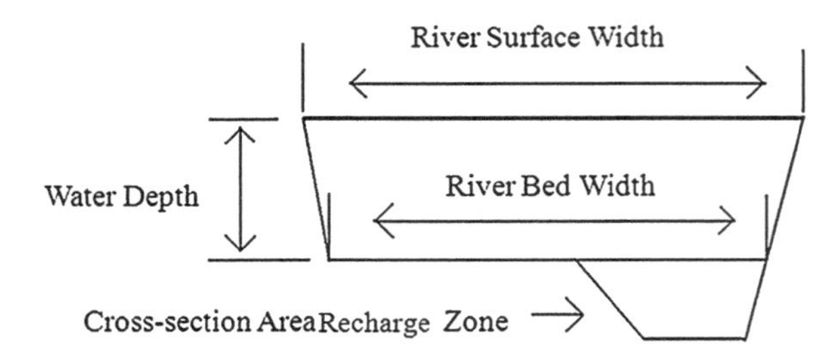

Fig. 42.1 Approximate shape of river cross-section

section characteristics of Hankou Hydrometric Station and the W-likeness shape of river bed, and also considering the deep groove in Wuchang and the scour pit in Hankou, the approximate shape of the cross section is shown in Fig. 42.1.

42.4.2 Flow Calculation Results

According to the model for the relationship of water depth with mean flow velocity and surface flow, data is detected with surface flow; combining the approximate shape of river cross-section, the flow result at a specific moment is acquired with cross-section flow estimate algorithm; to verify the effect of algorithm, the estimate result of flow result is acquired with three methods respectively. In the mean time, according to the hydrologic information released on Yangtze River hydrologic website each day, the actual water level and measured flow data of Hankou Hydrometric Station are acquired, and then the actually measured information and the calculation result are compared. Specific flow calculation numerical values and errors are shown in Table 42.1.

42.4.3 Analysis and Conclusion

The time slot of the radar data applied in the flow estimate calculation of this paper is corresponding to the time slot of the hydrologic information reports released on the Yangtze River hydrologic website of Hankow hydrological station, so as to easily make a comparison on the final results. Seen from the flow calculation table, it is found that the trend of the changes of different calculation methods keeps consistent though their calculated flow numerical values are different. Comparing the flow values calculated by the three methods, the value gained by the approximate method is the most approximate to the measured value under the current calculation conditions, and simultaneously its calculation error percentage is less than 5 %. However, the error of the numerical value directly calculated by the logarithmic method is higher than 10 %. The error of the numerical value calculated by the power method is just between the above two.

Table 42.1 Flow calculation values and errors

	8:00 PM, December 11	8:00 AM, December 12	2:00 PM December 13	8:00 AM, December 14	8:00 AM, December 15
Measured water level (m)	15.4	15.31	15.19	15.15	15.12
Measured flow (m^3/s)	10,000	9,880	9,670	9,600	9,840
Approximate method (m^3/s)	9,926	10,235	9,207	9,600	9,741
Error percentage (%)	0.73	3.6	4.79	0.1	1.0
Power method (m^3/s)	9,537	9,831	8,851	9,226	9,360
Error percentage (%)	4.63	4.9	8.47	3.89	4.87
Logarithmic method (m^3/s)	8,726	8,996	8,092	8,439	8,562
Error percentage (%)	12.74	8.92	16.32	12.09	12.98

The flow calculation results in this paper are restricted by hydrologic data preparation and experimental conditions, and also many approximate calculations are applied in the calculation process. These can affect the accuracy of calculation results. However, a large amount of values can be compared and calculated through the value range of hydrologic parameters, so as to find a method with minimum error. Also, the error of flow estimate can be made up through modifying the cross-section geometric model and other methods. According to calculation results, the method of using radar to detect surface flow data and acquiring cross-section flow is verified to be feasible. More accurate results can be acquired through further modifying a specific river cross-section model, and simultaneously the application range of the non-contact rivers measurement is expanded.

References

1. Han Y, Huang S, Wei J (2005) Comparative measurement and precision research of flow in inner yangtze river using ADCP. Autom Water Resour Hydrol 22(3):1–12
2. Shen W, Biyang W, Zili L (2007) UHF radar system tested on the bridge of yangtze river, radar systems IET Int Conf 22:1–5
3. Shen W, Biyang W, Ding F (2009) UHF radar designed for inshore wave watcher and ocean power application. Power Energy Eng Conf APPEEC 5:1–4
4. Grafton ST, Wenxian Z, Wan Z (1997) Fluvial Hydraulics. Chengdu Univ Sci Technol Press 12:281–286
5. Coleman NL, Alonso CV (1983) Two-dimensional channel flows over rough surface. J Hydraul Eng 109(2):175–188
6. Chen CL (1991) Unified theory on power law for flow resistance. J Hydraul Eng 117(3):371–389
7. Engelund F, Hansen E (1972) A monograph on sediment transport in alluvial streams, teknisk forlag. Copenhagen 6:62–67
8. Chow VT, (1964) Handbook of applied hydrology, vol 4. McGraw-Hill book Co. Inc. New York, pp 28–32

9. Xiwan G, Chen J, Zou N et al (2006) Research on stage-discharge relation of main hydrologic stations on middle and lower reach of the yangtze river. Yangtze River 37(9):68–71
10. Zefang C, Tong H, Yao L (2006) Analysis of river channel evolution for wuhan reach of the yangtze river in recent years. Yangtze River 37(11):49–50

Chapter 43
Study on Circulation of Right to Use Rural Curtilage

Zuwei Qin and Jiyu Tang

Abstract The ultimate goal of the balanced urban and rural development is to make a new pattern of integrating urban and rural economy and society realized. The limit to the circulation of the right to use rural curtilage is a reflection to the unequal rights that are enjoyed by urban and rural residents, can not adapt to the needs of the balanced development of urban and rural areas, gives rise to serious waste of land resources, goes against the benefits of peasants to make full use of curtilage resources, and also makes a large amount of "invisible markets" with hidden contradictions existed. Therefore, in terms of the right to use rural cartilage, it is necessary to ensure the equal rights of both urban and rural residents based on the ideas of legislation, build up a compensated obtainment and deadline use system and a strongly operational circulation system, and also make an improvement to the rural social security system that is the foundation of circulation.

Keywords Balanced urban and rural development · The circulation of the right to use rural curtilage · Innovation

43.1 Balanced Urban and Rural Development and the Circulation of the Right to Use Rural Curtilage

The balanced development of urban and rural areas is a new strategy that is proposed by Chinese government in the new historical period according to the imbalanced development of domestic urban and rural areas for over 50 years since the foundation of new China.

Z. Qin (✉) · J. Tang
School of Politics and Laws, Chongqing Three Gorges University, Wanzhou, Chongqing, China
e-mail: zuwei123t@126.com

W. Du (ed.), *Informatics and Management Science VI*, Lecture Notes in Electrical Engineering 209, DOI: 10.1007/978-1-4471-4805-0_43,

In 2010, "Opinions of the CPC Central Committee and the State Council on Exerting Greater Efforts in the Overall Planning of Urban and Rural Development and Further Solidifying the Foundation for Agricultural and Rural Development" clearly shows that it is highly necessary to take the balanced development of urban and rural areas as one of the fundamental requirements for constructing a moderately prosperous society in all aspects. It also explicitly indicates that the reform of rural land management system should be promoted with orders.

In the general conception of the integration of urban and rural areas, rural land circulation including the circulation of the right to use rural curtilage is the most important element beyond all doubts.

Rural curtilage refers to the lands that are legally acquired by rural residences with legitimate procedures and used for building up dwelling houses. The right to use rural curtilage is one of the core issues that have a direct relationship with the rights and interests of peasants in the real rights law system.

Along with the development of rural economy and the advancement of urbanization construction in China, the trading volume of rural houses increases with each passing day. In the mean time, because of integrated house and land, the circulation of the right to use rural curtilage emerges.

43.2 Drawbacks of Limiting the Circulation of the Right to Use Rural Curtilage

On the basis of the legislation idea that the right to use rural curtilage is protection-based non-real right and of non-circulation, it can be known that the circulation of the right to use rural curtilage is limited by existing relevant laws and regulations.

Investigation has proven that the limit to the circulation of the right to use rural curtilage has been unable to adapt to the needs of developing modern society and maintaining the rights and interests of peasants, and therefore presents multiple drawbacks.

43.2.1 Making the Rights Enjoyed by Both Urban and Rural Residents Unequal

Urban curtilage is allowed to be circulated together with the housing-reform houses of welfare. Although preferential buy right is given to urban units in general, rural curtilage of the same welfare is strictly limited in circulation and can only be circulated with houses within their collectives.

Therefore, it is obvious that such a two-dimensional urban and rural segmentation pattern is unequal.

43.2.2 Unable to Meet the Needs of Balanced Development of Urban and Rural Areas

The existing rural curtilage use legal system, which was established in the early liberation and simply oriented at protection-based residence right, has been far away from meeting the needs of the balanced development of modern urban and rural areas.

On the one hand, along with the development of the compensated state-owned land use system and the focus of the land estate system shifting from the ownership previously to the utilization currently, a great impact will certainly be exerted on the rural curtilage use system in which uncompensated use and circulation limit rules are implemented.

On the other hand, under the conditions of market economy, rural labor force, as production element, is required to flow, and therefore the fundamental and protection roles of rural land in the life of peasants are greatly reduced.

43.2.3 Giving Rise to Serious Waste of Rural Land Resources

Because acquisition and tenure of the right to use rural curtilage feature identity and non-compensation, Chinese peasants always hold an attitude of "trying to get more as much as possible" toward rural curtilage. Therefore, various illegal means are taken by them to make a constant expansion to the floor space of curtilage. This phenomenon is very common in China's rural areas at present, thus making the protection system for the survival of existing curtilage use system shake [1].

43.2.4 Going Against Benefits of Peasants to Make Full Use of Land Resources

The right to use rural curtilage as one of the most important resources of ordinary families makes peasants' financing channels greatly reduced and costs of life and production increased because of the weakening and circulation limit in its property rights.

If rural curtilage and dwelling houses are allowed to circulate, the peasants, who work in urban areas, can use the earnings from selling or rending houses and curtilage as the capitals of purchasing houses and managing business in urban areas. This is helpful for migrant workers to thoroughly transform to urban residents. Subsequently, the urbanization process can be accelerated in China.

43.2.5 Giving Rise to the Existence of Many "Invisible Markets" with Hidden Contradictions

For various reasons, vacancy or inefficient utilization of urban dwelling houses and curtilage is in existence universally in Chinese rural areas at present. However, in the mean time, a large number of urban residences, because of incompetence in buying a commodity house, shift their eyes only for the low-cost dwelling houses and curtilage of peasants. Therefore, great demands are generated, making many invisible curtilage markets with hidden contradictions existed.

According to the existing laws and regulations of China, the transactions of housing properties and land without registration in the transfer of real rights are not protected legally.

However, in fact, the rights of a large number of peasants to use rural curtilage have been circulated privately, but no efforts or measures can be taken by government for this phenomenon.

43.3 Innovation Ideas for the Circulation of the Right to Use Rural Curtilage

43.3.1 Protecting the Equal Rights of Urban and Rural Residents from Legislation Idea

The reasons for going against the circulation of the right to use rural curtilage are rooted in the traditional legislation idea.

First, it is thought that the current rural social security system is far from being sound, and a large number of peasants may be ultimately homeless if the circulation of the right to use rural curtilage is allowed legally.

Second, the protection of cultivated lands will be in an adverse state if the circulation of the right to use rural curtilage is allowed legally. Seen from the current situation, these reasons seem to have been unable to hold water. As a market main body, peasants have been very rational for the circulation of the right to use rural curtilage and will not easily sell their cartilages if they are not forced by actual circumstances even though the circulation is allowable in law.

Real Right Law of China explicitly shows that the equal legal status and development rights of all market bodies shall be protected legally. Specifically speaking, all equal legal status and development rights of Chinese peasants in curtilage shall be protected legally as well. The values of rural dwelling houses and curtilage property right can be added and embodied only by market allocation, and then a solid material foundation can be laid for the basic human rights such as the right to live and the right to develop [2].

43.3.2 Building up a Compensated Obtainment and Deadline Use System

Uncompensated and dateless policies are applied in China's existing rural curtilage use system.

As mentioned above, however, along with the development of society and the improvement of people's living standard, there are increasingly more drawbacks appearing in the uncompensated and dateless curtilage use system.

Therefore, it is necessary to build up a compensated obtainment and deadline use system in accordance with existing state-owned land use system. Fees shall be charged for the acquisition of curtilage in rural areas, and higher fees shall be charged for those wasted curtilages.

43.3.3 Formulating a Strongly Operational System for the Circulation of the Right to Use Rural Curtilage

As a basic law of adjusting property ownership relationship and property use relationship, Real Right Law of China has provided relevant principles for the right to use curtilage. The system for transferring the right to use curtilage shall be specifically provisioned in land administration law and administrative rules and regulations.

First, the scope of circulation objects shall be limited.

Second, the scope of transferees of circulation shall be expanded. Therefore, rural curtilage shall be allowed to open to all kinds of social objects, but not limited in the personnel who are in the collective economic organization and also meet the application requirement of curtilage. The personnel in the collective economic organization and also meeting the application requirement of curtilage can obtain rural curtilage through application under the current situation of acquiring rural curtilage without compensation. Under the same circumstances specified in rules and regulations, however, the fact should be that other members in collective economic organization shall own the preemptive right of purchase.

Third, circulation price and distribution system shall be made. On the one hand, to prevent rural curtilage sold with too low price and damage the interests of peasants, the minimum protection price can be provisioned by government in accordance with the economic developments of all local places. Therefore, the circulation shall be prohibited if the price is obviously lower than the protection price. On the other hand, a transfer fee distribution system for the circulation of the right to use rural curtilage should be established.

According to the interest distribution mechanism for the circulation of the right to use state-owned land, the principle of consistent rights and obligations and the principle of integrated investment and return, such an earning shall be reasonably distributed among the state, the collective, and the owner of curtilage, so as to protect the legal interests of the owner of curtilage and also prevent the problems of nonstandard transfer fee and power corruption [3].

43.3.4 *Perfecting the Rural Social Security System that is the Foundation of the Circulation of the Right to Use Rural Curtilage*

For a long time, there were no endowment insurance, medical insurance, and housing accumulation fund, various subsidies and minimum life guarantee system available for peasants in China.

For this reason, the right to use rural curtilage played an important role in protecting the survival and living of peasant in the past time.

However, there is no a parallel choice relationship constituted between the state's social security and land security: social security was originally supposed to be an obligation of some social organization (e.g., the state, society, community, family), but turns to a task related to some production element (land), and these seem to be of self-deception [4].

What Chinese peasants really need now is real social security, medical insurance and endowment insurance, but not nonconvertible rural curtilage.

Therefore, a complete social security system should be established, which can help fundamentally solve the subsequent troubles from the circulation of the right to use rural curtilage and also provide a stable social environment for the settlement of these troubles.

According to the situation that higher important is attached to China's urban social security system but rural social security undertakings are still seriously backward, Chinese government proposed that the security system for ensuring endowment insurance, medical insurance, and minimum life guarantee of rural people shall be explored and established in the places with conditions in the Report of the Sixteenth National Congress.

Therefore, it is necessary to make every effort to bringing true rural curtilage back to peasants and promoting it to be a real right freely handled by peasants by comprehensively impelling the construction of the rural minimum life security system, actively developing new rural cooperative medical system, and improving and innovating rural endowment insurance system, etc.

Acknowledgments This paper is a stage result of key project of Chongqing Three Gorges University—Research on Rights and Interests of Migrants Losing Lands because of the Construction of Three Gorges Project from the Perspective of Balanced Rural and Urban Development.

References

1. Liu J (2007) Study on rural residential land use rights system. J Southwest Univ Nationalities. Humanities Soc Sci 12(3):56–62
2. Yuan C (2009) Study on rural residential circulation legal issues under the background of urban and rural balanced development. Inf Econ Law 14(2):184–187
3. Long Y, Lin X (2009) Legislative suggestion to legal adjustment for right to the use of curtilage in china rural areas-and discussion on the solution to small property rights houses. Law Sci Mag 21(9):79–87
4. Qin H (2003) Peasant-oriented China: historical reflection and real choice, vol 12. Henan Publishing House, Henan, pp 45–50

Chapter 44
Study of Emotion Experience in Product Experience Design

Gang Wang

Abstract From an economic point of view, human history has gone through four stages: from agricultural economy times to industrial economy times, then to service economic times, and last to Experience Economic Era now. What is Experience? For the authors of Experience Economy, Experience is an activity to create a memorable experience. It is a memorable event around consumers using service as the stage and commodity as the properties. The appearance of Experience Economic Times illustrates people's humanized, emotional and personalized needs. The emotional experience design as an important form of experience design will become the future design trend.

Keywords Experience economy · Experience design · Emotional experience design

44.1 Emotional Experience

Based on Dewey's Experience Theory, Hekkert advances Product Experience Theory, which is defined by him as a series of results from the interaction between people and products, including all the sensory senses (sensory experience) [1], the meaning given to the products (meaning experience) and all emotions in the interaction (emotional experience) [2]. As shown in Fig. 44.1

G. Wang (✉)
School of Art and Design, Wuhan University of Technology, Wuhan 430070, People's Republic of China
e-mail: gangw49ang@163.com

W. Du (ed.), *Informatics and Management Science VI*, Lecture Notes in Electrical Engineering 209, DOI: 10.1007/978-1-4471-4805-0_44,

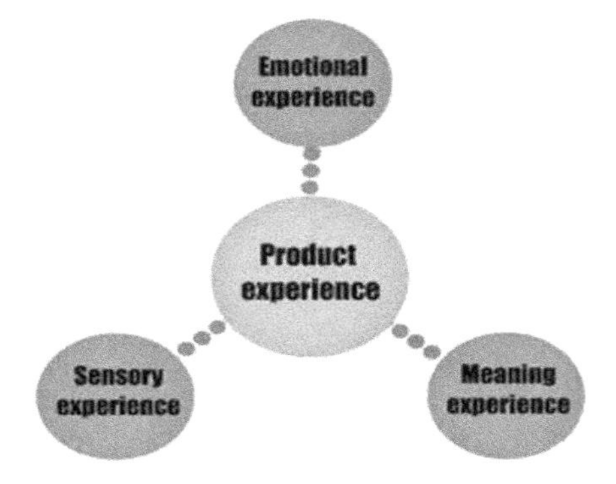

Fig. 44.1 Hekkert's product experience structure

Donald Norman points out in his book Emotional Design that there are three levels in design and the aim of design namely, instinct level (viscera), and behavior level (behavior), reflection level (reflective) [3]. As shown in Fig. 44.2. The level of behavior refers to the use process and to be entertained a sense of satisfaction and accomplishment in the process. The highest level is the reflection layer. It actually refers to the depth of emotion, the overall impression and meaning understanding generated by the user in mind, under the reflection of the first two levels. Reflection layer has great significance on the contemporary products design, which contains the users and the products' emotion and interaction relationship, and helps to improve users' brand loyalty.

44.2 Experience Design of Contemporary Product

Experience Design is generally referred to as User Experience Design. User Experience (abbreviated as UE) is a kind of purely subjective emotion experience which is produced in the process of using a product (or service). Because it's purely subjective, there are some uncertain factors. Individual differences also decide that each user's real experience cannot be copied or reproduced through

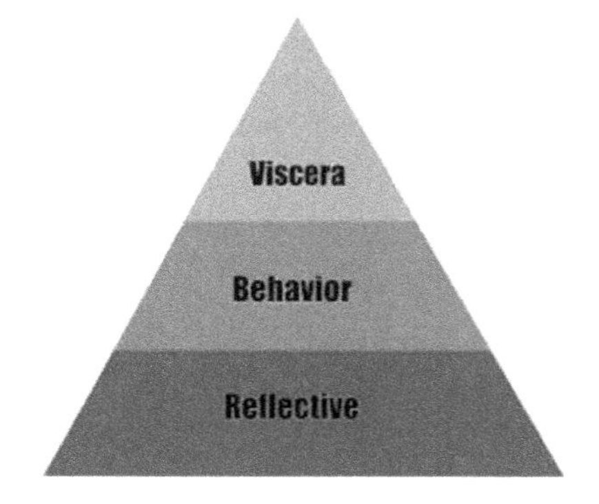

Fig. 44.2 Donald Norman's three levels of design and the design goal opinion

any ways. But for a specific user group, the common user experience can be realized through a good design experiment.

The UE mainly comes from the interaction of man-machine interface. In the early stage of the software design, man-machine interface was seen as a kind of decoration or package attached to the design of core function of the product and it did not get enough attention. The development of man-machine interface was separated from the development of the core functions, and very often it began in the end of the whole development process. This greatly limits the design of man-machine interaction. Thus, results in a high risk. Because revision in the final stage of the core function design will take high cost, sacrifice the design of man-machine interface is the only way out. Such practice is obviously doomed to fail to get a favorable user experience from the beginning of the development. As for customer service, it is also part of the user experience from a broad sense. Customer service is the media between products and customer. Whether this channel is obstructed or smooth directly affects the customer feelings. Good customer service can bring a customer into product related experience, feeling a sense of culture when using products. This is a higher level of design. So the design of the product itself and the UE cannot be divided. Customer service is more a service requires personnel quality, and it's difficult to change when a product has been completed and put into the market. There is also a theory that a good design can reduce the user's needs of the customer service. That results in reducing a company's cost in customer service, and also reduces the risk of customer loss due to customer service quality [4].

44.2.1 The Coming of Experience Economy

In the past, people often consider product function design from the perspective of technology or the product, considering only the factors of products but not users. When a designer design a product, he thinks firstly the product function factors, rather than the user experience of feelings. In other words, between technology and users, people pay more attention to technology, and ignore the users' psychology and emotions. In fact, a person is the designer of the product, and is also the final user. He is both the beginning and the end in the development of a product. So people should be the center of attention, the feelings of people should be the core of the product design [5].

With the experience economy era coming, experience design also comes into being. The changes of production and life style under experience economic conditions lead to the development of product experience design Information technology switches many substantial products and service to non-physical, virtual world in computer and network system. Those networking products we are familiar with, such as email, e-commerce, have become experience's media channels.

44.2.2 Users' Experience is the Core of Production

Product experience design can arouse the users' participation through thoughts and action, evoke memories of product users and form emotional resonance. In traditional product economic times, manufacturers tend to focus only on technology parameters to enhance the development and neglect experience and emotional factors.

Mobile phone, for example, the majority of mobile phone brands compete advantages and disadvantages of parameters such as the processor, screen size, pixel size, standby time, and do not concern about the emotional experience design of interactive human-machine interface, do not care about users' emotion changes. And a good product wants to convey to the target people, not just the look, performance or material, but rather to convey a way of life they should have. What they want to see, hear, what can move them, the product is sensitive to "prophetic vision", and once users reach some kind of emotional recognition, for its brand loyalty, while the parameters upgrade is not comparable. The reason why iPhone is sought after in the world, because it not only met all of the fashion element, but also bring a iPhone-style way of life and emotional experience—easy access, social networking, video games, music, web surfing, travel and navigation.

As another example, when the coffee is only as "goods" are sold, a small packet of coffee (10 g) only to sell 5–7 yuan (function value realization); and when the coffee is packaged as a fine product, a cup can sell for 25 yuan (aesthetic value realization); in the coffee shops, when it joined the service, a cup may be 35–100 yuan (primary experience of value realization); but if customers can experience the coffee wine and pleasant way of life, it can be sold for 150 yuan a cup or even a few hundred dollars (advanced experience to value). Starbucks, the first success use the depth "experience" to improve the value-added products, achieve a high profit return. From this we can see that user experience is a key to purchase behavior and brand building.

44.2.3 Technology is not for the Sake of Technology

Steve Jobs, the father of Apple never produces new technology for technology, and never adds various fancy functions to his products. He would rather reduce product complex function to achieve simplicity and easy use, because Steve Jobs keeps the philosophy that the most important decision isn't something you do, but you don't do anything. Therefore, a lot of Apple products are all designed from the angle os user experience. The success of the iTunes again proved the importance of the user experience. ITunes online music store launched in 2001, it was just the time when the Internet freely sharing music. Many people doubted how could an original music store compete against how piracy music. When people can get free songs, who will spend money in here to buy original music? Steve Jobs' answer is "user experience". To look for sound quality of a good song, users have to spend a lot of time by the file sharing network technology. In contrast, music lovers can log on iTunes online to buy high quality songs. The purchase process is very easy and quick. And more importantly, music quality and reliability is guaranteed. Steve Jobs said, "I don't believe we can convince people to leave piracy by language only, unless you can offer carrots-and not stick." "Our carrot is: We will provide you with better experience… and every song will only cost you one dollar." The same success of the user experience is also demonstrated in the success of the iPod.

The core of creativity in art and technology is the individual experience. In order to better grasp the pulse of the product emotional experience, what Jobs is always experience in person. Because he is not an engineer, and he has no any software engineering training previously, he has no college degree, either. So he can think like a layman. This makes him the best test platform of any apple products—Steve Jobs is an Apple ideal consumer. He himself is a user experience expert [6]. It is because of his precise understanding of users' emotional experience in the use of the products; in 2010 Apple successfully surpassed Microsoft, becoming the world's largest technology company.

In summary, product design is not only a simple function to meet and no longer a modeling beauty to demand. We pay more attention to product design experience. Stressing the harmonious interaction between man-machine interfaces, emphasizing the user experience has become a design trend.

44.3 Emotional Experience of Contemporary Product Design

Emotional experience design is humanization design; it reflects a kind of human spirit. It's a perfect harmany between people and products. The success of a designer lies in emphasizing 1 the easy use of the product of and at the same time, he also attaches importance to product to the user emotional needs.

44.3.1 Cracking the Technology Puzzle, Creating a Relaxed and Friendly Human-Computer Interaction

In modern society, material life is rich, life pace accelerates and high-tech products continue to emerge, undoubtedly it's a never-ending technology dilemma. Faced with a deluge of information, heaps figures and different colors buttons, helplessness, panic is the most profound feelings in information age. Therefore, the product design's basic functions have met the demand, the need for more attention to people's emotional needs and spiritual comfort. Friendship, compatibility, convenience, intelligence, these emotional cares can effectively alleviate the fear of high technology and enable people to build trust and rely on the product.

"The best technique makes people feel no technology". Steve Jobs returned to Apple and then he promoted "i Series" products (from the iMac to the iPod then iPhone), activated and enlarged the Apple design concept: the need of user is not technology, even not computers, but the maximization of interests and the optimization of experience [7].

44.3.2 Comforting Emotion, Building High-Tech and Balancing High Emotion

The rapid development of modern technology changes not only the material world, but also our spiritual lives. Behind extremely rich material life, people could not conceal the new anguish and anxiety, that is, emotional indifference, alienation and emotional imbalance. The reinforced concrete city blocks people's contact, pull away the distance of each other. In the high-tech society, people never desire to strike a balance that between high tech and high emotion, high rationality l and high humanity. Some emotional experience products bring us network communication, extend and enrich our lives. Social Network Sites (SNS) platform, mobile QQ, video phone makes our emotions have a certain comfort and balance.

Because of different people's life experiences, ethnic habits, different interests, the experience will be different. However, a successful products experience design can catch the users' emotional needs, trigger the users' "emotional point", and then generate the exchange collision between products and users.

44.4 Conclusion

In summary, emotional design will become the future design trends. Future designers should fully think about the users' psychological and emotional factors, pay more attention to emotional contact between human and products, strengthen the emotional experience of product design, so that human and object can exchange and make dialogue harmoniously.

References

1. Pine BJ, Gilmore JH (2008) Experience economy. Revised edition: (trans: Yeliang X, Wei L) vol 1(11). Engineering Industry Press, pp 1–5
2. Hekkert P (2006) Design aesthetics: principles of pleasure in design. Psychol Sci 48(2): 158–164

3. Norman D (2005) Emotional design (trans: Fu Q, Cheng J) vol 22. Electronic Industry Press, pp 435–439
4. Du XZ (2011) User experience design vol 11(2). Electronic Industry Press, pp 23-27
5. Kahney L (2008) Inside Steve's brain. In: Qiu JS (ed), vol 22. Portfolio, pp 81–86
6. Steve Jobs's Z (2009) 21st century business review. In: Bofan W (ed), vol 8. pp 11–17
7. Weifang W, Chaoran F (2007) Emotional design—contemporary product design and development trends. Ind Eng Learn Netw 05:383–388

Chapter 45
Study of FDI Location Choice in Mainland China Between Taiwan Area and South Korea

Yixian Gu

Abstract This thesis describes our investigation of 12 provinces or cities, including the top 10 Taiwanese and South Korean FDI locations, and presents our analysis of the province-level panel-data from 1990 to 2005 via logit model. Our results indicate that the two areas not only share investment location choice in mainland China in common, but also evidence their own characteristics; the two areas both have undergone agglomeration efforts in their investment location selections, and have invested proportionally to the basic establishment of the area; the principal distinction of the two is that Taiwan pays more attention to the market gravitation of areas, but South Korea focuses particularly on the labor resources of areas.

Keywords FDI · Location choice · Comparison · Demonstration

45.1 Study Background

FDI has recently evidenced some new characteristics and development trends, while its scale has continually expanded, especially with regard to factors including sources of funds, flow direction of areas, distribution of industries, and modes of investment. As a rapidly rising global socialist country, China amazed the world by manifesting sustained and rapid growth in the national economy, since its reform and opening, and also achieved a huge upswing in FDI.

Y. Gu (✉)
Suzhou Industrial Park Institute of Services Outsourcing, No.99 Ruoshui Road, Suzhou Dushu Lake Science and Education Innovation District, Suzhou, Jiangsu, China
e-mail: guyi20cxian@163.com

W. Du (ed.), *Informatics and Management Science VI*, Lecture Notes in Electrical Engineering 209, DOI: 10.1007/978-1-4471-4805-0_45,

A.T. Kearney, Inc. published their FDI Confidence Index 2002 report and showed that China, replacing the US for the first time, has become the most attractive FDI target country, whereas a decline in FDI has been observed in almost all the studied countries [1].

Korea and Taiwan, the areas immediately neighboring China, rose in the 1980s, being grouped into the category of "Asia's Four Little Dragons", and both began to invest in China in the 1990s. The two regions evidence considerable comparable factors, including geography, economic development scales, etc. Via the comparative analysis of FDI area selection in mainland China, this study will discuss the methods employed for area selection, and will summarize the experience and lessons learned in conjunction with FDI, in an effort to provide China with a more effective FDI strategy, especially in reference to Korea and Taiwan. Additionally, our results will provide a reference for Korea and Taiwan, by which they may improve and expand their direct investment efforts in China.

45.2 Empirical Model Specification

We have modelled the location decisions inherent to Korean and Taiwanese FDI in mainland China as a conditional logit problem, where the dependent variable is the province selected by each investor. In accordance with the methods employed by earlier researchers including Carton and Head et al., we exploit McFadden's result that logit choice probabilities may be derived from individual maximization decisions, provided that unobserved heterogeneity takes the appropriate form. We assume that each investor selects the province or city that is anticipated to yield the highest profit [2, 3]. Then, in order to evaluate the profitability of each investment, the profits of an investor in a province or city is dependent on the sets of independent variables, which are considered to influence the revenues and costs of investors [4].

On the revenues side, one variable is the GDP of every province. It reflects the market demand of the province, and is associated with the horizontal FDI. The greater the GDP is, the greater are the economics of scale, which translates to greater opportunity for industries and enterprises, and results in an increase in FDI; the GDP per capita is associated with the level of available development; the agglomeration effect is another variable affecting revenues. This is derived from the operating activities of nearby related enterprises, or from the FDI stock. For example, high-density manufacturing activities may attract more FDI in a province, as foreign investors serve existing manufacturers. Studies conducted in the 1990s also showed that relatively more FDI stock would attract more investment when area policies stabilized [5, 6].

With regard to cost, wages and interest are the principal factors. The province that costs least will be selected through investigation into the labour and capital markets. With regard to the labour market, wage level, labour quantity, labour quality, and employee-employer relations are extremely important factors; as for the capital market, one of the primary indices is the interest rate and adaptability of the capital market. Relatively high wage levels could limit the importation of FDI, which is referred to as vertical FDI [7, 8].

In addition, infrastructural construction is another factor that influences FDI. A well-developed transportation system could reduce operating costs via a reduction in the transportation costs of importing raw and processed materials and facilities, as well as exporting products. Additionally, preferential policies in specific areas may also influence FDI, as government policy is one of the most crucial factors in developing countries. For example, Chinese government granted preferential tax policies to foreign investment enterprises.

In order to conduct the quantitative analysis of regional factors of FDI, the following model was established via comparative analysis:

$$FDI = PGDP^{\beta_1} \times ROAD^{\beta_2} \times \cdots \times WAGE^{\beta_8} \tag{45.1}$$

We have dealt with the model logarithmically. There are several advantages to the logarithmic model: firstly, possible nonlinear relations can be transformed into linear relations; secondly, it could reduce the extreme values of variables; thirdly, it may avoid multiple contributions; finally, the coefficient of the regression equals the variable value of FDI in specific areas. Linear equation:

$$\ln FDI_{state} = \beta_0 + \beta_1 \ln PGDP + \beta_2 \ln ROAD + \cdots + \beta_8 \ln WAGE + \varepsilon \tag{45.2}$$

In the equation, FDI_{state}, as a dependent variable, is the gross FDI of a certain country which is actually utilized by every province; $\beta_{\mathrm{n}}(n = 0, 1, 2, \ldots, 8)$ is the coefficient of regression to be calculated; ε is a random disturbance variable; others are independent variables, temporarily referred to as area factors that influence FDI in mainland China.

45.3 Estimation Results

The data selected for data analysis in this study is the calculable data that may influence FDI. The data for China were obtained from the China Statistics Yearbook (1991–2006), the data regarding Korea FDI to mainland China were obtained from the Korean Export–Import Bank, and the data regarding Taiwan FDI to mainland China were derived from the Investment Commission, Ministry of Economic Affairs, Taiwan, China. Data selection included 12 provinces or cities of mainland China, (Beijing, Tianjin, Liaoning, Jilin, Shanghai, Jiangsu, Zhejiang, Fujian, Shandong, Hubei, Guangdong, Sichuan including Chongqing), which included the top 10 areas in which Korea and Taiwan have invested.

Using the up-conditional logit estimation, using the EViews 5.1 Standard Edition software, Table 45.1 provides the estimation results. The table shows that Korea and Taiwan evidence three points of homology. First of all, the agglomeration effect is statistically significant, and thus as the FDI appear to be positive and significant for the three models of Korea and Taiwan, they are considered to be highly pertinent to global FDI, and the significance for Korea is slightly higher than Taiwan.

Second, the IM-Exports are also positive, and the three models of Korea and Model 3 of Taiwan evidence significance, but are positive and non-significant for Models 1 and 2 of Taiwan. This means that Korean investment places a higher significance on the degree of the opening of the region.

Third, transportation development and the kilometers of railways in operation/hundred square kilometers by region were shown to be both positive and significant for both the three models of Taiwan and model 3 of Korea, but were non-significant for models 1 and 2 of Korea. It is interesting to note that the kilometers of highways in operation/hundred square kilometers by region were shown to be negative for Korea and positive for Taiwan. This issue is explained in more detail in Table 45.1, which depicts the amounts of investment by region from Korea and Taiwan.

According to the results of estimation, there are also significant differences between Korea and Taiwan. First of all, the developing levels of invested-in regions differ. The PGDP, which would be positive as has been shown in earlier studies, are negative for models 1 and 2 of Korea. As is shown in the appendix in Table 45.1, which present the regional distribution for Korea and Taiwan, it can be observed that the top 5 regions of Korean investment changed from 1991 to 1995s Shandong 33.39 %, Tianjin 11.89 %, Liaoning 10.41 %, Jiangsu 8.94 %, Beijing 7.53 % to 2001–2005s Shandong 26 %, Jiangsu 22.79 %, Beijing 13.72 %, Tianjin 8.15 %, Liaoning 7.14 %. At the same time, Taiwan's top 5 investment regions have not changed, and only distribution has changed: 1991–1995s Guangdong 31.35 %, Jiangsu 15.06 %, Shanghai 14.7 %, Fujian 13.79 %, Zhejiang 4.65 % to 2000–2005s Jiangsu 35.55 %, Guangdong 23.56 %, Shanghai 15.33 %, Zhejiang 8.3 % Fujian 7.34 %. This also shows that the regions in which Korea has invested are not the most developing regions of mainland China, because the surrounding Bohai Bay area is geographically closer to Korea. This also explains the issue mentioned in the third point above.

Second, the population was shown to be both negative and significant for model 1 of Korea and positive and non-significant for model 1 of Taiwan. Also, the employees–the number of staff and workers–are shown to be positive but significant for Korea, and non-significant for Taiwan. Future research should also be conducted on the regions in which they have invested. In mainland China, tens of millions of migrant labourers have left their hometowns in the central and western regions, for south-eastern coastal areas in which Taiwan has primarily invested, but the population variable shows only the number of people counted by the residence registration system. Thus, the population variable should not be considered to reflect the actual situation regarding where people live and work, and can only show where they officially belong. Also, the population by region variable is insufficiently accurate to account for quantity of labour. Thus, the employee variable is more useful in accounting for the quantity of labour. As compared to the Taiwan area, South Korea pays more attention to China's labour resources.

Table 45.1 Estimation results: Korea and Taiwan

	Korea			Taiwan		
Variable	Model 1	Model 2	Model 3	Model 1	Model 2	Model 3
C	17.13825 (1.02963)	−10.88827 (−1.88282)	−6.95419 (−2.09935)	−5.50431 (−0.32117)	−2.72626 (−0.43719)	−3.54073 (−0.92241)
Log $_{PGDP}$	−1.15293 (−1.27436)	−0.74116 (−0.88075)		0.58633 (0.60444)	0.73410 (0.80568)	
Log $_{ROAD}$	−0.19313 (−0.33223)			0.42321 (0.70248)		
Log $_{FDI}$	0.57950 (3.40144)***	0.66275 (4.03211)***	0.63676 (4.72194)***	0.58560 (2.84337)***	0.56998 (2.88260)***	0.62418 (3.39139)***
Log $_{IM\text{-}EXPORT}$	0.58864 (2.06251)**	0.53526 (2.03695)**	0.68209 (4.00676)***	0.35425 (1.18041)	0.44230 (1.64368)	0.45109 (2.45726)**
Log $_{RAILWAY}$	0.74029 (1.58074)	0.69596 (1.58970)	0.71827 (1.69062)*	0.79718 (1.68805)*	0.91052 (2.06211)**	0.94696 (2.19985)**
Log $_{POPULATION}$	−3.99813 (−1.76458)*			−0.06574 (−0.02997)		
Log $_{EMPLOYEE}$	1.62027 (2.08726)**	1.26661 (1.99824)**	0.86609 (1.94915)*	0.82941 (1.00831)	0.54047 (0.79487)	0.69736 (1.46121)
Log $_{WAGE}$	1.81863 (1.58580)	0.96060 (0.96650)		−0.46717 (−0.39761)	−0.64570 (−0.61726)	
R-squared	0.81210	0.80802	0.80683	0.79446	0.79374	0.79263
Log likelihood	−211.20749	−213.02393	−213.54509	−200.66634	−200.94888	−201.37899

Number of province = 12; Time = 15; number of observations = 180

Note t-Statistic are in the parentheses. Standard errors are included but not reported

*** significant at 1 % level

** significant at 5 % level

* significant at 10 % level

In the end, the wage variable, which shows the average wages of staff and workers by sector and region, was shown to be non-significant for both Korea and Taiwan—it was positive for Model 1 and 2 of Korea and negative for Model 1 and 2 of Taiwan. As has been demonstrated in earlier studies, the wage should be negative with the FDI, but Korean FDI in mainland China tells a very different story. There are two reasons for this: one is due to the geography as mentioned above, the other involves investment purposes. As has been previously determined, the primary investment purposes include the reduction of the cost of manufacture and trade, increasing the market share, bypassing the trade barriers, etc. On the basis of our research results, the principal purpose of Korean investment is to reduce the cost of manufacture and trade, and the wage level of the regions in which they have mainly invested is not the highest in mainland China. Specially, compared with the Korean wage level, mainland China's wage levels are only 10–20 % of the Korean levels, and many Korean investment enterprises utilize relatively higher wages to attract high-quality personnel. At the same time, according to research regarding the positive and significant relationship between IM-Exports and FDI, we can also show that a high share of products made by Korean investment enterprise has been exported. Thus, the level of wages is associated with Korea's FDI in mainland China. However, for Taiwanese investment enterprises, their principal investment purpose is increasing the market share in mainland China, and the majority of their competitors also make products in mainland China. Thus, it appears that Taiwanese investors pay more attention to wage levels than do Koreans.

45.4 Conclusion

We have researched the location choices of South Korea and Taiwan regarding FDI, and determined their characteristics. The points of similarity between the two are that the two areas both have undergone agglomeration efforts in selecting their investment locations, and have invested proportionally to the development of transportation in those regions. The principal distinction is that Taiwan considers more deeply the market gravitation of areas, but South Korea particularly emphasizes the labour resources of areas.

In this paper we selected 12 provinces or cities of mainland China to study, which include the top 10 regions in which Korea and Taiwan have invested. The logical extension of this study would involve more mainland Chinese regions, and would attempt to determine why the other regions (particularly the Western portion of mainland China) have such difficulties in attracting FDI, and attempt to formulate some suggestions for other regions which wish to attract FDI.

References

1. Kearney AT (2002) Global business policy council, vol 9. FDI confidence index, New York, pp 37–46
2. Carlton DW (1983) The location and employment choices of new firms: an econometric model with discrete and continuous endogenous variables. Rev Econ Stat 65:440–449
3. Head K, Ries J, Swenson D (1995) Agglomeration benefits and location choice: evidence from Japanese manufacturing investments in the United States. J Int Econometrics 38:233–247
4. McFadden D (1974) Conditional logit analysis of qualitative choice behavior. In: Zarembka P (ed) Frontiers in econometrics, vol 33. Academic Press, New York, pp 105–142
5. Wheeler D, Mody A (1992) International investment location decisions: the case of U.S. firms. J Int Econometrics 33:57–76
6. Huallacháin BÓ, Reid N (1992) Acquisition versus greenfield Investment the location and growth of Japanese manufacturers in the United States. Reg Stud 31(4):403–416
7. Barrell R, Pain N (1999) Domestic institutions, agglomerations and foreign direct investment in Europe. Eur Econ Rev 6:925–934
8. Timothy JB (1985) Business location decisions in the United States: estimates of the effects of unionization, taxes, and other characteristics of states. J Bus Econ Stat Am Stat Assoc 3(1):14–22

Chapter 46
Study of Capital Flows Effect on Regional Economic Development

Ke Hu

Abstract Speculativeness and unstable short-term capital flows will greatly influence the country's financial stability and economic development. Summarizes the impact "push–pull" China short-term capital flows are mainly: interest rate both at home and abroad, and expect to spread exchange, domestic capital market price and the domestic price in commodity markets, and access to the profound economic analysis. This paper the empirical study of China's short-term capital flows the relationship and the influence factors and monthly data from January 2004 to March 2009, we found a positive correlation relation communication, capital market interest rates, the price of commodity market price and short-term capital flows negative correlation with the expected, exchange rates and short-term capital flows, expected the greatest contribution of exchange rate is short-term capital flows.

Keywords Short-term capital flows · Regional economic · Sales price index of real estate

46.1 Introduction

In this unprecedented financial crisis, the amount of liquidity injection from the federal reserve, consistent, put the dollar exchange rate, and ultra-low interest rates in the United States, and create a perfect external environment, arbitrage. International deal with the appreciation of the renminbi continues to climb, and in China's rapid recovery crisis brought them the investment environment, the internal

K. Hu (✉)
ZhongZhou University, Zhengzhou 450044, Henan, China
e-mail: huke@cssci.info

W. Du (ed.), *Informatics and Management Science VI*, Lecture Notes in Electrical Engineering 209, DOI: 10.1007/978-1-4471-4805-0_46,

environment for arbitrage [1, 2]. From the combination of internal and external environment of the opportunity attracted a lot of international capital flows to China. Have all kinds of background money to have different ways of transfer capital controls into China [3]. The extending and deepening financial crisis in the world, has resulted in a loss of balance monetary and financial system, and at the same time, the pace of currency appreciation is slowing, domestic monetary policy has begun to turn, therefore, number quietly and direction and evolution of short-term capital flows trend in China began to face much greater uncertainty.

46.2 Economics Analysis and Theoretical Framework

It should be considered from abroad and domestic economic environment, the basic economic conditions the role of push and pulls the two problems [4], the first capital flows between developed and developing countries. The domestic factors increase risk in emerging markets, usually refers to the external factors have in the developed countries [5], leading to the global recession interest rates sharply, make profit opportunities in emerging market economies more attractive relative, so a lot of money to turn to emerging market and economic development more quickly find better investment opportunities.

46.2.1 Domestic and Foreign Money Market Arbitrage

Development mode (Mondale 1960 PTS); [1] based on the theory of interest rate parity a better explanation of problem of short term capital into China: people think capital flows is incomplete, or provide arbitrage capital has a limited nature, domestic and foreign interest rates would cause limited spread flow of capital [6]; At the same time, it is assumed that arbitrageurs are risk averse, they need some extra money before, they are willing to hold risk assets, that is, the amount of money into the domestic monetary growth function of the risk premium regulations, and the local currency-denominated assets a huge outflow of funds from the function is to reduce risk premium place currency-denominated provide assets [7]; Finally, assuming that while facing major macroeconomic policy changes, the rise (depreciated) value of the rate of expected exchange rate is given, that is expected exchange rate is static [8, 9]. In these assumptions, we can use interest rate parity Eq. (46.1) instead of the general uncovered interest rate parity equation, that is:

$$rd = rf + \triangle Ee + P \tag{46.1}$$

46.3 Exchange Arbitrage Profit

Dornbusch (1976) with the exchange rate determination model of sticky prices still has to prove that found that interest rate parity from a long-term perspective, in the short term the real exchange rate deviating from the equilibrium exchange rate, and gradually regresses equilibrium exchange rate [10, 11]. The model is based on the following core assumptions: (1) in product market price adjustment of the existing sticky; (2) prior purchasing power parity; (3) found that the interest rate parity. Assuming that (1), when a one-off currency, realizes the impact in product market price, and thus cannot instantaneously adjust the cleaning products market. Real exchange rate will be temporarily diverted from its long-term equilibrium level, then gradually "regresses" to the balance. Assuming that the "regression" process be described by the following Eq. (46.2):

$$E_t\left\{q_{t+k} - \bar{q}_{t+k}\right\} = \alpha^k(q_t - \bar{q}_t)(0 < \alpha < 1) \tag{46.2}$$

q_t is logarithm of the real exchange rate for t periods, defined as $q_t = \text{et} + pt^* - pt$, et is the logarithm of nominal exchange rate, while the $pt(pt^*)$ is the logarithm of the domestic (foreign) price marked by domestic (foreign) money [12]. The nominal exchange rate use direct quotation, that is say the units of foreign currency price in local currencies; while the real exchange rate is defined as relative price of a basket of foreign goods for valuation of a basket of domestic goods. q rise means depreciation of the real exchange rate of national currency; q decline means appreciation of the real exchange rate of national currency. E_t is conditional expectation based on available information in the t period; is equilibrium real exchange rate under a flexible price adjustment. Parameter $0 < \alpha < 1$ represents the adjustment speed of real exchange rate regressing to its equilibrium state, usually is a function of other exogenous parameters in the model [13] (Fig. 46.1).

The deviation between the real exchange rate and the equilibrium exchange rate in short term has led to speculative capital arbitrage mechanism and, thus, short-term capital flows.

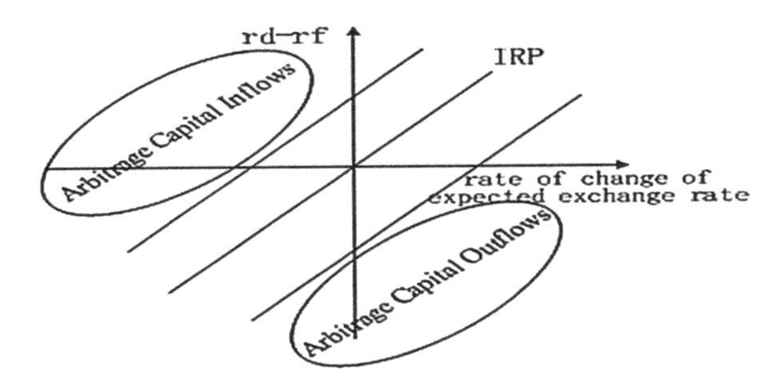

Fig. 46.1 Arbitrage capital flows with risk premiums and transaction costs

46.4 Empirical Analysis and Result

46.4.1 Stationary Analysis

In order to avoid the false regression with use of non-stationary variables, we use ADF (Augmented Dickey-Fuller) method for all variables to conduct a unit root test. The test results are shown in Table 46.1.

Be seen from Table 46.1, even in the significance level of 10 %, ADF values of *r* and LNE is not significant, indicating that these series have unit root. ADF values of first-order difference of all series reject the unit root null hypothesis at the 5 % significance level, so that all variables are I (1) series.

46.4.2 Co Integration

Based on the above unit root test, although the r and LNE are no stationary series, but the first-order difference variables are stationary. Linear combination of first-order integration of non-stationary variables may have co integration relationship, which can reveal the long-term stability equilibrium relationship between various variables.

We carry out co integration relationship between two or more non-stationary variables by Johansen co integration test. Trace test is used. Table 46.2 shows the final result.

Table 46.2 shows that trace test indicates 3 co integrating equations at the 0.05 level, for considering the relationship between variables should be in line with actual economic situation, the paper retains a co integrating equation. The equation shows the long-term stable equilibrium relationship among the variables. The corresponding normalized co integrating Eq. (46.3) is as follows:

Table 46.1 Augmented dickey-fuller test

Variable	(c,t,k)t	t-statistic	Piob.*
Scf	(c,0,5)	−4.5689	0.0005*
Δscf	(0,0,0)	−9.9901	0.0000*
R	(c,t,0)	−2.5341	0.3113
Δr	(c,0,1)	−3.1807	0.0261**
Lne	(c,t,0)	−0.9699	0.9405
Pc	(c,t,5)	−3.5864	0.0400**
Δpc	(0,0,0)	−3.3824	0.0010*
pa	(c,t,1)	−3.5020	0.0485**
Δpa	(0,0,0)	−4.4738	0.0000*

Note C, T and K, respectively, is constant term, trend items and lag-order number
*indicates significance at 1 % significance level
**indicates significance at 5 % significance level

Table 46.2 Unrestricted co integration rank test (trace)

Hypothesized no. of CE(s)	Trace statistic	0.05 critical value	Prob.**
None × 1	110.3118 6	69.81889	0.0000
At most 1 × 6	63.20894 4	47.85613	0.0010
At most 2 × 3	33.84333 2	29.79707	0.0162
At most 3 × 9	9.694092 1	15.49471	0.3050
At most 4 × 1	1.272457 3	3.841466	0.2593

**denotes rejection of the hypothesis at the 0.05 level; trend assumption: linear deterministic trend; lags interval (in first differences): 1–3

$$\begin{aligned} \text{SCF} &= 20062.79\text{R} - 3100807\text{LNE} + 32.89900\text{Pc} + 19.61726\text{Pa} \\ &\quad (5212.80) \qquad (530702) \qquad (13.2923) \qquad (18.9365) \end{aligned} \tag{46.3}$$

The results show that in long-term relationship [Standard errors in ()] of short-term capital flows and its influence factors, there is a long-term equilibrium relationship between interest rate differentials, rate of change of expected exchange rate, the domestic capital market prices domestic commodity market prices and short-term capital flows. There is a positive correlation between Chinese short-term capital flows and spreads, stock prices [14], property prices, a negative correlation between short-term capital flows and the rate of change of exchange rate. When the expected exchange rate is more than zero, the RMB against the U.S. dollar nominal exchange rate is greater than the real exchange rate. The RMB is undervalued at this time [15], there is pressure on appreciation of the RMB and the international capital substantially flows into China. From the coefficients we can compare that the impact of short-term capital flows from changes in prices of the domestic capital market should be greater than the impact from the domestic prices of commodity market.

46.5 Conclusions and Implications

Through the above, we can see that the empirical analysis of the exchange rate, the interest rate expectations diffusion, capital market price and the commodity market price and the short-term capital flows [16], this is a long-term equilibrium relationship. A positive correlation relation communication, capital market interest rates, the price of commodity market price and short-term capital flows; there is a negative correlation between the pace of change in the exchange rate of the expected and short-term capital flows.

In conclusion, we suggested adding diversity means of control and flexible financial market and macroeconomic regulation and control policy, perfecting capital market operating mechanism, to avoid as far as possible, change the dramatic changes of short-term capital flows in the smooth to reduce bad effects on the economy.

Acknowledgments The paper is supported by Henan province social science research program (Henan capital flows and regional economic development empirical research).

References

1. Mundell R (1960) The monetary dynamics of international adjustment under fixed and flexible exchange rates. Q J Econ 74(2):227–257
2. Fleming M (1962) Domestic financial policies under fixed and under flexible exchange rates. Int Monetary Fund Staff Pap 36(9):369–379
3. Dornbusch R (1976) Expectations and exchange rate dynamics. J Polit Econ 84(6): 1161–1176
4. Prasad E, Wei SJ (2005) The Chinese approach to capital inflow: patterns and possible explanations. NBER Working Paper 1:13–06
5. Bouvatier V (2007) Hot money inflows and monetary stability in China: how the people's bank of China took up the challenge. Working Paper 11:23–30
6. Kindleberger CP (2000) Manias, panics and crashes: a history of financial crises. Peking University Press 76:247–255
7. Cai F (2008) The factors influencing short-term international capital flow in China-An empirical study based on the data of 1994–2007. Finance Econ 31(6):312–345
8. Qu FJ (2006) Chinese short-term capital flows and statistical empirical analysis. Rev Econ Res 87(40):212–250
9. Wang Y (2004) China's capital flow between 1982 and 2002. Manag World 24(7):33–65
10. Cuddington J (1986) Capital flight: estimates, issues, and expression. Princeton Stud Int Finance 35(58):1–40
11. The World Bank (1985) World development report, vol 25. Washington, DC, pp 111–123
12. Genevieve BD, Shang JW (2008) Can China grows faster Education of China. 87:23–65
13. Zhang J, Yang H (2005) A Diagnosis on the fragmentation of the domestic capital market. World bank report, vol 3, pp 19–73
14. Feldstein M, Horioka C (1980) Domestic saving and international capital flows. Econ J 35(90):314–329
15. Leachman LL (1991) Saving, investment and capital mobility among OECD countries. Open Econ Rev 32(2):137–163
16. Sinn S (1992) Saving-investment correlations and capital mobility: on the evidence from annual data. Econ J 2(9):1162–1170
17. Wang Y (2004) Capital flows in china after 1994. Stud Int Finance 22(6):143–165
18. Wang Q (2006) An econometric study of determinants of capital movement in China. Stud Int Finance 75(6):320–347
19. Wang SH, He F (2007) Short-term international capital flow in China: current situation, channels and influencing factors. J World Econ 58(7):99–132
20. Zhang YH, Pei P, Fang XM (2007) An empirical analysis on short-term international capital inflow and its motivation in china: based on triple-arbitration model. Stud Int Finance 21(9):88–92

Part III
Education in Informatics II

Chapter 47
Analysis on Double-Sword Effect of Spreading of Western Culture and Traditional Chinese Culture Based on T-Test Law

Xiaomei Qi

Abstract Through the analysis of Western culture on the dual effect, this paper is mainly about traditional Chinese culture spread analysis, a culture of my people in the West festival conducted an investigation, from a quantitative point of view to reflect the traditional culture of Western culture on China spread the impact of the research data, descriptive statistical analysis of the epidemic and permeability analysis of Western culture and Western culture paired t-test analysis, a fuller analysis of the pros and cons, so that everyone aware of the need to strengthen the traditional Chinese degree of attention to a culture, continue to inherit and carry forward China's traditional culture.

Keywords Double-sword effect · Globalization · Quantitative analysis · Statistical analysis

47.1 Introduction

Culture is not only the core of country, but also an important carrier of the human soul [1]. Each country has its own culture, Chinese culture is profound. With the rapid development of globalization, information technology, make the western culture constantly into China, and also gradually into our culture spread among. This paper will discuss the cultural aspects of the festival according to the current research found that more and more Chinese people to accept more western festival, and as important festivals celebrating, and for the Chinese traditional culture

X. Qi (✉)
Dongying Vocational College, Dongying 257091, China
e-mail: xiaomei_qi21@yeah.net

W. Du (ed.), *Informatics and Management Science VI*, Lecture Notes in Electrical Engineering 209, DOI: 10.1007/978-1-4471-4805-0_47,

transmission is consciousness is not good enough. This clearly indicated that the impact of Western culture on the traditional Chinese culture, how to let everyone to keep the correct world outlook and values, the correct treatment as the Western culture to learn from the role, and make better use of the advantage of the west culture to achieve its essence, to the purpose of the dregs, protection of China's traditional culture and at the same time, continue to inherit and carry forward the traditional culture of our country [2].

47.2 Double-Sword Effect

47.2.1 *Positive Effect*

First of all, learning and understanding of Western culture, help to strength enervations with Western friends, can promote the harmonious coexistence, and better able to achieve cooperation and development [3]. Secondly, at the same time celebrate the Chinese traditional culture, constantly enrich the content of self-culture, to strengthen exchanges between China and Western countries, but also by Western cultural reference, and to continue to improve the culture of learning and development, For example, Western culture in a Thanksgiving Day, Father's Day, Mother's Day, Chinese culture is a good manifestation of so you can take this opportunity, Chinese and Western cultures of effective integration, promote feelings at the same time, rich traditional Chinese culture, more reflect the culture of good moral character, enrich the spirit of the people of civilized life.

47.2.2 *Negative Effect*

According to survey data show that 86.88 % of people think that now the atmosphere of Western culture more and more concentrated, indicating that the era of such a message in the current rapid development of Western culture, traditional Chinese culture spread a huge impact, all kinds of information and forms of Western culture, values will continue to influx we have to live and work. This is the Chinese traditional culture adversely, will dilute the spread of traditional Chinese culture and heritage, development, and reduce the demand for traditional Chinese culture is not conducive to learning and heritage of our people of their own culture. Too much attention to the Western culture, the loss of fine culture, is not conducive to establish a correct outlook on life and values.

For western cultures, and always keep a correct attitude to learn from each other, make full use of the advantage of the west culture to improve and enrich our culture, absorbing the essence of western culture. In-depth understanding of western culture, and we better go out of the country, to the world and lay a solid foundation. In learning and celebrate the western culture, to strengthen our own nation the protection of traditional culture and promote effectively, China's culture is really profound, we need to constantly to explore, inheritance, because we need in its own traditional culture on the basis of the study, come again with western culture, not to appear at the shadow and lose the substance of this kind of phenomenon, lost their own national excellent traditional culture, will be a huge loss, and at the same time we need to actively to maintain and develop the traditional culture.

47.3 Research Methods and Quantitative Analysis

In this paper, taken to a sample survey of school students, select 150 samples. In this paper the design questionnaire, and then fill in the investigation, effective questionnaire for 150 copies. This paper is aimed at the research of western festival celebration of the data of the western festival celebrate frequency of descriptive statistical analysis, and then a contrastive analysis of the differences of on western culture, using SPSS statistical software in quantitative statistical analysis, western culture of influenza and permeability of the Chinese and western culture and correlation analysis, paired t-test and variance analysis.

Art field and cultural industry are the nation of the generally accepted art performance means and the characteristic sign of the region [4].

They described the function of cultural entrepreneur and art business enterprise and took making money as a purpose and not with make money for purpose of production, the movie, television, book, music of allotment and consumption, drama, dance, sense of vision art, disguise, multimedia, animation etc. Cultural section isn't only a business realm, it is a space that appreciates beauty to be worth of with social spirit, suggest, establishment and the mental meaning of the representative and the pleasant sensation of the body. Group of islands developing country culture produced since the childhood of the angle sees, are the cultural identity that important realm and means of an investment propped up. This also contributes to in effort that single cultural economy diversifies to depend on too much narrow of tradition and non- tradition export base.

It is increasing the contribution to GDP, exit and employment in some cultural industries of developing countries. The islet island developing country participation world culture analysis economy suggests, with a big of trade unbalance of big part operation, abnormality drives Singapore. Descended form to provide to cultural merchandise trade export and the import data, especially those nations with some exit ability. But be called their arts and cultural industry, but still have a trade deficit.

According to the Fig. 47.1 descriptive statistics and analysis for the attention of all western festivals very tall, and to participate in the festival is more also, in addition to Halloween accounted for only 11.88 % besides, April fool's day and

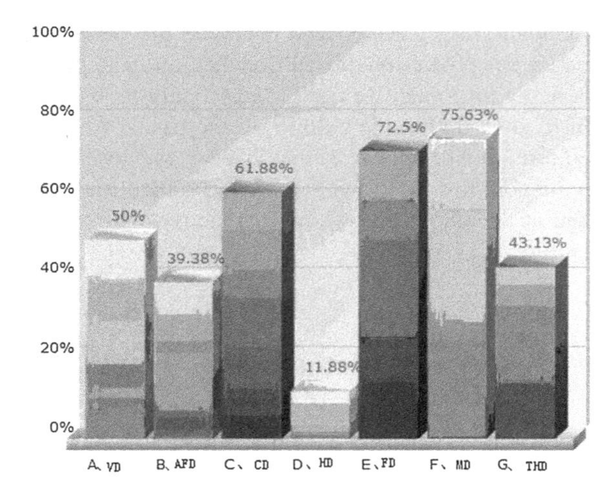

Fig. 47.1 Sample groups on the Western holiday celebration. *Notes VD* valentine's day, *AFD* april fool's day, *CD* christmas day, *HD* halloween, *FD* father's day, *MD* mother's day, *THD* thanksgiving day

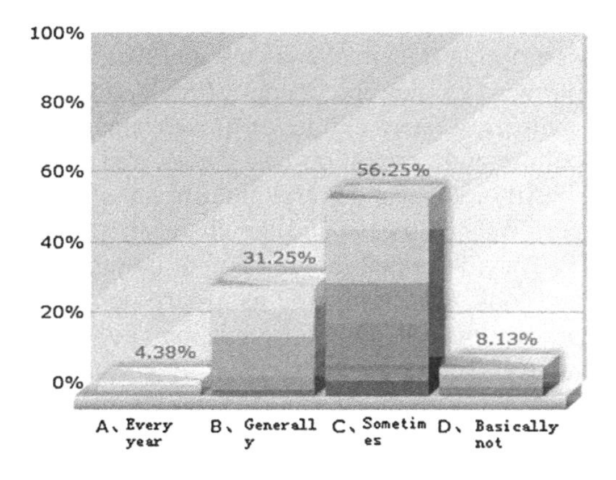

Fig. 47.2 Samples of the West festival to celebrate the frequency

Thanksgiving is at about 40 %, and Christmas, father's day, mother's day and were above 60 %, valentine's day also account for 50 % of the scale.

Descriptive statistical analysis in Fig. 47.2 can be drawn to a high degree of sample groups to celebrate the Western festival. Don't celebrate the festival of ratio is 8.13 %, and the account for every year than the minimum, only 4.38 %, are generally and the occasional respectively in the ratio is 31.25 and 56.25 %, so the two accounts for only 87.5 % of the proportion than reached [5].

Cultural industry is a problem that leads to go into a world to trade rule and manages and be subjected to huge concern. Cultural product kimono of passing duty United Nations teaches the section text the organization, the contesting of trade liberalism of its in many respect promotes cultural diverse sex. Many developing countries support its latent function to promote through IICD cultural diverse sex, and help more equilibrium cultural merchandise to trade, serve and intelligent property right. However, to what many developing country keys challenge is the convention don't produce right or the commitment for signing.

Table 47.1 The correlation analysis between epidemic of Western culture and permeability

		Epidemic	Permeability
Epidemic	Pearson correlation data	1	0.462**
	Sig. (2-tailed)		0.000
	N	150	150
Permeability	Pearson correlation data	0.462**	1
	Sig. (2-tailed)	0.000	
	N	150	150

May encourage more art productions at this meaning last convention, but it cannot promise to occupy a certain space market. Therefore, the main problem wants to ensure that the rule that be continuously changing is a basal bargain system, developing country can the vivid promote cultural business enterprise. The key here problem BE, pass applied a series of industrial and creative policy measure, developing country whether can meaningfully participate in economic boundary in the world of this kind of extends.

Under this background with ensure a competition ability and can keep on the existing strategy of development isn't enough. BE exactly under this kind of background; suggest protecting diverse sex of culture and promoting cultural industry. Pass to study traditional knowledge and cultural inheritance with, and cultural industry and start a business spirit, get is in cultural industry of some have the initiative realm.

From the Table 47.1 show, western culture of influenza and permeability in the level of 0.01, $p < 0.05$, and epidemic and permeability Pearson correlation coefficient is 0.462, said the investigation in the sample of most of the people of western culture more identity epidemic and permeability is of relevance.

From Table 47.2 can obviously, the conclusion of the Chinese and western culture in pairs of inspection Sig. (2-tailed) $= 0.001 < 0.05$, so it reflects the Chinese and western culture and significantly difference exists between the western festival, everyone for fresh degree is higher, on the western holiday has great interest.

From Table 47.2 paired t-test and Table 47.3 variance analysis can be concluded that the inspection, the Chinese and western culture in the sample survey of the population, especially among young people can get popular and spread to western culture or because the curious and as a result of freshness. Therefore also appeared to all of western culture advocate, for the Chinese traditional culture is weakened the relative. Cultural industry be the support that the center can keep on a development, on this foundation from interrogate to solve to reach of the meaning and fulfillment of cultural industry. This text outlined a cultural industry of can keep on to develop the target and value that the agenda has greater potential coherent, like social justice, renew through own efforts and ecosystem balance. Combine diverse sexual value from cultural, equal to formerly lift to inherit diverse sex is canning keep on an arguing of development. It makes sure with protection effective transition for target, can approve culture and promote cultural

Table 47.2 Chinese and western culture t test in pairs

	Differences in pairs					t	df	Sig. (2-tailed)
	The mean	Standard deviation	Standard error of the mean	95 % confidence interval of the difference				
				Low	High			
The fresh degree of western festival	0.460	1.383	0.134	0.185	0.716	3.392	150	0.001

Table 47.3 The variance analysis of Chinese and western culture

	The total variance	Df	Mean square deviation	F	Sig
In group	7.452	6	1.849	3.429	0.016
Out group	60.328	144	0.557		
In total	67.780	150			

industry, promoting the shape of cultural content and promoting among them can keep on a development. Cultural industry should drive the strategy for seeing to a key, have to have a foothold development foundation in the region, enhance special feature advantage, improve system environment, science establishment develops a target, so as to carry out cultural industry continuously to quickly develop.

47.4 Conclusion

In the face of the globalization of the international form, we want to fully grasp, strengthen and promote our country's own excellent traditional culture development, and at the same time effectively double edge effect of good western culture, do the disadvantages, and continuously learn from each other, take the essence and discard the dregs, constantly enrich Chinese traditional culture, in order to truly achieve our leapfrog development, can we get a chance to better understand the west, more learning, and use for reference, and finally achieve out of the country and the world and goals. In short, effectively grasp opportunities, will be the western culture double-edged blade effect adequately and hold of, so as to better promote the development of traditional Chinese culture, which can make our traditional culture more inheritance, spread and carry forward.

References

1. ChaoZheng Z (2008) Chinese culture and western culture self beyond comparison of. Guangxi Soc Sci 8(9):235–236
2. Fang L (2009) China traditional culture research status quo, orientation and development orientation. Jiangxi Soc Sci 4(5):131–133
3. Liu Y (2010) Analysis on Western cultural influence of the modern Chinese literature started cause. J Sichuan Educ Coll 12(16):380–384
4. Zhang L (2010) Under the background of globalization foreign culture on the impact of Chinese traditional culture. J Henan Sci Technol Univ 12(6):221–223
5. Sun L (2009) Chinese and western culture and values to the students' learning style influence. J Liaoning Educ Adm Inst 21(11):46–48

Chapter 48
Analysis of Contemporary Students Psychological Problems

Xuqian Zuo, Junling Wei and Jia Liu

Abstract Through anonymous questionnaire, we randomly selected college students to do the psychological questionnaires. At the same time, we selected 168 psychological counselling cases to analyze the psychological problems of the contemporary college students, and made analysis of personal reasons and social reasons for these problems, and finally we use evaluation school mental health open class to make teaching effectiveness evaluation. By analyzing we know that psychological problems mainly exist in the feelings of love, interpersonal relationships, and learning. Through the evaluation of teaching on college student's psychological effect, although colleges and universities offer this course, the role of the public class is not so optimistic.

Keywords Psychological dynamics · Game · Quantitative analysis · Team spirit

48.1 Introduction

In recent years, with the increase of the university campus emergency and malignant cases, people have to pay more attention to psychological problems, especially after the typical significance of the event—Majiajue incident, and the deal with psychological problems is regarded as an education research questions [1–3]. Types

X. Zuo (✉)
Chemistry and Environmental Engineering Department, Hebei Chemical and Pharmaceutical College, 050026 Shijiazhuang, China
e-mail: xuqianzuo@163.com

J. Wei · J. Liu
Polytechnic College, Hebei University of Science and Technology, 050000 Shijiazhuang, China

W. Du (ed.), *Informatics and Management Science VI*, Lecture Notes in Electrical Engineering 209, DOI: 10.1007/978-1-4471-4805-0_48,

of cases and problems of the entire campus, which is not difficult to find these two new cases: one is that a proportion of the contemporary psychological problems in a high group exhibition range; the other is that it is the psychological problems of the contemporary college students who are more likely to become evolution malignant cases, endangering personal safety. Due to these new features, we must take practical measures to solve the psychological problems of college students. With the intensification of the degree of modernization, market economy is in continuous rapid development and there is increasingly fierce competition in the community, so a variety of irreconcilable contradictions have emerged, such as housing problems in the employment problems after graduation and so on. College students as the main bearings of the Chinese society, the high expectations of community are on college students; college students as a special group of young people is at the core of interwoven contradictions, so they are in the face of the psychological pressure which becomes significantly higher than other social groups. How to improve the psychological quality of university students, and how to better the community healthy and stable development have become the needs of educators, and need to be resolved. On the basis of student's mental health status, we make analysis from the multi-angle of the psychological problems of the college students, and make analysis of college student's personal reasons and social reasons, so as to put forward constructive countermeasures to better solve the college students psychological problems, so that they can have a healthy psychology [4].

The psychological problems of the contemporary college students are mainly on learning, relationships, employment, and love. In recent years, problems such as suicide, network-dependent issues, and so on are increasingly prominent. Contemporary psychological problems have many reasons: the objective reasons are social, family and school environment, but also have their own reasons. Theorists generally believe that this is more important to student's own reasons, but their own reasons are more consistently summarized, including that the psychology is not mature, high expectations, lack of self-consciousness, weak, and so on, but behind a deeper level of value orientation, these reasons are the lack of research [5]. The anonymous questionnaire randomly selected college students in psychological questionnaires, made analysis of the main psychological problems of the contemporary college students by the cases of 168 psychological counselling cases, and made analysis of personal reasons and social reasons for these problems.

The mental health means the individual is able to adapt to the environment, has a perfect personality, and cognitive, emotional reaction, an act of will in an active state, and can maintain a normal ability to regulate and control. Some experts predict that: mental illness in the twenty-first century would seriously endanger the physical and mental health. World Health Organization surveys of many countries in recent year's studies have shown that every 2/5 persons have psychological problems in the world's population. Among the 18–22 years in China, the latest National Mental Health survey found those Chinese student's mental and behavioural problems of incidence up to 14.8 %. The relevant departments have also a sample survey of

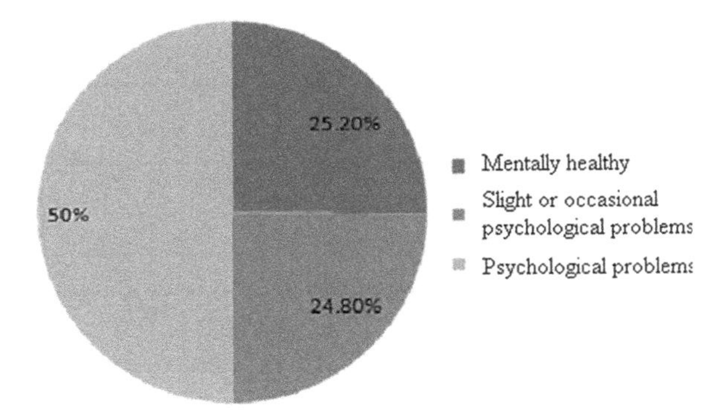

Fig. 48.1 Mental health of the college students

college students and found that among college students, about 3/7 have different degrees of mental disorder. These data suggest that in student's growth, their mental illness is more severe than adults.

48.2 Current Psychological Problems Investigation

48.2.1 Anonymous Questionnaires

We randomly selected six universities in Beijing and conducted a questionnaire survey; questionnaire takes anonymous form and questionnaire targets are college students; the contents of the questionnaire mainly focus on the psychological health, the main manifestation of psychological problems and so on. We use a random sample of 1,000 people, using an anonymous way to ensure real and effective questionnaire. Then selected another group of colleges and universities and conducted a questionnaire survey [6]. We repeated the above steps; the purpose of doing so is to exclude chance from the questionnaire data to make it more reliable and scientific. The findings of the survey show that nearly 75 % of college students are psychologically unhealthy, there being a variety of mental health problems, only 25 % of the students are mentally healthy, as is shown in Fig. 48.1.

In addition, nearly 75 % of the college students exist a variety of psychological problems; psychological unhealthy degree varies, of which the psychological problems of college students mainly show in six aspects, namely: psychological problems caused by employment pressure, psychological problems caused by pressure of study, the emotional trigger of psychological problems, a special family background causing psychological problems, interpersonal relationships lead to the psychological problems, and self-awareness causing the psychological problems. At the same time survey research is made on those in school student's psychological problems causes, and we analyze data, the ratio of psychological problems in college students are more common, and the main manifestations of psychological problems are also more diverse, such as

Table 48.1 Frequency distribution of visiting student for psychological counseling

Counseling frequency	Once	Twice to ten times	More than ten times
Number	115	47	6
Ratio (%)	68.5	28.0	3.5

Table 48.2 Visiting student's families economic situation

Whether or not poor students	Yes	No
Number	68	100
Proportion (%)	40.5	59.5

Table 48.3 Visiting the student's psychological problem type distribution

Consulting type	love	learning	Other aspects	Mental illness	self-image	Career
Number	48	24	14	12	11	8
Proportion (%)	28.6	14.3	8.3	7.1	6.5	4.8

depression and inferiority caused by psychological problems, and also irritability caused by psychological problems. Meanwhile, the main reason for college students with psychological problems are personal factors, school factors, social factors and family factors. Among them, personal factors accounted for 30 %, and social factors accounted for 15 %.

48.2.2 Counseling Case Studies

This survey data is from a counseling center record of 168 cases of college students in counseling, mainly targeted at college students, and they voluntarily take the initiative for counseling, the boys of 42, accounting for the overall number and 25 % of the girls are 126 people, accounting for 75 % of the overall number. The number of freshman is 50, accounting for 30 % of the total number [7]; 36 sophomore, which is 22 % of the total number; junior and senior students accounted for 48 %, a total of 82 people to participate in psychological counseling. By studying the case of psychological counseling, it can be found that the number of visiting students varies from person to person, as is shown in Table 48.1; visiting student consultation time shows a certain distribution; distribution of family economic status of visiting students is shown in Table 48.2; visiting student counseling psychological problems showing a certain type, as is shown in Table 48.3.

It can clearly be seen from the above chart that the number of visiting students varies from time. Visiting students also show a certain distribution difference: distribution of family economic status; visiting student counselling psychological problems are also different, showing a certain type.

48.3 The Mechanism of Psychological Problems in College Students

48.3.1 Environmental Changes Causing Them not Suited to the Environment

The survey shows that 30 % of people are more sensitive to changes in the weather; there will be various symptoms such as fatigue, forgetfulness, stars, depression, reduced work efficiency, poor sleep, and lack of concentration and so on. Therefore, the spirit of research scholars as the "weather" disease of modern society is a unique disease of civilization. If you master the law and patients can take appropriate measures, or in accord with nature, or make the best use or application of artificial confrontation technology to avoid or minimize the adverse effects of meteorological factors on the human body [8].

48.3.2 Competitive Pressures Bring Psychological Anxiety

Now China has more and more college students, so the natural pressure of competition is increasing, which for university students is a negative impact. The competition should be maintained by psychological stability to avoid the emotional ups and downs. Where there is competition, there is strong or weak, and the weak must withstand the blow of failure. Your failure in this competition does not mean you destined to fail but also in the future competition; your failure in this competition does not mean that everything is better people. You have to overcome inferiority complex, choose the direction of the effort, determined to catching up fishes. Give up on them is the idea intolerable, which leads to another type of loser defeat and hate and revenge, and has fully, exposed their selfish nature, and will put him to despair.

48.3.3 Disharmonies of Interpersonal Relationship

For the new university students, relationships become another major factor to affect their heart health, and there are points to note [9]: (1) You may be the first to leave the house to leave the parents feel helpless, and you need to overcome dependence, and learn to deal with a lot new transaction. If you did not do a good job it does not matter, and slowly will be good, you should be strong and independent, and allow others to more easily agree with you; (2) Living collective life to maintain the individuality, but at the same time do not forget to respect other people's will and the public will, and good personal accomplishment will help you to add points in interpersonal relationships; (3) After leaving their original environment, you may not be immediately understood by the people around by the original understanding of the characteristics with recognition and advantages may not be with recognition friends, teachers, students now, so give yourself and others some time to mutual understanding, not to quickly to any matters to conclusions; (4) It is possible that you are very good and the number one in a small circle of people, but as we get older with academic growth, the circle is growing, and now learn from the university but also from universities to society, you should understand that there will be a lot of the same or more excellent people; (5) Learn to appreciate others.

48.3.4 Lack of Psychological Qualities

Lack of psychological qualities such as self-awareness, one-sidedness, fragile emotional, impulsive, unstable, weak, cowardly, vain, cold, stubborn, lack of a correct outlook on life and positive attitude towards life, frustration and poor and do not understand mental health, and lack of the skills of psychological adjustment.

48.4 Evaluation of Teaching Effectiveness of the Student's Psychological Public Courses

Comprehensive evaluation method is an applied mathematics course with clear or unclear research with uncertainty. The phenomenon of expression is the gray system evaluation model form order differential equations to reveal the effect of the student's psychological public teaching. In this paper, the DPS mention software provides comprehensive evaluation system for the evaluation model GM (1, 1) and we make the analysis.

(1) We make an accumulated generating operation on data sequence (48.1) to get Eq. (48.2).

$$\mathrm{X}(0) = \{\mathrm{x}(0)(1), \mathrm{x}(0)(2), \ldots, \mathrm{x}(0)(\mathrm{N})\} \quad (48.1)$$

$$\mathrm{X}(1) = \{\mathrm{x}(1)(1), \mathrm{x}(1)(2), \ldots, \mathrm{x}(1)(\mathrm{N})\} \quad (48.2)$$

In the equation, $X(t) = \sum_{k=1}^{t} x^{(0)}(k)$.

(2) We construct the cumulative matrix B and constant vector YN, which is

$$B = \begin{bmatrix} -\frac{1}{2}(x^{(1)}(1) + x^{(1)}(2)) & \cdots & 1 \\ \vdots & \vdots & \vdots \\ -\frac{1}{2}(x^{(1)}(N-1) + x^{(1)}(N) & \cdots & 1 \end{bmatrix} \quad (48.3)$$

$$\mathrm{YN} = [\ \mathrm{x(0)(2), x(0)(3), \ldots, x(0)(N)]T} \quad (48.4)$$

Through public course of the student's psychological evaluation, it can be seen that university attaches great importance to student's psychological health problems, but investment in mental health education is not enough, and there is no effective practical solution to some college student's mental problems.

48.5 Conclusion

Mental health means the individual is able to adapt to the environment, has a perfect personality, and cognitive, emotional reaction, an act of will in an active state, and can maintain a normal ability to regulate and control. Mental health is long-term efforts in order to the progressive development of the establishment, do not look forward to by hypnotized, as if by magic, and quickly reached. In this unrest social, uncertainty filled with generations, so something putting on a false scientific cloak of religion, doctrine or activities are very easy to attract interest. In the past and at present, the revolution is specious book sensation selling which is indicative of the thirst of fast-food culture of the modern mind. Therefore, how to make our fellow citizens treat the transaction to achieve mental health step by step requires that we should work forward. By analyzing we know that psychological problems mainly exist in the feelings of love, interpersonal relationships, and learning. Through the evaluation of teaching on college student's psychology, although colleges and universities are offering this course, the role of the public class is not so optimistic.

References

1. Guo J, Deng X, Yang X (2007) 400 students' psychological problems investigation and countermeasures. SE Univ Med Sci 3(5):57–59
2. Hu Y (2010) Analysis on the status of college students from the 236 cases of psychological counseling. East Chin Jiaotong Univ 1(12):225–230
3. Li B, Kong F (2009) College students with financial difficulties and mental health and its educational countermeasures. Heilongjiang Higher Education Research 3(7):335–336
4. Qiu B, Gao L (2010) Poor college students and their mental problems. Med Educ 21(2):41–42
5. Wang T, Ma Y (2007) Research on the psychological problems in the past five years. Chin J Health Psychol 2(3):22–24

6. Wang X, Yu Y (2007) College Student's Mental Health Analysis and Countermeasures. Hubei Radio Telev Univ 12(07):55–57
7. Zhang D (2005) Students causes of mental disorders and Countermeasures. High Educ Res 13(26):137–139
8. Li L (2005) Positioning and Exploration of the Mental Health Education. Chin Educ 4:35–36
9. Ye B (2006) The influence of mode of exploration on the Chinese psychological counseling and treatment modalities. East Chin Norm Univ 12(2):29–34

Chapter 49
Study on Building of a Harmonious Campus Culture

Xue Yao, Wang Yan and Shanhui Lv

Abstract Harmony is the essence spirit of Chinese traditional culture and the campus is an important garden for the social training. It is also an important part for the society to build a harmonious campus culture, a stable school environment and a cultivate, so that more useful and better talent to the community can be cultivated, development and prosperity of the harmonious campus culture is one of the important feature to harmonious campus. This article is based on the campus culture content and effect to probing into the status of the campus culture building, and how to build harmonious campus culture.

Keywords Harmonious · Campus · Culture

49.1 Introduction

It is an extremely important part of school to build a harmonious campus, educational philosophy about comprehensive coordination, free, full development, interaction and overall optimization which based on Scientific Outlook on Development is of great practical significance to university. Meanwhile, the university possessing a strong radiation and leading effect and gathered a large number of highly qualified elite is a base to disseminate knowledge, heritage of civilization, personnel training, development of technology and the society services, it should set an example in the process of building a socialist harmonious society [1]. The building of harmonious campus culture is ideological and moral

X. Yao (✉) · W. Yan · S. Lv
HeBei United University, Tangshan, China
e-mail: yaoxue413@126.com

W. Du (ed.), *Informatics and Management Science VI*, Lecture Notes in Electrical Engineering 209, DOI: 10.1007/978-1-4471-4805-0_49,

basis to consolidate the harmonious campus. Therefore, colleges and universities with the goal of building a socialist harmonious society should use new concept, new ideas to build the university's campus culture and give full play to the effect of campus culture that create a harmonious environment for a harmonious society for talents cultivating [2–5].

49.2 Re-Understanding of Campus Culture

The so-called campus culture created in the process of teaching practice about the formation of the teaching and all other activities in schools to achieving its educational goals is material wealth and spiritual wealth. In essence, the campus culture spaced to campus, participated by teachers and students (mainly students) is ideological and cultural integration with moral values, behavior, ways of thinking, aesthetics and other aspects which is formed during a long-term; it is a common sense of identity, a spiritual atmosphere. It is the infiltration by campus culture that the students from different colleges but in the same conditions and professional show the different spirit and temperament. Campus culture is an atmosphere of the emphasis on education. It reflects strong colour of individual character, and follows the common characteristic because of common identity and cultural assimilation function. Mature campus culture can satisfy three requirements of the educatee's mental life. They are Beauty enjoyment, intelligence activities and personality ascension.

Content about the campus culture can be divided into material culture, system culture and spiritual culture.

Material culture which mainly refers to the school's physical environment is the hardware of campus culture. Campus physical environment condensed the subjectival creation of cultural is a cultural landscape; in turn it also has a great appeal to the educated. After extensive investigation, we found that students in the dirty, chaotic, and poor dormitory generally poor in performance, self-motivated and wilt in spirit. Visibly, the material culture on campus can not be ignored. On the other hand, certain construction of the campus, sculpture, and other attractions is the carrier of campus culture. Such as Oxford's library and Peking's Unknown Lake in some degree are symbols of the school.

Discipline culture is not only the written school rules, but also unwritten moral customs; it is the school's standardized objectives specimens on training. It is a culture system with a strong order, organization, and normative that should strictly to enforce and comply in campus-wide and it is consciously chosen by school.

Institutional culture should reflect the principles of equality, fairness and openness, its aims is to establish a behavior system that various factors coherence and growth under unified discipline also personal ease of mind, civilized, harmonious and orderly state. Therefor, under the constrains of well school system culture, the whole school's teaching, researching, management, life would be further standardized, democratic and the formation of harmonious orderly campus atmosphere can be strongly guaranteed.

Spirit culture which mainly include school's ideal climate, moral relations, seeking culture, artistic taste and other complex factors is an accumulation in the all members of the school and fill out in the entire campus. School spirit is both a sense of community and the heart and soul of the campus; it reflects the direction and essence of the campus culture. Although, we can see the material culture affection, and it can reflect mental state, the spirit culture is the core. It restricts the whole campus culture development level. Spiritual culture mainly reflects the value of the students, the psychological quality concept, aesthetic taste, etc.

In the main part of campus culture, material culture and system culture with the external, mechanical properties is at the low level, spiritual culture with higher level is at the core place. Plentiful culture and spiritual are guaranteed by the campus's physical facilities and strict scientific system, but the levels of physical facilities and the building of rules and regulations must reflect the spiritual and cultural significance. The campus cultural construction in the material should aim to make it become bear spirit the carrier of culture. Spiritual culture construction is in the material culture construction implied. It is the construction of campus culture, fundamental substantial parts of it.

49.3 The Role of Harmonious Culture in the Construction of a Harmonious Campus

At present, China is in the period of social transition during which one hand the university reformation continues to deepen, another interaction between universities and society is increasing and a complex relationship network formed in which they are mutual cooperate and mutual restraint, so came the culture collision and conflict and integration, it makes the university lose the lost quiet. Meanwhile, the rapid development in extension spawned a number of contradictions on the university campus. In such circumstances, the construction of a harmonious campus culture plays a pivotal role in the building of a harmonious campus.

49.3.1 The Role of Education Guidance

Campus culture is a kind of spiritual norms which possess a strong influence. It regulates the actions of the staff and students. Campus in the blowing of culture environment and spirit atmosphere, containing educational purposes, imperceptibly, intentionally or unintentionally influence and assimilate the people in the environment and consciously or unconsciously influence people, particularly it affect the values of students, the formation of moral character and the choices of lifestyle. In other words, it affects the student's consciousness, thinking and action. At the same time, its atmosphere and the formation of collective opinion embody the values, the honor sense of collective and the will of each member in the campus.

49.3.2 Incentive and Constraint

Colorful healthy campus culture provides the space to cultural and background to activities for the staff and students, it has strong appeal to the students on the faith of life, the ideals moral and rules of conduct, etc, so school can form an good atmosphere of exciting, vibrant and pioneering, a incentive system and environment of run and catch. Campus culture as the essence of the value system on the campus, a potential strength of achieving school development goals is undoubtedly a huge motivating factor and driving force to inspire members of the school to generate and maintain the positive motivation.

49.3.3 The Role of Unity and Cohesion

Campus culture as an organic combination on moral education is a cultural atmosphere including Moral thinking on education, philosophy, education intent. As a group culture, campus culture inspires the teachers and students a sense of community and identity after training on common values, so that schools truly become an organic integrity with cohesion and solidarity, form an atmosphere of unity and harmony, an good style of ethos on teaching and harmonious interpersonal relationships, by which the construction of "Harmonious Campus" can be leaded.

49.3.4 The Role of Social Radiation to Society

University is a place to spreading advanced modern culture, to converging and diverging health decorous culture, to guiding the social development and promoting social civilization. Currently, the campus culture is becoming more opening, it is not only regulated by the mainstream culture and influenced by the non-mainstream culture, but also radiate society through unstopping communication between university and society with it unique style (science, democracy, and beyond, critical), so that the special role for the campus culture to leading the cultural trends and spreading the scientific thinking can be demonstrated.

49.4 Status of the Construction of Campus Culture

The mainstream campus culture is positive and healthy. But, currently there exists many problems in the building of campus culture, prominently in the following two aspects:

Firstly, the campus cultural construction in certain extent weakens the Marxism theory. One survey about the campus cultural construction has showed: in the ideological and political theory teaching class see other books accounted for 66.7 %; think studying marxism is not very related to their success and adult accounted for 44.4 %; and to the university also open this class to feel bewildered accounted for 38.9 %. This suggests that a significant portion of the college students are lack of correct understanding and interest in the study of Marxist sense. Even have the rebellious attitude or negative attitude to it. This is a wake-up call for the ideological and political education workers.

Secondly, the quality of student organizations activities is declining. The taste of community cultural activities is not high in the core. Survey has found that people of 65.3 % think that some activities carried out by school is formalism in content and low-grade in culture; As to the quality of activities hold by various community organizations, students of 71.7 % feel it just ok, but 10.4 % feel poor, excessive rod sucks; only 17.9 % of the sample think "good, meaningful, and promoting school culture", that is the main reason why the school community activities can attract more students to participate in.

Thirdly, the world, life and values of college students have distorted in some extent. At present, some students is indifferent in concept of motherland and people; some individuals take the indolent, selfish, unscrupulous as nature of human and think it "reasonableness"; A considerable part of the students lack the spirit of arduous struggle, chase hedonism, do not take into account the actual situation of the family and do not understand the hardships of parents, just focus and enjoy the convenience bringed by material. According to surveys:

To the philosophy of that people should eat well, dress well and play well, students of just 37.4 % clearly express opposition, but as high as 62.6 % agree or partially agree.

Moral standards of the contemporary student overall in the whole society is at high level, but it exists contradictions and confusion in moral especially in value judgment and behavior. In recent years, the criteria of moral evaluation are losing, college students not aimless adore authority any more, but pay more attention to their feelings. Contemporary college students value their own interests, especially on self-development, desire to achieve oneself, desire to succeed, they just consider the individual's skills and personality development rather than take into national, collective and social's needs account. Faced with these phenomena, we have to admit that: the values of college students are facing serious challenges ever. Therefore, we should make the building of a harmonious campus culture as an opportunity to effectively establish a correct outlook on world, on life and values.

49.5 Push Forward the Building of a Harmonious Campus Culture, and Build a Harmonious Campus

49.5.1 Raise Awareness, Attaches Great Importance to Harmonious Campus Culture

Practice has proved that the campus culture as a specific society-cultural which has a huge impact on the students growth healthily. The core and essence of the campus culture is super-utilitarian, it is the carrier of culture which focuses on the spiritual construction, directly serve the overall development, its ultimate goal is to form an atmosphere for cultivating sentiments of students, improving student quality. University education should set a main concept of environment cultivating, strengthen awareness of the importance of campus culture, put all work into practice, no longer limited to just achieving the narrow purpose of management and entertainment and should fully understand it positive of building a campus culture for implementing CPC Central Committee and State Council "On Further Strengthening and Improving Ideological and Political Education" and has its own distinctive characteristics and harmonious campus culture.

49.5.2 Excellent Traditional Culture and the Essence of Outside Culture Nourish Campus Culture

China's long-standing traditional excellent culture is the national asset of priceless. The cultivation and growth for talents cannot leave the nutrients of traditional culture. Anyone is infiltrated by traditional culture in various degrees. We should endow traditional culture with new content, so the fine traditional culture of our campus culture can play a greater role in the campus' building. Meanwhile, the building of campus' culture is open, integrated national culture and alien culture, internal resources and external resources. But modern people should more carefully study the excellent culture lied in history of foreign educational building, it worthy of our learning to enhancing the quality of campus culture in educational philosophy, teaching methods, university's position and other aspects on development.

49.5.3 Building of Campus Culture Should Focus on All Sides

As to the promoting of building campus culture comprehensively and holistically, first, we should focus on harmonize development in material culture on campus, system culture, spiritual culture. Meanwhile, participants in the construction of campus culture also should be comprehensive, extensive, and form a atmosphere that everyone involved, everyone proud of, not just some people are concerned, few were satisfied, only leadership consensus, but should all school staff focus on.

49.5.4 Construction of Campus Culture Should Highlight Characteristics

With the popularization of higher education and university independent school is in strengthened ceaselessly trend, highlighting its characteristics become the urgent need to many schools, characteristics of campus culture is one of the very important part. Special campus culture reflects the specific historical tradition, the school spirit, pursuit of the goal and idea of running schools, while high grade campus culture can enhance the taste and reputation of a college, unconsciously influence the staff, improve education quality and efficiency and improve the social reputation to the university.

In short, the harmony campus needs harmony, healthy and positive culture. By virtue of it, can we nurture people, shape people, enrich the spirit connotation, enhance the cultural spiritual state and let we take a fine style of spirit, exhilarated state of spirit and high sense of morality, it also can strengthen the cohesion of the school to promote school unity stable and innovative development, and truly turn the building of a harmonious campus to reality.

References

1. CPC (2006) Central Committee's decision on building a socialist harmonious society and a number of major issues, vol 1. Beijing People's Publishing House, Beijing pp 124–133
2. Wang D (2006) Building harmonious campus and strengthening of building campus culture. Chin Youth Coll Polit Sci 3:23–27
3. Feng K (2010) The review In of the construction of harmonious campus. J Jilin Radio TV univ 1:37–39
4. Han Y (2010) Strengthening the education of gratefulness for university students promoting the building of harmonious campuses. Soc Sci J Coll Shanxi 1:80–82
5. Ma X (2009) Campus culture construction of higher learning education in harmonious society. J Huazhong Agri Univ Soc Sci Ed 2:110–113

Chapter 50
Firms Pollution Abatement R&D Investment Strategy on Tradable Emissions Permits

Yong Xi Yi, Shoude Li and Mengya Liu

Abstract Manufacturers pursuing profit maximization must decide whether and when to carry on the anti-pollution technology investment under the condition of tradable emissions permits and technical uncertainty, based on the real options theory we have constructed the pollution government technology investment strategy option gambling model and given the best opportunity of manufacturers to carry on pollution government technology investment. The study showing that in pollution treatment technology investment the competition between enterprises is not damage the options value. Rivalry between firms does not necessarily undermine option values. Instead the fear of sparking a patent competition may enhance the competitive effect, further raising the value of delay compared with the single-firm case.

Keywords R&D investment · Tradable emissions permits · Option gambling · Technical uncertainty

50.1 Introduction

Tradable emissions permits is according to the bearing ability of ecological environment in a region to determine the amount of sewage and then using a certain way to distribute it to the sewage source in this region, and then allows the

Y. X. Yi (✉) · M. Liu
College of Economics and Management of University of South China, Hengyang 421001, China
e-mail: yyx19999@126.com

S. Li
Antai College of Economics and Management, Shanghai Jiao Tong University, Shanghai 200052, China

W. Du (ed.), *Informatics and Management Science VI*, Lecture Notes in Electrical Engineering 209, DOI: 10.1007/978-1-4471-4805-0_50,

rights as commodity that can be purchased and be sold in order to control emissions of pollutants for the purpose of environmental protection. Because the emissions permit has the characteristic of ownership assignment and the actual use being not at the same time [1, 2]. Therefore, the emissions permits have characteristics of finance derivative product and stock or option. Developed real options gambling theory in recent years is a very good analysis tool for manufacturer decision-making of they should whether and when carried on the anti-pollution technology investment. This article based on the existing research results and established the pollution government technology investment option gambling model in the situation of manufacturers' research ability and investing cost are asymmetrical, and focus on analysis the impact of manufacturer own research ability and competitor's research ability to investing marginal value [3].

50.2 Fundamental Assumptions

Hypothesis 1, 2 risk neutral oligarchs manufacturers (i = 1, 2) product same product and in the process of production will discharge pollutants. The environmental managers supervise them strictly. If their discharging pollution exceeds the emissions permits, the manufacturer has to purchase emissions permits which it lack, and if it have surplus he can sell the surplus part [4, 5]. Therefore manufacturers must carry on the decision whether and when to carry on the pollution Abatement R&D investment [6].

Hypothesis 2, the two manufacturers is symmetrical. In the investing, the initial costs they have to pay are I and an on-going flow cost of C per unit time thereafter. The flow cost is incurred for as long as research activity continues, until a breakthrough is achieved by either firm [7, 8]. The investment success opportunity obeys to Poisson distribution. The investment success takes place randomly according to a Poisson distribution with parameter (or hazard rate) h > 0. Thus when firm i making breakthrough at first in a short time interval of length dt, given that it has not done so before this time, is $h_i dt$ and the density function for the duration of research is $h_i e^{-(h_i)t}$. The probabilities of success of each of firms are stochastically independent. Thus when both firms engage in research the density function for success is given by $h_i e^{-(h_i+h_j)t}$. $\theta_i \in (0, 1)$ Show the possible states of each firm for the idle and active states respectively. Each firm has a risk-free interest rate r [9, 10]. All parameter values are common knowledge.

Hypothesis 3, if one of the firm gets success at first, the firm use it in production can abate emissions quantity is D. Meanwhile, the price P of emissions permits per unit evolves stochastically over time. Therefore, the value that manufacturer gets success at first is $A = PD$. The investment is irreversible. P is a market demand factor which conforming to the geometry Brownian movement

$$P = uPdt + \sigma Pdz \tag{50.1}$$

In the (50.1), $u \in (0, r)$ is the expecting increment rate of P which is the emissions permits price, r is the non-risk interest rate, $\sigma > 0$ is the fluctuation rate of P.

50.3 Value Function and Investment Marginal Value

50.3.1 A Single Firm's Optimal Investment Timing

If in the case of lack of competition, the firm makes its investment decision unilaterally. By solving the following stochastic optimal stopping problem, we derive the single firm's optimal investment timing

$$v(P) = \max E\left[e^{-rT}\left(\int_T^{\infty} e^{-(r+h)}(hP_tD - C)dt - I\right)\right] \tag{50.2}$$

In Eq. (50.2), E is the expectation, T is the unknown future stopping time, the meaning of other variables are the same as having being indicated. Before manufacturer didn't get investment opportunities to invest, in this period the firm has no cash flows but may experience a capital gain or loss on the value of its option. Hence, in the continuation region, the Bellman equation for the value of the investment opportunity is:

$$rv_0dt = E(dv_0) \tag{50.3}$$

Using *Itô* lemma to expand dv_0 we get:

$$dv_0 = Dv_0'dP + \frac{1}{2}D^2v_0''(dP)^2 \tag{50.4}$$

Substituting from (50.1) and noting that $E(dZ) = 0$, we can write

$$E(dv_0) = uPDv_0'dt + \frac{1}{2}\sigma^2P^2D^2v_0''dt \tag{50.5}$$

From the Bellman equation we get following second-order differential equation

$$\frac{1}{2}\sigma^2P^2D^2v_0'' + uPDv_0' - rv_0 = 0 \tag{50.6}$$

Solving the Eq. (50.6) getting the following solution for the value of the option to invest in research

$$v_0(P) = A_0(DP)^{\beta_0} \tag{50.7}$$

In Eq. (50.7) A is a constant whose value is yet to be determined, and β_0 is the positive root of the characteristic equation:

$$\varepsilon^2 - \left(1 - \tfrac{2u}{\sigma^2}\right)\varepsilon - \tfrac{2r}{\sigma^2} = 0, \quad \beta_0 = \tfrac{1}{2}\left(1 - \tfrac{2u}{\sigma^2} + \sqrt{\left(1 - \tfrac{2u}{\sigma^2}\right)^2 + \tfrac{8r}{\sigma^2}}\right) > 1.$$

Next we consider the value of the firm in the stopping region. Since investment is irreversible the value of the firm in the stopping region $v(P_1)$ is given by the expected value alone with no option value terms:

$$v_1(P_1) = E\left(\int_t^{\infty} e^{-(r+h)\tau}(hDP_\tau - C)d\tau\right) \tag{50.8}$$

Since P is expected to grow at rate. We can get

$$v_1(P) = \int_t^{\infty} e^{-(r+h)\tau}(hPDe^{u\tau} - C)dt \tag{50.9}$$

From Eq. (50.8) get following expression for the expected value of the project

$$v_1(P) = \frac{hPD}{r+h-u} - \frac{C}{r+h} \tag{50.10}$$

The Bellman equation in the continuation region is:

$$\frac{1}{2}\sigma^2 P^2 D^2 v_0'' + uPDv_0' - rv_0 = 0 \tag{50.11}$$

The value-matching condition is: $v_0(P) = v_1(P) - I$. The boundary between the continuation region and the stopping region is given by a critical value P^* of the stochastic process. This condition requires the value functions $v_0(P)$ and $v_1(P)$ to meet smoothly at P^* with equal first derivatives $v_0'(P^*) = v_1'(P^*)$. Substituting it for the value functions from (50.7) to (50.10), getting the value matching and smooth-pasting conditions for the single-firm optimal problem:

$$B_0(P^*D)^{\beta_0} = \frac{hP^*D}{r+h-u} - \frac{C}{r+h} - I; \quad B_0(P^*D)^{\beta_0-1} = \frac{h}{r+h-u} \tag{50.12}$$

By Eq. (50.12), we can get the critical value P^* and the unknown coefficient B_0. Then we get the single firms' unilateral trigger point:

$$P_U = \frac{\beta_0}{\beta_0 - 1}\left(\frac{C}{r+h} + I\right)\frac{r+h-u}{h}; \quad B_0 = \frac{h(DP_U)^{1-\beta_0}}{(r+h-u)\beta_0} \tag{50.13}$$

50.3.2 The Value Function and Investment Marginal Value of Joint Investment Firms

We now consider the case of the firms can commit to a joint investment in same time. Prior to investment the value of each firm satisfies the Bellman equation (50.3).

The value of each firm in the continuation time is: $v_{0,0}(P) = B_j(PD)^{\beta_0}$. B_j Is a constant the value is yet to be determined. The value of firm i is the expected value of its investment project after investing by both firms, it can be indicted: $v_{1,1}(P) = E\left(\int_t^{\infty} e^{-(r+2h)\tau}(h_j P_\tau D - C)d\tau\right)$. Solving it getting:

$$v_{1,1}(P) = \frac{PDh}{r+2h-u} - \frac{C}{r+2h} \tag{50.14}$$

The value-matching condition at the trigger point P_j is: $v_{0,0}(P_j) = v_{1,1}(P_j) - I$. Solving the unknown constant B_j, getting:

$$B_j = (P_j D)^{-\beta_0}\left(\frac{hPD}{r+2h-u} - \frac{C}{r+2h} - I\right) \tag{50.15}$$

The value of the firm prior to investment at any arbitrary joint investment point is:

$$v_j(P; P_j) = B_j(PD)^{\beta_0} \tag{50.16}$$

By Eq. (50.17), we can find the joint investment optimal stopping problem at time T:

$$v = \max E\left(e^{-rT}\left(\int_T^{\infty} e^{-(r+2h)t}(2hDP_t - 2C)dt - 2I\right)\right) \tag{50.17}$$

The optimal joint investment trigger P_C is:

$$P_C = \frac{\beta_0}{D(\beta_0 - 1)}\left(\frac{C}{r+2h} + I\right)\frac{r+2h-u}{h} \tag{50.18}$$

The value of each firm in the continuation time is:

$$v_C = B_C(PD)^{\beta_0} \quad B_C = \frac{h(PD)^{1-\beta_0}}{(r+2h-u)\beta_0} \tag{50.19}$$

50.3.3 The Value Function and Investment Marginal Value of the Follower

As the follower i, who invests strictly later than its rival j. Because the rival has invested irreversibly, in any time interval dt the probability of leader get success is $h_j dt$ and the probability he/she has not yet got success is $e^{-h_j t}$ and this probability is independent of whether the follower has or has not invested. The optimal stopping problem of the follower can be indicated as Eq. (50.20).

$$v = \max\left[E(e^{-(r+h_j)T}\left(\int_T^{\infty} e^{-(r+h_i+h_j)t}(h_i DP_t - C)dt - I\right)\right] \tag{50.20}$$

In the continuation waiting for time, the follower holds the invest option, at the same time he/she also faces the possibility of the rival get success. Therefore, in any short time interval dt, the followers' gain is dv with probability $1 - h_j dt$ and with probability $h_j dt$. Its option expires which make the option no value by the rival get success. The Bellman equation is $(r + h_j)v_{0,1}dt = E(dv_{0,1})$, According to *Itô* lemma it may be unfold into Eq. (50.21):

$$\frac{1}{2}\sigma^2 P^2 D^2 v''_{0,1} + uPDv'_{0,1} - (r + h_j)v_{0,1} = 0 \tag{50.21}$$

The general solution of (50.14) is expressed by $v_{0,1}(P) = B_F(DP)^{\beta_1}$. The $B_F > 0$ is a constant whose value is yet to be determined, and β_1 is the positive root of the characteristic $\varepsilon^2 - (1 - \frac{2u}{\sigma^2})\varepsilon - \frac{2(r+h)}{\sigma^2} = 0$, and β_1 can be written as $\beta_1 = \frac{1}{2}\left(1 - \frac{2u}{\sigma^2} + \sqrt{\left(1 - \frac{2u}{\sigma^2}\right)^2 + \frac{8(r+h)}{\sigma^2}}\right)$.

Next to determine followers' the investment critical point. Denoting the follower's optimal investment trigger by P_F, in the point the conditions can be stated as $v_{0,1}(P_F) = v_{1,1}(P_F) - I$, and $v'_{0,1}(P_F) = v'_{1,1}(P_F)$. Solving the simultaneous system we get:

$$P_F = \frac{\beta_1}{\beta_1 - 1}\left(\frac{C}{r+2h} + I\right)\frac{r+2h-u}{h} \quad B_F = \frac{h(P_F D)^{1-\beta_1}}{(r+2h-u)\beta_1} \tag{50.22}$$

50.3.4 The Value Function and Marginal Value of Pioneer

The leader's payoff is affected by the action of the rival firm who invest later than it. Taking account of the follower's action the leader's post-investment payoff is:

$$v(P_t) = E\left(\int_t^{T_F} e^{-(r+h)\tau}(hDP_\tau - C)d\tau + \int_{T_F}^{\infty} e^{-(r+2h)\tau}(hDP_\tau - C)d\tau\right) \tag{50.23}$$

The Bellman equation for the leader is $v_{1,0} = (hPD - C)dt + e^{-(r+h)dt}E(v_{1,0} + dv_{1,0})$. Unfolding it according to *Itô* lemma and simplifying we obtain $(r + h)v_{1,0} = \frac{1}{2}\sigma^2 P^2 D^2 v''_{1,0} + uPDv'_{1,0} + hPD - C$. Solving it we get:

$$v_{1,0}(P) = \frac{hPD}{r + h + u} - \frac{C}{r + h} - B_L(PD)^{\beta_1} \tag{50.24}$$

In Eq. (50.24) $B_F > 0$ is a constant whose value is yet to be determined and $\beta_1 > 1$ is as previously defined.

According to value matched condition, it can be got: $v_{1,0}(P_F) = v_{1,1}(P_F)$. By which to solve B_F. We get:

$$B_L = (DA_F)^{-\beta_1}(hP_F DZ_1 - CZ_2) > 0 \tag{50.25}$$

In Eq. (50.25), $Z_1 = \frac{1}{r+h-u} - \frac{1}{r+2h-u} \geq Z_2$, $Z_2 = \frac{1}{r+h} - \frac{1}{r+2h} > 0$.

Now we can get an expression for the payoff to investing as the leader, denoted it as $v(L)$.

$$v_L(P) = \begin{cases} \dfrac{hDP}{r+h-\mu} - \dfrac{C}{r+h} - A_L(DP)^{\beta_1} - I, & P < P_F \\ \dfrac{hDP}{r+2h-\mu} - \dfrac{C}{r+2h} - I, & P \geq P_F \end{cases} \tag{50.26}$$

50.4 Simple Numerical Simulating and Analysis

We consider the following parameter values to simple numerical simulate and analyze. The drift parameter u of the geometric Brownian motion is zero. The risk-free interest rate r is 5 %. The research technology involves a hazard rate h = 0.2, flow cost C = 0.4 and set-up cost I = 1 and if the investment success the firm can abate emissions D = 1for each firm. The volatility parameter $\sigma = 0.15$.

The simulation results are as follows: The leader's trigger point is $P = 3.48$ while that of the follower is $P_F = 5.25$. For comparison, the trigger point of the single firm is $P_U = 5.20$. Hence in this case it is clear that the pre-emptive effect undermines option values to a significant degree, causing investment to take place sooner than in either the single-firm or the coordinated case.

50.5 Conclusions

In this study showing that in pollution treatment technology investment the competition between enterprises is not damage the options value. Rivalry between firms does not necessarily undermine option values. Instead the fear of sparking a patent competition may enhance the competitive effect, further raising the value of delay compared with the single-firm case. Thus in situations where both option values and strategic interactions arise, it is necessary to study the circumstances carefully before forming a view on whether the incentive to pre-empt or the option value of delay will dominate. In the competitive environment of investment behavior is very sensitive, some specific factors, such as the degree of uncertainty, the different of speed of finding new technology, the competition results will also appear. In some cases also have path dependence, investment and on the uncertain variable of the exact evolution. This paper studies focus the situation of symmetric two-firm case. If the company's research technology basis is different, the risk is different, the value of having got success is different, and so on, and these differences will influence the studying results study. In the next step, we will focus on the study of pollution control technology investment game situation between the asymmetric manufacturers.

Manufacturers pursuing profit maximization must decide whether and when to carry on the anti-pollution technology investment under the condition of tradable emissions permits and technical uncertainty, based on the real options theory we have constructed the pollution government technology investment strategy option gambling model and given the best opportunity of manufacturers to carry on pollution government technology investment.

References

1. Dixit A, Pindyck R (1994) Investment under uncertainty. Princeton University Press, Princeton
2. Weeds H (2002) Strategic delay in a real options model of R&D competition. Rev Econ Stud 69(3):729–747
3. Nielsen MJ (2002) Competition and irreversible investments. Int J Ind Organ 5(20):731–743
4. Lukach R, Kort PM, Plasmans J (2007) Optimal R&D investment strategy under the threat of new technology entry. Int J Indus Organ 25(1):103–119
5. Lundgren T (2003) A real options approach to abatement investments and green goodwill. Environ Resour Econ 25(1):17–31
6. Zhao JH (2003) Irreversible abatement investment under cost uncertainties: tradable emission permits and emissions charges. J Publ Econ 87(12):2765–2789
7. Mesbah SM, Kerachian R, Torabian A (2010) Trading pollutant discharge permits in rivers using fuzzy nonlinear cost functions. Desalination 250:313–317
8. Niksokhan MH, Kerachian R, Karamouz M (2009) Game theoretic approach for rading discharge permits in rivers. Water Sci Technol IWA 60:23–29
9. Ghosh S, Mujumdar PP (2006) Risk minimization in water quality control problems of a river system. Adv Water Resour 29:458–470
10. Vidyottama V, Chandra S (2004) Bi-matrix games with fuzzy goals and fuzzypay-offs. Fuzzy Optim Decis Making 3:327–344

Chapter 51
Efficient Vocal Music Education Scheme Based on Samplitude

Tingjun Wang

Abstract The paper is designed to study the application of voice recorder software—Samplitude in vocal music education. Compared with traditional education means, computer-software-aided vocal music courses have incomparable advantages. To apply computer software in vocal music teaching could bring students with visualized and vivid experience. Furthermore, it will greatly promote the innovation and development of vocal music education concepts. Currently, this technology has attracted teachers' enormous attention on vocal music teaching methods. We will reaching to the effect of professional skill improvement and properly usage of software by application Samplitude in vocal music teaching.

Keywords Voice recorder software · Samplitude · Vocal music

51.1 Introduction

With the development of society and the advancement of science, computer technology has gradually penetrated into music majors. The application of computer music software has played very significant and promotional role in the development of many music subjects. Vocal music education is a complicated and systematic task [1]. In traditional vocal music education, there are many problems hard or even impossible to be solved. As for this, this paper mainly discusses the following two problems: Firstly, students can't hear their real voice, for what students heard is actually a mixed sound—a combination of the sound transmitted through their skull and the sound transmitted by air around [2]. Secondly, students

T. Wang (✉)
Music College, Beihua University, Jilin 132013, China
e-mail: wangtingjun2013@yeah.net

W. Du (ed.), *Informatics and Management Science VI*, Lecture Notes in Electrical Engineering 209, DOI: 10.1007/978-1-4471-4805-0_51,

are lacking of practical experience in singing and recording, showing no self-confidence on their singing effect. By properly making use of computer music software, the aforementioned two vocal music problems can be perfectly solved. Firstly, teachers may use the recorder software Samplitude to record students' voice, and then to playback students' real voice for students to listen to. Meanwhile, they may as well make use of spectrum analyzer embedded in Samplitude to help students analyze defects in their voice [3]. Secondly, as students are lacking of practical experience in singing and recording, in teaching, digital multi-track recording software can be applied to provide students with the similar experience of being in sound-recording studio, so as to improve students' confidence in vocal music learning via modifications upon their voice [4].

51.2 Helping Students Set Up Correct Voice Concept

In traditional vocal music teaching, teachers' instruction and demonstration is the only source for students to learn voice production. The so-called "voice direction", "voice position" and other teaching terms given by teachers are too abstract for students to understand [5]. In practical voice music teaching, lower grade students often raise a question to their teachers: they feel that their voice is resonant and mellow, so why teachers would say that their voice is dry and lacking of penetrating power? In fact, the reason is that, what performers heard is actually a mixed sound—a combination of the sound transmitted through their skull and the sound transmitted by air around. However, what the teacher heard is real sound transmitted via air around [6]. This could easily explain why we should make use of voice recorder software to help students set up correct voice concept. In this way, teachers may make use of the recording function embedded in the digital voice recorder software Samplitude to recorded students voice. After that, by drawing support from the playback function, students are able to hear their real voice. Moreover, students may see the wave form displayed on the software at the same time [7]. By utilizing the super useful multi-track recording function of Samplitude, teachers may backup all students' voice in class. On this basis, they may select from the backups for several typical voices, and then put the selected voice on the same platform for comparison and analysis. In this way, teachers may find out the correct waves and wrong waves. Furthermore, they may also constantly make comparison, analysis and explanation by integrating the actual effect, so as to consolidate students' judgment standard in correct voice and wrong voice, as well as to help students better and efficiently understand the concept of correct voice. In each stage during vocal music learning, teachers are able to collect and preserve students' voices [8]. Then, comparative analysis can be performed with regard to students' voices in different stages, so that students may clearly understand their learning progress in each stage, to correct their voice, so as to better encourage students' interest and enthusiasm in vocal music learning, as well as to improve their self-confidence [9].

51.3 Application of Digital Voice Recorder Software Samplitude in Vocal Music Teaching

Samplitude is multi-track digital voice recorder software, widely applied in voice-recording studios. With Samplitude and by integrating advanced hardware equipments, users are able to produce excellent recording works [10]. In vocal music teaching, when recording students' voice, teachers may firstly import the accompaniment to one of the tracks, and then set the track at playing state. In the meanwhile, another track is set at recording state. As for the configuration for recording, instant replay of the recording effect is selected during the recording process [11]. As for this, during recording, students could hear the accompaniment, as well as their own voice at the same time. When the recording is over, both the two tracks will be switched to playing state, so that voice with accompaniment could be appreciated (Fig. 51.1).

In this paper, students will be divided in gender in vocal music teaching, with two typical students' voice spectrum curves as the examples to demonstrate the application of Samplitude in classroom teaching.

The frequency of male voice is normally located between 64 and 523 Hz, with overtone could be expanded to 7–9 kHz. In general, requirements on male voice include solidity, powerfulness, and clearness. Figure 51.2 shows a male students' voice spectrum curve gathered by the author with Samplitude. Analyzing by the

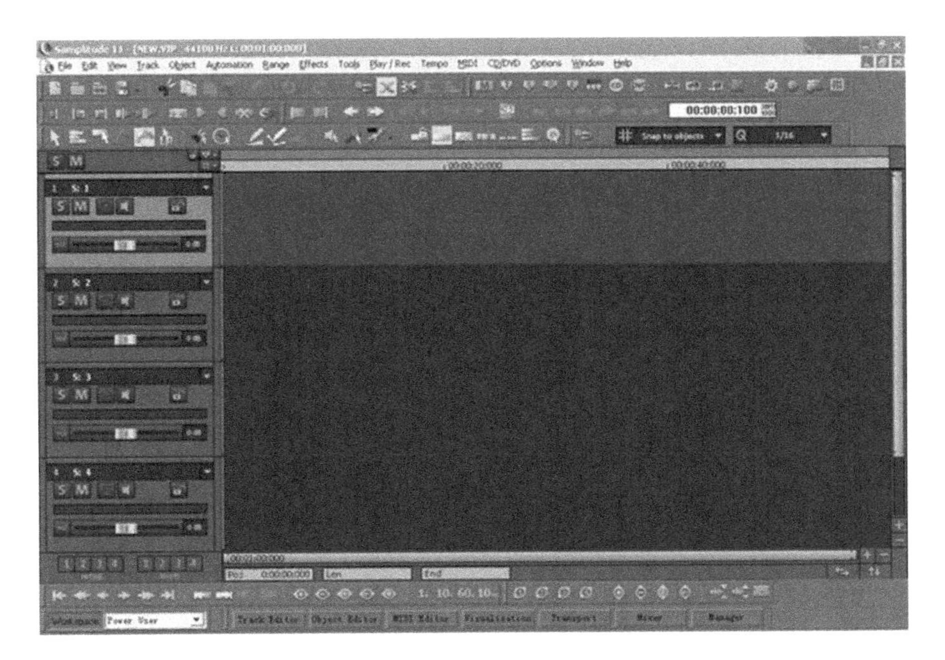

Fig. 51.1 User interface of samplitude

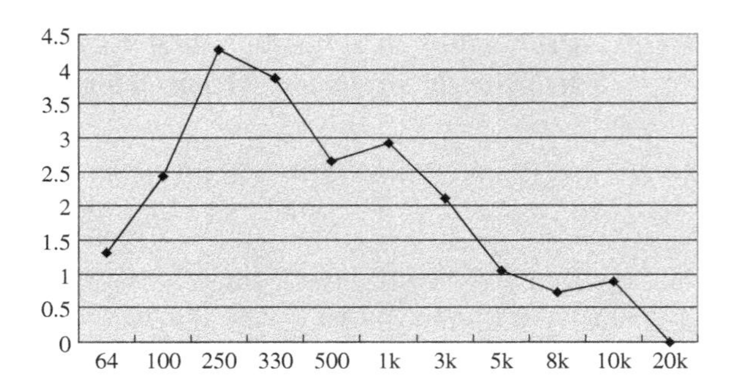

Fig. 51.2 Male voice spectrum curve

students' situation, the main frequency section of tone frequency is to be regulated and processed (Table 51.1).

The author has invited the student to analyze his own advantages and disadvantages, finding places needing to be improved and continued. In the meanwhile, he was required to learn to process voice with Samplitude. Therefore, firstly, the students needed to get some improvement in the area between 64 and 100 Hz. This area is male basso's range, requiring some powerful elements. The secondary emphasis was placed on frequencies between 250 and 330 Hz. This is the major frequency range of male fundamental tone, so that the student needed to improve the strength of voice, which was also frequently emphasized in daily trainings. In the end, as for the frequency range around 1 kHz and above 10 kHz, we may get some improvement with Samplitude, so as to ensure the performance of overtone, as well as the brightness of tone (Table 51.2).

The frequency of female voice is normally located between 160 and 1.2 kHz, with overtone could be expanded to 9–10 kHz. Figure 51.3 shows a female students' voice spectrum curve gathered by the author with Samplitude. Analyzing by the students' situation, the main frequency section of tone frequency is to be regulated and processed.

Still, the author invited the student to analyze her own advantages and disadvantages, finding places needing to be improved and continued. In the meanwhile, she was required to learn to process voice with Samplitude. Female voice needs to reflect features like mellowness, brightness and clearness. According to the spectrum analysis, the student needed no improvement in the range between 160 and 250 Hz. However, from 250 to 523 Hz, the student needed to be enhanced in

Table 51.1 Male voice tone spectrum data

Frequency Hz	64	100	250	330	500	1 k	3 k	5 k	8 k	10 k	20 k
Energy	1.32	2.44	4.28	3.87	2.65	2.91	210	1.05	0.73	0.89	0

Table 51.2 Female voice tone spectrum data

Frequency Hz	64	100	250	330	500	1 k	3 k	5 k	8 k	10 k	20 k
Energy	0	0.12	5.15	5.05	5.43	3.67	3.84	2.69	1.45	1.78	0.94

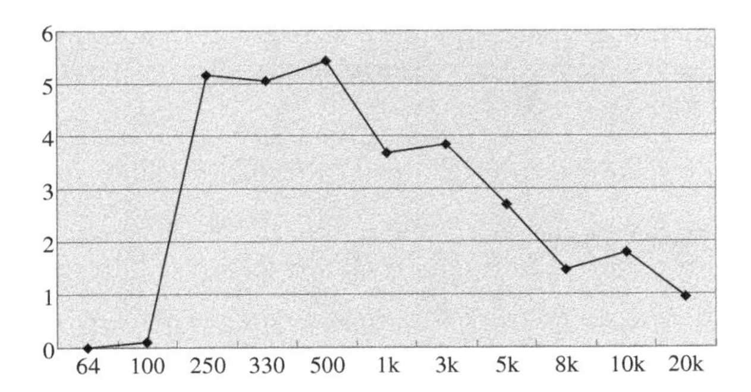

Fig. 51.3 Female voice spectrum curve

strength and plumpness, which should be especially noticed in female voice practice, for this area is the major tone range of female voice, as well as the medium and low pitch area. Secondly, the frequency section from 1 to 3 kHz should also be improved, so as to make the tone more beautiful, as well as to increase the brightness of timbre. Thus, in vocal music practices, the female student should try to endow overtone section in timbre structure with better frequency continuity. Related practices should be strengthened by integrating voice recorder software Samplitude. In the end, as for the frequency above 10 kHz, it needed to be further improved, or it can also be processed with Samplitude, to endow the timbre with better expressive force. Through processing, tiny and exquisite parts in timbre could be high-lightened and emphasized.

In addition to this, Samplitude is also applicable to opera duet practice. During duet, students may pay over attention on performance and neglect listening to others' or their own voice. On the other hand, if practicing in part, the overall effect of duet could not be fully reflected. Multi-track recording enables users to record voice of different parts in different tracks. By drawing support from this function, students are enabled to hear the overall effect through playback. Meanwhile, they may also listen to a certain part or several parts, to find out the defects and to improve their performance level.

In this research, the author had just used a few functions of Samplitude, and achieved very good teaching effect. As for this, the author hopes that Samplitude and other music software play better role in vocal music teaching. It is true that this needs further practice and exploration of colleagues in the same field.

51.4 Conclusion

Technology advancement has provided us with handful software like Samplitude. We should correctly and fully make use of them. In the meanwhile, we should specially notice that, these tools should never be utilized blindly. Teachers may take them as teaching tools. After students have used them for recording, they shall be guided to listen to and analyze their voice before processing. Once given correct guidance, students will be trained with correct voice concept, so as to timely rectify their voice, reaching to the effect of professional skill improvement and properly usage of software.

References

1. Pantano E, Servidio R, Viassone M (2011) An exploratory study of the tourist-computer interaction: the role of web site usability on hotel quality perception. JDCTA 5(5):208–216
2. Avargil S, Herscovitz O, Dori YJ (2008) Teaching thinking skills in context-based learning teachers challenges and assessment knowledge. J Sci Educ Technol 21(2):207–225
3. Xinhua Z, Yanshuai Z (2009) Teaching and thinking of multimedia. Adv Intell Soft Comput 116:809–813
4. Bani-Salameh H, Jeffery C (2011) Teaching and learning in a social software development tool. Soc Media Tools Platforms Learn Environ Part I 32(4):17–35
5. Ferreira MJM (2012) Intelligent classrooms and smart software teaching and learning in today's university. Educ Inf Technol 17(1):3–25
6. Hedin G, Bendix L, Magnusson B (2008) Teaching software development using extreme programming. Lect Notes Comput Sci 4821:166–189
7. Li Q, Myaeng SH, Guan DH, Kim BM (2005) A probabilistic model for music recommendation considering audio features. Lect Notes Comput Sci 3689:72–83
8. Chen Y, Chen M, Li Q, Zhang Y, Huang C, Sheng X, Zhou M (2011) Voxel deformation of virtual objects in augmented reality human computer interaction system. JDCTA 5(7):207–214
9. Nielsen F (2007) Music (and arts) education from the point of view of didaktik and bildung. Springer Int Handbooks Educ 16(2):265–285
10. Richardson C (2012) Narratives from preservice music teachers: hearing their voices while singing with the choir. Narrative Soundings Anthology Narrative Inq Music Educ 4:179–200
11. Bærendsen NK, Jessen C, Nielsen J (2009) Music-making and musical comprehension with robotic building blocks. Lecture notes in computer science. Learning by playing game-based education system design and development 5670:399–409

Chapter 52
Correlation Analysis of Occupational Stress and Family Support

Xinliang Ju and Xinming Qian

Abstract Objective to know the current status of occupational stress and family support in an enterprise and explore their correlation. Methods 914 production laborers who were on duty sampled by stratified cluster sampling were investing acted by Occupational Stress Inventory-Revised Edition (OSI-R) and Comprehensive Work Ability Index (CWAI). Results The Occupational Role Questionnaire (ORQ) score was lower than the product ion laborer group norm and had no statistic difference with the occupational stress norm of Southwest China. The Personal Strain Question naira (PSQ) score was lower than t he production n laborer group norm and higher than the occupational stress norm of Southwest China. The Personal Resource Questionnaire (PRQ) scare was lower than both 2 norms. 38.3 % of the investing acted laborers whose CWAI score was lower than 42. Both occupational stress and CWAI were influenced by many factors. CWAI correlated with several subscales of OSI-R, generally, nag actively associated with PSQ and ORQ and positively associated with PRQ. Conclusions Emphasis should be put on the frail group when intervening with occupational stress and CWAI, The work is still needed to be further strengthened in the work environment, task allocation, social support, health education, psychological counseling, family support and other aspects for Enterprises Occupational stress improvement should be conducted simultaneously with the CWAI import movement to obtain effective intervention outcome.

Keywords Family support · Occupational stress · Current status investigation · Correlation analysis

X. Ju (✉)
China Academy of Safety Science and Technology, Beijing, China
e-mail: xinliangju@126.com

X. Qian
Beijing Institute of Technology, Beijing, China

W. Du (ed.), *Informatics and Management Science VI*, Lecture Notes in Electrical Engineering 209, DOI: 10.1007/978-1-4471-4805-0_52,

52.1 Overview

Occupational stress comes from professional activities, it is the interaction of occupational factor and individual factor. When the individual's coping capacity can not meet the work requirements, the psychiatric or somatic anomalies reaction [1]. Its characteristic is:

- The long-standing;
- It is difficult to adapt to, often cause insomnia or excited state;
- If you cannot adapt to the working environment, can lead to serious consequences;
- The work of tension effect can be "spill" to other functions, such as affect occupation outside the family life;
- The tension effects of occupation will become harmful to health factors accumulation.

In recent years, along with our country society, the rapid development of economy, people's rhythm of life speeds up day by day, increasing pressure of work, occupation health and ability to work under some of the traditional physical and chemical harmful factors, occupation tension have also been researchers attention as a prevalent and harmful factors. We have a certain present situation investigation and analysis for technical workers of comprehensive work ability in a steel enterprise, in order to understand the investigation enterprise occupation hazards and workers and comprehensive work ability influence each other, so as to establish an effective health intervention measures.

52.2 Research Subjects and Research Methods

52.2.1 Study Object

We had carried on the questionnaire survey in workers out duty in a certain iron and steel enterprise, including iron, steel rolling (including steelmaking and steel rolling), the energy supply of three branch, a total of 997 copies of the questionnaire, to recover the questionnaire 967, rejecting unqualified questionnaire 53, and practical investigation techniques 914 workers. As shown in Table 52.1, iron making, steelmaking, continuous casting, covered rolling, energy production and supply, a total of more than 80 jobs.

Table 52.1 Investigation of basic characteristics

	Sex		Age			
Type	Male %	Female %	<30 %	30–40 %	40–50 %	> = 50 %
Proportion	85	15	10	50	35	5
	Education			Marital status		
Type	Junior middle school %	High school and technical school %	college %	Married %	Unmarried %	Divorce %
Proportion	20	52	28	86	8	6

52.2.2 *Research Methods*

Occupational stress inventory-revised edition, OSIR [2]. The scale consists of three subscales, a subset of fourteen items, one hundred and forty items. The three subscales is occupational role quest Bonaire (ORQ), personal strain questionnaire (PSQ) and personal resources questionnaire (PRQ), as is shown in Table 52.2. Each sub item was composed of ten items, each entry according to the five score, occupation and individual stress subscale higher scores are expressed in higher strain rate, individual coping resources subscale scores higher on nervous strain ability.

Comprehensive work ability was evaluated using the Zhang Lei system of comprehensive work ability index (Comprehensive work ability index, CWAI) [3]. CWAI consists of five items, seventeen items, and five items for work capacity

Table 52.2 Occupational stress inventory project description

Questionnaire (project)	Abbreviations	Number
Occupational role quest	ORQ	60
Role overload	RO	10
The task of discomfort	RI	10
Role ambiguity	RA	10
Task boundaries	RB	10
Responsibility	R	10
Working environment	PE	10
Personal strain questionnaire	PSQ	40
Service stress	VS	10
Psychological stress reaction	PSY	10
Reactions to interpersonal tensions	IS	10
Somatic stress response	PHS	10
personal resources questionnaire	PRQ	40
Leisure	RE	10
Self health care	SC	10
Social support	SS	10
Rational act	RC	10

evaluation (C1), physiological state (C2), psychological status (C3), social function (C4) and predictors of exercise capacity (C5). The lowest score of 11 points, the highest score of 56 points, scored higher work better, and <42 representation of comprehensive work ability abate [4]. The CWAI scale has good reliability and validity.

52.3 Results

52.3.1 The Surveyed Occupation Status of Stress

52.3.1.1 Occupation Stress Subscales and Often Die More Occupation Stress Subscales and Sub Item Scores and Technical Workers Often Mode Comparison

Occupation stress subscales/child scoring compared with the norm, Occupation stress subscales/child scoring compared with the production laborer group norm [5]. The investigation object and role overload, task discomfort, role ambiguity, interpersonal tension responses, coping resources, social support score, rational work scored lower; task conflict, working environment, personal stress reaction, stress response, total business psychological stress reaction, somatic stress response, entertainment and leisure, self-care higher scores. See Table 52.3.

Table 52.3 The compare of occupation stress questionnaire/sub score and norm ($\bar{x} \pm s$)

Component table /sub item	Score	Production laborer group norm	Stress norm of Southwest China
ORQ	162.4 ± 20.3	183.4 ± 22.0	162.9 ± 27.0
RO	25.7 ± 5.2	29.3 ± 5.7	29.0 ± 5.8
RI	27.7 ± 5.2	35.6 ± 4.9	30.3 ± 6.9
RA	24.1 ± 6.0	38.6 ± 5.4	28.2 ± 10.6
RB	27.6 ± 5.3	26.8 ± 4.9	24.8 ± 5.1
R	24.9 ± 6.6	24.6 ± 6.1	24.7 ± 6.3
PE	32.4 ± 7.1	29.0 ± 8.0	25.9 ± 7.4
PSQ	95.1 ± 22.1	97.5 ± 13.8	91.0 ± 17.2
VS	21.1 ± 6.4	22.6 ± 3.7	20.0 ± 5.1
PSY	25.7 ± 6.9	24.1 ± 5.3	23.7 ± 5.0
IS	24.9 ± 5.4	27.5 ± 3.8	25.4 ± 4.4
PHS	23.4 ± 7.1	23.3 ± 4.7	22.0 ± 5.5
PRQ	126.4 ± 9.2	128.9 ± 18.4	129.2 ± 7.7
RE	28.3 ± 5.7	27.7 ± 5.5	27.4 ± 5.5
SC	30.4 ± 5.7	29.8 ± 5.9	29.5 ± 5.7
SS	34.6 ± 6.9	35.6 ± 6.9	36.6 ± 6.5
RC	33.1 ± 6.5	35.7 ± 6.3	35.7 ± 6.0

First, confirm that you have the correct template for your paper size. This template has been tailored for output on the US-letter paper size. If you are using A4-sized paper, please close this template and download the file for A4 paper format called CPS_A4_format.

52.3.1.2 Occupation Stress Grading Composition

Occupation tension in accordance with literature [6], taking skilled workers norm as reference, the occupation tension coarse norm linear conversion, converted to a mean of 50, the standard deviation 10 score (T), for the group of applied research, that is Quantitative evaluation for the individual and organizational level occupation tension. According to the formula $T = 50 + 10 \times \frac{x - \bar{x}}{s}$ for conversion (Where x is the original score, for norm sample average, S is standard deviation). The questionnaire/child course changed into T table, According to the T table will be divided into four levels of occupation tension. For occupation tasks and Stress Reaction Questionnaire: T score larger than 70 is divided into highly occupation tension and stress response; 60–69 score is talked into middle occupation tension and stress response; 40–59 are divided into moderate occupational stress and strain response, Less than 40 is divided into the relative lack of occupational stress and strain of reaction; Personal resources Questionnaire: T score less than 30 divided into a high degree of lack of response to resource, 30–39 are divided into moderate lack of response to resource, 40 to 59 min of moderate response to resource, within the normal range; More than 60 points indicate a strong response resources. This survey, occupational tasks, moderate/relative lack of tension accounted for 97.9 %, moderate/relative lack of individual stress reactions accounted for 76.2 %, moderate/strong individual response to resource accounted for 80.7 %.

52.3.2 Impact of Occupational Stress, Factor Analysis of Population Characteristics of the Ability to Work

Occupational stress scores for each questionnaire and comprehensive ability to work an average of percentile spacing divided into three parts (cut-off point for the P33.3 and P66.7) to convert Polychromous response variables. First univariate analysis, second univariate analysis of $P < 0.05$ of the population characteristic factors and response variables, Polychromous response variable logistic regression for multivariate analysis. A group to be substituted into the equation between the independent variables have a strong correlation ($r > 0.7$, $P < 0.05$) phenomenon, select only the strong representation of the variable into the equation.

52.3.2.1 Univariate Analysis

Occupational tasks had higher scores for the population characteristics: male, age <30 years old, non-urban living, post-secondary and higher education level, BMI of 25.0, with a total length of service, is now the type of work seniority <5, steel rolling factory, weekly working time of 45 h. Noise exposure of <5.

The individual response to resource score lower population characteristics are: the economic situation is better or worse, the night shift interval <3 days, occupational exposure to noise, smoking and drinking.

Individual stress response score higher population characteristics: male, age <30 years old, non-urban living, unmarried, better economic conditions, the total length of service <5 a, steel rolling factory.

Work ability score lower population characteristics as follows: non-urban living, the economic situation is better or worse, the iron-smelting factory.

52.3.2.2 The Multi-Factor Analysis

The main factors affecting the occupational tasks: branch, sex, place of residence, BMI, educational level, weekly working hours;

The main factors to affect individual stress response: the place of residence, economic status;

Personal resources: economic status, exposure to noise, night work situation;

The main factors to affect the comprehensive ability to work as: residence, branch.

52.3.3 The Comprehensive Work Ability and the Correlation of Occupational Stress Analysis

To CWAI as dependent variable, the occupational stress for the independent variable sub-project (independent variables into the equation method using stepwise regression method), fitting multiple linear regression equation. Into the equation according to the independent variables on comprehensive ability, the order to influence body size nervous reactions, tasks, discomfort, rational, the stress response to, responsibility and work environment. The comprehensive work ability and the body nervous reaction, tasks, discomfort, the stress reaction and working environment is related to, and reason, sense of responsibility to a negative correlation. See Table 52.4.

Table 52.4 The comprehensive work ability and professional strain of multiple linear regression analysis

Variables	Partial regression coefficient	Standard partial regression coefficient	P value
PHS	−0.23	−0.27	0
RI	−0.27	−0.23	0
RC	0.19	0.21	0
PSY	−0.16	−0.18	0
R	0.11	0.12	0
PE	−0.05	−0.06	0.04

52.4 Discuss

This investigation of the steel enterprise, is attached to a large iron and steel group of listed company, is a collection of steelmaking and continuous casting, iron, steel rolling and other production links as one of the major steel joint enterprise. Although 97.9 % of respondents career missions moderate/lack the tension, and professional task tasks in the questionnaire, tasks, discomfort, overweight task and the individual nervous reaction fuzzy questionnaire interpersonal tension scoring a southwest reaction of working group, technical workers norms are low, individual, deal with resources in the questionnaire entertainment and leisure, self care score is two norms are high, 19.3 % still moderate/height lack of individual deal with resources, 23.8 % of people high/medium individual nervous responders, another career missions tasks in the questionnaire, working conditions and the conflict of individual nervous reaction in the questionnaire of psychological nervous reaction is scoring two norms are high, the individual deal with social support questionnaire resources, rational scoring two things were lower than the norm. Clew, survey in the work environment the transformation, the enterprise work task allocation, social support, worker health education and psychological counseling, family support and so on work still need to further strengthen.

Occupational stress factors by professional, individual factors and social support and the influence of many factors. Influence factors of occupational stress analysis showed that the crowd characteristics, research enterprise occupational stress weak groups (career missions, individual nervous reaction score higher and individual to score lower resources) characteristics are: male, age <30 years old, the total length of service or now work <5 a, length of service cultural degree diploma and above, the city living, overweight and obese, working hours per week, 45 h, noise contact <5 a, unmarried, economic situation is good or poor, night shift frequency is higher, smoking, alcohol consumption, steel rolling factory. First intervention (multiariate analysis results) of people for residence, economic characteristics factors in, sex, BMI, cultural degree, factory, working hours per week, noise contact, the night shift. Enterprise in occupational stress intervention must strengthen the weak link of the crowd, the attention, can achieve good effect of intervention.

The crowd features on comprehensive ability factors impact analysis shows that the living city, iron making plant, economic status a good or a CWAI dealing lower scores, CWAI intervention for the key crowd. Occupational stress can cause individual including physiology, psychology and behavior all aspects of reaction, and even lead to disease, no matter what nervous reaction will influence the worker's capacity for work. At the same time, bad working ability will also reduce the comprehensive worker career missions coping, increase individual professional nervousness. To this, has the domestic and foreign scholars research of the relationship between. The investigation shows, CWAI and body nervous reactions, tasks, discomfort, rational, the stress response to, responsibility and work environment more than occupational stress related sub-project, overall and individual nervous reaction, professional task a negative correlation, and individual deal with resources into positive correlation. Clew, CWAI intervention and occupational stress intervention should be at the same time, both can supplement each other.

References

1. Zhi-ming D (2008) Occupation stress assessment method and early health effects. Fudan University press, vol 12, pp 4–5
2. Lei Z, Ming-zhi W (2008) Development and evaluation of comprehensive work ability index scales. Chin J Indus Hyg Occup Dis 26(6):350–354
3. Xin-Wei Y, Ze-Jun L, Xing-Huo P (2007) Study of the occupational stress norm and its application in Southwest China. Chin Mental Health J 21(4):233–236
4. Rout Stress management for primary health care professionals (2002) Klluwer Academic Plenum publishers, New York, vol 19, pp 17–39
5. Tennant C (2001) Work related stress and depressive disorders. J Chasms Res 51(5):697–704
6. Holmes S (2001) Work related stress: a brief review. J Soc Health 121(4):230–235

Chapter 53
Research on Intelligent Recording and Broadcasting System in Classroom Teaching

Qing Dong, Fengting Jiang and Guangxing Wang

Abstract Though currently a large number of intelligent recording and broadcasting system of classroom teaching can proceed with transcribing and webcasting, the effects of the radio is far from ideal. It can't maintain the personalized and customized video recording. We researched and analyzed the content and form of the common classroom teaching, and then we designed a kind of personalized intelligent recording and broadcasting system of classroom teaching according to the pattern of Project Based introduction, at the same time, we adopt the DirectShow structure and C++ programming language.

Keywords Recording and broadcasting system · Directshow structure · Filter component technology · Video coding · Streaming media technology

53.1 Foreword

"The major points of Ministry of Education's work in 2011" points out that Taking the public video class as a breakthrough, exploring educational resource construction and new shared pattern, new mechanism. Video education resource wined pro-gaze of many learners with its vivid and visual characteristic. Education department of other areas started to develop video education resource construction energetically under the guidance of the policy. But traditional scheme of video education resource construction do have many shortcomings: it takes a mess of human resources to adopt professional DV to shoot scene, of course, to some

Q. Dong (✉) · F. Jiang · G. Wang
Information Technology Center of Jiujiang University, 332005 Jiujiang, People's Republic of China
e-mail: qingdong212@126.com

W. Du (ed.), *Informatics and Management Science VI*, Lecture Notes in Electrical Engineering 209, DOI: 10.1007/978-1-4471-4805-0_53,

degree, it has a bad influence on class teaching, and collecting and edition later takes a lot of time [1]. Though the common recording and paly system in classroom teaching can deal with the basic shooting problem and achieve recorded broadcasting automatically, it can not shoot the video that closes to professional cameraman's view. And it's hard to cope with the shooting work in classroom reaching such as seminars, practical courses, debate, etc. So we need to develop a new kind of personalized intelligent recording and broadcasting system of classroom teaching according to the content and form of the classroom teaching.

53.2 System Requirement Analyses

The classroom teaching is likely to be divided into 4 parts: Multimedia Courses mainly based on teacher's explanation; Classroom Teaching mainly based on showing the content of teaching; Theory Teaching based on blackboard-writing and explanation; Practical Courses such as discussing, debate and simulation program which gives priority to students [2]. Recorded video are required to be high-lighted, and it should reappear the classroom teaching activity with impersonality, fact city and lifelikeness. What's more, it can show adequately the teaching level and quality. So we should be niche targeting when we record video, obviously the common recording and broadcasting system does not conform to the requirement.

53.2.1 The Video Recording of Multimedia Courses Based on Teacher's Explanation

The multimedia courses based on teacher's explanation, that is, the teacher dictate the content of course by themselves, they stand on the platform to dictate courses with their rich knowledge reserves and vivid teaching posture; they seldom step down to interact with students, the position between teacher and students is relatively fixed. The whole process of classroom teaching is based on oral teaching. Therefore, when recording the video, it's required that the picture is mainly focus on the teacher [3]. When needing Multimedia courseware or picture of students, they will switch the video.

53.2.2 The Video Recording of Classroom Teaching Based on Showing Teaching Content

This kind of classroom teaching is a teaching method which teachers teach by multimedia courseware, visual materials, pictures and words etc. There are several common classroom teaching such as <<Art Appreciation>>, <<Analytical Application of Literary works>>. The teacher's position of the course is relatively fixed, they teach by showing teaching content with computers and oral explanation. The recording of this kind of video is relatively easy. It's mainly based on showing the teaching content and teacher's explanation, it rarely use the student's picture.

53.2.3 The Video Recording of Blackboard-Writing and Explanation

This method, namely, what we often mentioned "blackboard class". It mainly consists of blackboard-writing and teacher's explanation. Maybe the teacher will explain first, and then write on the blackboard; or he will write on the blackboard first, and then explain, sometimes, he does both at the same time. In the process the teacher will ask students to answer questions, maybe students answer collectively to response the teacher, sometimes it needs individual to answer, and the teacher will use teaching aids such as chalks, erases, triangular rules etc., they move between blackboard and teacher's desk. The position of the teacher is not fixed, so it's not easy to record the video of this kind of course. The common automatically recording and broadcasting system is not qualified to the job. Therefore, the video recorder will consult with the teacher in advance to fix the sphere of the activities of the teacher, the relative position of the blackboard-writing and orientation of the teacher and so on. The personnel director is required to have rich experience and foresee ability to record through artificial directing.

53.2.4 The Video Recording of Practical Courses Based on Students

The practical courses mainly contains discussing, debating and scene simulation program. The students are supposed to be the main parts. The teacher plays a role of guidance, dispatch and direct. The main teaching contents are accomplished by students. This kind of classroom teaching video id required to focus on students, lens scheduling changes frequently, the picture should be linked up compactly, it needs to cohere with the teacher before recording in order to make it easier to shoot. In addition, recording and broadcasting system can record appointed OPED automatically and webcast, it can reedit, that is non-linear editing function, it can also record automatically.

53.3 Design and Realization of the System

Personalized recording and broadcasting system in classroom teaching should have the function of auto conducting and customized locate function, it can also be easy to operate and be able to live broadcast and edit in later stage. So the intelligent recording and broadcasting system we designed has the following function module: The Teacher Positioning Module, The Student Positioning Module, and Intelligent directing Module, Video Recording Module, Live Online Module, and Post-production Module [4]. The picture of system structure frame is as shown in Fig. 53.1.

Teacher positioning module: the recognition and tracking of the teacher's image, thus realizing the aim of location.

Student positioning module: the recognition and tracking of the student's image.

Intelligent directing module: the switching between anchor picture and backup images, require the picture and frequency of the teacher, have the function of adjusting and revising.

Video recording module: to realize frequency and video recording of anchor picture.

Live online module: the videotape of the classroom teaching can webcast through the system.

Post-production module: to realize the function of editing in later stage and exporting.

Camera positioning system can ensure the relative position between teacher and student, positioning camera feedback to the main camera (also called teacher camera) and auxiliary camera (also called student camera) through location module and infrared communication. Intelligent directing module obtains the main camera, auxiliary camera and the picture of teacher's computer to record personnel to watch and adjust, and then the record personnel choose the anchor picture, and artificial direct or automatically direct according to the parameter defined in advance, the video recording module gain the picture of the anchor picture to save. Live online module gain the anchor picture to webcast. Post-production invoke the

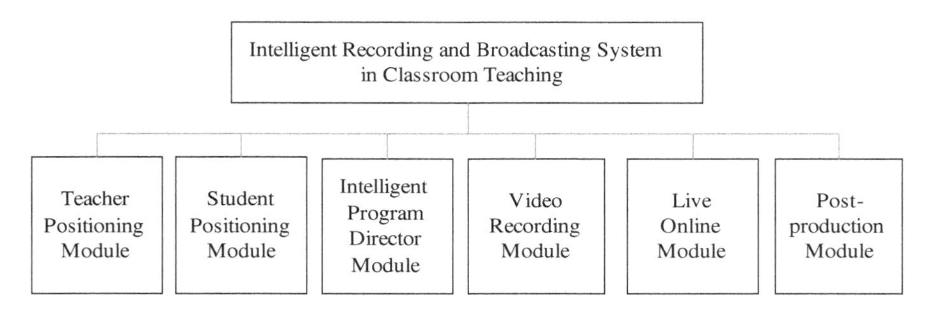

Fig. 53.1 Structure of classroom teaching video intelligent recording and broadcasting system

transferred video to proceed with post-production work. According to the function module, the topological structure picture of the designed system is shown in Fig. 53.2.

53.3.1 The Realization of the System

Intelligent recording and broadcasting system in classroom teaching adopt Filter Graph Construction technology of the DirectShow structure and C++ language program due to the reason that DirectShow provide perfect total solution to Windows platform to deal with all kinds of high-powered media. What's more, it's the new generation Streaming media processing SDK (software developer's kit) based on COM. It provides a brand new multimedia data processing model, takes the Filter component as the core concept, and packages a series of arithmetic like collection, condensation and decompression. Developers not only can make use of Filter component's own services, but also can customize Filter component according to their own needs, thus making multimedia application system development such as Video Surveillance, video conference more convenient and faster. Another advantage of DirectShow structure lies in the separation between

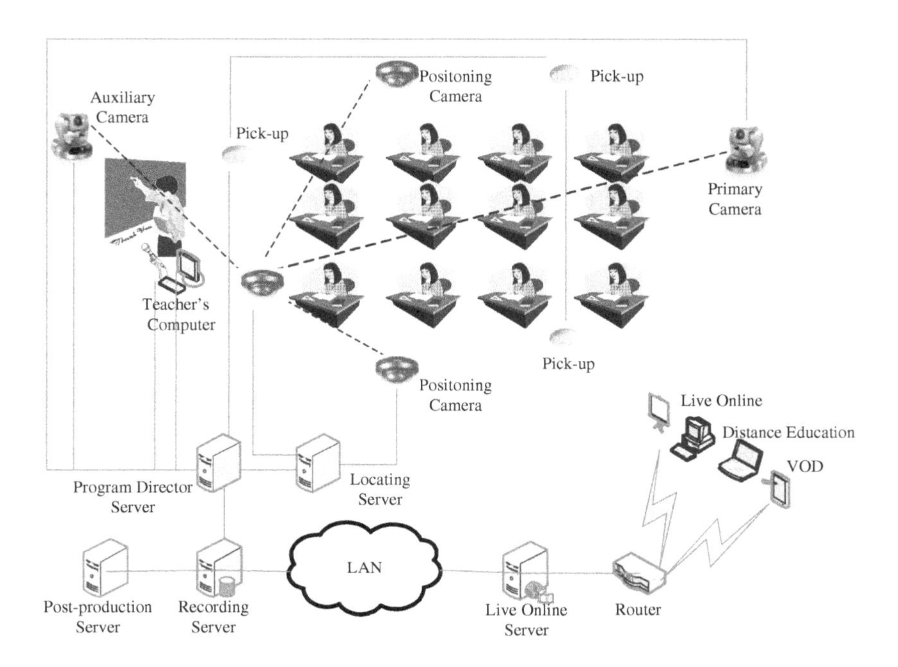

Fig. 53.2 Network topology of classroom teaching intelligent recording and broadcasting system

hardware and software. The developers do not need to concern for implementation details of the underlying hardware. All what they need to concern is the software, thus increasing efficiency. The picture of DirectShow system structure is shown in Fig. 53.3.

From the figure we can see that, DirectShow is made up by Filter and Filter Graph Manger, application program set up corresponding Filter Graph according to certain uses, then control the whole data processing through Filter Graph Manager. DirectShow can collect all kinds of events when the Filter Graph is running and it sends to application program by message. In this way, it realizes the interaction between application program and DirectShow system.

Talking the Filter Graph constructive process in video recording module for example, we introduce briefly the realization of the system and key of C++code. The steps of other module are similar to it.

Step 1: Define video recording component interface parameters //define audio flapping catch interface
Step 2: Add and deploy video mixture of Filter
Step 3: Add and deploy video source connect to video mixture of Filter
Step 4: Add and deploy word mixture Filter
Step 5: Accomplish the Filter Graph establishment and save the video //to realize the coding and save of the video
Step 6: Add and connect VMR Filter //monitor video picture and control

Among others, the Filter Graph saved includes video scaling, coding, and transferring. Here the key part is the video coding configuration; the C++ coding is as follows:

```
//set up WMV coding Filter
CComPtr<IBaseFilter>pVideoDMOFilter;
Hr = CoCreateInstance(CLSID_DMOWrapperFilter, NULL, CLSCTX_ALL
```

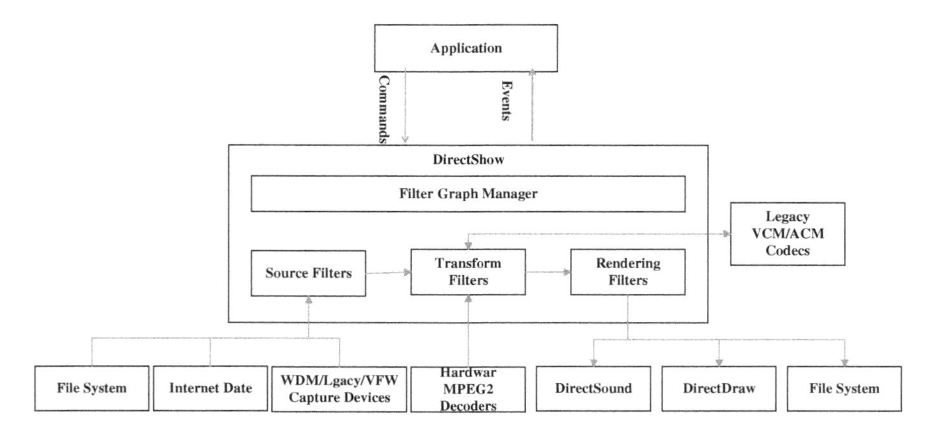

Fig. 53.3 Direct show system structure

```
IID_IBaseFilter, (void**)&pVideoDMOFilter);
WINDY_hr_JADGE(hr);
CComPtr<IDMOWrapperFilter>pVideoDmoWrapper;
Hr = pVideoDMOFilter->QueryInterface(
IID_IDMOWrapperFilter, (void**) &pVideoDmoWrapper);
WINDY_hr_JADGE(hr);
//initialization encoder
hr = pVideoDmoWrapper->Init(
CLSID_CWMVEncMediaObject2, DMOCATEGORY_VIDEO_ENCODER);
WINDY_hr_JADGE(hr);
//add coding Filterhr =
m_pVideoGraph->AddFilter(pVideoDMOFilter, L"WMV9 DMO");
WINDY_hr_JADGE(hr);
// connection video scaling Filter 与 WMV coding Filter
hr = CGraphUtility::ConnectFilters(
m_pVideoGraph, pSaveVideoSizeFilter, pVideoDMOFilter);
WINDY_hr_JADGE(hr);
// WMV encoder configuration
bRet = ConfigWMVDMO(
PVideoDMOFilter, dmoVideoMediaType, m_dwSaveKBitRate,
m_dwSavePrivateData);
WINDY_b_PASS_EX(bRet,"Fialed to ConfigWMVDMO");
m_VideoMediaType = dmoVideoMediaType;
```

The configuration VC video Coder is realized by Config WMVDMO, the content of configuration includes computational complexity, code quality, code stream and so on, it will produce private data based on WMV coding, it's usually made up by 4 bytes of data. It records coding information, the data is needed in writting files and live decoding. The coding above, after configuration, will save the private data in m-dwSavePrivateData variable. After the coder deployed, it adds and connects video to transfer Filter through network to save and webcast, the C++coding is as follows:

```
// add network to send Filter, save the video
hr = CGraphUtility::AddFilterByCLSID(m_pVideoGraph, CLSID_NetSend,
L"Video Send", &m_ pVideoSaveFilter);
WINDY_hr_JADGE_EX(hr,"NetSendFilter.ax no register!");
// configuration network send parameters
CComPtr<INetSendInterface2>pNetSendInterface;
hr = m_ pVideoSaveFilter -> QueryInterface(
IID_INetSendInterface, (void**)&pNetSendInterface);
WINDY_hr_JADGE(hr);
hr = pNetSendInterface-> SetDestIPPort(
CCommonUtils::IPNumToStr(m_dwVideoIP), m_wVideoPort);
WINDY_hr_JADGE_EX(hr,"Net Send Filter param setting failure!");
```

```
// connect to encoders and live broadcast transmissions Filter, proceed with live broadcast
hr = CGraphUtility::ConnectFilters
(m_pVideoGraph, pVideoDMOFilter, m_pVideoSaveFilter);
WINDY_hr_JADGE(hr);
```

Through the six steps above, we finished the development of core function in video recording module. The video coding and arithmetic invoked are packaged in DirectShow structure, the developers do not need to consider it, just invoke it.

Through this system, the video recording member can install in intelligent directing module and locationing module and go on with niche targeting recording work in classroom teaching. You can choose proper personalized recording type according to the pattern of classroom teaching.

53.4 Conclusion

Intelligent recording and broadcasting system in classroom teaching adopt both DirectShow and Windows Streaming media technology, it realized the functions such as AVCap, AVS, AV save, AV editing and webcasting. It has been tested that it achieved 1024*768 resolutions and 25 FPS video of the coding efficiency, the screen is clear and smooth, it sounded good and covers basic 4 kinds of recording work that we mentioned above. Of course, system diversity in picture needs to be improvement, we will move forward to explore.

References

1. Jianggong S, Jizhong Z (2011) The research and application of intelligent recording and broadcasting system in classroom teaching. Mod Educational Technol 20:44–48
2. Yijun J, Wenming W (2011) Exquisite course recording and broadcasting system function analysis and prospect. China Education Inf 3:83–85
3. Ji Y, Wang Y (2009) The design and application of automatic recording and broadcasting system in University Exquisite course. Mod Chinese Education Equip 10X:44–46
4. Andrade EL, Khan E, Woods JC, Ghanbari M (2003) Description based object tracking in region space using prior information. Electron Lett 39(7):600–602

Chapter 54
AHP-Based Teaching Evaluation Index System of Weights

Liang Qin

Abstract Teaching evaluation is an important part of the teaching quality monitoring system. It's important to improve the quality of teaching and promoting teaching reform. But the traditional teaching evaluation index system could no longer meet the new teaching model's evaluation which teaching both working and learning. How to use the analytic hierarchy process to establish a scientific and rational, adapted to the new concept of teaching under the new situation of index system of weights is the focus of this paper.

Keywords Index system · AHP · Weight

54.1 Introduction

Teaching, as the core of higher learning schools, is one of the most important ways of making personnel training realized. How to effectively monitor and evaluate the teaching quality of teachers is an important measure for guaranteeing the quality of personnel training [1]. Today, vocational education has been under a vigorous development, and also a fundamental change has taken place in teaching model and teaching methods. Therefore, under the combination of working and learning and the integration of theory and practice, a teaching evaluation index system is urgently expected to appear, so as to more scientifically, reasonably and effectively make evaluations on the teaching quality of teachers [2, 3]. Also, result from the teaching evaluation index system is of certain significance for guiding target evaluation of teachers and educational reforms.

L. Qin (✉)
Sichuan Vocational and Technical College of Communication Chengdu, 611130 Sichuan, China
e-mail: liangqin1232121@126.com

W. Du (ed.), *Informatics and Management Science VI*, Lecture Notes in Electrical Engineering 209, DOI: 10.1007/978-1-4471-4805-0_54,

54.2 Concept and Characteristics of AHP

In 1980s, Analytic Hierarchy Process (AHP) was founded by Thomas L. Saaty, famous American operational research expert and Professor at University of Pittsburgh. It is a powerful method for system analysis and operational research, and plays a considerably effective role in multi-factor, multi-standard, and multi-plan comprehensive evaluation and trend forecast. According to hierarchically structured decision analysis problem composed by "solution layer + standard layer + target layer target layer", a complete set of method and process was provided. The primary advantage of AHP is processing qualitative and quantitative multi-factor decision analysis problem, importing subjective judgments and policy experience of decision makers into model and then processing them with quantification [4]. The main characteristics of AHP can be concluded from the four aspects below.

First, in the comprehensive evaluation on an integral problem with a hierarchical structure, layer-by-layer decomposition is used for changing it into one with multiple independent evaluation standards. Then, the comprehensive evaluation can be made based on multiple independent standards.

Second, to solve the processing and comparability problems of qualitative factors [5, 6], Saaty suggested that comparison on "importance" (weight in mathematics) could be used as a united processing format, and comparison results could be quantifiably measured with 1–9 levels according to important degrees.

Third, the transitivity of comparison chain is inspected and adjusted, namely the acceptable level of consistency is inspected.

Fourth, matrix set collecting all comparative information is processed with the theory and methods of linear algebra, for the purpose of digging out deep and substantial comprehensive information as decision support.

54.3 Application of AHP in the Teaching Quality Evaluation Index System for Teachers in Higher Vocational Colleges

54.3.1 Determining Teaching Quality Evaluation Hierarchical Model

According to the characteristics of higher vocational education and the actual conditions of school, teaching quality evaluation is made from three aspects, namely students' evaluation on teaching, teaching and research office's evaluation on teaching, and department's evaluation on teaching; the hierarchical model of teaching quality evaluation is set up as shown in Fig. 54.1.

According to the different natures of courses and the theory and practice integration idea in higher vocational education, students' evaluation on teaching is classified into two aspects (liberal art, and theory and practice integration). In this

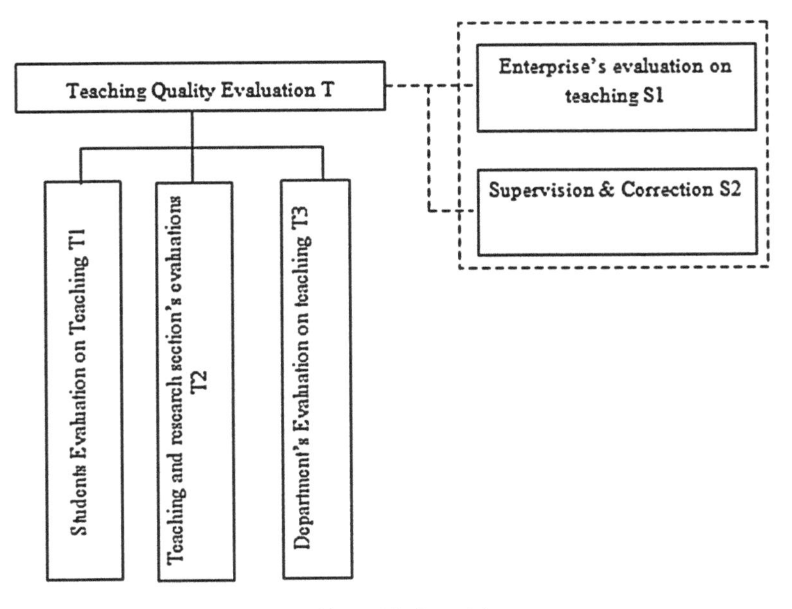

Fig. 54.1 Teaching quality evaluation hierarchical model

paper, the application of AHP in the index system is introduced from students' evaluation on teaching (theory and practice integration). In the concrete teaching process, teaching and learning behaviors of teachers and students are diversified, and also the relationship between the two sides is complicated. For this reason, the factors playing an effect on teaching quality evaluation are various. In general, students' evaluation on teaching are conducted from six aspects, namely professional dedication, professional ethics, professional skills, guidance in experiment practice, evaluation and review of experiment practice and report, and communication ability. Therefore, the Teaching Quality Evaluation index system (theory and practice integration) can be divided into two layers and six modules, and the established hierarchical model is shown in Fig. 54.2.

54.3.2 Weights of the Teaching Quality Evaluation Index System

Each index in the teaching quality evaluation index system (theory and practice integration) plays a different influence on the teaching quality of teachers. Therefore, it is necessary to set the weights according to the degrees of influence of index evaluation contents on teaching quality. For some important indexes, it is necessary to give a larger proportion to them, and therefore their weights should be set highly. However, for some less important indexes, it is only necessary to consider a smaller proportion, and therefore their weights should be set lowly. The

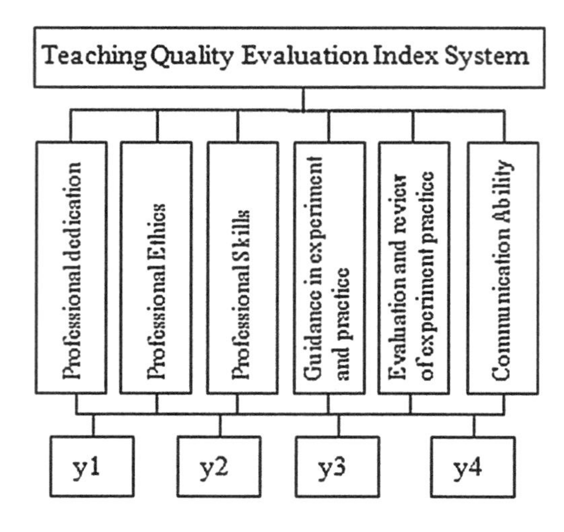

Fig. 54.2 Teaching quality evaluation hierarchical model

setting of weights can be determined with the commonly-used AHP, experts meeting method, statistics analysis method and Delphi method, and the set comprising of weights is referred to as weight set. In this paper, the weights of the teaching quality evaluation index system (theory and practice integration) are set with AHP, and the established system is shown in Table 54.1.

54.3.3 Rating Standards of Determining Evaluation Indexes

In the index system, the grades of evaluation index A_{ij} are divided into "excellent, good, average, and poor", which are corresponding to 95 points, 85 points, 75 points, and 60 points respectively.

54.3.4 Organizing Evaluators to Participate In Evaluation and Gaining Evaluation Matrix

$$D = \begin{pmatrix} d_{111} & d_{112} & d_{113} & \dots & d_{11k} & A_{11} \\ d_{121} & d_{122} & d_{123} & \dots & d_{12k} & A_{12} \\ d_{131} & d_{132} & d_{133} & \dots & d_{13k} & A_{13} \\ d_{111} & d_{112} & d_{113} & \dots & d_{11k} & A_{14} \\ d_{211} & d_{212} & d_{213} & \dots & d_{21k} & A_{21} \\ \dots & \dots & \dots & \dots & \dots & \dots \\ d_{231} & d_{232} & d_{233} & \dots & d_{23k} & A_{23} \\ \dots & \dots & \dots & \dots & \dots & \dots \\ d_{611} & d_{612} & d_{613} & \dots & d_{61k} & A_{61} \end{pmatrix} \quad (54.1)$$

Table 54.1 Teaching quality evaluation index system (theory and practice integration)

Primary index	Weight	Secondary index	Weight
Professional dedication A1	α1	Carefulness of teacher in reviewing your homework and report A11	α11
		Patience and responsibility sense in answering your questions A12	α12
		Teacher does not do things not related to teaching, such as answering mobile phones and chatting with other idly A13	α13
		Not be late for class; dismissing no class early; leaving classroom not randomly A14	α14
Professional ethics A2	α2	Be loyal to their duties; suiting action to word (e.g. adjusting no class randomly; allowing no others to take over his class; keeping promise for students) A21	α21
		Respecting and fairly treating students, and criticizing students never with radical words A22	α22
		Teacher behaves himself well, and has no uncivilized behaviors at classroom and training rooms (e.g. cigarette smoking, chewing gum, spitting) A23	α23
Professional skills A3	α3	Familiarization on work flow of enterprises and industries A31	α31
		Skilled and standard operation 32	α32
		Professional qualification certificates related to enterprises and industries A33	α33
		Satisfaction with professional and practice operation skills of teachers A34	α34
Guidance in experiment and practice A4	α4	Giving accurate Explanation and demonstration to experiment principle, purpose, content and method A41	α41
		Proficient in experiment contents and skills; able to help students get rid of problems and errors in experimental process A42	α42
		Giving explanation patiently to questions raised by students; paying attention to the reactions of students, and also providing timely helps A43	α43
		Managing experiment class orderly A44	α44
		Treating experimental teaching with chariness and responsibility; never leaving classroom randomly for a long time A45	α45
		Giving concentrated explanation to contents questioned by students frequently in the experimental process A46	α46
		Gains from practice A47	α47

(continued)

Table 54.1 (continued)

Primary index	Weight	Secondary index	Weight
Evaluation and review of experiment practice and report A5	α5	Requirements of teacher on experiment report A51	α51
		Reviewing experiment report seriously; pointing out problems in reports of students and also providing answers A52	α52
		Normative evaluation standards and methods available A53	α53
Communication ability A6	α6	Communicating with students in break A61	α61

Evaluation index weight sets are determined with analytical hierarchy process as follows
Evaluation index A_i weight set $B = (\alpha_1, \alpha_2, \alpha_3, \alpha_4, \alpha_5, \alpha_6) = (0.125, 0.125, 0.250, 0.175, 0.075, 0.250)$
Evaluation index A_{1i} weight set $B_1 = (\alpha_{11}, \alpha_{12}, \alpha_{13}, \alpha_{14}) = (0.300, 0.300, 0.200, 0.200)$
Evaluation index A_{2i} weight set $B_2 = (\alpha_{21}, \alpha_{22}, \alpha_{23}) = (0.300, 0.400, 0.300)$
Evaluation index A_{3i} weight set $B_3 = (\alpha_{31}, \alpha_{32}, \alpha_{33}, \alpha_{34}) = (0.200, 0.300, 0.200, 0.300)$
Evaluation index A_{4i} weight set $B_4 = (\alpha_{41}, \alpha_{42}, \alpha_{43}, \alpha_{44}, \alpha_{45}, \alpha_{46}, \alpha_{47}) = (0.150, 0.200, 0.200, 0.100, 0.100, 0.100, 0.150)$
Evaluation index A_{5i} weight set $B_5 = (\alpha_{51}, \alpha_{52}, \alpha_{53}) = (0.300, 0.500, 0.200)$
Evaluation index A_{6i} weight set $B_6 = (\alpha_{61}) = (1.000)$

$$D = \begin{pmatrix} 95 & 85 & 85 & 95 & 95 \\ 85 & 95 & 95 & 85 & 75 \\ 75 & 75 & 60 & 60 & 75 \\ 95 & 95 & 95 & 85 & 95 \\ 85 & 85 & 85 & 95 & 85 \\ \cdots & \cdots & \cdots & \cdots & \cdots \\ 85 & 95 & 85 & 95 & 75 \\ \cdots & \cdots & \cdots & \cdots & \cdots \\ 85 & 75 & 75 & 85 & 75 \end{pmatrix} \tag{54.2}$$

There are k students assumed to participate in the evaluation. According to the contents of the evaluation index system and the grades of evaluation index, all evaluators give points to the teaching quality (theory and practice integration) of a teacher. According to the evaluation tables of k persons, the evaluation matrix for the teacher is gained.

d_{ijm} is points made by the mth evaluator on index A_{ij}. For easy demonstration and calculation method, demo calculation is conducted under the condition that evaluators participate in evaluation, and evaluation matrix D_1 is gained.

54.3.5 Whitening Quantification of Indexes

According to evaluation grades, the number of gray evaluation types is set to be 4, and the set of serial numbers of gray types is $e = (1, 2, 3, 4)$. Then, whitening function is calculated for all qualitative indexes.

1. For the first grey type (excellent), $e = 1, \eth_1 \in [d_1, \infty)$, namely whitening function of $\eth_1 \in [95, \infty)$ is

$$f_1(d_{ijm}) = \begin{cases} \frac{d_{ijm}}{95}, & d_{ijm} \in [0, 95) \\ 1, & d_{ijm} \in [95, \infty) \\ 0, & d_{ijm} \notin [0, \infty) \end{cases} \tag{54.3}$$

2. For the second grey types (good and average), their function forms are the same, namely $e = 2, 3, \eth_e \in [0, d_e, 2d_e]$, in which d_e = [85,75], and the whitening function is

$$f_e(d_{ijm}) = \begin{cases} \frac{d_{ijm}}{\mathrm{de}}, & d_{ijm} \in [0, de) \\ 2 - \frac{d_{ijm}}{\mathrm{de}}, & d_{ijm} \in [de, 2de) \\ 0, & d_{ijm} \notin [0,\ 2\mathrm{de}) \end{cases} \tag{54.4}$$

3. For the fourth grey type (poor), there is $e = 4, \eth_4 \in 0, d_4, 2d_4]$, namely $\eth_4 = [0, 60, 120]$, and the whitening function is

$$f_4(d_{ijm}) = \begin{cases} 1, & d_{ijm} \in [0, 60) \\ 2 - \frac{d_{ijm}}{60}, & d_{ijm} \in [60, 120) \\ 0, & d_{ijm} \notin [0, 120) \end{cases} \tag{54.5}$$

54.3.6 Calculating Grey Evaluation Coefficient

For evaluation index A_{11}, the eth grey evaluation coefficient $\chi_{11} = \sum_{k=1}^{5} f_e(d_{11k})$ can be gained, and then $\chi_{111} = \sum_{k=1}^{5} f_1(d_{11k}) = f_1(95) + f_1(85) + f_1(85) + f_1(95) + f_1(95) = 1+1 + 1 + 0.895 + 0.895 = 4.789$ can be gained. Similarly, $\chi_{112} = 4.647, \chi_{113} = 3.933$, and, $\chi_{114} = 2.417$ can be gained. Based on this, the total grey evaluation coefficient of evaluation index A_{11} is solved, namely $\chi_{11} = \sum_{k=1}^{4} x_{11k} = 15.786$.

54.3.7 Calculating Grey Evaluation Weight Vector and Matrix

Through above deduction, the grey evaluation weight vector of evaluation index A_1 can be gained, namely

$$\begin{aligned}\gamma_{11} &= (\gamma_{111}, \gamma_{112}, \gamma_{113}, \gamma_{114}) \\ &= \left(\frac{\chi_{111}}{\chi_{11}}, \frac{\chi_{112}}{\chi_{11}}, \frac{\chi_{113}}{\chi_{11}}, \frac{\chi_{114}}{\chi_{11}}\right) \\ &= (0.303, 0.294, 0.249, 0.153)\end{aligned} \tag{54.6}$$

With the same method above, the grey evaluation weight vectors in the following can be gained:

$r_{12} = \{0.283, 0.287, 0.260, 0.170\}$, $r_{13} = \{0.216, 0.249, 0.282, 0.253\}$, $r_{14} = \{0.316, 0.293, 0.256, 0.145\}$, $r_{21} = \{0.291, 0.310, 0.267, 0.132\}$,..., $r_{23} = \{0.205, 0.302, 0.348, 0.145\}$,..., $r_{61} = \{0.245, 0.274, 0.279, 0.202\}$

Thus, grey evaluation matrixes R_1, R_2, R_3, R_4, R_5, R_6 of all evaluation indexes A_1, A_2, A_3, A_4, A_5, A_6 for all grey types can be gained.

54.3.8 Comprehensive Grey Evaluation

A teacher's R_1, R_2, R_3, R_4, R_5, R_6 is evaluated comprehensively, gaining the primary comprehensive evaluation results H_1, H_2, H_3, H_4, H_5, H_6:

$H_1 = B_1 \cdot R_1 = (0.3354,\ 0.3369,\ 0.2094,\ 0.1442)$, ……, $H_6 = B_6 \cdot R_6 = (0.245,\ 0.274,\ 0.279,\ 0.202)$

According to primary comprehensive evaluation results H_1, H_2, H_3, H_4, H_5, H_6, the total grey evaluation weight matrix $R = (H_1, H_2, H_3, H_4, H_5, H_6)$ of the teacher can be gained, and $H = B \cdot R$ can be established.

According to the comprehensive evaluations of primary and secondary evaluation indexes, grade-point vector of each grey evaluation index is C = (95, 85, 75, 60), and comprehensive evaluation value is W = H·CT = 84.72.

54.4 Conclusion

In this paper, a teaching quality evaluation hierarchical model (theory and practice integration) is scientifically and systematically established under the combination of working and learning and the integration of theory and practice; through the development B/S-based software, the paper-based traditional manual evaluation has been transformed to computer-based evaluation, thus effectively improving evaluation efficiency and statistical and analysis functions. With the purpose of making evaluation data more reasonable and accurate, the teaching quality evaluation results (theory and practice integration) are processed with the multi-level grey evaluation method, which can promote students not to subjectively evaluate teachers with malevolence. Through the processing by grey evaluation method, qualitative indexes can be scientifically and reasonably converted to quantitative analysis, thus making it easy to fairly evaluate the teaching quality of teachers. Also, evaluation results can be more valuable as reference for evaluation on teachers.

References

1. Chen B, Wang Y, Yang H, Yang X (2008) Evaluation of higher vocational educational quality and development of evaluation system. Southwestern University of Finance and Press, Chengdu
2. Deng Y (2004) Application of multi-level grey evaluation method in the evaluation of classroom teaching quality. Sci Technol Prog Policy 12:143–144
3. Shi G, Chen X (2000) Research on the AHP model used in evaluation of undergraduate teaching quality. J Wuhan Univ Technol 2:74–75
4. Zhang X (2006) Exploration on the teaching evaluation in higher vocational colleges. Education Vocation 6:508–509
5. Wang S, Xu Y (2003) Establishing the model of teachers' teaching quality evaluation. China Education Info, 22(04):116–118
6. Li P, Gao Y (2004) An improved method for cadre assessing based on gray fuzzy synthetic evaluation. Sci Technol Manage Res 3:103–105

Chapter 55
Study on Construction of Excellent Teaching Team in Higher Vocational Colleges

Guangyan Liu

Abstract Excellent teaching team is an important guarantee to cultivate the high quality skill type talents. The construction of the teaching team should pay special attention to the cultivation of the double-qualified teachers, who are specialized in particular fields. It should as well pay attention to the construction of the professional teaching team echelon. In addition, it should place heavy emphasis on the improvement of the levels of the professional teachers. At the same time, it should hire the part time teachers who are skillful in the practical skills, which can serve as the forceful supplement of the excellent teaching team construction.

Keywords Higher vocational colleges · Teaching team · Construction

55.1 Introduction

With the gradual deepened of the constructions of higher vocational colleges, which demonstrates the mainstay of the country, it has been an important task in the face of us on how to cultivate a teaching team with double-qualification teachers, who are not only able to adapt to the needs of the industry and the specialized posts, but also has clear characteristics of their own [1]. The quality of the talent cultivation in higher vocational colleges is not only able to be related to the development of the college itself, but also has some contact with the whole situation of the national

G. Liu (✉)
Binzhou Polytechnic, Binzhou, 256600 Shandong, China
e-mail: gyliu201032@126.com

W. Du (ed.), *Informatics and Management Science VI*, Lecture Notes in Electrical Engineering 209, DOI: 10.1007/978-1-4471-4805-0_55,

economic construction and development. Fastening the constructions and cultivation of the high quality skill type talents has been an urgent task for the higher vocational education. There is not a moment to be lost thus.

55.2 The Meaning of the Teaching Team Construction of the Higher Vocational Colleges

Teaching team construction is an important link for the educational and teaching resources team constructions of the higher vocational colleges. Teaching team construction is as well an important measure to improve the teaching quality and to realize the sustainable development for higher vocational colleges.

55.2.1 Be Beneficial to the Professional Development of Teachers

Teachers in higher vocational colleges face the new situation of the professional courses with the continuous adjustment of the market needs. The professional development is far from enough merely by the individual learning and explorations of the teachers themselves. It as well requires the realization of knowledge communication and share of the teachers through team learning, which can promote the growth of the teachers. New and old teachers, as well as the professional and part-time teachers can reach the goal of mutual development in drawing on each other's merits and raise the level together as well as mutually communicate [2].

55.2.2 Be Beneficial to Improve the Comprehensive Strengths of the Teaching Team

Compared to the teachers that are specialized in the compilation of teaching materials, the preparation of the teaching schemes, the lecture of courses and the assistance and answering the questions, the teaching team can largely expand the radiation effect of the team leaders and thus form the comprehensive strengths. It can efficiently promote the optimization of the entire teaching process system [3].

55.2.3 *Be Beneficial to the Improvement of Talent Cultivation Qualification*

In the contemporary society, the development speed of the higher vocational colleges cannot be exceeded by any time in history [4]. It is beneficial to the improvement of the talent cultivation quality. However, the construction of the teacher team recourses is far from the expansion of the students' scale. The proportion of the students and teachers is so high that it has become a universal problem in face of the higher vocational colleges. In terms of the efficiency, teaching team construction is able to improve the entity efficiency under limited teacher resources through the internal communication, integration, and the team members can share skills and experiences at the same time. When the teaching team is in the process of lecturing professional knowledge, the teaching team is able to fully realize the horizontal interaction of the team between the teachers and teachers, the students and the students as well as the teachers and the students. In this way, a good teaching effect is able to be achieved.

55.3 The Ways to the Constructions of the Excellent Teaching Teams

The construction of the excellent teaching team should surround the mainstream of the "double-qualification teachers" structure and the "double-qualified teachers" qualification. In this way, it is able to outline the center of the students. Take the ability to the post as the emphasis and cultivate the high-skilled talents with certain excellent professional theoretical foundations and the relatively strong professional operation abilities.

55.3.1 *The Cultivation of the "Double-Qualified Teacher" Qualifications in the Professional Teaching Team Construction*

55.3.1.1 Send Some Teachers to Do Research and Exercise in the Enterprises

Doing research and exercise is to improve the operation skills of the professional teachers in the actual production environment. In this way, the professional teachers can grasp the core competence in order to satisfy the needs in the actual posts in the enterprises. The construction of the teaching team should pay special attention to the cultivation of the double-qualified teachers, who are specialized in particular fields. It should as well pay attention to the construction of the professional teaching team echelon. In addition, it should place heavy emphasis on the improvement of the levels of the professional teachers. At the same time, it should hire the part time teachers who are skillful in the practical skills, which can serve as the forceful supplement of the excellent teaching team construction. In the first place, the professional and specialized teachers can make some practice in the enterprises in the corresponding posts. In the second place, through the practices in the corresponding posts, the professional and specialized teachers are able to determine relevant kinds of jobs in the major and thus be aware of the talent cultivation specifications and standards needed in the posts. In the third place, the professional and specialized teachers can formulate the professional cultivation standards and specifications according to themselves. In the fourth place, set forth from the professional cultivation schemes and deepen the educational teaching reformation.

The professional and specialized teachers should go to make researches and have exercise in the enterprises of relevant industries for more than half a year. The colleges should follow the teachers' exercise and learning circumstances. It as well requires the realization of knowledge communication and share of the teachers through team learning, which can promote the growth of the teachers. The human resources department in the enterprises and the practice post administrators can give evaluations and opinions. These opinions can be serving as an important foundation for the new teachers to be official and enter compilation and determine the posts in this case.

55.3.1.2 Go Studying In the Nationwide "Double-Qualified Teacher" Qualification Training Foundation

Based on the "On the strengthening of vocational (high end) college teachers team construction opinion" announced by the ministry of education, it should stick to the needs of the excellent teaching team construction and send professional teachers to the "double teacher qualification" training foundation to get further study in the national higher vocational colleges every time at regular times. In this case, the teaching level and the professional practice ability of the teachers can be improved and the quality of the "double-teacher qualification" of the teaching team can be further strengthened.

55.3.1.3 Encourage the Professional Teachers to Participate In the Professional Skill Level Examinations and Acquire the Related Professional Qualification Identification

Actively encourage teachers to participate in the national occupation proficiency exam, obtain professional qualification certificate. Obtain the professional qualification certificate of professional teacher award, and the column for the promotion of the assessment project. Make professional teachers continuously improve the occupation skill level, so as to improve the professional quality of "double-qualified teacher".

55.3.1.4 Encourage the Professional Teachers to Take Part-Time Job in the Enterprises

Encourage the professional teachers to take part-time job in the enterprises. Professional teachers to the organization and personnel department, office of academic affairs for the record, to fill out "professional teacher enterprise industry part-time job application form", in addition to a part-time business signed opinions shall be affixed with the official seal, should at least be in 1 with the specialized related enterprise scientific research item or technology, equipment, technology reform project. The college is thus improved, without affecting the normal teaching tasks under the premise, is able to meet the needs of the professional teachers according to business time required.

55.3.2 Construct the Excellent Teaching Team Combining both Professional and Part-Time Teachers

In terms of the student post ability analysis in the future, excellent teaching team must have a group from the industry, proficient in the production operation technology, master Post Occupation Skills of the professional and technical personnel to serve as part-time teacher, must ensure that sufficient quantities from the industry professional and technical personnel to augment the teaching team, structure to special combination, proportion for each 50 %.

55.3.2.1 Build Part-Time Teachers Resource Database

On the basis of analyzing the professional post ability investigation, go to the enterprise, and from research enterprises, find a number of skilled craftsmen operations proficient in professional skills and the technical backbone. With the approval of the enterprise human resources department and the agreement of the individual, record the information in the Information Institute Department of organizational. After the approval of the college through the audit identified as optional part-time teachers, and according to its specialty, identify as teachers or practice, training teachers.

55.3.2.2 Hire the Part-Time Teachers

We should break the original five working days, adopt flexible teaching arrangement time, Saturday, Sunday and evening self-study time as a part-time teacher centered teaching time. Strict control of part-time teacher appointment link, every alternate part-time teacher in storage before must pass through the interview, practice link. By the human resources department, office of academic affairs, professional teaching three parties involved throughout, and in the "part-time teacher qualification approval sheet" signed opinions and archive.

55.3.2.3 Manage the Part-Time Teachers

The Department mainly focuses on part-time teaching quality, teaching effect for monitoring during the entire process of part-time teachers, the teaching methods, teaching means, teaching equipment records. To the professional competence of student's growth degree as a part-time teacher evaluation is the main basis. According to the curriculum standards, curriculum must master the professional ability of decomposition, and develop the course student's professional ability appraisal standard.

55.4 The Reformation and Practice of the Construction Model of the Excellent Teaching Team

55.4.1 Cultivate Excellent Professional Teacher Vocational Team

55.4.1.1 Improve the Vocational Ability of the Teachers

Organize the teachers seriously to study the teaching concept, change their ideas, and take the lead in the reform. All of the young teachers should have the vocational education level test. Select the backbone of young teachers to Australia to study TAFE vocational education ideas, the introduction of advanced vocational teaching methods, enhance the comprehensive quality of vocational education.

55.4.1.2 Improve the Service Operation Level of the Teachers

Through and industry experts to "friend" in a good method to solve this problem, and business experts affection exchange process we got a business information for enterprises reform and work-study combination teaching provides support, understanding of the foreign trade enterprise's new foreign trade dynamic in time for the students update necessary knowledge, the combination of the actual significance.

55.4.1.3 Improve the Certificate Training Abilities of the Teachers

Improve teachers' research training ability is the essence of work-integrated learning quality and professional ability raise. The three China World Trade Center professional occupation qualification certificate is the future jobs, by sending backbone teachers to participate in the national foreign trade salesman qualification certificates in teacher training, the national foreign trade clerk qualification certificates in teacher training and the national foreign trade merchandiser qualification certificates in teacher training, teachers' research training skills and levels have been greatly improve.

55.4.2 Build an Excellent Foreign Trade Manager Professional Team

Outstanding foreign trade manager occupation troops, including small and medium-sized foreign trade enterprises, the backbone of the business department manager and general manager, according to their service school degree of enthusiasm and personality preference to participate in the whole process of personnel training. A good manager of foreign trade occupation team is to cultivate qualified foreign trade occupation people important indispensable pole, in combination of some of the key areas in need of support. The amount of the team members should be relatively stable and keep growing.

55.4.3 Dig Various Key Elements in the Teaching Team and Giving the Promotion Role of the Double-Qualified Teachers Team into Full Play

In the double construction of teaching team, teaching team enzymes plays an organization coordination and catalytic effect, can reduce the construction cost and improve efficiency. Teaching team teaching team enzymes include internal rules and regulations, organizational culture, shared vision, including members of the team's desire to learn, teach aspiration as well as members communicate cooperative desire.

55.5 Conclusion

Improve the teaching qualities of the double-qualified teachers teaching team should on the one hand stimulate the teachers to take part in all kinds of enterprise practice activities in the forms like "make practice in the enterprise" and "go to the industry" and so on. Improve the teaching qualities of the double-qualified teachers teaching team should on the other hand attract the post technical employees who are excellent in the enterprises. This is especially the case for the post technical employees who have medium to high level of posts. Invite the experts in the industry to take part in as the part-time professors or the client guests or the honor leaders and so on.

In the construction process of the double-qualified teaching teams, only by giving all kinds of factors in the team into full play can we strengthen the cooperation mechanism of the team members and form the team cohesion. In this case we can improve the teachers' team levels and the talent cultivation quality and thus improve the entire teaching level of the team. In this case we can reach the goal of double-qualified teacher construction.

Acknowledgments This paper is the research achievement of Shandong Province educational science "the Twelfth Five-year" planned project, "The Research and Practice of the Teaching Team Building of "Double-Competency Construction" in Higher Vocational Colleges" (Project Number: 2011GG405).

References

1. Ye C (2007) Analysis and reconstruction of teachers with double certificates in our country's vocational education. Exploring Education Dev 10:35–39
2. Yan-qi M (2007) The target orientation and strategic analysis of the teaching team construction in colleges and universities. Chinese Higher Education 11:121–126

3. Ping-li X (2008) An interactive integration of industry and academy: a new approach towards reconstructing the system of faculty with industrial experiences in higher vocational education. Exploring Education Dev 3:54–58
4. Zhao DY (2010) Department of higher education higher vocational education office. Year higher occupation education the basic requirements of national level teaching team. 15(03):121–125

Chapter 56
Study on Computer Technology Teachers Training Based on Systematic Method

Yan Zhao

Abstract In this paper, we investigate the professional ability training and evaluation for computer technology teacher training. First, we propose the basic structure of a computer teacher's professional abilities. Then, based on that, we discuss the purpose and main contents of the professional ability training as well as the associated approaches. A suggested curriculum for teacher is presented after that. And at last, we propose a set of criteria for professional ability evaluation of the computer technology.

Keywords Teachers training · Professional ability training · Evaluation · Computer technology

56.1 Introduction

With the rapid development and wide application of information technology in the problem of China's development, how to train high quality computer primary and secondary school teachers has been more and more attention [1]. We know, in computer science graduate students of normal universities from the main source of computer teacher provide in elementary school and middle school in our country. To be a qualified teacher, grasp the system and solid theoretical knowledge, in computer science is not enough. High teaching skills and capabilities are also important. So how to design a practical and effective training system for teacher is important and meaningful.

Y. Zhao (✉)
Shandong Medical College, Jinan, 250002 Shandong, China
e-mail: yanzhao123sw@126.com

W. Du (ed.), *Informatics and Management Science VI*, Lecture Notes in Electrical Engineering 209, DOI: 10.1007/978-1-4471-4805-0_56,

56.2 Basic Structure of a Computer Teacher's Professional Abilities

The professional abilities that a computer teacher should have can be structured as Table 56.1 shows.

56.3 Training of Professional Abilities

Because good oral English is a high quality the necessary of a teacher, mandarin oral English training is the basic requirement of teacher's education [2]. According to our normal undergraduate course, oral Chinese language course form to provide new lectures and strong practice. The second grade students, young and seniors to improve their pronunciation and oral English ability is mainly through the self-practice. The teacher encouraged Chinese reading and communication in their daily life, and establishes a training organization oral practice mandarin. In pedagogy curriculum and discipline education courses, we pay special attention to in teaching and the ability to use the language of instruction.

Table 56.1 Basic structures of a computer teacher's professional abilities

Basic abilities	Abilities in handwriting	Ability in chalk writing
		Ability in pen writing
		Ability in chinese brush writing
	Abilities in expressing	Ability in oral expression
		Ability in written expression
	Abilities in application of modern educational technology	Ability in operating modern teaching equipments
		Ability in collecting and processing multimedia materials
		Ability in the design and making of multimedia CAI courseware
		Ability in the design and development of online courses
Applicable abilities	Abilities in teaching	Ability in teaching design
		Ability in classroom teaching
		Ability in experimental teaching
		Ability in after-class coaching
	Abilities in educating	Ability in education and management of students
		Ability in communication and coordination
Personal development abilities	Ability in educational research	
	Ability in educational evaluation	

Good written expression of basic communication and leadership, students and their parents. So this is a must for teachers in quality. In order to improve the written expression ability of teachers, ordinary education course name will be comprehensive writing is the third term teach them the basic skill of the general types of writing and common actual writing. In the related specialized courses and subjects of education courses, we can put the course of essay writing another way is to strengthen the training of the performance evaluation of the written expression ability. In addition, in writing papers may also help guide students improve their writing skills a comprehensive development of the students.

The modern education technology extensive application does not mean that writing is not important. According to our normal undergraduate course, course, handwriting for teaching, put forward the new form, lectures and intensive training. After that, they also need to practice, sometimes in the supervision and guidance of teachers. The blackboard to each of the dormitory for their teacher practice chalk to write freely [3]. All kinds of handwriting showed, contests held regularly to create an exciting atmosphere, and work hard.

A qualified teacher should have the ability to organize teaching resources in the teaching design, can make, the choice of appropriate media or lever or software in classroom teaching, and on the basis of teaching aim, teaching content as well as the characteristics of his students. Therefore, the teacher must know information retrieval, processing and use, and the method of selection, use and development of the most commonly used by modern education technology teaching media research. They are also in the course of eia, teaching media, the material, the learning process, the teaching effect, must pass all the experiments step by step. After class, encourage the students to develop competitive or exhibition courseware making courseware online or offline.

Teachers should in creative thought general theory knowledge to study the teaching design system, familiar with the basic method, and through some courses, such as teaching theory and computer science teaching activity design. Invite experts lecture implementing new course standard, and to improve students to the new curriculum concept of teaching reform, the importance of it. Under the guidance of tutor, college students learn how to analyze the computer teaching materials for primary or secondary students; learn to set the teaching goal. They can also be arranged in groups or together practice teaching course design unit number rate and comment on design or mentor each other in small groups. During the internship, they will get more practical experience in teaching design.

Often, they are arranged in the actual school course to teaching practice and observation, gain experience. Course, college students also need to explain his own teaching practice technology design and comment on each other's performance in the group. In addition, intensive training under the guidance of the micro-standard teaching, the teacher invited elementary school and middle school.

Experimental teaching is the computer teaching is an important part, in elementary school and middle school. The teacher must be familiar with the content and requirements for computer course of experiment teaching in elementary school and middle school. They can learn the basic rules, procedures and operation requirements of computer experiment teaching, through the teaching practice and internship observation, familiar with. The computer lab and other existing resources can be applied to let the teacher doing the experiment and the teaching content, in the school textbook. In addition, they are also encouraged to develop and maintain campus network and web sites.

After-class coaching covers rich contents to a computer teacher, including course problem solving, homework instruction and correction, individual study guidance for top students or poor students, National Olympiad in Informatics (NOI) coaching, professional practice tutoring and so on. So, as computer teachers before employment, the computer technology should take enough training in coaching. Obviously, the specialty courses in computer science and technology are most important for them to systematically master the knowledge of computer software and hardware, as well as information retrieval, transformation, processing and application, which will serve as a thorough and firm grounding that coaching is based upon. On the other hand, the computer technology is also taught the basic knowledge and experiences about how to coach through the causes like Pedagogy Basics and other discipline educational courses. And intentionally, some primary and secondary school teachers are invited to give lectures to them about after-class coaching. In addition, the teacher is encouraged to actively join in the practices of various forms of after-class coaching in their spare time.

The teacher must have learning theory and the education method of the basic courses, such as psychology, pedagogy through the foundation, moral education and the class management, education the investigation and statistics and education evaluation. Through a variety of ways, such as the analysis and discussion of the real case teaching, learning thought of famous educator, with outstanding educator in the primary and secondary school education survey from the lecture and practice, the teacher can gain practical experience in education, learning goals and requirements of the elementary school and middle school education rights and daily behavior standards of school students. In the education internship experience, familiar with practice, teachers are encouraged to help in the work of the head teacher, observation and study real instances of how teachers guide and management students, to participate in the preparation methods, design, organization, implementation and management of class activities, try to make friends with students and help them deal with problems in study and life.

The teacher not only to get along together with students, and their colleagues, leadership, the parents of the students and other people social work. How to strengthen communication skills and interpersonal good this is what all the teachers have to face. A qualified teachers should have good communication and coordination ability, create harmonious atmosphere, education. Therefore, the teacher must learn and master the basic theory and knowledge of education policy and management courses, teachers' professional development, etc. The lecture interpersonal communication ability and psychological is carried out in a planned way. The teacher also allowed touching and holding education, investigation, observation or practice. In the process, they are encouraged and inspired to deal with problems they face, so that they can gradually understanding and the understanding of the methods and skills of dealing with various kinds of types of interpersonal relationship from personal experience.

The teacher education research is the ability to his ability and creativity and he did research in the teaching and education. According to our normal undergraduate course, the course of education research methods, investigation, statistics and education of teachers' professional development to provide help students understand the fifth semester basic education theory and method of the research. At the same time, teachers encourage students to attend college students' scientific research and innovation program or to some of the research group in computer science professor. Internal or external experts were invited to teach students in a planned way scientific research, in order to arouse their interest, to do research. In addition, some topics can be designed for the high level of undergraduate study around and do further research.

In the school, the teachers are not only objects, and the main body of education evaluation, can help stimulate growth of the teachers and students. Students will learn the system of basic theory and methodology of education evaluation through the education evaluation process. In the familiar experience of practice, teaching training or practice, we can lead them to do the experiment of education evaluation specific activities. And some experts in related areas can invite class must according to the actual needs of the students.

56.4 Curriculums for Teacher

Course, is the core of the teaching activities, the main approaches to achieve our education goals. In our experience, the course system for teacher can by the big three modules, that in general education foundation course module, module in computer science major courses, and the normal education course module.

The basic course module in general education course widely-ranged from various aims to improve students' comprehensive cultural knowledge and personal quality level.

This module contains the entire professional course group in the computer science, such as common core curriculum group, the software foundation course group, software engineering course group, hardware course group, the group and the multimedia teaching network course group.

In practice, the normal education course module can be divided into four parts, namely basic education of the normal curriculum, teaching skills classes, discipline education courses and teaching practice.

Two compulsory courses and the number of elective courses to provide basic education course of normal. The two requirements is psychology and pedagogy foundation course foundation. And elective courses can education the investigation and statistics, the teaching activity design, and the development of school-based curriculum, education evaluation, education philosophy, education policy and management, the thought of the famous educator, analysis of case teaching, education research method, etc.

Teaching skill course covers four required curriculum, this is oral English teaching, calligraphy teaching, modern education technology, computer teaching skills training.

56.5 Evaluation of Professional Abilities

The professional ability testing and evaluation for teacher is in the form of a pass/fail test during the pre-service teacher education, which aims to improve their basic qualities to be a computer teacher. The test should be easy to organize, easy to operate and easy to assess.

Table 56.2 Professional ability evaluation criteria for teacher

	Module	Factors
A	Common teaching skills	A1. Teaching language
		A2. Blackboard writing and drawing
		A3. Teaching manners and suit
		A4. Preparation of teaching aids
B	Teaching design skills	B1. Analysis of learning conditions
		B2. Interpretation of curriculum standards
		B3. Analysis of textbooks
		B4. Design of teaching goals
		B5. Planning of the teaching process
C	Classroom teaching skills	C1. Introduction skill
		C2. Interpretation skill
		C3. Questioning skill
		C4. Communication skill
		C5. Learning guiding skill
		C6. Concluding skill
		C7. Organizing and coordination skill
		C8. Presentation skill
D	Practical teaching skills	D1. Skill for CAI courseware making
		D2. Operating skill of conventional electric education media, e.g. projectors, video recorders, CD/VCD/DVD players, microphones, power amplifiers, etc.
		D3. Skill of using popular software tools
		D4. Experimental preparation skill
		D5. Experimental presentation skill
		D6. Experimental guiding skill
E	Extended teaching skills	E1. Skill of explaining one's teaching design
		E2. Skill of observing a lesson
		E3. Teaching evaluation skill
		E4. Teaching research skill

Here, we propose a set of criteria for professional ability evaluation of computer technology. They can be divided into five modules, i.e. common teaching skills, teaching design skills, classroom teaching skills, practical teaching skills and extended teaching skills. Each module is composed of a few related factors. For each factor, corresponding evaluation criteria are established, against which the testers can give out different grades or levels, such as A, B, C and D. Finally, the decision of whether an undergraduate can pass the test or not is made on the overall evaluation of all the grades he/she has derived. Note that in the computation, those factors can be given different weights according to their significance. The five modules and their associated factors are listed in Table 56.2.

56.6 Conclusion

In this paper, we mainly study questions ability training and evaluation professional major in computer science teacher. First, we put forward the basic structure of the computer teacher professional ability. Then, we discuss the main contents and related professional ability training method. A suggestion of course teachers put forward. Finally, we put forward a set of professional ability evaluation standard of computer technology.

References

1. Tan Z, Duan Z (2010) Pre-service teacher education modes abroad: comparative research and enlightenment. J Jiangsu Univ High Educ Study 27(4):26–30
2. Chen H (2010) Teacher education in US and its implication on Chinese education reform. Int Forum Teach Educ 4(3):35–39
3. Qin L (2003) Comparison of the talent cultivating modes in higher teacher education between America and China. Training Res-J Hubei Coll Educ 20(6):94–96

Chapter 57
Efficient Teaching Scheme Based on Multiple Intelligence Theory

Huanxin Jiang

Abstract Administrators, especially in the field of ESL and EFL, should seriously consider Multiple Intelligences Theory as one of the policies to enhance English language teaching. The results in this study encourage English teachers to integrate Multiple Intelligences Theory in teaching and learning activities to improve students' learning. There is a consensus among researchers and scholars that attitude is very important in successful learning. The findings of this study proved that MITA Syllabus encourages students to have a positive attitude in learning. Consequently, this study stimulates teachers to incorporate Multiple Intelligences into their teaching.

Keywords Multiple intelligences teaching approach · Syllabus · English teaching

57.1 Preface

As a result of globalization, the National Education Act, promulgated in 2004 focused on developing the whole system of education in Thailand. Many educational policies were declared by the Ministry of Education (MOE) to support globalization [1]. According to this tremendous change, many new and unprepared universities encountered many problems because of the lack of readiness in many aspects such as lack of its own curricula [2], lack of motivating and interesting approaches in teaching, and paying less attention to the individual in teaching.

H. Jiang (✉)
Changsha Aeronautical Vocational and Technical College, Changsha, China
e-mail: huangxinj234@126.com

W. Du (ed.), *Informatics and Management Science VI*, Lecture Notes in Electrical Engineering 209, DOI: 10.1007/978-1-4471-4805-0_57,

Gardner's Multiple Intelligences (MI) Theory addresses how the brain deals with information, stating that there are nine different ways of thinking, solving problems, and learning [3]. Even though it is a theory and has yet no specific application method or instructional approach, it does offer a structure by which to develop a model for teaching [4]. Multiple intelligences theory suggests that there is not just one concrete measure of intelligence and by implication not just one single way of teaching. Hence Gardner suggests that learning and teaching can be understood and practiced through many avenues. In 1983, he started with seven intelligences but his research has now described nine intelligences. The first is logical/mathematical intelligence which refers to the ability to reason deductively or inductively, to recognize and manipulate abstract patterns and relationships, and to use numbers effectively [5]. The second is verbal/linguistic intelligence which refers to the ability to use language effectively and to communicate in both speaking and writing. The third is visual/spatial intelligence which refers to the ability to comprehend mental models, manipulate and model them spatially and draw them in detail.

57.2 Purposes of the Study

To develop an effective MITA Syllabus for teaching English Conversation 1 course based on multiple intelligences theory based on the 80/80 standard level.

To compare the students' English proficiency that was taught through the MITA Syllabus and those via a traditional method.

To compare students' Multiple Intelligences both before and after learning through the MITA Syllabus.

To explore students attitudes towards learning the English Conversation 1 course via MITA Instruction.

57.3 Research Methodologies

The present study is a mixed-method research consisted of both quantitative and qualitative data analysis. The experimental group was taught by the researcher via the MITA Syllabus on English Conversation 1 course, whereas the control group was taught by the researcher through the traditional method which focuses on lectures by the teacher. However, the two groups were taught in the same content of English Conversation 1.

57.4 Subjects

The subjects in this study were 66 mixed-ability fourth year students at Rajamangala University of Technology Isan Kalasin Campus. They were studying English Conversation 1 course in the second semester of the 2009 academic year. The subjects were divided into two groups: 33 students in the experimental group and 33 students in the control group.

57.5 Research Instruments

57.5.1 Multiple Intelligences Teaching Approach Syllabus

A MITA Syllabus for teaching English Conversation 1 was developed by the researcher under the framework of Multiple Intelligences Theory to promote students' autonomous learning and improve students' learning achievement. It was tried out with 27 students. The result is shown in Table 57.1.

57.5.2 Traditional Teaching Method

The traditional teaching method was employed for teaching the control group. Most learning activities focused on giving lectures and doing the exercise at the end of each unit.

57.5.3 Rubrics Assessment

Rubric assessment for the English Conversation 1 course was adapted from Webber [7]. This instrument was employed to collect the students' scores from their individual and group work based on the criteria of each grade indicated in the rubrics [6]. The rubric assessment was revised according to the comments and suggestions of experts. The followings are the example of A and B grade rubrics assessment:

Table 57.1 Effectiveness of MITA syllabus

Effectiveness	Number of students	Total score	X	S.D.	Percentage
E1	27	70	56.46	2.746	80.04
E2	27	30	24.35	1.896	80.46

* El = Effectiveness of the process (total scores of learning activities)
* E2 = Effectiveness of the learning outcomes (scores of test)

57.5.4 Criteria for a Grade Individual Work

Collects information from at least six resources to create the individual work
Shows a deep understanding of research
Uses logical sequence of ideas and easy to follow
No errors in spelling, grammar, punctuation, and capitalization.

57.5.5 Group Work

Collects information from at least six resources to create the group work
Shows deep understanding of knowledge presenting
Well organized presentation with knowledge in a logical sequence which audience can easily follow
Maintains eyes contact and does not refer to notes
Uses a strong and clear voice, good pronunciation, tone, pitch, and stress with appropriate pacing and gestures English Proficiency Test
The English proficiency test was created by the researcher, level of difficulty (p) = 0.333–0.630, level of discrimination (r) = 213–0.663, and the reliability (KR 20) = 0.82.
Student's Multiple Intelligences Inventory
To obtain students' predominant intelligences for grouping students in learning, the multiple intelligences inventory adapted for Thai context from Students' Multiple Intelligences Inventory [7]. The Coefficient Alpha of Cronbach of this inventory was 0.83.

57.6 Data Analysis

The data obtained from different research instruments was analyzed and interpreted in two ways; quantitatively and qualitatively.

57.6.1 Quantitative Data Analysis

Quantitative data analysis includes the data obtained from a rubrics assessment, test, and students' multiple intelligences inventory.

Table 57.2 Student's English proficiency of learning through MITA syllabus

Group	Independent-samples T-Test				Sig. (2-tailed)
	Mean	SD	t	df	
EPT					
EG	32.15	2.84	26.648	64	0.000
CG	17.1	1.57			
LAS					
EG	70.46	5.28	8.325	43.434	0.000
CG	62.12	2.27			

EPT English proficiency test, *LAS* learning activities score, *EG* experimental group, *CG* control group

57.7 Results of the Study

The three main results of the study are presented below:

Table 57.1 presents the result of effectiveness of this MITA Syllabus, including MITA Lesson Plan, Rubrics Assessment, and MITA Guide for Students on English Conversation 1 Course was 80.04/80.46 which means that it was effective in teaching English according to the 80/80 standards.

According to Table 57.2, it is obvious that there are highly significant differences between students' English Proficiency Test and students' scores of learning activities of both experimental and control groups at $p < 0.05$. This indicates that the students who studied English Conversation 1 Course via the MITA Syllabus had better English Proficiency Test and learning activities scores than those who studied the English Conversation 1 Course by the traditional method.

Table 57.3 shows that there are highly significant differences between students' Multiple Intelligences before and after learning through MITA Syllabus at $p < 0.05$. These indicate that learning through the MITA Syllabus could improve all nine Multiple Intelligences of students.

As shown in Table 57.4, the mean score of students' attitude toward learning through MITA Syllabus was 3.50 which higher than 3.26. It means that the students have a very positive attitude toward learning through MITA Syllabus on English Conversation 1 Course.

57.8 Discussions

According to the results of the study illustrated above, they revealed that the MITA Syllabus was effective in language teaching. The evidence for this effectiveness of the MITA Syllabus is discussed below.

The MITA Syllabus was proven to be effective according to the 80 1 80 standards, where both scores from learning process and test were 80.04/80.46. This might be related to the following processes involved in conducting the MITA Syllabus.

Table 57.3 Results of students' multiple intelligences of before and after learning through MITA syllabus

MI	Independent-samples T Test			
	Mean	SO	df	Sig. (2-tailed)
V				
Pre	8.73	0.63	49.530	0.000
Post	11.42	1.15		
L				
Pre	7.52	0.71	49.587	0.000
Post	9.15	1.30		
Vs				
Pre	7.88	0.78	50.011	0.000
Post	9.67	1.41		
B				
Pre	8.21	0.89	49.472	0.000
Post	9.94	1.64		
M				
Pre	8.64	0.93	54.352	0.000
Post	10.06	1.46		
Ie				
Pre	8.30	0.64	51.272	0.000
Post	10.91	1.10		
Ia				
Pre	8.39	1.12	63.429	0.000
Post	11.03	11.03		
N				
Pre	7.97	0.73	50.805	0.000
Post	10.52	1.28		
E				
Pre	8.33	0.817	58.485	0.000
Post	11.15	1.12		

MI Multiple Intelligences, *V* Verbal/Linguistics, *L* Logical/Mathematics, *Vs* Visual/Spatial, *B* Bodily/Kinesthetic, *M* Musical/Rhythmic, *Ie* Interpersonal, *IA* Intrapersonal, *N* Naturalistic, *E* Existential

This might be related to the processes of development of MITA Syllabus. Firstly, it was developed by the researcher step. By step under the principles of Instructional System Design (ISD) and it was revised according to the comments and suggestions of experts.

Secondly, the content in the MITA Syllabus was based on the results of needs analysis of teachers and students which supported their needs and interests. Moreover the activities in each step promote cooperative learning.

Finally, the rubrics assessment for MITA Syllabus helped students to get their expected learning outcome for their individual work and group presentation according to their language proficiency.

Table 57.4 Student's attitude toward learning through MITA syllabus

Statements	X	S.D.	Level of satisfactory
1. Learning through the MITA Syllabus helps me understand my real abilities	3.46	0.588	Very positive
2. Learning through MITA Syllabus helps me succeed in english conversation course	3.42	0.654	Very positive
3. Learning through the MITA Syllabus gives me a variety of academic information from many sources	3.42	0.654	Very positive
4. Learning through the MITA Syllabus gives me a chance to acquire and express knowledge in multiple ways	3.54	0.588	Very positive
5. Learning through the 3 MITA Syllabus enhances my independence and cooperative learning	3.50	0.659	Very positive
6. Learning through the MITA Syllabus lets me make use of all my language skills	3.54	0.588	Very positive
7. Learning through the MITA Syllabus enhances learning discipline	3.42	0.654	Very positive
8. Learning through the MITA Syllabus enhances me to get expected grade	3.54	0.588	Very positive
9. Learning through the MITA Syllabus enhances critical and logical thinking and improves problem-solving skill	3.67	0.702	Very positive
10. Getting immediate feedback through the MITA Syllabus highlights any weak points which need to be improved	3.50	0.590	Very positive
Total	3.50	0.626	Very positive

The results show that the students' English Proficiency m the experimental group taught through the MITA Syllabus was significantly different from the students' achievement in the control group taught via a traditional method.

These results indicate that the MITA Syllabus encourage students to learn effectively. The learning objectives stated and introduced clearly in the first period of each unit help the students to understand the purposes of each unit; and they know exactly what they have to learn to reach the goal of each unit. This helps students to stay on the correct track of effective learning.

As mentioned earlier, Multiple Intelligences Theory needs new and multiple ways for evaluation. That is why rubrics assessment was implemented in this study. Rubrics assessment helped students to get high scores in both individual work and group presentations. When the students knew exactly how assignments would be graded, they helped each other to prepare the group work very well for a satisfactory score.

References

1. Gardner H (1983) Frames of mind: The theory of multiple intelligences, vol 23. Basic Books, New York, pp 87–89
2. Graves K (2010) Designing language courses: A guide for teacher, vol 6. Heinle and Hinele Publisher, Canada, pp 78–82

3. Green F (1999) Brain and learning research: Implications for meeting the needs of diverse learners. Education 4(119):682–687
4. Kim IS (2009) Relevance of multiple intelligences to CALL instruction. Reading Matrix 25:79–84 (Available Online at)
5. Nitko A (2010) Educational assessment of students, vol 15. Upper Saddle River, NJMerri, pp 11–17
6. Shore R (2009) Investigation of multiple intelligences and self-efficacy in the university English as a second language classroom. Doctoral dissertation, George Washington University, USA 13:36–39
7. Webber E (2000) Five—Phases to PBL: MITA (Multiple Intelligences Teaching Approach) model for redesigned higher education classes. Retrieved 22:06–11

Chapter 58
Penetration of Moral Quality Education on University Life Sciences Courses Teaching

Mei Chen, Yunlai Tang and Duan Ning

Abstract With the rapid development of life science and technology, and the increasing influence of life science and technology on human life, the moral quality training of professionals in the field of life science must be valued seriously. This paper discusses the necessity and feasibility of the penetration of the moral quality education in the university life science courses teaching, and is designed to provide reference and reflection for the training students with high moral qualities and professional knowledge in universities and colleges.

Keywords Biology science courses teaching · Moral quality · Education

58.1 Preface

Humanistic quality education is the basis of scientific and cultural education, and is an important part of the national quality education. Education without humanistic education is not complete education, and one without the soul. Albert Einstein once said: It is not enough to teach man a specialty. Through it he may become a useful machine but not a harmoniously developed personality [1]. The moral quality is a fundamental part of the humanistic quality. The moral education objective in universities and colleges is to foster graduates with good moral character, strong will, a high professionalism and sense of social responsibility. Professional teachers bear inescapable responsibility for moral quality education

M. Chen (✉) · Y. Tang · D. Ning
School of Life Science and Engineering, Southwest University of Science and Technology, Mianyang, China
e-mail: meichen182m@126.com

W. Du (ed.), *Informatics and Management Science VI*, Lecture Notes in Electrical Engineering 209, DOI: 10.1007/978-1-4471-4805-0_58,

of students. The phenomenon that teacher considers knowledge be more important than humanistic and moral quality is still common. The majority of teachers even think that humanistic and moral quality education has nothing to do with their professional courses teaching. The teaching of professional courses, however, account for most of the time of the students. If professional courses teaching cannot be integrated with humanities and moral quality education, we will lose the opportunity and time given to students with humanities and moral education, which would be difficult to in-depth and long-lasting. Therefore, penetration of the humanities and moral education into the courses teaching is an important part of the training high-quality talent. This paper aims to discuss the problem of the penetration of the humanities and moral education into the courses teaching.

58.2 Life Science Professionals Training Urgently Need to Focus on the Moral Education

In recent decades, many achievements have been made, and life science technology brings human with health, prosperity and convenience, but it also brings confusion, problems and even disasters to the human. Moral quality of life sciences scientists or professionals is becoming more and more important.

58.2.1 The Brilliant Achievements in the Field of Life Sciences and Their Contribution to Mankind

Life Science is a nature science to study the phenomenon of life and life activities, the nature, characteristics and laws, as well as the various biological relations between biology and environment. Since the second half of the twentieth century, human beings have made a series of tremendous achievements in the field of life sciences and its application technology: In 1978, the English bred the world's first test-tube baby, causing the widespread concern of the world's; In 1985, mankind realized the automation of the sequencing of DNA base pairing, marking life science technology to further mature. In 1997, the cloned sheep "Dolly" was born, and human cloning became possible etc. [2]. With these achievements, life sciences and technology are impacting on human beings lives. As a scientist once said, the rapid development of the life sciences has opened up tremendous prospects for the improvement of the health of individuals and mankind as a whole [3]. In a sense, the life sciences scientists or professionals are playing the role of God: They not only changed the problem that some organic diseases cannot be cured by the technology of organ transplantation, but also can create a nature original species through transgenic technology.

58.2.2 The Development of Life Science and Technology May Bring Human Confusion and Disaster

Everything has two sides, and the same is true of advances in life science and technology. Development of human history has proved: science and technology is a double-edged sword. Science and technology is more developed, and it brings us more positive and negative effects [4]. Development of life science bring human with health, prosperity and convenience, but it also brings confusion, problems and even disasters to the human, and life science and technology are bringing new challenges to human ethics and jurisprudence. With the development of life science and technology, life science and technology crime has been the attention of scholars in related fields. Life Sciences criminal, which is an abuse of modern life science and technology, is a serious departure from the basic purpose of the development of life science and technology, which should have been taken for the benefit of human society.

58.2.3 Moral Quality of Life Sciences Scientists or Professionals is an Important Thing

The life science research is maybe driven by a wide range of value targets, including academic value, economic value, moral values and aesthetic value. Moral value is one of the goals of the important value of the life science research. History tells us that scientists, who work for the benefit of the people and promote the harmonious development of human-oriented, can make research for the real benefit of mankind. Therefore, the scientific conscience of professionals is at stake, because they have special knowledge of the background and make it authoritative in the field of knowledge, and they need to have a strong sense of social responsibility.

As mentioned above, the life science has indeed brought people a lot of hidden dangers and threats, and the moral quality of life science professionals, grasping the life science and technology, has a significant impact on this. There is life science major in the China's Majority comprehensive universities, and life science professional culture is an important part of higher education. In the era of highly developed life science research, life science major students should be excellent both in morals and studies, having social awareness, sense of mission and responsibility.

58.3 How Moral Quality Education is Integrated into the Life Sciences Courses Teaching

Life science, which is a nature science, is vested in the science areas. But the harmonious development of man and nature is the ultimate life science research as far as the eye could. From this point, life science is a human colure. Following the many years of teaching practice, in this paper the author will talk about the penetration of the Humanities and moral quality education into the life sciences courses teaching.

58.3.1 Improve Teachers' Moral Qualities and Impress Students with the Power of Teachers' Personality

As the saying goes, example is better than precept. Teachers' personalities have a profound effect on students, and this effect is even lifetime. A teacher, who has charisma, will win the respect and affection of the students and promote the formation of students' personality. So teachers should pay attention to inspiring students with their own charisma in the process of teaching and education. The students recognize the following things about their teacher very clearly: Whether the teacher has a high degree of social responsibility or not, whether the teacher deals with his own education and scientific research with a high sense of responsibility or not. Because the teachers' personality will affect the growth of the students, so a university professional teacher should be strict demands on themselves, and strive to improve his human and moral qualities.

58.3.2 Help Students Build up a Scientific Morality and Way of Thinking by Giving Full Play to the Role of Scientist Deeds and the History of Science

The life history of science includes not only the ingenuity of scientists, but also scientists' rigorous scientific attitude, perseverance, the spirit of the pursuit truth and the sense of social responsibility. There is a saying that the power of example is infinite. The young students always are deeply affected by famous scientists. During the life sciences courses teaching, teachers can tell the students some stories of scientific discovery, revealing how the natural mysteries was uncovered, encouraging students to engage in the science research, helping them to establish a noble scientific ethics, exercising the right way of thinking. Albert Einstein once said in memory of Marie Curie: "The first-class figures of the times and historical significance of the process, in its moral Quality, perhaps more than pure talent will be even more significant achievements, even if it is the latter, they depend on the degree of character, perhaps more than usually perceived as". The infinite charm of the great scientists is a valuable material for scientific moral quality education.

58.3.3 *Cultivate the Students' Noble Morality and Social Responsibility by the Combination of Social Problems with Teaching Process*

Scientists who engage in some life science researches, which are scientific experiments related to the human health, should also consider whether the techniques and results will bring the human being with hazards, which is different from other disciplines. Scientists must comply with appropriate laws, regulations and ethical boundaries. Human beings have the ability to control and guide the development of technology to the desired direction. Human beings, however, have grasped lots of knowledge, which give human beings the ability to control the nature. Therefore, the desirability of a variety of technologies must be judged carefully, insuring such a powerful force must be used for noble purposes [5]. In order to culture the social responsibility of students in the process of life sciences course teaching, teachers cannot preach simply, but combined with the actual social problems, so that students have practical experience. For example, in July 2007, a local media in Xingtang county Hebei province china reported that a mentally ill beggar was murdered and his organs were sold. From the adverse consequences of these events, the students should learn: we learn the scientific knowledge in order to benefit mankind, and to solve the problems for mankind. If scientific research and technology is misused in order to get financial gain, it will bring mankind and society harm.

Academic fraud and academic corruption are currently serious problems that Destruct the social prestige of the life science professionals. As we all know, Hwang Woo Suk, a disgraced cloning expert from South Korean, fabricated data in a paper published in the journal Science. When teaching stem cell research progress, teachers can tell students this disgraced story. From the cloning scandal, students will have these insights: Scholars must have lofty academic integrity, and academic fraud is the direct damage to the academic character, the collapse of the academic research, leading to the academic bubble and unnecessary consumption of manpower and resources.

58.3.4 Guide Students to Use the Scientific Knowledge in the Social Practice, Deepening the Moral Quality Education in Practice

Students' social practice education is an important part of college quality education. In course teaching, teachers should tell students to emphasis on social practice, and grasp the opportunity to exercise their own social practice. Units and departments of social practice should be consistent with the student's major n principle. Through course teaching, students have mastered the theoretical knowledge, but the knowledge cannot be elevated to the ideals and beliefs of students, unless the knowledge is integrated with his emotional experience. On the other hand, the social practice can make up for the imbalance in the search for self and understanding of the mentality for students, and can communicate with others and self, self and society, as a result of improvement of their own personality traits. Therefore, from this perspective, the social practice is a strong complement of the moral quality of education, and it will promote the harmonious development of humanistic quality of the college students.

To summarize what has been mentioned above, with the development of life science, life science and technology are exerting more and more influence on human social life. The life science teachers who engage in the higher education should establish the education view in which a cultural and moral qualities, scientific literacy and innovation capability are integrated. University or college provides students not only the academic knowledge, but also the noble moral qualities. Such education can cultivate a genuine talent with a solid life sciences expertise and noble human and moral qualities, making life science research results service the benefit of mankind.

References

1. Einstein A (1952) New York Times, Concerning Education and the Humanities. Three Rivers Press, New York 58–62
2. Guo Z (2002) Legal and ethical issues of biomedicine. Peking University Press, Beijing 163–168
3. John TH (1984) Science, technology and environments. Popular Science Press, Beijing 16–22
4. Liu C (2009) Phenomenon of crime caused by the modern life science and technology and its Criminal Law. J Zhengzhou Univ Light Ind Soc Sci 12(1):27–31
5. Lin D (2000) Science and technology philosophy and the future destiny of human. Science Technology and Dialectics, Beijing 11–12

Chapter 59
Study of Capital Construction Project Financing Mode for Local Universities

Lijun Fan

Abstract In recent years, commercial banks tighten money for the capital construction of local universities. Some universities' new districts under capital construction are facing lack of money supply for construction. Using a typical inland local university-A university as an example, this paper analyses the success story of A university in the financing mode of its new district's capital construction, proposes its directing thoughts and basic idea and addresses the financing channels in university A's financing modes. University A's financing modes for the capital construction of its new district consists of several financing channels organically and is a systematic, virtuous and sustainable financing mode. University A's financing mode for capital construction is a feasible and extendable and successful mode, which can serve as a guidance and reference not only for local universities, but also for the capital construction of originally subordinated universities for the central government and other types of universities.

Keywords Local university · Capital construction · Financing mode

59.1 Introduction

From the end of 1990s of last century, our nation's higher education has seen a tremendous developing period that never appeared before since the deepening of the reform of our nation's education system and the need of the popularization

L. Fan (✉)
Department of Engineering Management, Luoyang Institute of Science and Technology, Luoyang, 471023 Henan, China
e-mail: lijunfanx4983@126.com

W. Du (ed.), *Informatics and Management Science VI*, Lecture Notes in Electrical Engineering 209, DOI: 10.1007/978-1-4471-4805-0_59,

situation of higher education [1]. Colleges all over the nation expanded largely. The gross enrollment rate of higher education has increased from 4 % or so in the 1980s to 24.2 % by 2009, and is expected to increase to 40 % by 2020. In order to improve the conditions for running universities and ensure the quality of higher education, large scale expansion will definitely require universities to invest more funds for capital construction.

Where do the funds for capital construction come? The investment by the government into the construction of local universities is far from satisfying the current need and it's not feasible to rely on promoting tuition fee. Therefore, both the government and universities agreed undesignedly on lending money from banks [2]. Because of too much relying on bank loans, some universities suffered from very large scale of bank debts, and was badly influenced in its sustainable development. In 2004, the country entered into relevant policy documents to give policy instructions concerning college loans[3]. Meanwhile, the country cut down the scale for commercial banks to lend money to universities. Particularly, commercial banks reduce money supply to local universities in recent years. Under such a circumstance, how is universities' (especially local universities) capital construction fund guaranteed and how the funding channels for capital construction is extended through diversification of financing modes to ensure the investment on capital construction and the sustainable development in later periods have become a task of top priority for most local universities to achieve.

59.2 Success Story of the Financing Mode of the New District's Capital Construction of University A

A University was an ordinary local training school which occupied a land of less than one hundred mu and a school area of 30 thousand square meters with less than 2,000 students [3]. Through ten years of development and through successful adoption of financing modes for new district's capital construction, by utilising a virtuous and sustainable financing mode which consisted of several financing channels organically, A University finished the capital construction of its new district favorably and realised a great-leap-forward development with an apparent improvement of the scale and conditions for running a school and an ascending education level [4]. It has developed from a frequently yellow-carded college by the ministry of education into an undergraduate school with 10,000 students at school and with a land of over one thousand mu and a school area of 300,000 square meters and with modern and complete infrastructure. Not only so, the total loan of A University's assets doesn't exceed 100 millions. With a low debt ratio, a good debt structure and a good distribution of time span, A University's assets are entering a benign circle and the prospect of A University's development is gratifying [5]. The financing mode for the capital construction of A University is based on the characteristics of the capital construction projects of universities and has broken through the financing channels of the past conventions and macroscopic views, providing a feasible idea and approach for the development of universities' capital constructions. Compared with the approaches that rely too much on bank loans to do capital construction by many universities, the financing mode of A University is successful and merits discussion, learning and reference.

59.3 Guiding Thoughts and Basic Idea of a University's financing Mode for Capital Construction

59.3.1 Guiding Thoughts

Universities' capital constructions have a huge investment, a tight time constraint, a long period and a strong industry sense and societal cooperativity. It's apparent that it's very difficult to do a good construction by merely relying on schools to invest manpower, material resources and money [6]. It's a definite trend for schools' capital construction project to involve socialised and marketized operation. The guiding thought for the financing model of A University's capital construction is: to go a way of socialization for universities' capital constructions based on the characteristic of universities' capital constructions; to extend financing channels respectively according to the economic properties and running characteristics after the completion of different projects; this is a trend of the development of universities in the future (Table 59.1).

59.3.2 Basic Idea

According to the guiding thoughts of the financing of universities' capital construction and based on the characteristics of universities' capital constructions, universities' construction projects are classified into three types from the economic property of the running revenues [7]: public project (e.g. library and teaching building), projects of narrow profit margin (e.g. students' apartment and students' canteen) and projects for profit (e.g. front building for business use and external reception center). The realization paths and extents for socialised financing are different according to different projects. A University adopted certain types of financing channels based on different extents of marketization. Various financing channels joint organically and complement one another, establishing the specialised capital construction's financing model for A University. The details are listed in the following table.

Table 59.1 Illustration of the financing model of A University' capital construction

Mode	Channel	Applicable mode	Specific project	Characteristic
The financing mode of A University' capital construction	Asset exchange	Public	Library, teaching building	Led by government, no compensation, as starting capital generally
		Fundamental	Prophase, land, way and pipe network	
	BOT and TOT	Narrow profit margin	Students' apartment and students' canteen	Incomplete marketization
		For profit	Shopping Store, book store, natatorium and gymnasium	Complete marketization
	Trust financing	For profit	External inception center, factory run by university	Complete marketization
	Bank loan			Assistance, turnover and supplementarity capital on a base of a certain amount of capital

59.4 The Implementation Details of the Financing Mode of A University' Capital Construction

From the economic attribute of the running revenue, it classifies universities' capital construction projects into three types: public projects, narrow-profit margin projects and for-profit projects. The realization paths and extents of extended social financing are different for different types of projects. A University adopted the following types of financing channels with different extents of marketization.

59.4.1 Asset Exchange

Suits for public projects invested or led by the government. The so-called asset exchange of A University is a method that the capital for the capital construction of the new district is funded by operating its own assets, including transferring the original district, ceding tangible or intangible assets of the university, opening oriented market, renting facilities, etc. Take the selling of the original district's property as an example. The geographical position of the original district of A University is excellent, with an elegant environment and complete service facilities and possesses an extremely high business value for development. The location of the new district needs not only to consider environmental factors, but also to obey the principles of economic benefit. By choosing lands with a low price and with a vast area, a differential land rent between the original and the new district is produced and thus provides sufficient capital for the capital construction of the new district through exchanging the original and the new district with the cooperating company.

59.4.2 BOT Mode and TOT Mode

Suits for for-profit projects run by A University via marketization, including completely marketized for-profit projects and incompletely marketized low-profit margin projects. The BOT mode, namely build-operate-transfer mode, refers to transferring the capital construction project to a project management company which develops the capital construction, puts into operation and enjoys the operation right, using right and income right during the contracted period. After the contracted period, the project company transfers back the project unconditionally. TOT mode, namely transfer-operate-transfer, refers to that A University cedes the completed capital construction project (or idling land resource) in order to obtain rent and invest into other fundamental projects.

59.4.3 Trust Financing

Trust financing transfers the warrant or claims of A University's building project to investors through trust plans. Trust separates investors and the project managers of A University, whereby investors enjoys the income right of operating the project in the future, but the project is still managed by A University, who enjoys the property right and control right of the capital construction project. The substance is a project financing mode that A University uses the future income of the new capital construction project as a warrant to finance through issuing income warrants in the internal of external capital market of the university.

59.4.4 Cooperation Between Banks and the University

Refers to under the condition that A University has approval of the government, the bank issues A University a certain amount of debts with a certain deadline directly after the university has passed the evaluation of commercial banks according to relevant program. It is an extension and supplement for university's financing mode for capital construction, which may ease the tension of capital construction capital during the peak time of the use of funds. It may realise a win–win outcome between university and banks.

Besides, there are financing modes such as setting up investment fund for advanced education, issuing lottery tickets of advanced education and capital operation, etc.

59.5 Effect of the Implementation of the Financing Mode of a University

Under the condition the government gives policy support without increasing investment, the successful implementation of the financing mode of A University makes A University gradually walked out distress, realised the extension of a new campus successfully and realised a leap-forward development through asset exchange to efficiently use state-owned assets, activating incremental capital by reserve capital and conducting socialised financing through modes such as attracting social investment and oriented trusting on the basis stated.

A University's extension project of the new district is classified into three periods. Period one project started capital construction from 1997 and finished in the year end of 1999, during which the university implemented a move entirely. Period two project finished in the summer of 2007 which makes the university begin to take shape occupying a land of 1,045 μ with 300,000 square meters of school buildings and relevant associate works. Period three project is now under capital construction. According to the standard issued by the Ministry of Education to local universities, the current hardwares of A University surpass the scale satisfying 10,000 students' need.

The university has had a leap-forward in quality no matter concerning harwares and soft environment through the extension of A University's new district, changing the dropping-behind aspect that was shown yellow card yearly, with a land not exceeding 100 μ, only 2,000 students, lagging in teaching and researching facilities and lagging in education level. Not only so, A University's education level was promoted from local higher vocational and academic college to an undergraduate institution.

59.6 Conclusion

Starting from the characteristics of university's capital construction projects, the financing mode for the capital construction of A University broke through the traditional and macro-scopic financing channels and offered feasible ideas and approaches for the development of university's capital constructions. In these channels, there are widely verified by universities and mature ones such as asset exchange and BOT. There are also ones that are newly adopted by university financing but have been adopted by enterprises such as trust financing and TOT. Besides these, there are also ones that are controversial such as cooperation between banks and universities. These channels occupy different positions and exert different effects in the financing mode of A University's capital construction. Asset exchange is mainly to start the foundational capital, while BOT and TOT is to develop capitals, trust financing is an important resource and bank loans are auxiliaries and supplements on the basis of the application of the above channels or on the basis of owning a certain scale of capital already. In summary, the financing channel for the capital construction of A University's new district is a systematic, healthy, and sustainable financing mode composed of several financing channels. A University's financing mode is successful and merits extension. It has guidance and reference values not only for local universities but also for centrally run or other types of universities.

Acknowledgments Fund Project: 2010 research topic of the Union of Social Sciences Circle and Union of Economic Community of Henan Province: Extended Research of Financing Channels for The Capital construction of Local Universities—Based on The Financing for The Capital construction of The New District of University A (No. SKL-2010-3047). This project was directed by the author and was finished in April 2011 with a first prize honor.

References

1. Cao S (2004) On the economic property, publicity of education, the non-for-profit property of university and the marketization reform of education. Theory Pract Educ 9:27–33
2. Liu R (2003) Discussioin of the fundamental methods of the financing marketization of advanced education. Coll J Shandong Fiscal Inst 13(2):32–37
3. Dai X, Anbang X (2004) Marketization of advanced education. Beijing University Press, Beijing 4–8
4. Fang F (2003) Investment and financing of engineering projects. Shanghai University of Finance and Economics Press, Shanghai 214–219
5. Jian S, Zou S (2003) Investigation of the current extension paths for China's diversification of financing of advanced education. Adv Educ Coals 21(5):153–156
6. Li Y, Ding Y, Zhang W (2004) Attracting folk capital to participate in university's industry development by using BOT. Coll J Jiangxi Educ Ins 34(6):88–94
7. Tang Z, He G, Liu X (2004) Some thoughts about several problems in the investment system's reform of China have advanced education. Concurrent Educ Forum 25(8):78–83

Chapter 60
Research on Cooperation Spirit and Training Methods in University

Zhonghua Li

Abstract Cooperation means the joint action of members. Reasons for college students' lack of cooperation spirit are the following: First, the influence of Chinese traditional agricultural society. Second, the influence of social factors. Third, Colleges and universities don't stress sufficiently the importance of cooperation spirit. To cultivate students' cooperation spirit, we must cultivate their interpersonal communication skills.

Keywords Chinese colleges and universities · Cooperation spirit · Reasons analysis · Cultivation methods

60.1 Introduction

"He Zuo" is "Cooperation" in English. It comes from the Latin, meaning the joint action of members. Market economy is an economic form which stresses competition. People tend to look at market economy as an economy of fierce competition, ignoring the importance of cooperation in market. In fact, there is no contradiction between competition and cooperation. Cooperation is the most basic human survival mode, including the co-operation between individuals, the cooperation between an individual and a group, and the cooperation between groups. The more intense competition is, the more we need people to have a sense of cooperation. People can innovate in the process of cooperation; can survive in a

Z. Li (✉)
Department of Humanity and Social Sciences Henan Institute of Engineering,
451191 Zhengzhou, China
e-mail: z2hualiu@126.com

W. Du (ed.), *Informatics and Management Science VI*, Lecture Notes in Electrical Engineering 209, DOI: 10.1007/978-1-4471-4805-0_60,

strong team of competitiveness. Compared with world's developed countries especially with the United States, China falls behind in the level of scientific research. This is because historical reasons. But the key point is that there is a big gap in the sense of cooperation between researchers of China and the United States [1]. In today's science and technology, researchers must have the spirit of cooperation. Many research projects and scientific problems require a huge investment of material and energy inputs. One person alone cannot finish the research work. Many research projects and scientific problem require the joint efforts of an enterprise, a research unit, or even the whole society, the whole country [2]. In addition to scientific research, it often requires the joint efforts of many people to bring an enterprise to prosperity, to make a business unit develop rapidly. In these areas, the spirit of cooperation is also needed. In the university campus, close cooperation among students is also needed to build a good class atmosphere and a good studying atmosphere, to carry out class activities and school activities, to finish a research project, to do an experiment. It can be said that many students do not lack personal skills and talents, but they lack cooperation spirit and cooperation ability. In 2005, Heilongjiang Institute of Technology finished a survey on college students; the result showed that many students believe the spirit of cooperation is not important. Many students lack the sense of social responsibility. They put more emphasis on personal struggle. They have a strong sense of social competition. But they seldom contact with others. They lack the sense of unity and cooperation with others. They lack the collective concept. Some students lack the ability to communicate with others. Some students even have communication difficulties. These phenomena show that college students' cooperation awareness and cooperation ability don't meet the requirements of modern society. The lack of cooperation ability will not do any good for the future development of students when they go into the community. It has been proved that students who have a strong sense of cooperation, communication skills, and leadership ability would adapt to society shortly after their graduation. These students have good prospects to achieve greater success. Otherwise, students who lack cooperation ability, despite their better individual achievements in colleges, often have difficulties to adapt to society. Their prospects are usually not as good as the prospects of students with strong cooperation ability [3].

60.2 Reasons for College Students' Lack of Cooperation Spirit

Generally speaking, we cannot be optimistic for college students' cooperation ability. The reasons are the following [4]:

First, the influence of Chinese traditional agricultural society.

In Chinese traditional smallholder economy, farmers provided for themselves, commodity economy was underdeveloped. Every peasant family was a production unit. The biggest task is to plant their own land well. They had little chance to contact with the outside world. Their motion range was very small. So they were bound to be self-centered and cooperation spirit was scarce in them. Almost all of the college students are of peasant origin. Since childhood, they have been immersed in traditional ideas. So their values and ways of thinking are inevitably influenced by the traditional culture.

Second, the influence of social factors.

One of the major reasons for the lack of cooperation spirit is individual selfish departmentalism. Traditional Chinese society was a mess. People lacked cooperation spirit in the old times. In modern period, the slogan of collectivism had been shouted for many years, but it has not been really established in many people's minds because of a serious disregard of individual rights and personal interests. And some people received the adverse influence of extreme individualism, their personal selfishness swelled to extreme. Some students worry about their own interests, don't care about the interests of the collective at all. "Why misfortune always comes to me?" "Why the leaders don't remember my good? Why they always remember my bad clearly?" "He is not as good as I in many ways, why is his job better than mine?" "everyone is equal, why should I obey his command?" These complaints were the performance of individual selfish departmentalism.

Another reason for the lack of cooperation spirit is the examination oriented school education. In the fierce competition of college entrance examination, the score is the lifeblood of students. Under the baton of college entrance examination, competitive learning dominates in primary and secondary schools, and students in their more than ten years of study and career, are used to competitive learning, not cooperative learning. They extremely lack cooperative learning experience.

Third, Colleges and universities don't stress sufficiently the importance of cooperation spirit. Many university teachers also believe that the 21st century is a highly competitive era. The economic competition and technological competition will be very fierce. The fierce competition requires students to have a strong sense of competition. Students should say, "you can, I can do better than you; you are good, I am better than you". The sense of competition should be enhanced, and cooperation will naturally relegate to second place. Cooperation is not as important as competition. This idea is reflected in the teaching style. Teachers do more instilling, but less dialogue and discussion with students. In many colleges, students don't take part in management; their rights to participate in college management are ignored. Even teachers don't cooperate with each other in scientific research. As a result, teachers seldom impart the knowledge of cooperation to students. There is no platform for students to practice cooperation. It is difficult for students to realize the importance of cooperation. So they cannot cultivate their cooperation awareness consciously. And in colleges and universities, there is little practical cooperation education. The first thing to cultivate students' cooperation spirit is to cultivate their cooperation awareness. But the more important thing is to give them chances to practice cooperation. Teachers should give students opportunities to practice cooperation. To get happiness in mutual assistance and to develop appropriate moral habits is an important way to cultivate students' cooperation spirit. At present, china is giving more opportunities for students to practice cooperation than before, but compared to the United States and the U.K., China has not created an effective practice environment for students, and China's college teachers have not used a variety of teaching methods to foster co-operation spirit. Because students lack the necessary co-operation relations practice, their cooperation concept and cooperation behavior are backward.

In china's colleges and universities, there is no mechanism to protect and evaluate students' cooperation spirit. We have taken too many quantitative indicators into consideration in awards appreciation, but we did not form an effective incentive mechanism to encourage cooperation between students. For example, some students who have no cooperation spirit can also be rated as outstanding graduates. This made some students think cooperation spirit not important. They have no strong cooperation sense, no mental ability to adapt to a cooperation circumstance. In interpersonal relationships, the majority of students choose to pay attention to their own moral uplift without thought of others.

60.3 Ways to Cultivate College Students' Cooperation Spirit

Colleges should do the following two things to improve students' interpersonal communication skills. The first is to impart interpersonal knowledge to students by opening elective or required courses. Interpersonal communication is students' practical needs, institutions of higher learning must look at it as a task to train students. Students should not gain communication skills automatically after they go into society and encounter many setbacks and failures. Courses such as public relations, management, lecture and eloquence, public relations, etiquette, mental health education, and so on play an important role for students to understand themselves and the society, to improve students' physical and mental health, to get more knowledge, to promote their communication skills. Because these courses are very useful, they should be used as elective or required courses for students to attend. In college years, it is important to learn professional knowledge, but to develop communicative competence can never be ignored. Many students are good at expertise, but poor at interpersonal communication skills. When they go into society in the future, this defect will show up. Students with good communication skills will be more successful than the eccentric students. College students have little social experience. Many of them failed to recognize the importance of communication skills. At this time, teaching them communication knowledge will have the effect of getting twice the result with halt the effort. It will be too late to make up the deficiency of communication knowledge when students get into society and meet many setbacks.

The second way to cultivate students' communication ability is to provide students with the chances to practice. There should be many collective activities and social opportunities in campus, and teachers should encourage students to participate in interpersonal relationships. It's important for students to gain knowledge, but if knowledge cannot be used in practice, it will remain at the theoretical level, cannot be internalized into skills. So the actual exercise of interpersonal exchanges has more evident effect. In Chinese universities, there is a wide variety of student organizations, such as student unions, Communist Youth League, literary agency, news agency, English society, lectures Association, the Reporters Association, etc., which provide a platform for students to practice interpersonal communication. In these societies, associations, students have to work with others to achieve a certain purpose. They will have to keep in touch with their fellow students, and improve their communication skills and cooperation capabilities. Inferiority, shyness, alienation, jealousy and other undesirable psychology will be overcome. In these community activities, students can also acquire a certain amount of social skills.

The third way is to learn from foreign countries, especially the American experience, and vigorously carry out the cooperative education in vocational colleges.

The slogan of the college-enterprise cooperation has been exalted in China. But China's the college-enterprise cooperation is carried out mainly between the officials of colleges and enterprises. The cooperation is mainly high-tech cooperation. China's college-enterprise cooperation does not pay enough attention to cooperative education. It does not play its due role for the cultivation of students' ability to cooperate. While cooperative education dominate in American college-enterprise cooperation, and play a major role in the cultivation of students' cooperation spirit.

Based on the American experience, the Central Government of China and local governments at all levels should attach great importance to cooperative education.

The federal government of the U.S.A does not directly run the cooperative education of every state and every college. But this does not mean that the federal government turns a blind eye to higher education. There are three ways for the federal government to influence higher education. The first is to use legislative means, the second is to assess, and the third is the use of funds means.

Legislative means. The U.S. federal government encourages all states to strengthen the management of the cooperative education, encourages all enterprises to participate in higher education so as to make higher education a cause of the whole society. The U.S. government has made a number of bills to develop the U.S. cooperative education and the effect of this policy was very good. In June 1991, in order to help colleges understand how to reform the curriculum and teaching content to allow students to obtain efficient skills required for workplace success in the future, the U.S. Department of Labor established The Secretary's Commission on Achieving Necessary Skills (referred to as SCANS). The Committee stressed that colleges must enable students to learn to survive. U.S. Department of Labor published a report on the requirements of workplace, asked colleges, parents and businesses to help students acquire the necessary skills in the current and future workplace, these skills includes the ability to dominate a variety of resources, interpersonal skills, information capacity, and the ability to analyze comprehensively and systematically, and so on. This report reflects the requirements of the employers on students' learning. The report suggested that the capacity should be obtained in the actual practice, and therefore cooperation between colleges and enterprises became the best choice. This report has greatly promoted the cooperation between U.S. colleges and enterprises.

Assessment. In the organization and management of cooperative education, the relevant regulatory agencies play an important role. In 1962, the United States established the National Cooperative Education Committee and Cooperative, Education Association. These two agencies were responsible for coordinating the nation's colleges and universities to carry on cooperative education. They played a positive role in promoting the smooth development of college-enterprise cooperation. They secured the support of the Advertising Association, used various media to advertise for cooperative education, and improved public understanding of the cooperative education. Its publicity was large-scaled; the advertising time value reached $150 million.

Funding means. The financial support is the key to the development of cooperative education. Funding of the successive U.S. administrations greatly promoted the development of cooperative education. During the Clinton administration, the United States gave a lot of funding on cooperation in education, so cooperative education developed rapidly. The Obama administration also tried to give financial support to the development of cooperative education, but Congress refused.

China's has seen 40 years of reform and opening up which has greatly increased China's national strength. China has become the second largest economy in the world. With the increase of national strength, China's investment in higher education is also increasing. We should innovate management system to give more policy support, management support and financial support to China's cooperative education. Cooperative education will play an important role in improving college students' cooperation spirit.

Acknowledgment This is the mid-term result of the soft science research projects of Henan Province—"Research on the Cultivation of College Students' Cooperation Spirit on the market economy conditions". The item number is 112400450226.

References

1. Chen Y (1999) On the cooperation spirit and cooperation education. Jinan J Jinan Univ 4:28–35
2. Li W (2001) Thinking on the cultivation of College Students' cooperation spirit Nanning. Guangxi Univ 1(6):90–99
3. Sun X (2006) Ability to cooperate—college students' key ability when they go into the community. Beijing Sci Technol Inf 9:09–15
4. He Y (1999) Mental health and psychological counseling. Shenyang Liaoning Univ Press 1(2):012–017

Chapter 61
On Philo-Semitism

Yanming Lu

Abstract Philo-Semitism is an interest in, respect for, and appreciation of the Jewish people, their historical significance and the positive impacts of Judaism in the history of the western world, generally on the part of a gentile. The motivation of Philo-Semitism consists in religious, economic, and secular cultural factors. Under some circumstances, the survival of a Jew relies on Philo-Semitism among his neighboring gentiles. Apparently Incompatible, Both Philo-Semitism and anti-Semitism are sentiments and behaviors formed upon the same condition.

Keywords Philo-Semitism · Christianity · Holocaust

61.1 Introduction

In the long historical process of Jewish civilization, the clouds of anti-Semitism are always shrouding in the head of the Jewish nation for most of the time [1]. It is precisely because of this that, a number of other equally important issues have often been obscured or not been taken seriously enough attention.

Since the Second World War, pro-Jew's re-emergence has made some people reconsider the Jewish history.

Philo-Semitism or Judeophilia is an interest in, respect for, and appreciation of the Jewish people, their historical significance and the positive impacts of Judaism in the history of the world, in particular, generally on the part of a gentile.

Harvard University scholar and the controversial author of the book in Hitler's Willing Executioners, Daniel Goldhagen considers that pro-Jew often goes hand in

Y. Lu (✉)
Department of Religious Studies, Nanjing University, Nanjing 210093, China
e-mail: Yminglu210@126.com

W. Du (ed.), *Informatics and Management Science VI*, Lecture Notes in Electrical Engineering 209, DOI: 10.1007/978-1-4471-4805-0_61,

hand with anti-Semitism. Norman Finkelstein, who disagrees with him on many issues, agreed with this point. Pro-Jew in history is a reaction of anti-Semitism, such as the thinker Nietzsche calling it as "anti-anti-Semitism".

61.2 Motivations for Philo-Semitism

There are all kinds of motivations for the existence of Philo-Semitism.

First of all, of course, it is on the religious and cultural motives. As two monotheistic religions closely connected in the aspect of the doctrine, Judaism and Christianity have gone through love and hate of nearly two thousand years. It is undeniable that the acts of persecution and discrimination against Jews have been prevailed in the Christian community for a long period of time.

The early Christian after all, is separated from the ancient Judaism. These kinship ties of the relationship cannot be severed by the religious anti-Semitism. Dietrich, Bonhoeffer, the famous modern theologians and Germany famous theologian in Simpson Church said that: "Jesus Christ is the Messiah promised by Israel-the Jewish people. Therefore the family of our ancestors can be traced back to the nation of Israel, who appeared before the Jesus Christ. According to God's will, not only from the perspective of genetics but also from the experience that never truly interrupted, the history of the West and the nation of Israel are inseparable. Jews put forward the problem of Christ the Redeemer. He is the symbol who freely and kindly selects and soothes the wrath of God. It can be seen that the mercy and severe of God" (the Bible, Romans 11:22). Jews' expelling from the West will certainly lead to Christ's expelling, for the reason that Jesus Christ was a Jew.

On the aspect of the Catholic society, French novelist Francois Mauriac sent a letter to Oscar de Ferenzy, the editor of the magazine La Juste Parole in 1937. In the letter, he cried solidarity with the Jewish nation which was facing with the unprecedented disaster: For Catholics, the anti-Semitism is not just the flagrant violation of mercy. We are integrated to Israel as a single entity, whether we want it or not [2].

After the war, Pope John Paul II's visit to Israel in 1986 on behalf of the Catholic Church official has further proposed that we have relationships with the Judaism that we would never have with any other religious relations. You are our dear brothers, to a certain kind of sense, you are our brother [3].

Jewish religion culture has gone deep into every corner of Western civilization. As a result, whenever the dregs of anti-Semitism appeared, there are always non-Jews fought back.

Another motive cannot be ignored of Philo-Semitism is the economic factor. For example, in 1492, King Ferdinand II of Spain ordered the expulsion of all Jews. The Ottoman Empire quickly accepted these Jews, and placed in the provinces of the empire, and also ordered to treat the Jews well, and offenders would be sentenced to death. Yitzhak Sarfati, Chief of Edirne, said: "Turkey has it all," "doesn't life living under the rule of Muslim stronger than live living under the Christian rule?" [4].

In the late twentieth century, "multiculturalism" roused in the Europe and the United States, which was one of the driving forces of the Philo-Semitism in the contemporary society. Philo-Semitism also has the interactive factors coming from the Jews communities.

61.3 Philo-Semitism from Ancient Times to Early Modern Period

Jews often become the object of curiosity and ideological insinuation by the intellectuals and those in power. Philo has taken note of the situation that strong interest of the Romans towards the Judaism has even caused the situation that they abandoned the original polytheism. Khazar people's converting to Judaism has fully proved the charm of the latter.

However, in addition to the Khazar people, these kinds of people who are for Philo-Semitism are rarely interested in the Jews as a real nation.

It is very representative that Mather Luther King swung between anti-Semitism and Philo-Semitism.

Although there are pitiful voices of Philo-Semitism in the early Protestant church, they still exist. Calvin's main rival Sebastian had the courage to stand up against the persecution of the Jews [5].

Due to the development of linguistics, ethnology and anthropology in the eighteenth century, the British began to find some of the historical ties and the early confluence between the ancient Hebrews and the British, as well as the Hebrew race, language and religion's impact on the shape of the characteristics of the United Kingdom [6].

To the late eighteenth century and early nineteenth century, only fewer and fewer antiquities and linguists still adhered to the idea that Hebrew was taken as the human primary language [7].

With the end of an era of Enlightenment, the once very popular antiquarianism has become gradually out of people's vision. In addition, it was replaced by the science of archeology. The traditional culture Philo-Semitism began looking for new backing.

61.4 American Spirits and Philo-Semitism

Philo-Semitism really carried forward in the New World. The Hebrew spirit is inherent in American culture. It is the inherent tradition and an integral part. Puritans compared themselves to the ancient Israelites wandering in the land of

Canaan, and the New World was the new Israel in their minds. The United Kingdom was compared to Egypt, while Washington has been likened to Moses. Puritans and Scottish-Irish immigrants have created a Jewcentric American society and political culture, although the United States did not have a lot of Jews then.

John Adams, the second president of the United States, once said, "I think that Hebrew people have made far more contributions to the world civilization than the other nationalities" [8].

The text of the early American history was filled with the rhetoric of the Jewcentric. For example, Benjamin Franklin and Thomas Jefferson (and John Adams) recommend the use of the Exodus story in the Continental Congress when commissioning the design of the national emblem in July 1776. President Johnson said to the visiting Israeli President Xia Zhaer in 1966 that "... we grew up in the Hebrew culture as well. ... This is not only our heritage, but also your heritage." [9].

Philo-Semitism permeates every aspect of early American literature and politics. Among the literature works, the most famous of which is the letter of George Washington in Newport, Rhode Island, Jewish synagogue. In the letter, he put forward the principle of religious tolerance and freedom of his worshipers.

Mark Twain wrote in an essay entitled Concerning the Jews, saying: "The Egyptian, the Babylonian, and the Persian rose, filled the planet with sound and splendor, then faded to dream-stuff and passed away; the Greek and the Roman followed, and made a vast noise, and they are gone; other peoples have sprung up and held their torch high for a time, but it burned out, and they sit in twilight now, or have vanished. The Jew saw them all, beat them all" [10].

It is noteworthy that, before the rise of the Zionism movement, the Americans have voices of helping Zionists. Abraham Lincoln has once said in 1863 that: "Let the Jews restore their homes, which is a noble dream jointly owned by many Americans" [11].

61.5 Holocaust and Post Philo-Semitism

From the early nineteenth century to the beginning of holocaust, European Jewish has the choices to acquire personal space or to continue as the victim in a larger society. However, Nazi has deprived of their rights to select. Holocaust has created a new kind of Philo-Semitism.

With the end of the Second World War and the publicity of Holocaust atrocities, the hostile attitude of the Catholic Church towards the Jews has first become the object of broad public accusations.

From the year 1962 to the year 1965, the court held a second Ecumenical Council in Vatican, issuing a series of important reform documents and decrees. The documents and decrees have included the "Nostra Aetate", Church's Universal Declaration of religious attitudes towards non-Christian. In addition to relations with other religions, this Universal Declaration of the Ecumenical Council specially mentioned Judaism, which was the first time that they clearly stated: "The Church not only opposes the persecution of any person, but also commemorates the shared heritage with the Jews. It by no means was the political factors. While it was pressed by the gospel and loving religious grounds, denounced all the hatred, persecution, and the measures in any of the times and the people waged anti-Semitic by any person [12].

Once available, Nostra Aetate has won the praise of world. "It is very noticeable in the history of Christianity.... Christianity has taken an important step in correcting its attitude towards Jews and Judaism" [13]. The whole world heard the first public praise by the Catholic Church, which is praising the religion of the Jews:

Church cannot forget that it is infinite mercy once by God of the Old Testament with the meaning knot voters to accept the revelation of the Old Testament. At the same time, the wild olive branches of the Gentiles were taken up in the good olive root and received nutrition. The faith of the Church of Christ is our peace. He relies on the cross and makes Jews and Gentiles be good. In this case, they become one in him [13].

The Declaration recognizes Christianity and Judaism that they "share such a great spiritual heritage", and "promote and encourage both sides to know each other and respect each other". These are the positive attitude that has never been in history [13].

61.6 Conclusions

Philo-Semitism is not the opposite of anti-Semitism. Instead, they are two sets of emotional and behavioral patterns that are formed basing on the same conditions. Philo-Semitism does not mean that non-Jews love the Jews as individuals or as groups merely because they long for Jewish property. Only when the Jewish religion, history, culture and characteristics become the center of the non-Jewish social consciousness and discourse can Philo-Semitism exist. Even if the pro-Jew sentiment is high, anti-Semitism will not disappear.

References

1. Barnes KC (1999) Dietrich Bonhoeffer and Hitler's Persecution of Jews in Betral German Churches and the Holocaust. In: Robert P Erickson and Susannah H (eds) Minneapolis, Fortress Press 67:780–786
2. Birnbaum P (1992) Anti-semitism in France: A political history from Léon Blum to the present. Oxford 2(3):183–186
3. Pope John PauL II (1995) Spiritual pilgrimage texts on Jews and Judaism 1979–1995, The crossroad publishing company, New York 5(4):63–67
4. Lewis B (1984) The Jews of Islam. New York 2:135–136
5. Martin B, Schulin E (1985) Die Juden als. Minderheit in der Geschicht 1(2):129–136

6. Joseph L, Irish O (2004) A Literary and intellectual history. Syracuse NY 12(3):80–88
7. Blake W (1967) Poetry and Prose, vol 112. In: Geoffrey Keynes (ed) Trianon press, London p 463
8. Adams J, Adams CF (1854) The works of John Adams, Second president of the United States: with a life of the author. notes and illustrations, Brown, Little 12:609–614
9. Wang SM (2004) Cultural origin of pro-Jew-ism in contemporary America. J Soc Sci 11:120–127
10. Twain M (1898) Concerning the Jews. Harper's Magazine
11. Oren M (2007) Power, Faith, and Fantasy. W.W. Norton, New York 3:215–221
12. Translated by china catholic bishop college (1992) Declaratio De Ecclesiae Habitudine Ad Religiones Non-christianas Nostra Aetate (NAE)
13. Bokser BZ (1968) Vatican II and the Jews. The Jewish quarterly review new series 59(2):387–396

Chapter 62
Optimization of Curriculum System of Higher Vocational Education

Juntao Mei

Abstract The curriculum system of higher vocational education should be established based on professional properties, and simultaneously attaches high importance to the training of personnel professional ability. The core of personnel training lies in the construction of curriculum system. In this paper, the current situation of the curriculum system of higher vocational education is discussed, the problems and weaknesses in the system are introduced, and also the idea of necessarily making an optimization after the construction of the curriculum system is demonstrated if personnel with professional skills are intended to train successfully. Therefore, a conclusion is reached that the key to optimizing the curriculum system lies in teaching innovation, promoting the reform of vocational education courses, and propelling vocational schools to attain a healthy development in the training of personnel professional ability.

Keywords Vocational education · Curriculum system · Optimization

62.1 Introduction

Higher vocational education refers to the education, which is provided to train students to acquire knowledge, skills and attitudes on the basis of high school education, aiming at promoting them to meet the needs of an occupation post or business area [1].

J. Mei (✉)
Jiangxi Vocational & Technological College of Electricity, Nanchang, Jiangxi, China
e-mail: juntaomei138@126.com

W. Du (ed.), *Informatics and Management Science VI*, Lecture Notes in Electrical Engineering 209, DOI: 10.1007/978-1-4471-4805-0_62,

The core of personnel training lies in the construction of the curriculum system [2]. The training objectives of higher vocational education are achieved through the teaching of courses [3].

62.2 Current Situation of the Curriculum System of Higher Vocational Education

The curriculum system of a majority of higher vocational colleges is the specialized disciplines course system, which is established mainly drawing on the experience of general higher education disciplines curriculum system [4]. The characteristic is that the curriculum system is implemented according to three kinds of courses, namely foundation courses → major fundamental courses → major technical courses. In this model, high importance is attached to the comprehension of students on what they have learnt, the evaluation on their ability in analyzing professional theories, and the fundamentals of discipline theories; the practice courses are a separate system, which mainly focuses on verifiability, and also is provided with independent courses.

According to the Requirement of Deepening Educational Reform proposed in the Decision about Greatly Developing Vocational Education of the State Council, great numbers of useful reform attempts have been conducted by all higher vocational colleges on the construction of the specialized disciplines course system. For example, a large number of foreign vocational education cooperation projects such as the "dual system" curriculum model of Germany, the "MES" course model of the International Labor Organization and the "CBE" courses model of Canada have been absorbed by domestic higher vocational college.

In the mean time, on the basis of drawing on the experience of the construction of the foreign courses system, a good many vocational schools have made numerous advantageous attempts in construction of the curriculum system according to China's national conditions.

However, because biased educational objectives and insufficiently explicit educational ideas still exist in all higher vocational schools and also the "application of vocational technologies" defined in higher vocational education and the "essential and enough" principle in theory are simply and unilaterally comprehended by them, the specialized disciplines course system and curriculum form continue to be used by some higher vocational colleges for the implementation of professional training plan.

62.3 Problems and Weaknesses in the Curriculum System of Higher Vocational Education

62.3.1 *Attaching Importance to the Construction and Innovation of the Curriculum System of Higher Vocational Education*

Overall, high importance is attached by all higher vocational colleges to the construction and innovation of the curriculum system of their higher vocational educations. They think that the construction of the curriculum system of higher vocational education gives specific reflection to the "service" and "oriented" relationship between disciplines and employment [5].

The problem is that the main mechanism of personnel training is naturally thought by schools to be imparting knowledge in the construction of the curriculum system. In other teaching mechanisms, however, top priority is given to the "knowledge points" of courses.

62.3.2 *Undefined Connotations Provided for the Training of Practical Ability and Ability Structure of Students*

The connotations of the training of practical ability and ability structure of students are not clearly provided by higher vocational colleges.

More specifically, the components of the practical ability of students at higher vocational colleges and the basic training framework have not been clearly defined; the guiding strategy for the training of practical ability, based on inconsistent consensus, makes all higher vocational colleges implement the training of the practical ability of students in accordance with their own understanding and ideas [6].

Therefore, it is very difficult to investigate whether there is a real improvement achieved for the practical ability of students.

62.3.3 *Separation Between Professional Knowledge Courses and the Training of Professional Technical Abilities*

Separation between professional knowledge courses and the training of professional technical abilities is another long-standing problem in vocational education courses.

Generally, vocational education courses can provide the students with all knowledge in the theory of professional techniques, but is unable to provide them with the "professional technical work ability" and basic work experience, which receive the most attention from enterprises. The relationship between the professional technical learning opportunities and the practical professional technical work ability, which is provided for students, is indirect.

Therefore, it can be seen that the key of the construction of the curriculum system of higher vocational education is how to make the establishment and compilation of learning contents out of the fence of the disciplines system after the completion of the construction, but not making clear the macro-structure of the courses [7].

The overall quality of China's vocational education can be fundamentally improved only if the reform and attempt of the vocational educational courses can jump over this fence, but only stay at the amendment and improvement of the original discipline courses.

62.4 The Optimization of the Curriculum System Lies in the Innovation of the Personnel Training Model

It is necessary for higher vocational education colleges to carry out majors setting, courses arrangement, teaching plan adjustment and curriculum system optimization in accordance with the real needs of the personnel markets.

However, how do higher vocational education colleges optimize their curriculum systems? For schools, the core lies in promoting the change of educational courses as well as the teaching innovation.

62.4.1 Teachers Used to Apply the Linear Thinking Way in the Previous Education

In the previous education, teachers used to apply the linear thinking way, namely teaching or making up what students were short of. In such an education, teachers were in shortage of dialectically analyzing the education or teaching problems, and hence often sunk into an awkward situation (namely, only bringing about a temporary solution to a problem).

From the point of view of teaching evaluation, the real reason why such a situation occurred was that the learning interest and confidence of students are not fully stimulated by teachers in the teaching process, and it was difficult to play teaching effect only relying on teaching contents. "Interest" and "confidence" should be equally regarded as two core elements of higher vocational education.

In teaching, therefore, it is necessary for teachers to possess their own independent thinking on teaching contents. At classroom, teachers should allow students to get experience not only in cognition and also in emotion.

62.4.2 Setting up a Teaching Idea Centered at Students

A teaching idea, which is centered at students, should be established. In the teaching activities, the subject activities of students play a decisive role in the implementation of teaching values.

In the courses teaching of higher vocational education, teachers in different majors should carry out the "project teaching" as much as possible in accordance with the requirements of major courses teaching.

In the mean time, teachers can help students make and implement work and learning plans and also provide them with evaluations through the design and development of the applicable major teaching projects and multiple auxiliary teaching methods such as multi-media and learning guiding materials.

The project teaching method can give expression to the subjective characteristics of students very well, allowing students to receive, analyze and complete tasks as a whole in the project teaching process, and simultaneously promoting them to initiatively construct a professional knowledge structure in the process of completing working tasks.

62.4.3 Adjusting the Existing Courses Structure, Attempting to Break the Original Professional Curriculum System and Giving Students Multiple Course Selections

It is necessary for higher vocational colleges to make an adjustment to the existing courses structure, attempt to break the original professional curriculum system, and provide students with more course selections. In this aspect, it is necessary to pay attention to two points:

1. Increase more selections through the setting of selective courses;
2. Improve more selections of professional courses with a gradual improvement.

62.4.4 Innovating Ideas

It is necessary to bring forth new ideas, transcend the difficult choice between theoretical teaching and practice teaching, and integrate theoretical teaching and practice teaching into the implementation process of higher vocational education.

Therefore, higher vocational education should be the educational system, which is centered at the educational ideas bringing the creative ability of students into full play.

62.4.5 Teaching Thinking Quality is Important for a Teacher, and the Thinking Way of Teachers Directly Influences the Educational or Teaching Effect of Higher Vocational Education

Teaching thinking quality is an important quality of a teacher, and the thinking way of teachers directly influences the educational or teaching effect of higher vocational education.

Therefore, teachers in courses teaching should learn the advanced knowledge to effectively teach students. It is necessary for a teacher in higher vocational education to make an expansion and in-depth to learning contents and range. This can be mainly reflected from three aspects.

First, higher vocational education requires students to possess practical ability. Therefore, teachers should learn knowledge not only from books and also from the people who own practical ability, not only pay attention to knowledge and also learn how to solve actual problems with knowledge, and not only know how to organize students to participate in practice activities and also personally demonstrate and guide them to increase practical ability.

Second, it is necessary for teachers to be adept in making scientific researches on all sorts of problems in teaching, namely teachers should possess the ability in scientific education or teaching research.

Third, teachers should not limit the learning of the courses they teach within the update of what they teach at classroom, but should make an enhancement to the learning of the leading-edge and interdisciplinary knowledge, thus improving their own comprehensive knowledge level and ability.

62.5 Conclusion

The optimization of the curriculum system of higher vocational education is a long and difficult task, and has a close tie with the success of the reform of higher vocational education, the improvement of teaching quality, and the implementation of personnel training objective.

For this reason, it needs all higher vocational colleges to research and practice with joint efforts. Only in such a way, the courses in higher vocational education can change into a flexible interface between schools and society, and the connection between the training of the professional skills personnel provided by schools to meet the needs of society and the application of skills personnel needed by enterprises.

References

1. Jiang D (2007) Research theory of vocational education. Science and Education Press, Beijing 22:38–44
2. China's Vocational and Adult Education (1998) Principles of vocational and technical education The publishing house of economic science, Beijing
3. Shengchang T (2011) Innovation of personnel training model: from thinking to action. People's Education 11:920–928
4. Xu G (2007) Several key problems in vocational education project curriculum. Chinese vocational and technical education 4:28–36
5. Li Yao LY (2011) Teaching Autonomism. Teaching Study 2:939–947
6. Wu S (1997) Education quality theory. The Publishing House of Economic Science, Beijing 12:388–394
7. Liming Y (1998) Analysis on the Advantages (CBE) and Disadvantages of the Education System based on Ability. Mechanical Vocational Education 8:99–105

Chapter 63
Efficient Teaching Scheme of Economics Courses Based on Teaching Method Reform

Honghui Wang

Abstract In this paper, as course education reform cored at thoroughly changing and improving teaching quality is comprehensively implemented in an all-round way, it is particularly important to carry out a discussion on how to reform teaching methods and improve teaching effect. The methods improving classroom teaching effect of economics courses based on the perspective of teaching method reform are hereby proposed by the author from attaching importance to participatory teaching, heuristic teaching and case teaching, and applying diversified teaching method and modern teaching means.

Keywords Economics · Teaching effect · Heuristic teaching · Experiment teaching

63.1 Introduction

Classroom is the main battlefield of school education; teaching effect at classroom directly plays an important influence on the education quality, but classroom teaching methods also directly decide teaching effect. Therefore, with the purpose of improving teaching effect, it is necessary to take the reform of teaching methods as entry point and breakthrough [1, 2]. In this paper, based on the perspective of teaching method reform, some ideas about improving teaching effect are briefly discussed, hopefully receiving corrections from experts [3].

H. Wang (✉)
School of Economics and Management, Changchun University of Science and Technology, Changchun, 130022 Jilin, China
e-mail: honghuaw12w@126.com

W. Du (ed.), *Informatics and Management Science VI*, Lecture Notes in Electrical Engineering 209, DOI: 10.1007/978-1-4471-4805-0_63,

63.2 Guiding Students to Participate in Classroom Teaching, and Promoting Heuristic Teaching

In contrast to the teaching of other disciplines, the teaching of economics needs an interactive education model [4]. Cyxomjnhcknn, educator in former Soviet Union, used to say: "there is a fundamental need in the bottom of people's hearts, namely hoping them to become a discoverer, researcher [5], or explorer". Therefore, it is necessary for the teaching of economics to include a method arousing the interest of students in cognition.

Heuristic teaching is a guiding thought for teaching methods, but not a specific teaching way or method. Supported by such a teaching method, the rising of the twenty-first century should not leave the basic teaching model achieving teaching effect with learning, and the key lies in recognizing what to teach and how to teach starting from the principal position of students in learning.

Aristotle, Greek philosopher in the ancient times, thought that thinking began from wonder and doubt. Without doubts, it is difficult to induce and stimulate desires in learning [6, 7]. Then, students will not feel the existence of doubts, and naturally will not think. Also, learning can only stay at the surface level or form. Therefore, the key to effective heuristic teaching lies in the reasonable setting of classroom questions. It is universally advocated by us that students should raise more questions than teachers at classroom, but knowledge of students is limited. Thus, questions can be suggested by teachers for their students at classroom.

63.3 Necessary to Carry Out Case Teaching in Combination with the Features of Major Courses

Case teaching is teaching method, in which the situations in the real world are typically processed based on the needs of educational objects and trainings, forming cases (usually in written form) for students to think, analyze and judge, and improving the abilities of students in analyzing and resolving problems.

In general, case teaching is one of effective means to stimulate creativity. In case teaching, it is necessary for teachers to leave enough space for students to do creative thinking; teaching materials should keep close to practice, promoting students to think problems in the place of professionals and also mobilize all of their knowledge and potential to solve problem. Therefore, the distance between teaching and practice is greatly shortened, and also the psychological preparation, knowledge structure and operation skills that are necessary for students to meet the needs of future work are possessed at school. With the purpose of achieving a good case teaching effect, it is necessary for teachers to possess the ability in applying theory and the rich experience in practices. Thus, it is necessary for schools to actively provide practice venues and opportunities for students to practice.

63.4 Paying Attention to Cultivating the Ability of Students in Solving Problems and Strengthening Interpretations on Research Methods

Learn to be, which was published by the United Nations Educational, Scientific and Cultural Organization (UNESCO), lays a stress on "illiterate people refer to those who do not learn how to learn, but not those who do not recognize characters". The teaching of the traditional economics in China attaches high importance to learning results and acquisition of knowledge itself. In such a teaching model, teachers imparted knowledge to students who were only necessary to accept it, and then teachers checked teaching effect and saw if knowledge was solidly remembered by students. Reform is to change such a traditional teaching model, convert knowledge impartation to learning how to learn, just as the saying goes that it is always better to teach a person who are hungry to fish than to give him some fish. Shengtao Ye, educationalist of China, used to say that teaching was realizing teaching became unnecessary for students and allowing students to learn how to learn with research methods. The research methods in economics can be concluded from four aspects: abstract analysis, empirical analysis, standard analysis and mathematical analysis.

63.5 Diversifying Classroom Teaching Models and Modernizing Teaching Means

63.5.1 Appropriately Using Experiment Method of Economics and Increasing Students' Understanding of Professional Knowledge in Economics

Great theoretical achievements have been achieved with experiment method of economics, but also some practical problems have been solved. Not only existing economic theories can be tested by it, and also new economic laws can be discovered with it. Therefore, designing scientific experiment of economics is a primary issue that researcher has to face up with to make studies with the experiment method. From experimental process standardization, monetary incentive significance, unbiased experimental language, clearly-defined comparison benchmarks and consistency with realities, discussing how to design scientific experiment of economics is an important direction of constructing the experiments.

63.5.2 Trying to Use Game Teaching Method According to Teaching Contents

Teaching through lively activities is the top state of teaching, but also an effective teaching method meeting characteristics of economics-oriented major. Game teaching method was first proposed by American economics and education Association, and then was widely applied in Japan. Hicks, who used to acquire the Nobel Memorial Prize in Economic Sciences, said: "Economics is an intelligence-oriented game in essence". Teaching through lively activities is the top state of teaching but also an effective teaching method meeting characteristics of economics-oriented major. It can make the principles of economics deeply rooted in the minds of students, and can help improve the interests of students in learning. Therefore, it is a very efficient way of teaching.

63.5.3 Improving the Comprehensive Ability of Students by Organizing Personal Speech, Evaluating Speech, and Other Teaching Activities

The purpose of making speeches is to allow students to appreciate the style of leaders, and generate a self-consciousness changing from audience to speaker.

Making speeches is not only a comprehensive conclusion on what students have learnt, and more importantly is the best learning way for students to give speeches and exchanges between each other. In general, a satisfactory speech is an accumulation of what students have learnt at classroom and extracurricular place for many years.

Evaluating speeches is to require audiences to draw up a conclusion on speeches, mainly including two aspects.

First, contents in speeches can be evaluated, including if contents are to the point, discussion topics are new and clear, evidences are full and speeches have powerful influences ad persuasion force.

Second, temperament and appearance of speakers can be evaluated.

63.5.4 Making Full Use of Modern Teaching Methods

Multimedia teaching system has entered higher learning schools for many years, and also is applied by increasingly more schools. This plays a very positive role in promoting teaching reform and improving the quality of teaching. The enthusiasm of students in learning can be mobilized by multimedia teaching system to the maximum, and also the teaching style and characteristics of teachers can be fully reflected by it, thus making contents taught at classroom leave an in-depth impression in the minds of students.

In making courseware of economics, it is necessary for teachers to make sure the interesting and beautiful characteristics of courseware in addition to ensuring the integrity and accuracy of teaching content. Thus, students can be attracted by teacher along with interesting explanation, and also a good teaching effect can be obtained. In addition, the Internet can be utilized fully for strengthening the contact between teachers and students. Teachers can help students apply for a free email address as public class mailbox, opening up a space for both teachers and students to exchange and operate. Then, lecture notes, courseware or some typical exercises can be sent by teachers to the email address, convenient for student to review after class. Also, research direction or economic questions interested by teachers can be proposed and then discussed with students in E-mail. Furthermore, teaching suggestions can be asked from students through E-mail.

Therefore, it is necessary for teachers to exchange with students and absorb suggestions from them on teaching, for the purposes of teaching students in accordance of their aptitude an acquiring both teaching and learning effect.

In the mean time, students can take advantage of this space for putting forward questions, or expressing their opinions.

63.6 Conclusion

From above analysis, it is necessary for teachers in the teaching of economics to apply new teaching methods rich in heuristic and helping students develop intelligence.

In the mean time, it is necessary to change the traditional teaching method, keeping its reasonable elements and transforming its negative factors.

Therefore, teaching methods meeting China's national conditions can be researched and created by teachers.

References

1. Liu H (2010) Several problems necessary to stress in selecting western economics teaching method. China Collective Economy 16(4):466–475
2. Zhou Z (2010) Discussion on case teaching process of western economics. Estate Sci Tribune 16(13):56–65
3. Lian X (2010) How to play the principal role of students in the teaching of economics courses. Business China 3(1):34–39
4. Guo W, Luo J (2007) Discussion on the teaching reform of economics courses. J Taiyuan Normal Univ 3(3):87–95
5. Tan J (2009) Analysis on the factors affecting the teaching effect of western economics and the countermeasures. The South of China Today 10(8):456–464

6. Lei L (2010) On the application question-oriented teaching method in classroom teaching. J Chongqing Univ Sci Technol Soc Sci Educ 11(7):477–483
7. Bian S (2010) Exploration on the application of inquiry teaching method in the teaching of western economics. J Henan Inst Sci Technol 12(11):98–106

Chapter 64
Research on Science and Moral Quality Education

JinRui Zhao and XinYing Zhao

Abstract The engineering students' education is to train both rich theoretical knowledge and the ability of strong operating engineer type, education of skilled personnel. To improve engineering students' level of education, it is necessary for the education of engineering students' characteristics and students' actual targeted education, especially in the era of knowledge economy, to enable students to have a very good career prospects, it is necessary in science and technology. The moral quality education efforts, it is necessary to enable students to "three" quality.

Keywords Engineering students' education · Quality education · Three

64.1 Introduction

"Vocational skills education is a means to cultivate the grassroots level of theoretical knowledge and practical ability-oriented, production-oriented, service-oriented, management-oriented first line of the occupational status of practical, skills-based expertise for the purpose of career education" [1]. Innovative thinking sex education includes an important part of science, moral quality education. Vocational skills education is a professional general education content of this paper is how college

J. Zhao (✉)
Department of Ideological, Shaanxi University of Science and Technology,
Xi'an, Shaanxi 710021, China
e-mail: jinruizz22c@163.com

X. Zhao
School of Information, Northwestern Polytechnical University,
Xi'an, Shaanxi 710072, China

W. Du (ed.), *Informatics and Management Science VI*, Lecture Notes in Electrical Engineering 209, DOI: 10.1007/978-1-4471-4805-0_64,

students a quality education from the point of view of science, moral education. Technology ethical point of view, the engineering students' students a quality education is necessary to attach importance to the student "three" training.

64.2 Engineering Students Must First be a "MAN"

In many modern enterprises, in the end what kind of talent? In research done on this issue, get the most the answer is that the students should have basic moral qualities. Expertise is of course essential, it is also important, but there is a more important quality than the professional skills is the individual's character qualities. Professional skills not learned in the campus, especially good, but go to work, in social practice, is also able to quickly learn to adapt to a person's moral character qualities is the process of a long-term cultivation of mold not be able to develop overnight. If a student is left on the job, do not know the specific method of operation of the machine, but as long as there are knowledge of the heart, seriously study the old masters, help in the field, I believe, will soon learn to use and master the essentials. Hacking knowledge of the heart, diligent belongs to the scope of the technology and moral qualities. A serious, earnest heart belongs to the scope of the ideological and moral education, if a student does not have these basic ideological and moral qualities, even if it have good professional skills, in future work it is difficult to progress. He did not know ahead and improve. Therefore, the moral education of the college is in place; largely determine the students' future direction.

Before university education, knowledge and ability education is the most important, but as the problems in the universities, such as by some of the negative phenomena in society and the impact of bad ideas, part of the Students Knowledge and Practice, in the thinking of the value of life a lot of confusion, individual students or even the wrong outlook on life. This has led to the thinking of the China Higher Education.

2004, the CPC Central Committee and the State Council issued an important document of views on further strengthening and improving the political and ideological education of college students, from the issuance of documents shows that our mode of education in continuous improvement, teaching the focus of attention in the idea of "people-oriented" laid down more harmonious. Therefore, in the Engineering College Students in Educational Administration, especially its ideological and moral education should take active and reasonable manner, to correct the problems of many students.

Move people to emotion. The most perfect representation of humanity is the human emotion. Emotional education in the educational environment in China is the eternal subject. Now, faced with new situations and new problems emerging in the rapid development of society and students, the time has given a new meaning to emotional education. As a teacher, should Germany high division, the body is Fan. "Morals", "only to convince the people" is important, but if left "emotional" education is very difficult to receive the desired results. Because education is the cause of an emotional, loving, and not sentimental education is the education of feeble. In the teaching process, the formation of a harmonious, democratic teacher-student relationship is an emotional education can be realized factors. “Trees, takes a hundred years”, which is saying passed down since ancient times so far. He said point to understand the ideological and moral education is a long process, but also illustrates the importance of the ideological and moral education. Therefore, in education, it should be impressed by emotion. Emotional education than criticism the absolute system and the rational way of education to enable students to accept, absorb, and use [2].

At the same time to respect their dignity. College students are at the period of the formation of an independent personality in this period, the teacher’s guiding role is very important. Built on top of the boot should be respected, there is no respect for personality is difficult to form a healthy and positive attitude. Therefore, as a teacher in the teaching process must respect the independent personality of the students. Give students the right to speak, so that they learn to form their own judgment, to express their views, the only way to into the community, will always be able to speak. In the usual education, the teacher should be promptly observed the advantages of students for recognition. Of course, criticism should be taken to criticism under the premise of respect for the students, so that it can embrace and be corrected. Respect for the process of the students’ personality, is a subtle process of education. Allow students to learn to respect others, respect for society, and the formation of conscious awareness [3].

64.3 Students will Learn to Make

The most useful knowledge in the world will learn. At present, China is a new era of reform and opening up, and also into the new era of knowledge explosion in the worldwide explosion of knowledge is not only reflected in the expansion of knowledge turnover rate on the amount of knowledge is also reflected in the expansion of knowledge dissemination channels, according to Internet reports the amount of knowledge in the past 30 years is equal to the sum of the amount of knowledge over the past 2000; 2050, the current state of knowledge only when 1 % of the total knowledge While these data are correct or not difficult to trace, but it does feel today’s new science and technology emerged in the amount of knowledge of the aging speed, each person can only continue to learn in order to survival and development and the success of engineering undergraduate course construction and reform stressed that form the basic development of students in this era of rapid change, System of the course, the emphasis is no longer the accumulation of knowledge, but knowledge as a carrier to guide students through the learning process, learning to learn, develop students ‘interest in learning, stimulate students’ desire for knowledge, training students problem-solving skills, ability to recognize ability and innovative thinking of things [4].

Rapid technological development, the intense shock of the economic structure, the one hand, people feel more and more to stick to a mature skills are no longer able to guarantee life-long career with no fear, and therefore pay more attention to learning, constantly acquiring new knowledge, new skills; another aspects of occupational division of labor is increasingly flat, a career change in the degree of frequency and intensity. Phenomenon has gone years ago, the kind of life in a profession. In the face of new situations, how to calm effectively respond to new work, to learn new knowledge, new issues would require graduates in new work situations, beyond the existing knowledge, methods and work experience to continue learning, innovation and problem-solving approach and flexible to effectively adapt to new situations, to achieve self-development.

Soviet educators said: Sukhomlinski, when students taken the school gates, let him take away the spark of the thirst for knowledge, and let his lifetime endless burning down. The students' learning ability in the survival and development of the new era of knowledge explosion and the basis of success, it affects the speed of development of students' career, the depth and breadth. The success of lectures should enable students to deeply feel their wisdom indeed been developed, the potential to arouse, a steady stream of release, tasted the knowledge, tasted the joy of learning, resulting in power to continue its efforts to study hard.

64.4 Innovation of the Three Let the Student Union

The core of education is to nurture talent, creative thinking and innovation capability, and good at finding problems and ask questions is a prerequisite for innovation. But the traditional classroom teaching to the teachers in the teaching, teaching and learning is separated from each other. There is no teacher-student interaction. Under the influence of this mode of teaching, Chinese students are reluctant to ask questions, reflecting the weak students questioned awareness in China and lack of innovative capacity deficiencies. Its root lies in our education focus too much on the common, paid insufficient attention to the personality of the students, this education out students, and the inevitable lack of independent thinking ability and creative spirit. Social development requires innovative education, advocacy in the field of engineering students' education to train a large number of innovative talent that can truly meet the needs of modern society, compound talents with innovative spirit and ability, thus, the innovation ability of institutions for the implementation of the engineering students. The soul of quality education is to achieve the requirements of the engineering students' institutions for educational purposes. Engineering undergraduate education due to the impact of traditional educational values, in many ways, disadvantages, is not conducive to the cultivation of students' creative ability, the school trained personnel cannot really adapt to the needs of the community, therefore, socio-economic development of the urgent needs of the diverse talents called for reform of the existing education, to carry out the spirit of the times, with the vitality of personalized education, played the main theme of the educational reform and development. Personality development of students can be said that a self-fulfilling, self-creation.

The personality is the formation of relatively stable individual character in a certain social environment and education mode, is a person different from the others at. Good sense of individuality and personality are strongly promote the development of individual creativity is the key to successful [5].

Personalized education is an effective way to cultivate innovation capability. Personalized education mainly includes two aspects: First, it emphasizes people as the starting point and destination of education, focusing on shaping and improving the overall quality of the students 'personality through education to promote the structural optimization process of the quality of students, quality education to develop students' personality. Second, to emphasize the person's self-realization and personal advantage, which means to promote the development of students' creative potential? Creativity and personality are closely linked, only to give full play to the students' personality, to develop their creative abilities [6].

Personalized education must deal with the four relationships: First, the development of personality and the strengthening of the relationship of the "double base". The second is to cultivate the relationship between personality and the development of innovative ability. Individualized Education must implement the knowledge, skills, education, and foster creativity trinity of education, the ability of emphasis on innovation as a core feature. Should cultivate individuality and creativity development, give full play to the personality of the students, so as to better train students' ability to innovate. Similarly, the full attention of innovation ability, but also will promote the students' character development. The third is the development of personality and culture "collective spirit". "Collective spirit" in the development of the personality at the same time, we must focus on training, teamwork, learn to cooperate, to contain the spirit of serving the people wholeheartedly, to respect the spirit of social norms and laws and regulations and others something else heterogeneity, diversity, tolerance the heart. Fourth, we emphasize personalized and relationship oriented. Students starting from an international perspective, both to maintain the personality of our traditional culture, but also a deep understanding of the ability of the multiculturalism of the superior personality; is based on the personality of their own culture, but also the mind of the world.

Personalized education, teachers should make an effort to encourage students to divergent, to develop the good habit of questioning. Personalized and creative thinking is usually linked with questioning, questioning is an innovative starting point, did not question the thinking is superficial thinking will not help in the formation of personality. Teachers should be good to continue to raise questions in the classroom teaching to the students to create scenarios to stimulate students' creative thinking. We should actively explore the personalized training program. Cultivation of diversified individual talent, bold and innovative educational models and programs, and actively explore and improve the credit system, the establishment of student transfer, transfer program to turn professional, elective courses and Flexible Credit management system as well as the choice of sites for learning timing provide greater choice of space and time, for students personality development, and fully reflects the "people-oriented modern educational thought. Classroom teaching, to encourage and organize students to participate in practical activities, both by observing the experience to form the perception of society, combined with the specific requirements of each course projects practice, understand and meet the needs of the many. More critical is that students in practice to identify opportunities for open entrepreneurship and innovation, inspiration, shaping a strong interest in entrepreneurship, and to lay a good foundation to take the entrepreneurial path. Therefore, the engineering students' institutions should provide practice spaces and opportunities for the students to build internship bases outside the school. Implementation of individualized education, innovation and application of a variety of individualized teaching mode. Discuss teaching and other teaching methods such as heuristic teaching method, consider the individual differences of students, and develop students' individuality and innovation capability [7].

64.5 Conclusion

In short, the engineering students' education is to develop a practical, skills-based education of expertise. To improve the level of education of engineering students' education, it is necessary for the education of engineering students' characteristics and students' actual targeted education, especially in the era of knowledge economy, to enable students to have a very good career prospects, it is necessary to quality education efforts, it is necessary to allow the students will be quality. Cultivate and foster adapt to the socio-economic development needs of complex, diverse and creative talents, and culture must always seize the core of the students 'innovative spirit and ability to constantly improve the quality of education of engineering students' objectives and teaching content training of high quality of the plasticity of talent.

References

1. Liu D (2006) Ministry of education and comprehensively improve the teaching quality of higher vocational education in view of teach high. J High Edu 23:(2)16–23
2. Shi W (2006) Attention and innovation: key to the effectiveness of engineering students' college students' ideological and political education work. High 3:83–85
3. Ma M (2006) Lee's ultra-ideological and political education and innovative talent cultivation. Beijing educ high educ ed 4:53–55
4. Wei L (2008) Analysis of the construction of university courses and countermeasures of higher engineering education research. Shaanxi Univ Technol 1:38–45
5. Feng J (2004) Personalized education theory. Educ Sci 2:29–37

6. Deng Z (2000) Personalized teaching theory, vol 12(1) Shanghai Education Press, Shanghai, pp 47–57
7. Tang D (2005) Promotion of vocational education students' personality development strategy. Sci Technol Monthly 7:280–286

Chapter 65
Study on Catholic Church Attitude

Yanming Lu

Abstract Since the end of the Second World War and the unveiling of the Holocaust committed by Nazi, the Catholic Church has begun to examine its standpoint and sought methods to improve the relationship with other religious groups including Jewry. The change of the Catholic Church's attitude towards Jewry is a progressive process. The post-war Christian democratic movement in Catholic countries is important force, which drives the change of the Catholic Church's attitude towards Jewry. During the period of the Second Vatican Council, great pressure from public opinions was carried by the Catholic Church. Therefore, Nostra Aetate became a hot topic among people all over the world since its emergence.

Keywords Nostra aetate · Holocaust · Reconciliation

65.1 Introduction

With the end of the Second World War and the unveiling of the Holocaust committed by Nazi, the Catholic Church is forced to encouter great crisis, because a policy of appeasement was carried out by it for Nazi during the Second World War. For this reason, the Catholic Church has to examine its standpoint again for seeking methods to improve its relationship with other religious groups.

From 1962 to 1965, a series of documents and laws oriented at reform were promulgated by the Catholic Church, including the Nostra Aetate, which is a great declaration about the attitude of the Catholic Church towards the non-Christian

Y. Lu (✉)
Department of Religious Studies, Nanjing University, Nanjing 210093, China
e-mail: luyanming@guigu.org

W. Du (ed.), *Informatics and Management Science VI*, Lecture Notes in Electrical Engineering 209, DOI: 10.1007/978-1-4471-4805-0_65,

religions. In addition to making efforts to improving the relationship with other religions, the Catholic Church made a special mention of Judaism in the Nostra Aetate. This was the first time for the grand duke of the Catholic Church to mention that the Church not only objected making persecution on everyone, but also was willing to commemorate the common heritages that Jewry owns. At the same time, the Church disagrees with any enmity, persecution and measures taken by any people in any times for fighting against Jewry under the compelling of religious gospel and kindheartedness. All these new changes are not driven by the political reasons [1].

From the history of the Catholic Church, it can be known that extremely high attention had been paid by people to the declaration Nostra Aetate. Therefore, the Catholic Church has made a highly important progress in the change of its attitude towards Jewry and Judaism [2]. Nostra Aetate is a declaration, which is really not so easy to be owned. However, the important significances of Nostra Aetate can be reduced enormously if people in the latter times read it without considering its concrete historical background.

65.2 Continuation of the Anti-Jewish Traditions: Kielce Pogrom

It is a gradual process for the Catholic Church to change its attitude towards Jewry. Before the declaration Nostra Aetate about the Catholic Church's attitude towards the non-Christian religions was issued for 20 years, the frictions between the Catholic Church communities and the Jewish communities occurred with high frequency. The most extreme example was the Kielce Pogrom in Poland.

In July 1946, 80 Jewish people were killed by poles in the Kielce of Poland, and also a quarter of the adults in the city were involved.

The Kielce Pogrom became the largest European event in which many innocent Jewish people were cruelly massacred after the Second World War. This violence was caused by the blood libel, which could only be seen in the middle ages. However, the Catholic Church set their feet in it [3].

The anti-Semitism in Poland had been notorious for a long time before the Second World War. As early as February 1936, August Hlond, from the Cardinal of Poland, openly declared: "we should reject the harmful influences from the Jewish people on morality, keep far away from their anti-Christian culture, and especially need to boycott the newspapers and relevant publications of Jewry" [4].

Before and after the Kielce Pogrom in 1946, the Catholic Church indulged the anti-Semitism again dishonestly. Czesław Kaczmarek, bishop from the Catholic Church in Kielce, used to say: "There has nothing to say about the Jewish people as long as they can be absorbed in their own private affairs, but they insult the national sentiment of the people of Poland if they begin to get involved in the political and public life of Poland" [5].

The presentation of Kaczmare was just a centralized reflection on the attitude of the monks and priests groups of the Church.

In addition, during the Second World War, the people or organizations from the Catholic Churches of all European countries carried out a very detailed and systemic distinction between the "converted Jewry" and the "common Jewry".

When the Second World War came to end, the Pope of the Catholic Church made a decision, allowing the inquisitions to proclaim the bishops all over the Europe not to return the Jewish children having received baptism in the Catholic organizations. In the mean time, the Pope also permitted to keep the Children, who had not received baptism and also had no family members to recognize them [6].

65.3 First Signs to Call on Reconciliation

A great number of the tenets in the declaration of Vatican Council II had been advocated as far back as the 1930s. Some prestigious French Catholic thinkers in the 1930s, such as Jacques Maritain, Emmanuel Mounier and Francois Mauriac, and the influential daily La Croix, were opposed to the anti-Semitism of the Catholic Church [7].

After the Second World War, among the multiple factors forcing the Catholic Church to make a change to its attitude towards other religions, the Christian democratic movements in the Catholic countries might be the most important driving force [8].

Christian democratic movement emerged in the European continent in the middle of the nineteenth century or latter. Along with the end of the Second World War, Christian democratic movement welcomed its development peak.

All Christian Democrats believed that they could seek a compromise road between capitalism and socialism according to the religious doctrines and encyclical spirits of the Catholic society.

The change of the Catholic Church's attitude towards Jewry happened first in Italy. In January 1943, De Gasperi, who led the Italian government after the Second World War, established the Italy Christian Democratic Party.

The Christian democratic movements in France had positive changes, which were of the same significance. However, the People's Republic movement that the anti-Catholic members established changed into one of the most important powers during the fourth republic period.

After the Second World War, the development achievements, which were made by the Christian democratic movement in German, were the most significant.

In August 1945, CDU was founded. The goal of the CDU was to establish an alliance of religious sects, not only bringing Catholic liberal and conservative sects together and also accepting new Protestants and the people from other religious faiths.

65.4 Questions Inside and Outside Catholic Church

In 1963, German dramatist Rolf Hochhuth described the Pope Pius XII as hypocrite in play, because he kept silence about the holocaust. In the critically-academic Catholic Church and Nazi Germany of Guenter Lewy, the spear was directed at Vatican as well [9].

All these criticism sounds were attacked firmly from Vatican, but the influences that they exerted were great in the real world.

Criticism vioces were not only from the world outside the Church, and also the discontents inside the Church grew. All these dissatisfactions were mainly directed at the negative attitude of the Pope Pius XII towards Vatican Council I movement [10].

In 1958, Pope Pius XIII provided a new historical opportunity for Vatican to make an amendment to its conservative route. Also, Pope Pius XIII became highly known within the Church for a very long time due to his assertions towards freedom and reform.

65.5 Vatican Council II and Nostra Aetate

On October 11, 1962, ecumenical council I was unveiled at St. Paul's Cathedral of Vatican, and ended on December 8, 1962. Because of the argument between the innovation and conservative routes, there were no formal documents agreed at the Vatican Council I. Therefore, Pope Pius XIII still released the general spirits in April, setting a tone for the second-half of the council.

In the process of the Church to issue the declaration about its attitude towards the non-Christian religions, Augustin Bea, general secretary of the Christian tribunal secretariat, played a very important role.

In 1959, John XIII named Augustin Bea to prepare the draft of a document. After Vatican Council II was held, the formal drafting of the declaration began to be implemented with tension. At the Council, Alfredo Ottaviani, leader of the conservative sect of the Church, made a fierce debate with Augustin Bea concerning about religious freedom. Although Ottaviani also gave his supports to the religious tolerance, he still was opposed to the separation of religion and politics and the endowment of all equal religious rights, and claimed it was necessary to suppress the public expression of the non-Catholic religions as far as possible. The draft, which was submitted to the Council, received the opposition from the majority of the representatives. For this reason, the Council decided to rewrite the resolution. In such a way, the Catholic Church held a more flexible attitude towards many issues, and also it was possible for the secular party to generate a larger influence on the Church [11].

However, unfortunately, John XIII was unable to see the end of the Vatican Council II. New Pope Paul VI was necessary to set prestige in the public when he was just the successor.

In November 1964, before the third meeting of the Vatican Council II, the topics about the religious freedom received fierce discussions under the request of Paul VI.

The fourth meeting of the Vatican Council II was held from September 14 to December 8, 1965. On October 26 of this year, the declaration Nostra Aetate was issued in the world.

The Church can't forget, it is voter of the Old Testament that the God made with infinite mercy and accepts the enlightenment of the Old Testament, and at the same time the wild olive branches of the foreign people were connected to the fine olive tree roots, absorbing nutrition. It is our peace when the Church fells a faith on Christ, and the cross makes the Jewry and foreign people peaceful and friendly with each other, becoming an entirety.

In the declaration, it is admited that there are great common spiritual heritages between Christian and Judaism, and also the two religions are advocated and encouraged to know each other with esteem. All these are positive attitudes towards Jewry, which could not be seen in the history.

65.6 Several Questions

The questions on the declaration about the change of the Catholic Church's attitude towards the non-Christian religions are never stopped since its emerngance in the world.

Question 1: Does the declaration treat Jewry equally?

The fundamentally unequal tenet for the Christians and Jewry forms a "gray zone" in the consciences and moral obligations of the Christians actually. However, in the declaration, Jewry are still regarded as the objects that do not receive the gospel and need to be converted (this can be proved in the Holy Bible that most Jewry did not receive the gospel and were restrained from the communication of the gospel when the Jerusalem did not know them).

Question 2: Is it right to publicize the conversion of Jewry?

From the perspective of the devout Christians, making Jewry converted can be regarded as a religious obligation and the complete kindness no matter how the environment is severe. However, all these "good" things are questioned by Jewish communities.

Question 3: Is there rescue out of the Catholic Church?

This is a very important question for the Catholic Church. In a traditional sense, it was understood that there was no rescue out of the Church. However, the declaration of the Catholic Church towards the non-Christian religions aroused a question (is there rescue out of the Catholic Church?).

Beyond all doubts, these questions cast a shadow over the declaration Nostra Aetate. The responses of the Jewish communities to the declaration are different as well.

However, in recent 40 years, the relationship between the Catholic Church and the Judaism has been improved constantly. This has a very close connection with the release of the Nostra Aetate and the commitments of the Popes to scrupulously abide by the religious tolerance.

Nostra Aetate is the core for people to get a real understanding of the relationship between Christianity and Judaism after the Second World War. Today, the Nostra Aetate has changed into a model for all religious groups to communicate with each other.

References

1. Translated by China Bishop Secretariat. Documents of Vatican Council II: attitude declaration of the Catholic Church towards the Non-Christian Religions
2. Bokser BZ (1968) Vatican II and the Jews. Jewish Q Rev New Ser 59(2):455–472
3. Times Correspondent (1946) Anti-Jewish Riots in Poland. The Times Retrieved 19(2):79–88
4. Porter B (2003) Making a space for antisemitism The Catholic Hierarchy and the Jews in the Early Twentieth Century. Polin Stud Polish Jewry 16(13):976–1011
5. Aleksiun N (2005) The Polish Catholic Church and the Jewish Question in Poland, 1944–1948. Yad VaShem Stud 33(10):643–652
6. Friedländer S (2007) The years of extermination: Nazi Germany and the Jews 1939–1945. Harper Collins, New York
7. Birnbaum P (1992) Anti-semitism in France: a political history from léon blum to the present. vol 45. Blackwell, Oxford, pp 5–14
8. Von Beyme K (1985) Political parties in Western Europe. vol 12. Gower, Aldershot, pp 290–298
9. Marchione M (2000) Pope Pius XII: architect for peace. Paulist Press, New York
10. Guokun H, Longyun J (2008) Historical inspection on Vatican Council II. Collected papers of history studies, 9(5):75–85
11. Williams PL (2003) The Vatican exposed. Prometheus Books, New York

Chapter 66
Study on Reconstruction of Life Education in Universities

Mingming Liu and Rongting Qin

Abstract According to some factors that affect life education in universities in our country, this thesis explains the necessity and urgency of carrying out life education. It puts forward some suggestions of reconstructing universities, such as deepening understanding to change the education concept, paying attention to guidance to widen connotation of education, strengthening training to improve the quality of teachers, establishing the course system of life education, combining with families to develop life education, appealing the society to participate in life education and so on. These suggestions can ensure the implementation of life education in universities.

Keywords Universities · Life education · Lack · Reconstruction

66.1 Introductions

The current high education mainly serves for cultivating the standard construction-centered talents. Students are faced with an increasing high standard of professional skills, which have been the first and only measurable criteria of the students' ability while life education comprised of the ideological exquisiteness, cultural enrichment, psychological quality and the personal integrity is over neglected, inevitably leading to bruise or suicide. Life is precious as well as fragile, and loving life tends to be the most significant value orientation in Chinese contemporary education [1]. Therefore, the life education has become an urgent issue needing to be given an in-depth research in high education.

M. Liu (✉) · R. Qin
Chongqing University of Arts and Sciences, Yongchuan, Chongqing, China
e-mail: liumingming@guigu.org

W. Du (ed.), *Informatics and Management Science VI*, Lecture Notes in Electrical Engineering 209, DOI: 10.1007/978-1-4471-4805-0_66,

66.2 Analyses of Influencing Factors and Reasons Concerning Universities Life Education

66.2.1 Part of Students Holding a World-Weary Attitude

Contemporary universities students are faced with challenging psychological problems. What results in university students' pessimistic attitude is as the following. One thing for certain is that the university students are frustrated facing failures and even unable to get through setbacks, easily tending to suicide in these cases. Investigation on part of the university done by the Nanjing Crisis Intervention Center shows university students' suicidal rate rises in recent years; another thing is lack of independence and self-confidence prevents quite a few of students from making decisions on their personal things. The loss of autonomy makes it easy to cultivate a dependent psychology. Under this circumstance, students are often at a loss and chose extreme ways to solve problems when facing setbacks; the third thing is the majority of students can be well aware of loving life and cherishing life, but still not take a particularly positive attitude; the fourth thing is a variety of students ignore the true meaning of life, and hold false values, false outlook on life, false world outlook, such as Ma Jiajue, a Yunnan University student, created the appalling killing classmates case; Liu Haiyang, Tsinghua University students, spilled sulfuric acid to bear.

66.2.2 Professional Teachers Deficient: Overall Planning Still

At present, Chinese university life education does not kick off in this true sense, even if there is the so-called life education, done by instructors and the psychological consulters. Professional teachers are relatively scarce. Through a few universities' carrying out life education, we can see from the teaching content aspect that a system structure with a organic link between the theoretical education and actual application is nowhere evident; in terms of the teaching arrangements, the phenomenon such as repetition, dislocation still exists in learning period; in addition, the relevant classes get no guarantee and are kind of unworkable; from the curriculum aspect, there is no denying that content is monotonous and too much emphasis are attached on knowledge acquisition, while little attention is put on cultivating students' aesthetic appreciation and promoting moral development.

66.2.3 Standing Problems in Education System

For a long time, graduation rates, high scores, good grades and all these utilitarian goals digest and becloud the important and profound life education targets, which are beneficial to improve students' life quality, thus life education losing the inner Connotation and obtaining no due effect. What's a worse, university launch the expansion action, bringing in a sharp rise of the students population. Even though it has a certain extent reform, university education is still far away from the fundamental change from exam-oriented education to quality education. If too much emphasis is focused on the professional knowledge and skills and so on, the bottlenecks of traditional consciousness will restrict the university from effectively implementing and developing life education.

66.2.4 Family Factors

The family is the first class for each individual, adverse factors in family can simply result in the students' unsound personality, mind twist, extreme trends, producing a "earthshaking" light-weight move. Tracing its reasons, there are three factors lying behing the phenomenon. Firstly, it is these thoughts deeply rooted in the heart that make some parents set a high standard for the children regardless of the specific conditions; secondly, it is single family influences including lack of care for their children, some even the parents' responsibilities that misguide the children vulnerably to the wayside; thirdly, it is the insufficient communication and exchange with their children, due to the family internal impeded communication, that estranges children from parents, and obtains no emotional interaction and emotional satisfaction, eventually diluting the family.

66.2.5 Social Problems

The modern society is a rapid evolutive, multicultural society. Especially, the market economy of our country is in the formation and consummation stage with the changes reaches every corner of social economy. Involved in such a dynamic, multi-choice social context, students can be easily become vulnerable and blank. During this extraordinary period, some regain the balance, while others' were broken. Faced with various unhealthy trends, highly competitive external environment, boost material temptation and adverse publicity situations, university students challenge themselves to psychological adjustment and withstanding ability. However, large portions of the university students grasp an inproper scale to distinguish right from wrong. And in this case, it will be inevitably not surprising to produce deviant behavior if there is not timely guidance.

66.3 The Necessity and the Urgency of the Life Education

Life, the most precious gift for everybody, is prerequisite of the existence of all significance and values. Needless to say, no life, everything will be out of the question. The life formation is a very complicated process, and hard-earned, a variety of elements in nature happening to meet and form at a certain time and a certain space through complex changes. Still its existence is also very dangerous, the space it exsits has a number of factors endangering its existence all the time. With little attention, it will die away. In this sense, cherishing and loving life are the fundamental conditions and necessities to ensure life meanings and values. The reasons for university students overlooking life are from various fields, including society, school, family. But from the universities' level, it is undeniable that our education has been short of life education, leading to university students' despising their life and others [2]. For this reason, paying more heed and strengthening the life education should be the top priority for the universities.

66.4 Thinking's About the Reconstruction of Life Education System

66.4.1 Enforcing Awareness: Changing Educational Views

Life education is not just a simple admonishment and conviction, but the science education we essentially master through the course of life. The existing life education, although has eased students' psychological pressure and corrected some deviant behaviors in a certain degree, is very often playing a fire brigade role, putting out a fire when it occurs. It is just a temporary solution and not a permanent cure. Therefore, we should be fully aware of the urgency to establish and consummate the life education, improve understanding of educational department's leaders at all levels, actively implement and promote life education, explore the rule of life education fundamentally. Through the above actions, desirable results of life education can be achieved eventually.

66.4.2 Attaching Emphasis on Guidance: Expanding Educational Content

First of all, life education should acknowledge and cherish value of life noumenon' existence. The correct understanding of life is the premise of loving and treasuring life. The value of life based on the existence and perpetuation of life, it is undoubtedly ridiculouos to realize the value of life independent on life noumenon existence and sustaining process. Students should be guided to fully recognize and respect the value of human life. Any behavior, harming yourself and others, is profaning and tramplng over person's value and life. From individual micro level, people's life inherites from history and extends the future. Only cherishing the real life, can continued vitality be provided for obtaining eternal life [3]? What's more, the life education should emphasize understanding of life and harmony life. Life is the basis of life, life is manifestation of life. No life, no lives. Lives are out of question without life. Whether it is wealth, power status or beauty, they are melting into dust finally. The characteristic of life is pervasive and historical. Only from my life to universal life, from the immediate life to the eternal life, will it eliminate the tension between life and lives. Furthermore, life education should attach importance to life's ultimate concern. Life's ultimate concern is the premise of life education, and is the highest pursuit for life education. Students should rationally grasp the limited life with mastering the spirit of transcendent and no repeating the material life and building the ultimate life beliefs, through which timeless faith of life can be established.

66.4.3 Strengthening Training: Improving Professional Level

Whether life education is able to smoothly develop in the school depends largely on relevant professional teachers. Universities and colleges should take powerful measures to strengthen the training of the staff. For all the teachers, they should be equipped with basic knowledge's of life education; for subject teachers, they must strengthen humanities cultivation and bioethics training; for the psychological counseling teachers, they had better enhance the understanding of life science; for counselors, the professional skills of life education guidance and psychological counseling should be mastered to improve teachers' quality and educating ability.

66.4.4 Constructing Life Education Curriculum System

Compared with foreign universities, our life education curriculum system appears a poorer elasticity. When launching life education in schools, a scientific curriculum system should be wholly constructed based on the students' physical and mental development and the education law; overall planning education content sequence so as to organically link the education content including "understanding life, cherishing life, respecting life, loving life" in every age, every learning phase, and wholly and systematic establishing a scientific and reasonable, standard new curriculum theory system step by step.

66.4.5 Carrying Out Joint Family Life Education

The family has been called the first class, and the parental education is the most crucial part. The family is the children's first point of contact, and is life's first school, the parents are children's first teachers, the advantages of family education are so obvious to be taken off. In the eyes of children, parents are the most reliable family members, family environment and parental words and deeds play a subtle role in their children's development. Parents, as the guardian of children, should set up the correct life values and educate children to harbor the heart of self-love, respect for life, reverence for life, and to be well aware of the value of life, to let their children better understand that limited life is not the private property. Everyone's life is the source connecting the parents' life, every life is correlated, interdependent and they should cherish each other. Besides, parents should use the right method to educate, exchange, communicate with their children frankly with inner thoughts, positive emotion so that the children can truly feel the warmth of family, promoting their the healthy growth.

66.4.6 Calling for Social Participation in Life Education

Life education is involved in every aspect of the society, and each corner of society should be filled with love for life. Only through a variety of ways, such as undertaking the education of life and health, safety, growth, and value towards students; helping and guiding the students to correctly handle the relationship between the individual, group, social and natural; making students learn to master necessary survival skills, knowledge, can students understand the meaning and value of life, cultivate respect for life, learn to appreciate and love life with the correct world outlook, life outlook and values outlook, develop a scientific spirit and humanistic quality of a new generation. In addition, supervision needs to be strengthened for propaganda Medias including the newspapers, films, televisions, the internet and so on. Forbidding the content of the promotion, showing no respect for life, from erasing student's thoughts, and building social atmosphere, where everyone protect life, cherish life, so as to construct a safe and harmonious environment in the students' growth process.

66.5 Endnotes

Education is out of question if life is not well proceeding. Life education is life-long and is the base of education as well. With all-around life education and joint efforts by family, school and society, students are able to form the correct life values, and they will reap the benefits for a whole life and make life more valuable.

In short, the life education is a systematic project, involved with a variety of macroscopic and microscopic factors. We proceed from the reality, dare to try, reform and explore life education with no sticking to one pattern from the characteristics of university life education, thus promoting education effect, further improving the level of life education in our country, laying a solid foundation for university students' security, physical and psychological health and happiness.

References

1. Huang B (2006) Theory and the practice concerning life education. Learn Monthly 4(2):25–26
2. Chen J (2004) New inquiry about meanings, contents and methods of university students' life education. J Guangdong Univ 4(1):63–65
3. Gao J (2003) Research about university students' life education. J Wuyi Univ 3(2):8–11

Chapter 67
Research on Cultural Knowledge and Awareness in Education

Junling Wang

Abstract How to develop the cultural knowledge of teachers and how to improve the students'cultural awareness. It is very important in culture teaching in our country. In order that students acquire cultural competence culture teaching in foreign language teaching must have many things to do. This article puts forward the ways of developing the teachers' knowledge of second culture and the ways of improving the students' cultural awareness.

Keywords Culture teaching · Communicative competence · Second culture cultural differences

67.1 Introduction

Culture is a way of life. Culture is the context within which we exist, think, feel, and relate to others. It is the "glue" that binds a group of people together. Culture might be defined as the ideas, customs, skills, arts and tools that characterizes a given group of people in a given period of time. But culture is more than the sum of its parts [1].

Many experts both abroad and at home have talked much about culture teaching. Brown [1] discusses how to learn a second culture; While Stern [2] expounds the cultural syllabus design and the goals of culture teaching. Chinese scholars, represented by Professor Hu Wenzhong, have presented a lot of papers

J. Wang (✉)
School of Foreign Languages, Shandong University of Technology, Zibo, 255049 Shandong, China
e-mail: wangjunjin2371@126.com

W. Du (ed.), *Informatics and Management Science VI*, Lecture Notes in Electrical Engineering 209, DOI: 10.1007/978-1-4471-4805-0_67,

on intercultural communication [3]. However few treatises systematically deal with culture teaching pricinples. In order that students acquire cultural competence, culture teaching in foreign language classes must have definite objectives and teaching principles through which the objectives are realized.

Learning to communicate in second language may be difficult and frustrating. It will offer a look at cultures from every part of the earth, which will reveal cultural similarities and differences that we had never noticed in the past.

Culture differences constitute such serious difficulties for English-Chinese learning. Apart from knowing the two languages well, learners should deeply understand the two cultures from different aspects, such as history, region, custom and brief which always hinder Chinese receivers from understanding the meaning of English context correctly and result in parralax on the Chinese receivers.

67.2 The Situation of the Culture Teaching in China

The goal of English teaching is only to equip students with communicative competence [4]. One of the important compnents of communicative competence is the ability to select a linguistic form that is appropriate for a specific situation, or to use English appropriately in social interactions [5].

As Chinese culture is not the same as that of the English speaking countries, the rules for using Chinese are, in some respects, different from those for using English. There is equally good evidence that Chinese students may transfer their mother tongue references of language use to their English performance and fail to communicate effectively. Misunderstanding caused by cross-cultural pragmatic differences is an important source of cross-cultural communication breakdown and much attention should be paid to them in English culture teaching.

Culture establishes for each person a context of cognitive and affective behavior, a blueprint for personal and social existence. But we tend to perceive reality strictly within the context of our own culture; this is a reality we have "created," not necessarily "objective" reality, if indeed there is any such thing as objectivity in its ultimate sense [6].

It is apparent that culture, as an ingrained set of behaviors and modes of perception, becomes highly important in the learning of a second language. A language is a part of a culture and a vulture is a part of a language; the two are intricately interwoven so that one cannot separate the two without losing significance of either language or culture.

Studies of culture teaching only have a short story, but systematic research in this field is deepening. However, how to conduct culture teaching well in language classes is an expanding problem for language teachers.

In recent years, culture teaching as well as foreign language teaching has got more and more attention than ever before, for more and more foreign language teachers are aware that second language learning is often second culture learning and that cultural competence is an integral part of communicative competence.

67.3 The Ways to Develop Teachers' Cultural Knowledge

Learning will not happen when the teacher insists on teaching rather than letting the students learn for themselves. Nor will it happen when the students do not know how or are not willing to learn for them. It is not enough simply to be willing to teach linguistic conventions, rules and norms of different cultures. It is clear that a language can only be learned and not be taught. A teacher can at any time be quite sure that he is teaching, but it is not at all easy for him to make sure that learning does take place. There are some ways for them to develop their knowledge of the second culture:

67.3.1 Teachers Should be Well Learned in the Target Culture in Terms of English Teaching

They should analyze the students' linguisitic mistakes due to the interference of cultural factors so as to enhance their cultural sensitivity and make them realize that communicative competence can by no means be acquired by the mere mastery of language forms.

67.3.2 Teachers Should Know Much More About the Foreign Language Culture Than He or She Will Teach the Students in Culture Teaching Class

Thanks to the development of science technology, we can know the culture of that language through the webs, e-books, watching video-tapes and reading materials about it.

67.4 The Ways to Improve the Students' Awareness of Culture

Culture should be included in our foreign language program. This is especially relevant to the foreign learning context in China, where the learner learns a non-native language in one's own culture with little immediate and widespread opportunity is to use language within the environment of one's own culture. In learning English, the learner has to learn how to use and interpret the socio-linguistic rules of English. To achieve this, the learner must develop "an awareness of areas in which the socio-linguistic system of his native language differs from that of English" [7]. Teachers should seize every opportunity to teach culture: in reading and in listening class.

1. Students should be aware of cultural differences from the start (intercultural awareness should be made one of the goals in English education).
2. Students should have easy access to visual aids, such as English original films and videotapes, and if possible, English native speakers. They should not learn English just to pass the exams, but to form the communicative competence, to satisfy the requirement of society.
3. Our school gives every student (not confined only to the students of the English major) opportunities to appreciate original films several times a term.
4. The teacher can provide an isolated item of information on the target culture per lesson and over time encourage the students to arrange the related capsules into a cluster and dramatize it. Such as how to set the table, how to behave at table, how to get to have dinner with Americans, and the family together at the table can be organized into a cluster of family meal for the students.
5. Assign the students poem recitations, news reports, or playing popular songs, widely reading the famous English original novels, just like Shakespeare's and Mark Twain's works. It's a good way for the students to improve themselves' awareness of second culture.

67.5 Conclusion

After all, the goal of English teaching is only to equip students with communicative competence [4]. During the stage of learning English, the teachers are a great help to the students. To improve the students' awareness of cultural differences, they should select the appropriate ways to improve it. To develop the teachers' knowledge of culture in second language, there are many ways to get. Teachers and students should make most use of the facilities offered at the university to teach culture through lectures, reading, personal contact and guided discussion. This is particularly important in a Chinese setting where the actual use of English is geographically and psychologically far removed from our classrooms.

References

1. Brown HD (1980) Principles of language learning and teaching. Prentice Hall, New Jersey 23:102–109
2. Stern HH (1992) Issues and options in language teaching. Oxford University Press, London 4:49–57
3. Hu W (1988) Intercultural communication—What it means to chinese learners of english. Shanghai Translation Publishing House, Sanghai 55:299–303
4. He Z (1984) Cohesion in adult language-teaching texts. J Lang Teach Res 22(03):59–68
5. Hymes D (1981) In vain i tried to tell you: essays in native American Ethnopoetics. University of Pennsylvania Press, Philadelphia, pp 5–6

6. Condon EC (1973). Introduction to cross cultural communication, (1). Rutgers University, New Jersey, pp 17–19
7. Holmes and Brown (1977) Language and culture research. Cambridge University Press, Cambridge, pp 73–74

Chapter 68
Study of Vocational Technical Education in Promotion of Chinese Rural Labor Transfer

Feng Wang

Abstract This paper has made explorations on the connotations of vocational technical education and its status and function of the promotion of rural area labor transfer. It has put forward some suggestions and countermeasures for the development of the vocational technical education. In this way, it can better serve the stable and well-regulated transfer of Chinese rural area labor.

Keywords Vocational technical education · Rural labor · Status · Function

68.1 Introduction

According to statistics, the current Chinese rural surplus labor has reached 2.5–300 million. The unitary transformation to achieve the dual economy the central issue is the transfer of the agricultural sector labor force. Economic developments in different countries around the world, labor transfer path, direction, patterns are not the same. Learn from foreign experiences and lessons of rural labor, rural labor transfer to explore the direction and path, which will help promote the orderly transfer of rural labor force. International experience shows that to achieve the dual economy the central issue of the unitary transformation is the transfer of the agricultural sector labor force. The higher proportion of agricultural labor, the lower the level of agricultural modernization, the more obvious characteristics of the dual economic structure; the lower the proportion of agricultural labor, the higher the level of agricultural modernization, the more obvious characteristics of the dual economic structure, or even completely disappear. On the whole, these

F. Wang (✉)
Shaanxi Radio and TV University, Xi'an 710068, Shaanxi, China
e-mail: fengwang02s@126.com

W. Du (ed.), *Informatics and Management Science VI*, Lecture Notes in Electrical Engineering 209, DOI: 10.1007/978-1-4471-4805-0_68,

countries non-agricultural and urbanization are basically synchronous. As a result of its economic, historical, cultural, institutional and environmental factors, the transfer of rural surplus labor force in these countries the specific model is still distinctive. Developed countries to understand and study the transfer of rural surplus labor transfer process and the model for China, a majority of the rural population is still developing a modern road, an important inspiration and reference. This paper has made explorations on the connotations of vocational technical education and its status and function of the promotion of rural area labor transfer. It has put forward some suggestions and countermeasures for the development of the vocational technical education. In this way, it can better serve the stable and well-regulated transfer of Chinese rural area labor.

68.2 The Current Situation of Chinese Rural Labor Transfer

The transfer of rural surplus population of the colonial state is the British transfer of rural surplus labor force is another feature. Britain was the world's largest colonial powers. The existence of surplus labor in the countryside and transfer of agricultural labor to non-agricultural sectors is a common phenomenon in economic development. Supply of rural surplus labor will change with the transfer of labor [1].

By calculating the difference between the agricultural labor force and current demand of laboring agriculture, this paper seeks to estimate the total amount of China's rural surplus labor as identified by age, gender and education. Currently, China's rural surplus labor supply falls short of non-agricultural industry demand. In the first place, it is the knowledge difficulty. In the second place, it is the household register difficulty. In the third place, it is the land difficulty. In the last place, it is the organization difficulty.

In the above difficult situations, the transfer of rural surplus labor force is not effective in promoting agricultural growth. In a sense, the British transfer of rural surplus labor, the expense of agriculture at the expense of the British in the process of urbanization of the population needed food and agricultural products as raw materials, mainly from abroad. From the transfer point of view, to absorb the rural labor force is the largest city department of industrial and service sectors. From the overall quality, it is still quite low [2].

68.3 The Position and Function of Vocational Technical Education in the Promotion of Chinese Rural Labor Transfer

Technical education is the formal education designed to provide knowledge and skills underlying production processes with a wider connotation than vocational education at secondary or higher level. Vocational education is the formal education designed to prepare for skilled occupations in industry, agriculture and commerce, generally at secondary level. In recent years, as China's rapid economic development, vocational and technical education has been gradually recognized by the society. As highly educated people, vocational and technical education graduate is an important force in the development of vocational education. But in reality, their employment orientation deviate from the trend of the profession and only a small part of them do related work after graduation. Combined with the current development status of vocational education, the article analysis the reasons of this problem and give policy recommendations, create a good environment for vocational and technical education talented people [3].

The ultimate goal of rural labor transfer is to change farmers to citizens. This requires a long period of time. Since reform and opening, China's agricultural labor force to non-mass transfer, and to some extent, promoted the development of urbanization, China's first industrial employment structure but still lags far behind the industrial structure changes. The transfer of rural surplus labor force in developed countries the pattern for the transfer of rural surplus labor force to provide a valuable experience.

The functions of vocational technical education have shown in the followings:

1. Vocational technical education can enable labor transfer to have the basic knowledge and skills and get good preparation for employment.
2. Vocational technical education can promote labor transfer to realize professional development and change the potential labor to be real labor.
3. Vocational technical education is beneficial to realize the labor transfer citizenship and quality modernization. It can increase the city belonging.

In a sense, the main task of China's modernization is to a large number of rural surplus labor transfer out for the effective transfer of rural surplus labor force path and pattern, has become our country's sustainable economic and social development.

68.4 The Countermeasures and Suggestions of Vocational Technical Education in the Promotion of Chinese Rural Labor Transfer

68.4.1 Fasten the Vocational Technical Education That Faced the Rural Areas and Ensure Each Work to Practice

In 2011, the ministry of education and the national development reformation committee have published the "Opinions on fastening the development of vocational education faced rural areas". Can say how fast the pace of industrialization, how large, the transfer of agricultural labor force is as big as the speed and scale. By the late industrialization, the development of tertiary industry, absorbing more of the labor force is moving fast. Early stage of economic development, these countries mainly through the rapid development of industries to solve the problem of rural surplus labor, and economic development depends mainly on the later development of tertiary industries to absorb surplus rural labor. The development of education means increased investment in human capital and improving the quality of labor, which makes Japan the rural labor force to non-agricultural employment opportunities with good adaptability, thus contributing to the rapid transfer of rural surplus labor. This can build a better condition for everyone.

68.4.2 Carefully Implement the "Rural Area Labor Transfer Training Plan" and Build Characteristics and Brand

Because people will be less, although there are a large number of rural surplus labor transfer, but not high concentration of land, remained small-scale production. How the process of industrialization in the optimal allocation of agricultural resources, requires a combination of its own national conditions, in accordance with their endowment of resources to develop appropriate land policy. From the dynamic point of view, people each year in rural areas by adding mobile force, the decline in comparative advantage due to agricultural and urban–rural differences, regional disparities widened, the surplus agricultural labor force is accelerating inter-provincial, cross-county market liquidity. Thus, at present about half of the rural labor force is in absolute or relative to the rest of the remaining state, to non-transfer of a lot of pressure, which requires us to further adjust the industrial structure, and explore suitable for China's agricultural surplus labor transfer road.

68.4.3 Strengthen the Overall Development and Application of all Kinds of Teaching Resources and Strengthen the Abilities of "Agriculture, Rural Areas and Farmers"

From our practice, the transfer of surplus agricultural labor force has the following characteristics. First, the rural labor force into the city, long-term stability and poor regional transfer completely inadequate. As the agricultural sector are inextricably linked, their roots still in the rural areas, and to the public by the farmers did not complete a thorough transformation. Encouraged farmers to leave their homes not encouraged farmers to leave their homes into the factory without entering the city; a phenomenon in the transfer of surplus labor in agriculture is widespread. Second, the quality of population and cultural point of view, the transfer of surplus labor out of agriculture is generally low educational level. Migrant is mainly engaged in industry, construction, commerce, catering and other labor-intensive physical exertion of the primary service industry, which is generally not high quality of their culture, lack of professional skills closely related. Statistics show that a considerable part of the labor force reached only primary school education. In addition, priority is often transferred out of high school and those with higher education level of the labor force, and those with junior middle school education level of employment, low educational level of the remaining surplus agricultural labor force cannot meet the non-agricultural industries, urbanization, development, and transfer with a certain degree of difficulty. Third, the population flow from the point of view, the main agricultural surplus labor from rural to urban flow, flow from the mainland to the coast, from north to south. China's agricultural surplus labor is a complex factor, almost covering the political, economic, social, geographical, cultural and other aspects. Therefore, the path to find solutions to problems despite some reliance, but fundamentally, it requires the joint efforts of the whole society in order to achieve a fundamental orderly transfer of rural surplus labor. At this stage, the transfer of surplus labor in the agricultural and urban development in the process of coordination should note the following aspects.

68.4.4 Strengthen the Policy Support and Mechanism Construction and Ensure the Well-Regulated Transfer of Rural Labor

Generally the higher quality of labor force, the transfer speed is faster than the lower quality of labor, while the transfer level is higher. Unprecedented fierce competition in the market today, China's rural labor force by virtue of years of education per capita reached only primary school level, how to get rich? Levels of government should the rural labor force skills training, transfer of surplus labor as the current and future service for a long period during the most important task, do a good job skills training for existing rural labor force work for the transfer of rural surplus labor to create the conditions for urban and rural areas do a good job of education, so that future generations of farmers to become knowledgeable, literate modern citizen, is China's farmers to bid farewell to poverty, backwardness of China's rural areas to bid farewell to the root of the road. Accelerated urbanization is a large-scale infrastructure development as the guide, and this large-scale infrastructure, the transfer of surplus agricultural labor force is bound to provide the preconditions. To develop urban transport, housing, education, water and electricity supply services, information consulting, property management for huge investments in these industries to absorb labor costs low, you can create a lot of jobs; help solve the problem of surplus agricultural labor force.

References

1. Jianmin LI (2010) Chinese labor market. Nankai University Pressm, China, vol 38(3), pp 479–485
2. Fang M, Xi-tong W (2011) Rural labor transfer and what can vocational education do. Chin Educ J 22(9):73–87
3. Nianhong Z (1988) Encyclopedia of education. Chinese Rural Scientific Press, Beijing, vol 3(3), pp 514–519

Chapter 69
Network Model of Practice Education in the Major of Art Design

Yankun Liu

Abstract In views of the problems that there is a lag in the network model of practice education in the current course system and a dummy in social practice activities for the undergraduates, the author thinks that in the network model of practice education in the major of art design, the network model of practice education should be coordinated with the social practice activities, which means that the professional practice education in the course system should be socialized and the social practice activities should be professionalized and informatization.

Keywords The major of art design · Network model of practice education · Social practice activities for undergraduates · Professionalized · Informatization

69.1 Introduction

The major of art design has experienced the stagnation and gradually recovery from the 60s to the early 80s in the last century, and until the late 80s, it stepped into the comprehensive development stage and got remarkable achievements [1]. But take the model of the western developed countries as a reference, the education of art design is still in the primary stage, and have not really set and been mature. As the economy of our country developing in high speed in recent years, there is a phenomenon that on one hand, our country is lack of the talents for design, on the other hand, the undergraduates in the major of art design tend to divert to other professions because they can't find jobs with matched major [2].

Y. Liu (✉)
Chongqing Normal University, Chongqing 401331, China
e-mail: 6706069@qq.com

W. Du (ed.), *Informatics and Management Science VI*, Lecture Notes in Electrical Engineering 209, DOI: 10.1007/978-1-4471-4805-0_69,

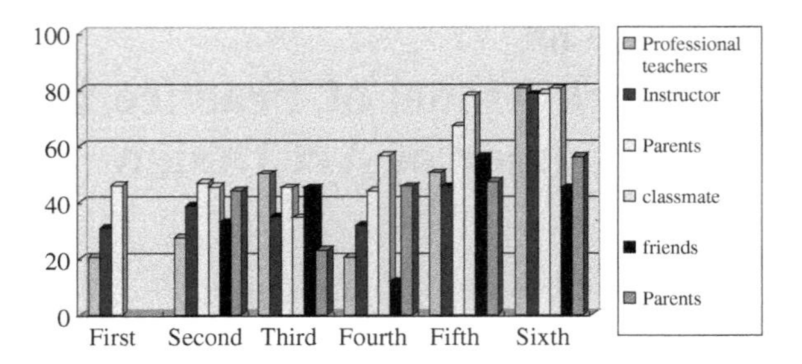

Fig. 69.1 Various factors affecting the proportion

It reflects a problem that why those talents cannot be fully endorsed by the society under the great investment from both the country and the students in financial and material resources and energy, besides, it will also cost a lot for the companies to give a pre-job training [3]. To deal with the problems above, the author thinks that in the model of practice education in the major of art design, the practice education should be coordinated with the social practice activities, which means that the professional practice education in the course system should be socialized and the social practice activities should be professionalized (Fig. 69.1).

69.2 The Professional Practice Education in the Course System Needs to be Socialized

The college practice education will be the real practice education only through the real social practice activities rather than the imitating practice education under a perfect state [4]. Social practice is the necessary approach to realize the goal of tertiary education, the complements of classroom teaching and the efficient method to coordinate and utilize the education resources. Along with the development of the society, the content of social practice is constantly extended and enriched, which has become the key session to enhance the professional skills and the social adaptation abilities for the undergraduates.

The art design education in its transition period is facing the most complex circumstances than ever before. The teachers and scholars of art design education around the country are all doing theoretical researches and explorations. Held by the Ministry of Education, the National Seminar on Practice-Oriented Education on Art and Design was born at the right moment, which specifically discussed the practice education in art education. Pan Lusheng, the dean of Shandong University of Art and Design, summarized the five-session model in the practice of art design education as the course practice teaching, the creation practice teaching, the program practice teaching, the industry practice teaching and the social practice teaching. "If the industry practice teaching is a little commercialized, then the social practice teaching will fully cultivate the ethics

Table 69.1 Student sample structure

Name	Category	Frequency (%)
Gender	Male	42
	Female	58
Subject categories	Engineering	13
	Bachelor of science	24
	Literature	46
	Art Studies	17
Professional type	Normal	61
	Non normal	39

and sense of worth including the self-discipline for design and the social responsibilities" The practice teaching in different levels aiming at by adopting dynamic and overall cultivation, to make the design talents qualified to requirements of the position and the society, integrated into the industry as well as the cultural development, observed the design ethics, undertaken the design tasks and served and led the lives. Pan Lusheng put the social practice teaching at the leading position in the entire practice education.

The exploration the practice education will finally end up in the coordination of the understanding and the requirements of the society, whose basic approach is to study and grow in the society and the market. To be more specific, it can be summarized into five models, which are the professional studio model, the institute model, the program model, the learning and working model and the government-led model, and all of these models aim at the social practice of the undergraduates. The school of Bauhaus in Germany emphasizes that the hands should work under the leadership of brain, that is to say, the practice exercises should be combined with the theory. Only if the practice teaching is linked up with the social practice activities, the students can truly understand the market and the society, and the vitality of the practice education can be stimulated (Table 69.1).

The platforms of the courses of the art and design practice education at different levels are consisted of the teaching of basic knowledge, the training of basic skills, the knowledge of primary materials and the series of courses for the cultivation of the students' professional practice skills, all of which are implemented according to the objectives in talent cultivation and the needs of the students. The schools should guide the teachers and students to collect, clean up and unearth the programs in the market or offered by the cooperated companies, seek for the combination point of the technology researching and developing for the companies and the objectives of talents cultivation of colleges, coordinate the talents resources of colleges with the advantages of the companies including the brand, the fund, the market and the programs, establish the practice teaching base in and out of the colleges and standardize and improve the regulations of practice teaching.

69.3 The Social Practice Education for the Undergraduates Needs to be Professionalized

The social practice activities held by the art institutes should enhance the professionalization to adapt to the mode of cultivation of the art students, so the direction of social practice is put forward as professionalization, which means the social practice education for the undergraduates needs to be professionalized.

69.3.1 Define the Importance of Social Practice

The social practice should be altered from the pure political education into the organic part of the professional education, which should be incorporated into the compulsory subject of the talent training schemes with proper credits, for example, the innovation credit system, that is to say, only if the undergraduates get complete corresponding credits identified by the colleges, should them got the qualification for graduation. To ensure the status of social practice and to strengthen the identification of the importance of it by the colleges, the teachers, the students and the parents are the method to make it the regular work of talent cultivation (Fig. 69.2).

69.3.2 Coordinate the Resources to Unified Plan the Social Practice Activities

To guarantee the social practice can be developed in a long term and get real efficiency, the colleges should make a unified social practice plan, such as the social practice base, the scientific research plan, scientific research fund and some coordinated sets of measures. From the aspect of the colleges, the academic system and the faculty system should be cooperated to bring the networking and professional advantages including stimulating the students' questioning awareness, designing and researching work and writing papers to summarize their social

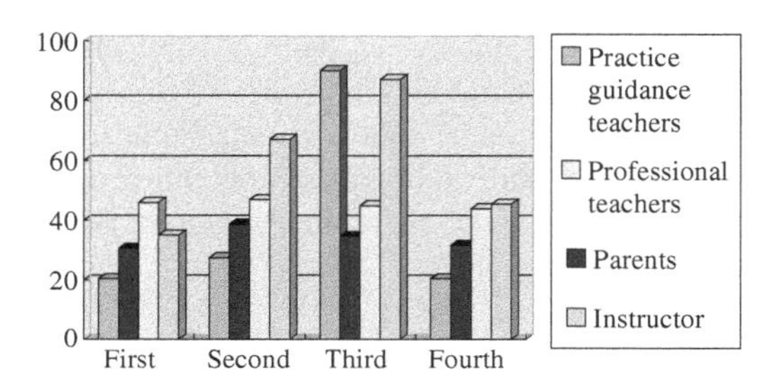

Fig. 69.2 Network mode of practice teaching responsibility

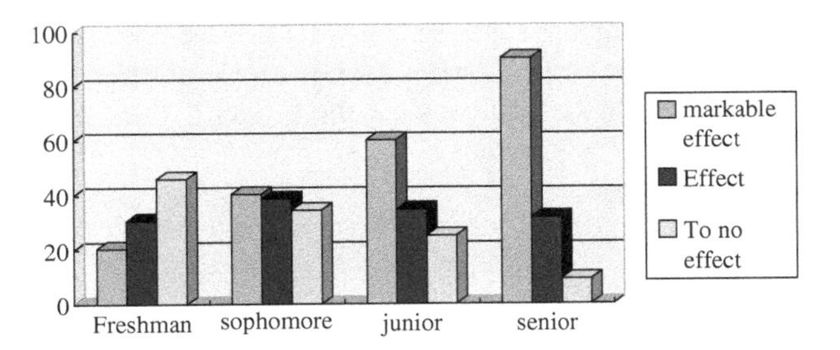

Fig. 69.3 Network mode of practice teaching

activities. The faculty should positively corporate with the organization of the activities, the setting of proper evaluation system and the promotion of the achievements. "The academic system together with the faculty system to implement the social activities", it is the necessary condition to bring the social practice into the professional education, as well as the key point of the social practice to act well (Fig. 69.3).

69.3.3 Arouse the Enthusiasm of the Teachers

The social practice activities are in urgent needs for the guidance of the experts of the original subject and the participation of the professional teachers, so that we should take full use of the teachers' theoretical knowledge and professional abilities to strengthen the theoretical study through social practice and to implement the dual-supervisor-system with both of the professional tutors and the instructors. To reinforce the construction of the team of the professional tutors, we should improve the appraising system of the teachers, give them corresponding workload and offer them equivalent treatment to arouse their enthusiasm.

69.3.4 Emphasize the Service Functions of Social Practice

Through practice activities, the students can fully understand the society and get into real life broke the walls between the colleges and the society, experience the beat of the city and thus find the fertilized field of art creation. They should take full use of their spare time to do the researching, collecting and creating work to promote their independent thinking and problem solving abilities, strengthen their employment and business power and refine their art-building. Through the practice activities, we can solve the traditional problems that the art education is broken away from the society and become an individual and closed type of teaching.

It can also carry out the thought that art should serve the society and the people. We should cultivate the talents with social responsibilities and innovation thoughts, who can find, create and promote the beauty.

69.4 Conclusion

The social practice is a systemic program of talent cultivation, which needs elaborate organization of the colleges, the guidance from the teachers, the attendance of the students and the support of all circles of the society to make it into a well-acted mechanism. The professionalization of the social practice is neither disposable nor efficacious forever. The developments of the society and the social practice itself require it to be a continuous and endless activity. We should constantly open up new thoughts and explore new approaches to alter the social practice into an exoteric and innovated type.

References

1. Liang S (2010) The exploration and practice of the model of institute for the cultivation of the major of art design. Decoration 8:32–39
2. Wang Y (2010) The research and practice of the combining learning with research and production for the major of art design. Chin New Technol Prod 3:25–29
3. Judith (2002) Bauhaus. China Light Industry Press, Peking, vol 15, pp 255–261
4. Liu J (2006) The analysis of the innovation credits in colleges. J Taiyuan Norm Univ 5:85–96

Part IV
Sports Management and Application III

Chapter 70
Transmission Mode Analysis of Sports Tourism Resources Based on Microblogging Platform

Lin Zuo

Abstract As a new network communication tools in today's society, Microblogging brought the explosion of development, and it is now in a rapid development stage, which is more and more popular with the customers and favor. People can understand each other not met by the Microblogging. In view of the Po platform, we can take a new visual angle for the sports tourism resources. Microbloggingost can promote the development of the local sports tourism industry. In this paper, we took the sports tourism resources in east island of Guangdong province for example by using the analysis method of SWOT. On the basis of Microblogging platform, the author put forward the corresponding propaganda pattern in order to promote sports tourism industry in east island.

Keywords Microblogging platform · Sports tourism resources · SWOT analysis · Transmission mode

70.1 Introduction

Sports tourism in China is a new tourism product, and the corresponding sports tourism is a new subject. Sports tourism is the combination of sports and tourism fitness methods [1]. Sports tourism resources as a natural part of the tourism resources, is a complex set of cultural tourism resources, the concept of sports tourism resources in nature and human society is that, for all the sports tourism development, can produce human movement tourist attractions, can produce economic benefits and social benefit and environmental benefits of all kinds of

L. Zuo (✉)
Wuyi University, Jiangmen 529020, China
e-mail: lin_zuo88@163.com

W. Du (ed.), *Informatics and Management Science VI*, Lecture Notes in Electrical Engineering 209, DOI: 10.1007/978-1-4471-4805-0_70,

things and combination of factors. Guangdong province is currently occupying the first in the national economy, the Pearl River delta island for an important development strategic geographical position. East island of Guangdong province is located in the southwest of the eastern Leizhou Peninsula, history; SuiXiXian state jurisdiction belongs to ray. Sports tourism is sports and tourism mutual fusion crossover part, it emMicrobloggingdies the social sports and tourism social [2]. Sports tourism belongs to the social sports a branch of the industry, tourism is an important component of the special tourism is a, is a kind of human social life of the emerging tourist activities, and the concept have broad and narrow the points. From the broad sense, sports tourism is to point to the tourists in the tourism of Microbloggingdy and mind, engaged in entertainment exercise, competitive race, exciting adventure, rehabilitation care, sports watch and sports cultural exchange and tourism, tourist enterprise, sports enterprise and society the sum total of the relationship between; Speaking from narrow sense, it is to meet and adapt to the tourists various special sports demand to sports resources and some sports facilities for conditions, the tourism commodities form for the tourists in the course of travel and provide harmony fitness, entertainment, leisure, communication is equal to one of the service, the tourists to get harmonious development in Microbloggingdy and mind, is promoting social material civilization and spiritual civilization development, social and cultural life of the rich purpose a social activities [3].

Guangzhou successfully held 2010 Asian games, for Guangzhou and the Guangdong area of sports and tourism industry in the breadth and depth of level cooperation, fast promoting the development of sports tourism offers a great opportunity. First, from the time at state level, the Asian games will promote the sports tourism increase and sustainable development. Second, from regional space at state level, the Asian games will promote the sports tourism to realize regional linkage type development. Third, from industry at state level, the Asian games will promote the sports tourism industry more competitive and quality optimization. In addition, the Asian games for the Guangdong area tourism exchange also can have direct influence. In short, sports tourism has a broad development prospects. The Asian games held in Guangzhou, Guangdong, for promoting and pan-pearl river delta region and the entire Chinese sports tourism development in south China is a very good opportunity, no matter for the area of sports tourism image or tourism products and service quality of ascension is of profound significance.

The Microblogging is a through the attention of real-time information sharing mechanism short GuangMicrobloggingShi social network platform. The Microblogging as the Internet's new things, having very strong penetration and attractive user groups, has the widespread fame and influence, make it become the influence thought and behavior of a powerful network power, it's happened and pop influences people's consumption idea, way of life, accept information habits and channel, meet all kinds of the diversity of the crowd requirements and personal demands, through the construction of the new media, the implementation of the sports tourism ideology, public opinion guide and Microblogging platform and realize the docking, through the mutual infiltration and gradual fusion effect, cleverly propaganda sports tourism industry promotion quickly.

70.2 Microblogging Platform in the Promotion of the Influence

Microblogging, namely the Blog (Micro Blog) for short, is a relationship based on users of information sharing, communication and access platform, the user can through the WEB, WAP and various client component individual communities to aMicrobloggingut 140 words updated information, and realize the real-time share. The earliest and most famous Microblogging is the United States' twitter, according to relevant public data, this product in the world already has 75 million registered users. In August 2009, China's biggest web portal sina.com launch "sina micro Microblogging" closed beta version, become the first portal sites provide a Microblogging service web site, the Microblogging formally enter the Chinese Internet mainstream people view [4].

The Microblogging platform has a great influence publicity flexibility, and content quality highly relevant. Based on the influence of the existing users by the number of "focus". The user release information's appeal, the stronger the news, the users interested, concerned aMicrobloggingut the number of users, the more the greater influence. In addition, the Microblogging platform authentication and recommend the itself also help to increase the number of "concern".

Influence coefficient Fj formulas for the formula:

$$F_j = \sum_{i=1}^{n} bij \Big/ \frac{1}{n} \sum_{i=1}^{n} \left(\sum_{j=1}^{n} bij \right) \tag{70.1}$$

Among them, $\sum_{j=1}^{n} bij$ is reported for the husband of fathers inverse matrix of the first j listed the sum $\frac{1}{n}\sum_{i=1}^{n} (\sum_{j=1}^{n} bij)$,

For listed the Cardiff inverse matrix column and reported the average.

According to the survey on the network, the random questionnaire, get as shown in Fig. 70.1 shows the Microblogging platform visibility of the change.

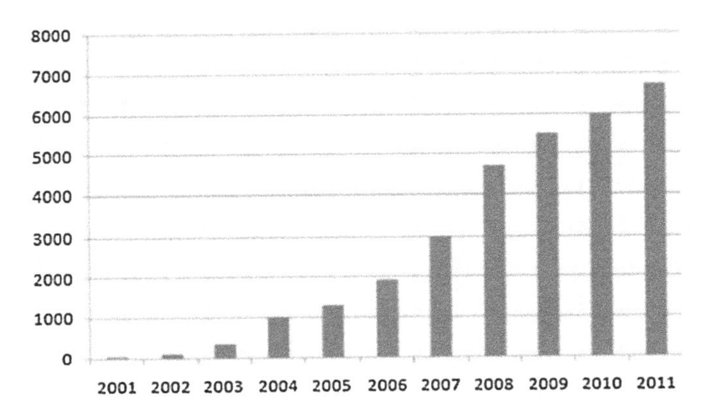

Fig. 70.1 2001–2010 Microblogging platform visibility

We can see from the aMicrobloggingve graph, in Microblogging increasingly development today, the Microblogging user is more and more, and Microblogging platform visibility also more and more big, therefore, in the network science and technology development today, the Microblogging platform as a means of publicity, it can lead to strong benefits. Use the Microblogging platform to promote the Guangdong province east island of sports tourism industry, can play a very significant role.

70.3 The SWOT Analysis on Sports Tourism Resources on East Island

SWOT model that situation analysis is often used for enterprise strategy, analysis of competitors and so on. Focus primarily on analysis the advantages and disadvantages of the strength of the enterprise itself and its and the comparison of the competition, opportunity and threat analysis focus on changes in the external environment and to the enterprise may affect [5]. Figure 70.2 shows SWOT model schemes.

70.3.1 The Advantage and Opportunities of Sports Tourism Resources on East Island

East island tourism resources are very rich, region has more on the beach, hot spring tourism scenic spots, and environment is appropriate, has four seasons selectivity, is a good place to be on a vacation the traveling. At the same time, with convenient transportation east island with conditions [6]. East island of Zhanjiang bay sea-crossing bridge is the bridge after another major cross-sea Bridges highway engineering, including east island of sea-crossing bridge, the Guangdong head to wei law at port of the highway engineering, connection has built up the traffic Zhanjiang dredging port highway, Chongqing Zhan highway, 325 national

Fig. 70.2 SWOT model schemes

highway and line, 207 lines and provincial road 373 line, 374 line connected together to form around Zhanjiang urban areas, Zhanjiang port area and east island of iron and steel base of the modernization of the highway network. East island resort beach water clean, the sand moderate thick soft, clean, and will have a variety of beneficial to human Microbloggingdy minerals, sand bath can cure various skin diseases.

East island as the sixth largest island of Guangdong province, and on the island of have not only high quality, scale, the number of on the beach and island LinZiRan reserve shape, and the rough places, winding mountains, rich and many different colored stone cultural landscape resources, landscape resources combination advantage. In addition, the east island area of the pearl river delta area is the best seaside resort, in the domestic and international sports tourism market has broad prospects, so has the rich island east island of cultural development prospects.

70.3.2 The Disadvantage and Challenges of Sports Tourism Resources on East Island

East island of Guangdong province in the tourism, transportation and communications sides main facilities investment, less western Guangdong coastal highway project has not been completed, the serious influence consumers to east island in sports tourism; In addition, wide bead light rail project is still in construction, inside short time cannot meet the needs of a large number of times ZiJiaYou tourists; And, in the east of the main island of sports tourism attractions, also have no city sightseeing bus to the station and tourist attractions, attractions and the attractions of train travel between only also lack, still need government support.

The hotel (hotel) restaurant industry position setting is not reasonable. Low-level hotel (hotel) position in the city common, to bring inconvenience passengers, can't let passengers in-depth understanding of overseas compatriots understand China's unique culture, ornamental birds in bird paradise best time should be early or late, but no nearby hotel, only a few of the cheap hotel, it is difficult to make the passengers in the sports tourist attractions enjoy oneself, such natural reduced the number of tourists. So suggest the government take active measures to promote overseas Chinese characteristics of construction activities of the hotel, and overseas Chinese unique culture to adapt to each other, make passengers appreciate overseas Chinese culture. In addition, the lack of sports tourism culture east island of entertainment. The government and folk to often run some size of local culture and recreational activities, such as lion dancing, dragon Microbloggingat, floating color, the Lantern Festival, the organization and local style of game performance, improve the urban culture art grade, and not every year of the dragon Microbloggingat places, such as special cultural activities to hold such entertainment activities. Sports tourism entertainment activities can greatly stimulate the curiosity, and to attract tourists to the east island of sports tourism activities on.

East island first traveling scenic area almost no corresponding overseas Chinese unique tourism commodities, therefore, should make some culture value-added tourism souvenirs, has the very high science and technology content, especially in kaiping castle, should develop the Marine tourist commodities products and local characteristics of commodities; Because of the backward marketing method, the domestic and foreign traveling in the fierce market competition, the lack of enough promotion means and strong marketing means adapt to market changes the grim situation, caused the city's overall image not outstanding, in the international, domestic popularity is not high, in addition to mining and play east island of sports tourism resources, development with local characteristics and competitive products. Although east island of sports tourism image and tourism environment present situation of the gap, but can through the integration of sports tourism resources to shape in the international, domestic image.

70.4 The Transmission Mode of Sports Resources Based on Microblogging Platform

70.4.1 The Cooperation Based on the Development of Sports Tourism Resources in Guangdong

Economic tourism integration, the integration of tourism, on the basis of east island of Guangdong province as a whole tourism development integration is the strategic target and key measures to improve east island of the core competitive power of the tourism industry in the development of sport in the realization of the economic integration, in close cooperation with the economy of Guangdong province, and establishing an effective partnership in the tourism management organization, coordination, control and prevent the common management, three places of interactive exchanges consultation, main is the transportation system, not only to make traffic travel unimpeded, in "sports tourism economic integration", logistics and information sharing the realization, and in the east of the island tourism, based on the construction of the famous international efforts into sports tourism resort.

70.4.2 The Transmission Mode of Sports Tourism Resources

The Microblogging platform mainly through the post of publicity east island of sports tourism industry, in the post process, a valid key emotion is especially important. In order to be able to accurately the extraction of emotion information in connection with the product before the need to set the mood of the mood of the main analysis, the word is mainly to users of emotion and attitude of the words, in addition to the common emotional vocabulary, specific products and corresponding professional vocabulary, this need professionals to participate.

Satisfaction of the market is said, east island of sports tourism product in Microblogging platform under the Xi, Wi said in a period of time to collect the post of the sports tourism product interest information, when this post express satisfaction with Xi to 1, when antipathy to 1 Xi, not on the clear attitude Xi is 0. The period of time of the east island of sports tourism product satisfaction for:

$$\frac{\sum_{i-1}^{N} xi \times wi}{\sum_{i-1}^{N} wi} \tag{70.2}$$

Attention is on the market of said, users east island of sports tourism product made of the proportion of the evaluation. For a b said in a period of time, for sports tourism products to collect interest for resent the number of posts, h for satisfaction of the number of posts, N for the sports tourism product information in the number of posts, then at that time, the east island to the attention of sports tourism product for:

$$\frac{b+h}{N} \tag{70.3}$$

The characteristics of the need for the Po platform, attention to concern said sports tourism product of netizens ratio, satisfaction for the east island of said sports tourism product of interest. Guangdong province and to estimate island of sports tourism resources market conditions. Finally, through the computer simulation, the Fig. 70.3 can reflect the net friend to the responsiveness of sports tourism industry.

70.5 Conclusion

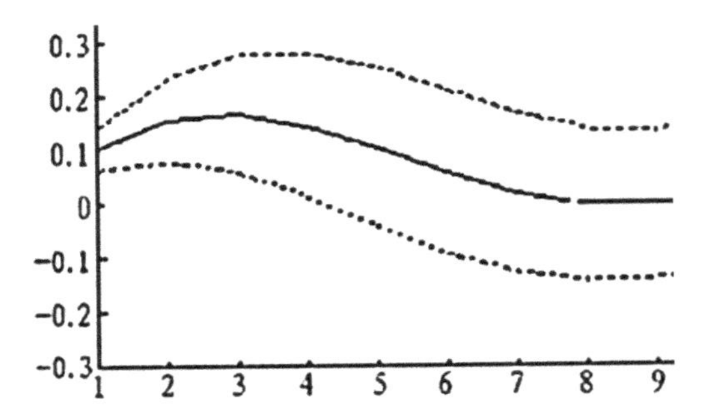

Fig. 70.3 The function of impulse response

Sports tourism is the mutual fusion crossover part of sports and tourism, which can bring the social sports and tourism social some favors by Microblogging. Sports tourism belongs to a branch of the social sports industry, and tourism is an important component of the special tourism, which is a kind of human social life of the emerging tourist activities. Through analyzing the propaganda pattern of sports tourism resource about east island of Guangdong province s based on the sis the Microblogging platform, we can promote the development of sports tourism industry on east island of Guangdong province, so that more of the masses of the visitors enjoy the fun of sports tourism brings, and for the tourists in the course of travel provide harmony fitness, entertainment, leisure, communication, which is equal to one of the service. It can make the tourists and mind get be harmonious development in the Microbloggingdy.

References

1. Fugao Jiang (2003) The status and sustainable development of town residents' sports tourism in Shandong province. Qufu Norm Univ 12(3):187–193
2. YuSuMei (2007) Some advice on sustainable development of sports tourism in our country. J Sports Cult 2(9):59–60
3. Minhui X (2005) The development research on Hainan sports tourism, Beijing Sport University Press, Beijing 12(2):73–78
4. Zhou H (2007) The planning analysis on the development of sports tourism resources. Sports Sci 18(2):47–49
5. Zhou L, He Y, Wang J (2008) Emotions analysis. Comput Appl 2(28):25–28
6. Zhang F (2010) The high level game let the sports "prosperous" up. Yulin Daily 06(27):29–37

Chapter 71
Analysis of Anaerobic Ability Basketball Player in Vertical Leap

Xiaorong Mi

Abstract In Ideological and Political Education, there are two forms of characteristics: conscious and unconscious, but in view of the community and academics' increasingly concern about the ideology and its nature, we can promote the understanding and practice of the ideology, which has great significance. Concern about the ideology can give full play to the ideological function. In this article, based on the concept of political ideology, we determine the methods measuring the ideology, make an analysis of existing measures of political ideology and improvement measures; according to the IRT model, we analyze the structure of the political ideology, and make clear measures based on the potential problems while giving its guiding significance our effective analysis.

Keywords IRT · Political ideology · Quantitative methods

71.1 Introduction

Ideology is a general awareness of the phenomenon in the society as a whole, which has a certain structure. In the construction of socialism, we need a profound and practical understanding for the ideology to deepen self-awareness, and to contribute to the research carried out and the ideology practical play to enhance its utility [1].

Given the current political polarization properties, it has been renewed people's attention and focus on the political ideology. In the history, ideology and partisan

X. Mi (✉)
School of Marxism Studies, Southwestern University, Chongqing 400700, China
e-mail: xiaorong_mi@163.com

W. Du (ed.), *Informatics and Management Science VI*, Lecture Notes in Electrical Engineering 209, DOI: 10.1007/978-1-4471-4805-0_71,

have close relations, and by view of the community and academics' focus on the nature of political ideology, but surprisingly, political ideology has so little actual structure and meaning, which is still a "black box". What implications dose awareness of self-placement has? Whether it is rooted in the problem of choice, which is often assumed or largely symbolic as the ideological label? To what extent has awareness measure shaped our understanding of the impact of allocation and policy?

71.2 Political Ideology and Its Measurement

Political ideology is about historical stage people live in; it is people's awareness of the political system, political actors, as well as life, relationships, etc., including these organizations, the groups and individuals' position in the political system the role played by and between each other and the comprehensive relationship [2]. The basic composition of the political ideology often leads people to think of political theory, political ideology, moral, legal, point of view, and education.

Ideology is a social co-existence in a system; it co-exists with economy, political structure, and interaction in the social structure. At different stages, different environment, culture, etc., will have a different political ideology system, and we define ideology as a relatively independent form of study, and thus it's better able to play its overall effect. Therefore, from a functional perspective, it is divided into basic theory, ideological era architecture, the concept of consciousness form the core value as well as external. Social existence through ideology can be effectively reflected, but it will in turn also have a very large role in the existence of society. Political ideology has a strong class nature and the economic base determines the superstructure.

We mainly through doing research on those thoughts easy to be generally accepted to study the structure of political ideology, mainly from the emotional point of view, and rational analysis perspective to analysis of the ideological structure; make joint analysis of the community system of thought. As for community thinking, ideology usually requires two specific rational and emotional understanding and analysis [3]. This mainly reflects the ideological consistency and completeness in the theory and practice, and these ideological systems are generally recognized by everyone, with no logic errors. Ideology is associated with the era in every stage, so people in different stages of the era should from carry out a comprehensive evaluation from different political, economic and other aspects, and they should also analyze and consider various viewpoints and attitudes. Analysis of the ideological theory should be linked with the background and environment; so with comprehensive analysis of its characteristics, we can get better feedback on the current outlook, values and philosophy.

71.3 Measurement of Thinking

Previous studies the ideological orientation is usually dependent on one or two measures: (1) self-placed in the conservative liberalism; (2) based on the attitude of a number of issues, usually create factor analysis, and to provide improved measures of ideology. By IRT sequence model analysis, we will use problem-based ideology scale [4].

First, we discuss existing measures' limiting ideology. (1) Self-placement measures: survey questions usually come from the following: When it comes to politics, you usually think you are very loose, free, slightly loose, moderate, slightly conservative, conservative, very conservative, or do not you think that you are the redundant one? This suggests that the conservative and classic political science liberalism label may reflect the specified ideological groups may recognize or just show a matter of preference, rather than the candidate. However, recent studies tend to assume ideological self-employment in voting behaviour, which fully reflects the public's policy preferences. It is an open experience in considering the more polarized political environment preferences to confuse ideological self-placement problem. (2) The ideological issues are clear and prominent. According to the survey sample design, use the idea in pre-election to measure the modal split and the question wording split. Although the various modes of reaction around the distribution are not very different, there are major differences in the branch and response scaling. There is the possibility of considerable difference in which it was thinking to their labels. In the original survey, 22.2 % of people who answer "do not know"; problem with the original branch, 33.9 % chose this response [5]. In fact, the reaction of the individual level comparison before and after the survey (only applies to the size format of the response in both studies) found that only 43 % is consistent with the scale of the response; needless to say, it is difficult to figure out how to interpret and understand these different ideas of self-placement problem. For example, how do we explain a large proportion and we do not know the response Table 71.1.

In fact, the study found that one dimension is a quite sufficient characteristic of political forms, accurately predicted by a single ideological dimension. However, the public has long been considered to be a better description of ideology that requires at least two aspects, such as the first contrast ideology that can be traced back to the New Deal politics, focusing on the economic well-being, including

Table 71.1 Ideology survey data

Percentage	Independence (%)	Slight independence (%)	Moderate (%)	Slightly conservative (%)	Conservative (%)
0–21th	41.8	19.8	29.4	5.0	8.3
22–40th	23.5	17.4	39.3	11.7	13.7
42–60th	13.4	13.5	41.2	18.5	18.0
62–80th	8.3	7.4	35.5	23.5	28.4
82–100th	3.3	4.4	17.4	23.8	55.1

health care, unemployment, social security; and the second aspect, in nearly a decade, involves cultural or value, such as abortion, school prayer, and homosexuality. There is no doubt that these dimensions are folded into a single dimension of the party's elite, but the existing research makes people have reason to suspect a dimension sufficient to characterize the policy preferences of the public, even in the contemporary political context. If the preferences of the people are two-dimensional, then how can we explain the ideology of self-placement as well as some personal and economic policy, and social policy preferences are the preferences of others Table 71.2.

Alternative measures relying on of the ideological construction scale, and the preferences problem in academic research is built by basic researchers. The classic factor analysis of the underlying purpose is to acquire understanding correlation between the latent construct, which may account for the various problems. Although the factor analysis is a common size and has a variety of ideological, it has some limitations of the method, especially that other data cannot effectively express their political ideology. First, the classical factor analysis is designed for continuous variables, binary discrete variables, procedures, or other types of technology, which can produce incorrect results. Second, the traditional factor analysis cannot easily handle missing data and factor structure is sensitive, especially for missing data Fig. 71.1.

In this paper, we estimate the improved potential ideological problems are issues-based measures. IRT method allows the use of item response (nominal,

Table 71.2 Two dimensions ideology survey reference data

	Difference between the parameters	95 % HPD casting range
Aid to poor consumption	1.483	(1.308, 1.648)
Government services (branch)	1.601	(1.312, 1.843)
Government services (scale)	1.489	(1.298, 1.710)
Assurance (branch)	1.317	(1.091, 1.556)
Assurance (scale)	0.991	(0.830, 1.168)
Health insurance (branch)	1.228	(1.001, 1.445)
Health insurance (scale)	0.928	(0.777, 1.093)
Public school spending	1.283	(1.087, 1.452)
Social security expenditure	0.880	(0.754, 1.022)
Tax cuts from the surplus	0.399	(0.298, 0.508)
Welfare spending	1.062	(0.941, 1.214)
Affirmative action	0.920	(0.789, 1.057)
Environment (branch)	0.919	(0.733, 1.126)
Environment (scale)	1.136	(0.944, 1.334)
Gun control	0.966	(0.813, 1.084)
Abortion	0.454	(0.341, 0.553)
Abortion with parental consent	0.393	(0.266, 0.505)
Death penalty	0.460	(0.352, 0.582)
Gay adoption	0.851	(0.710, 1.018)
The role of women (scale)	0.651	(0.502, 0.836)

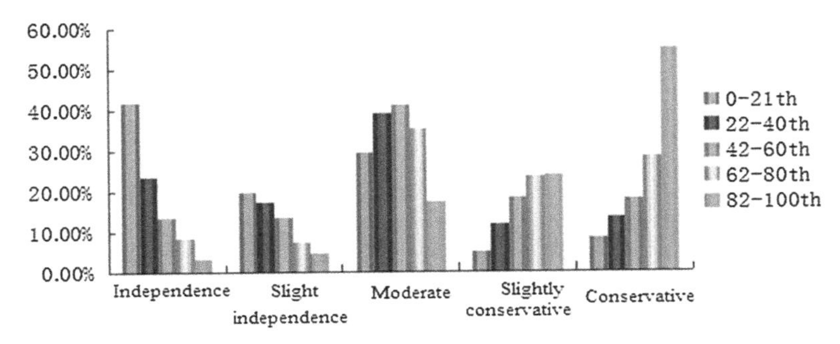

Fig. 71.1 Performance ratio of ideology people

binary, ordered, continuous) and missing data, only the construction of potential measures and the existing data. The most striking is that the IRT model, the potential measurement ideas directly modelled, and therefore by inference, just like any other model parameters. Therefore, we can directly estimate the individual's latent ideology score and uncertain assessment score. With this uncertainty, then we can consider potential measures included in the multivariate model and our vote choice.

71.4 Data and Methods

We use analysis of ideology to construct measures based on ideological issues. Through surveys, we capture the important issues of theory which may be based on personal thoughts. In view of this data structure, IRT method is ideal, because it calculates the potential scores for each respondent to provide information.

We propose IRT models relative factor analysis and it has the benefits, so we now turn to the more formal representative of our experience. The idea of building potential ideological measures as a function of individual preferences can be observed through the ordered item response model. Set i = 1..., N means index individuals and j = 1..., M means index problems. Let K = 1..., KJ index (ordered) response categories J; in our data, KJ is ranging from 2 to 7. So, our model is:

$$\begin{aligned}
\Pr(y_{ij} = 1) \quad &= F(\tau_{j1} - \mu_{ij}) \\
\Pr(y_{ij} = 2) \quad &= F(\tau_{j2} - \mu_{ij}) - F(\tau_{j1} - \mu_{ij}) \\
\cdots \quad &\cdots \\
\Pr(y_{ij} = k) \quad &= F(\tau_{jk} - \mu_{ij}) - F(\tau_{j,k-1} - \mu_{ij}) \\
\cdots \quad &\cdots \\
\Pr(y_{ij} = k_j) \quad &= 1 - F(\tau_{j,k_j-1} - \mu_{ij})
\end{aligned} \tag{71.1}$$

Two dimensions questionnaire is calculated by modeling and data investigation Fig. 71.2.

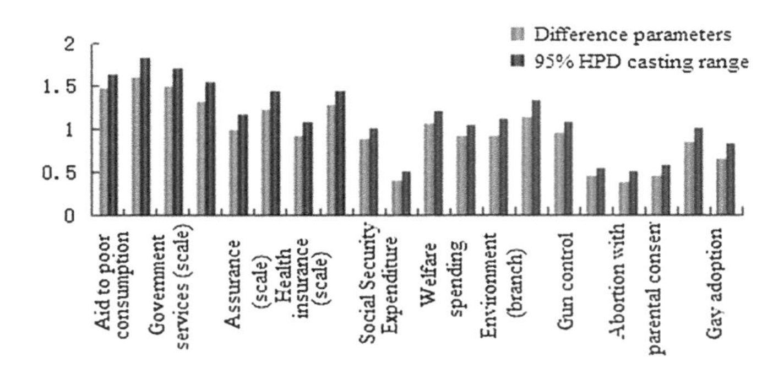

Fig. 71.2 Two-dimensional ideology characteristic scale

71.5 Conclusion

Through the conscious and unconscious characteristics of the two forms of research and ideological and political education, we can promote the understanding and practice of the ideology, which has very important significance. Through the determined ideology measuring, we make analysis of existing measures of political ideology and improved measures of the ideology. Based on IRT models, the structure of the political ideology is analyzed and we make clear measures based on the potential problems and also make effective analysis of its political significance, which helps to focus people's attention on ideology in order to give a better play of the function of ideology.

References

1. Xiao G (2009) The characteristics of innovation on Chinese ideology from transformed political science. J Zhejiang Province 4(4):11–14
2. Wang X (2010) Ideology of misunderstanding and its realistic enlightenments. J Zhejiang Normal Univ (soc sci) 23(2):32–35
3. Shi S (2010) Theory of ideological and political education and the unity of ideology. Explore 3(2):76–78
4. Liu H, Zhang J (2008) Function and social ideological trend lead of ideological and political education. Theory J 31(7):30–37
5. Liu J (2007) Discussion on the basic structure of ideology. The ideological Polit Work Res 3(12):47–49

Chapter 72
Research on Skill Level of Basketball Free-Throw Based on the Viewpoint of Biomechanics

Fenglin Dong, Sufei Yang and Peng Pu

Abstract From the basic mechanics theory, taking the angle of sport biomechanics as the foundation, this paper analyzed the principle and method of basketball technology. After researching on the percentage of biomechanical factors of basketball free-throw, six angles are high, medium and low grade three technical foul shot skill level quantitative analysis from the free-throw accuracy, free height ratio, center of gravity, trunk inclination ratio, pitching angle and pitching speed. The purpose of this paper is designed to combine the principle of biomechanics knowledge and improve on the technical ability level of basketball free-throw, which could effectively help basketball players increase the hit rate of free-throw.

Keywords Biomechanics · Basketball · Skill level · Influencing factors

72.1 Introduction

In the basketball game, as the key element, shooting free-throws already caused social all circles and education circles attention. But during the game there are basketball free-throw percentage decline, mainly by the two major categories of motor skill control task. First of all, the first task required accuracy penalty.

F. Dong (✉) · P. Pu
Sports Education Department, Teda-Tianjin Economic-Technological Development Area Polytechnic, Tianjin 300457, China
e-mail: fenglin_dong@163.com

S. Yang
Department of Physical Education, Hebei University of Technology, Tianjin 300130, China

W. Du (ed.), *Informatics and Management Science VI*, Lecture Notes in Electrical Engineering 209, DOI: 10.1007/978-1-4471-4805-0_72,

Second, it requires the basketball athlete's pitching speed point. This study analyzed the selected different skill players hit rate of free-throw in the biomechanical parameters [1]. Most discussions of the hit rate of free-throw biomechanics qualitative observation or mathematical derivation based on experimental evidence compare. There are numerous researchers agree with the "high" and "secondary" throw angle shot angle, however, there are a lot of people in favor of "low" free-throw angle [2].

Through this study, explore new program under the guidance of basketball training course, compared with other sports curriculum of sports on College Students' physical quality acquired role, as well as the psychological health factors of college students is active guide, stimulate on physical education interest, fosters the lifelong physical education habits, so that the students' physical health, personality, mental state is good, in the basketball training to achieve certain positive to promote physical and mental health purposes [3]. In this paper, according to the characteristic of basketball sports, from the train body stress ability, alleviate the psychological tension, improve emotional self-control, exercise a strong will and improve interpersonal relationship and other aspects, demonstration of basketball exercise on mental health of college student's role in promoting. The basketball movement to develop physical stress ability. The basketball movement is smooth, the need for strong physique, quick to judge and quick response, so it can cultivate students ability to stress, conducive to cope with the emergency events come unexpectedly. Basketball can alleviate the psychological tension. The basketball movement to increase the emotional self-control. Mood and emotion is the human to the objective things whether conforms to the subjective and psychological experience is accompanied by specific physiological responses and the external manifestation of a psychological process. They will directly affect the body's physical and mental health. The mood and emotion, human psychology can crab and physical energy exchange formed a significant interaction between. In basketball, the outcome of the game will give participants or spectators to produce strong, rich and varied emotion [4]. Against the complexity, diversity and strong, easy to make college students often appear a variety of emotional state. Therefore, the basketball movement process must know how to control their emotions, to overcome the various internal and external factors of the interference field. This paper firstly analyzes the students' existing mental health problems, these problems are mainly anxiety, loneliness, weak-willed, communication ability and low self-esteem and other aspects; then, from the training of physical stress ability, alleviate the psychological tension, improve emotional self-control, exercise a strong will and improve interpersonal relationship in five aspects, discusses basketball exercise can effectively promote the development of College Students' psychological health; finally, according to the Hunan province Xiangtan county first middle school teaching cases show the effectiveness of the method. Conclusion: basketball in college students' psychological health has a good role in promoting [5].

72.2 Influencing Factors of Basketball Free-Throw Based on Biomechanics Analysis

The basketball players' hit rate of free-throw biomechanical factors mainly contains six factors, namely, free-throw accuracy penalty height ratio, centre of gravity, trunk inclination ratio, pitching angle, and pitching velocity. In the process of penalty, penalty hit has a very important influence factor, namely the pitching stability or accuracy. Accuracy is dependent on good balance, but to balance the need to focus in on a support base. There is also discussion of trunk tilt variables to maintain stability. Be a head, back, hips, vertical alignment. Cautious to lean forward, can improve the hit rate of shooting. The four women's basketball team players, for the strength limit the hit rate of free-throw players maximum sustained winds near the need of production. The characteristics of high-speed throw conclude the following two aspects [6]:

(1) Focus shifted back, then forward displacement;
(2) Upper body bending, stretching.

The expert thinks of high consistency between the release is desirable. Because it reduces the minimum distance of projection angles, the minimum velocity projection, it increases the error. Suggestions on how to realize the includes [7]:

(1) The use of more flexion at the shoulder;
(2) The more extended elbow;
(3) To release the ball long range.

These include three elements: pitching angle, speed and accuracy; trunk tilt, the centre of gravity of the three process elements.

72.3 The Experimental Method

The three college students of mutually exclusive female as the research object, as the study task. Including high skills in the nine member of the group; secondary skills group: seven non scholarship in the school basketball team players; low skill teaching class of 9 members.

Each participant's testing protocols includes [8]:

(1) Control the warm-up period;
(2) The 20 precision testing: throw test;
(3) Exercise performance marking characteristics;
(4) Additional warm-up time;
(5) Hit three free-throws to test records for analysis.

The test records from the main body on the right side of the 23 m line extension. The camera is the speed of 64 frames per second times and 4 ms exposure time. Point on the edge of basketball and ends with digital pioneer motion

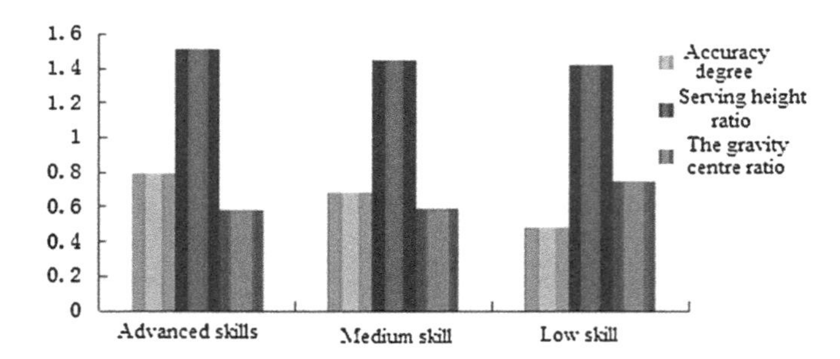

Fig. 72.1 The comparison chart of accuracy degree, height and gravity center ratio

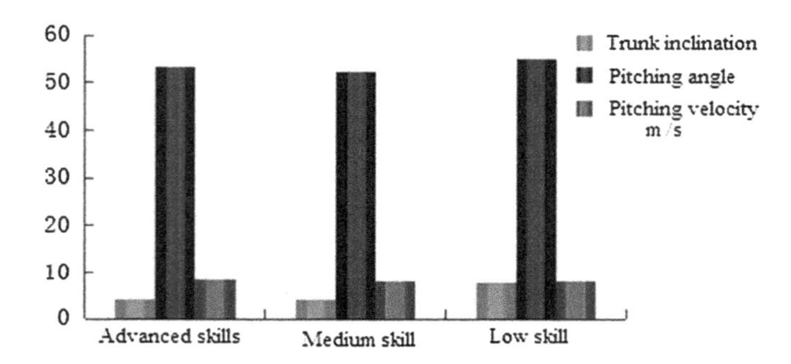

Fig. 72.2 The comparison chart of trunk inclination, pinching angle and pitching velocity

analyzer. Analysis on basketball projection characteristics it is necessary to know the location of the center of the ball. One three angle measurement method, using three coordinates of data point's ball edge ball, find the centre (Figs. 72.1, 72.2).

72.4 The Experiments and Structure Analysis

It is found in the horizontal and vertical components of the displacement of the center of the ball, passing the time between frames, the equation of motion. The resulting from the assembly speed, ball speed calculation. Free-throw angle, the angle speed and level (Fig. 72.3).

It can be seen from the figure, the sphere of the horizontal velocity: $V_h = V_0 \cos\theta$

Into the basket for the vertical velocity:

$$V_v = \sqrt{V^2 \sin^2\theta - 2gh} \tag{72.1}$$

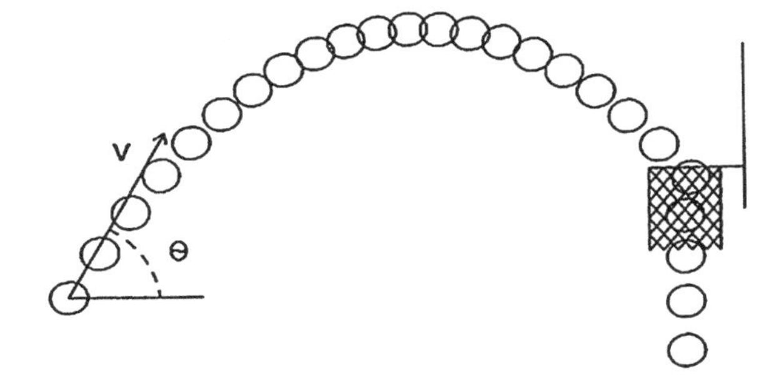

Fig. 72.3 Schematic diagram of the basketball free-throw angle

As is known, free-throw distance in the distance, the height of the release point throws the pitching angle and speed must be reduced, to ensure that the hit rate of free-throw.

Trunk inclination, verticality measurement is O, forward is. The base was defined as the support from the rear horizontal distance ahead of the toe ankle. Distance from the center of gravity is in the wake of the ankle support base length, with production center is divided into weight ratio. Calculation of the release rate highly release height divided by the ball center height shooter. Variance analysis and 0.05 level of significance were used to determine if the similar execution parameters.

Each skill group members held during the preparatory phase of the first time the ball backward shot. High skills, a member of the group, with extreme hip and shot in the initial part of the flexion of the knee. In 0.2 s before release, her right thigh is inclined 8° below the horizontal. Each skill set of mean value and standard deviation of the estimated for each selected 6 variables are listed in Table 72.1. Statistical data shows, at different skill set free-throw percentage accuracy have significant differences.

Once upon a time table can be seen, the subjects of anxiety factor is reduced considerably, and several of the students physical quality, low exercise capacity, low exercise capacity of students in general anxiety level rises apparently, through exercise, healthy feeling led to positive changes in heart function, exercise students' anxiety, students mental disorders were significantly decreased after exercise, self-concept has significantly strengthened, the exercise level, mental health level in the presence of strong interaction. College students, depression, anxiety, hostility, learning pressure, psychological balance, emotional balance, mental health, and academic performance and comprehensive quality through 6 months of basketball games, are a significant change has occurred, that basketball training and the improvement of students' learning promote each other, basketball training to promote the mental health of the students and improve the comprehensive qualities. Through the teaching of basketball skills and knowledge, change to teach skills oriented teaching mode, cultivate more knowledge Identification and adaptive sports participants, so that students can be more widely understood basketball knowledge, and in the pursuit of the process of this kind of understanding is applied to the system of basketball training, stimulate students' sports

Table 72.1 The average selected estimation error with standard variables

Variable	Advanced skills	Medium skill	Low skill
Accuracy (%)	79 ± 9	68 ± 14	48 ± 13
Serving height ratio	1.51 ± 0.05	1.45 ± 0.06	1.42 ± 0.07
The centre of gravity ratio	0.58 ± 0.08	0.59 ± 0.08	0.75 ± 0.30
Trunk inclination	4.0 ± 3.1	4.0 ± 1.8	7.8 ± 8.1
Pitching angle	53.4 ± 5.7	52.5 ± 4.8	54.9 ± 5.2
Pitching velocity of m/s	8.32 ± 0.62	8.04 ± 0.40	8.05 ± 0.53

interest, promote the intelligence development and comprehensive quality, and cultivate students' consciousness of lifelong sports.

Another significant difference, found in the center of gravity of the ratio. High skills and skills in middle member groups are fairly stable ratio of 0.48 and 0.49, respectively. Low skills group scored an average of 0.65, less stable at the release point in fact, low skill is a member of the group of center of gravity is greater ratio 1, marked in an unstable state.

Each trunk tilt amount had no significant differences. The significance of the possible interpretations of the extreme variability is due to the low skill set. Hypothesis, low skills group members in trunk tilt test comparison between subject variability of uniform.

Multivariate statistical analysis, compare these data with advanced theory the ideal parameter. Each shooter target point distance to calculate:

(1) The success of minimum angle;
(2) The penalty at the angle of 45°;
(3) The smallest angle speed penalty.

Because the angle prediction reported the lowest velocity projection, higher or lower angle shot requires more ratios the minimum speed. Therefore, the absolute value of deviation from the recommended penalty angle is analyzed. Low and moderate skills group missed angle of 3 and 4° missed the high skill set of the minimum speed. Although highly skilled group of projection features appear to be from low and moderate skills group variability is very high.

72.5 Conclusion

From the following six angle such as free-throw accuracy, free-throws, gravity center height ratio, pitching, tilt, and pitching velocity, this paper has done the quantitative analysis on the high, and low level of technical ability level and discussed all aspects of the free-throw biomechanical properties combined with ability level. Then draw some conclusions about many influence factors arising from the free-throw shooting from different angles. We can get some conclusions as the following:

(1) Greater stability (i.e., balanced center of gravity and vertical trunk tilt) can enhance the ability of basketball skills;
(2) The bigger weight proportion is, the slower the angle and speed of the ball in the shot is, which can improve the hit rate of free-throw;
(3) Adopting the independent, but irrelated skill level of pitching angle and velocity.

References

1. Liu M (2004) On improving basketball players shooting. J Yuncheng Univ 4(05):46–47
2. Zhou X, Zhao Fang (2009) Stop jump shot in basketball sport biomechanical analysis. J Beijing Sport Univ 41(9):23–25
3. Huang S (2004) The application of the spinning ball in shooting. J Sports Correspondence 11(02):85–88
4. Chen D (2005) Analysis on basketball cast ball rotation. J Wuhan Sports Inst 21(04):76
5. Yuan F (2007) Modern basketball shooting technique structure mechanics analysis. J Henan Univ Nat Sci Ed 02:55–57
6. Wu J, Gavin L (2003) The main factors affecting the rate of shooting. J Shaanxi Normal Univ Nat Sci Ed 1(01):35–37
7. Feng H (2006) Discussion on how to improve preliminary shooting. J Jiangxi Normal Univ Nat Sci Ed 23(02):71–72
8. Wang H, Hu X (2010) The analysis on the influenced factor of the hit rate of free-throw. Sports World Acad Ed 11(10):53–55

Chapter 73
Research on Athletes Psychological Dynamics in University Sports

Yong Wang and Bogang Huang

Abstract At present, the psychological momentum in sports and team spirit, a considerable number of studies, but researchers and theorists are still divided the concept of vague and hold different views of the situation. And its two joint study does not. Through the early literature theory, research, psychological dynamic characteristics of the athletes in university, qualitative surveys to a more comprehensive understanding of the concept that you want to achieve requires quantitative methods to analyze test the psychological dynamics of the athletes to explore from the qualitative and quantitative research specific cognitive, emotional and behavioral changes, combined with the perspective of the economic game analysis of the impact of teamwork in sports competitions in the athletes' psychological dynamic, effective assessment of the conceptual model of the psychological power.

Keywords Psychological dynamics · Game · Quantitative analysis · Team spirit

73.1 Introduction

The modern sporting event is characterized by many factors against fiercer than ever, the success or failure of the competition. Participate in sport, to specialized sports physical, technical skills, tactical quality athletes need to maintain a good emotional state, with high morale, tenacious style, and perseverance, which requires plenty of spiritual strength input. According to statistics, sports competitions is compatible with the body movement, the appropriate mood and mentality

Y. Wang (✉) · B. Huang
School of Physical Education, Jiujiang University, Jiujiang 332005, China
e-mail: yong_wang21@126.com

W. Du (ed.), *Informatics and Management Science VI*, Lecture Notes in Electrical Engineering 209, DOI: 10.1007/978-1-4471-4805-0_73,

of the generation of a good atmosphere and team are inseparable, confidence and determination of the help and encouragement of the coaches, athletes game victory or defeat will have a direct the impact of movement normal play and super technology to play with appropriate emotions, stable ideas, a good attitude, focused attention, quick thinking and reaction has a close relationship. These mental activities are subject to the direct impact of the inner psychological atmosphere and style. So, for an athlete, team spirit is not optional, it can give full play an important guarantee for the players' technical level and the team overall strength.

Social competition is so fierce, the players in the competition facing greater pressure, how best to give full play to their strengths so that they can with ease, and have obtained outstanding results for the game, not only with the need to focus on athletes' skills and physical fitness enhancement, and more to be concerned about the psychological quality and competition of athletes in teamwork. This plays a vital role in sport. The sports psychologist also provides a very attractive field of study [1–3]. Potential linkages to enhance mental function and performance, and consolidation of the phenomenon, to achieve a clearer understanding of the psychological dynamics and the importance of teamwork. However, the psychological dynamic appears to be an elusive concept is a very challenging study of teamwork is a complex research.

Earlier studies involving psychological dynamic to enhance mental strength, may affect sports performance, and is a two-way [4]. Positive psychological dynamic seems to reflect the shift of power and its attendant psychological cognition, affect the physiological parameters, and therefore affect performance; negative psychological dynamic is expected to cause the reverse effect. Rowing crew, for example, only in a 2,000 m race have more than one in the middle part of the 25 s will be considered as the momentum (positive momentum), while leading the crew, though still leading but will lose momentum (negative momentum).

The sports team is a special social group. It is the combination of certain social relations collective joint activities is the basic unit of social competitive sports organizations. To complete the training and competition task entrusted by the sports organizations, to meet the members of a variety of social needs, such as security, ownership, communication, self-esteem and achievement needs. In addition, sports team also has a social function cannot be ignored. Sports team members are young children, they are at learning, interaction and personality the best period of their long-term living, learning, training in teams, in addition to master specific sport, their world view, outlook on life the formation of values and aesthetics, ideals, ethics, basic life skills and social norms of behavior, learning, etc., are closely linked within the team environment by the coaches, team culture, the psychological atmosphere of group pension and other aspects of impact.

73.2 The Conceptual Model

Factors affect the athletes' psychological dynamic, athletic ability in sport conceptual model of the following aspects: First is the psychological momentum model. From changes in the cause and effect analysis of motivation, control, optimism, energy and synchronized view of the power of cognitive psychology toward a moving target [5]. Cause or effect of the difference between these models, the psychological momentum to admit defeats, the performance changes. The nature of the task is also an important consideration. Namely the psychological dynamic is likely to lead to a high level of awakening, which may contribute to the awakening of performance at a high level task [6].

Followed by multi-dimensional momentum model. The model of an incident or series of events, leading to the performance of athletes are different, subjective norm, athletes began to generate power, trigger change in cognitive, emotional and physical changes, which in turn affect the behaviour of the players, the performance and the final result of an event. The change in momentum led to corresponding changes in cognition, but not persistent.

In summary, this paper, the psychological dynamic model for the dynamic characteristics of athletes' psychological analysis that athletes can be effective and comprehensive analysis to consider from the perspective of qualitative and quantitative binding assay, fully teamwork used in which, to common analysis of the performance of athletes in the game, better analysis of the psychological dynamics, psychological momentum and momentum, and cognitive, emotional, etc.

73.3 Psychological State of Affairs of the Model Analysis

Despite the psychological momentum of the three competing and a lot of research models, the concept remains controversial. Through the analysis of evidence that in the competition or lose out, the psychodynamic view does exist and steering response. In addition, theorists proposed the concept of psychological momentum through the performance of cognitive and affective processes (that is optimistic about the sense of control, mediation, motivation, and self-efficacy), physiological factors, such as physiological arousal, changes. However, many of the recommendations mediators have not been empirical test, but still speculative psychological momentum to explain. For example, Adam Christie cognitive mediator unaudited cognitive interpretation of the observations of the psychological situation in the billiard players [7]. Adams data involved in the game of pool statistics, but because of the individualistic point of view a semi-structured 10 players to prove difficult to follow this model analysis results.

1. The research object

Taking the school sports students of Jiujiang University, Nanchang University, and Jiangxi Normal University as subjects, randomly selected 600 students in basketball, football, volleyball classes, including 300 boys, 300 girls (18–24 age) for the survey.

2. Research methods

The use of survey methods by the documents and files. According to the purpose and content of the paper, designed in accordance with the principles of the design of the questionnaire survey on teamwork status quo of the sports university students. In order to ensure that its research representative, 600 questionnaires were issued according to plan (300 boys, 300 girls) with filtered 596 valid questionnaires, the response rate of 99.3 %. Six questionnaires were excluded invalid. A total of 590. Effective response rate of 98.3 %.

In order to ensure that the investigation reliability, the questionnaire to the effective monitoring and inspection. Distinguished professor of 12 sports disciplines, experts, an associate professor, lecturer for evaluation of the validity of the questionnaire, including four professors, four associate professor, four lecturers. Use of the Delphi method evaluation, the results shown in Table 73.1.

Through effective inspection we can get a conclusion that the questionnaire easy to understand, to reflect the theme and reasonable structure.

Through questionnaires, and use the momentum of the scenarios and the scale of measurement is usually cognitive psychology. Written score assuming the configuration of a set of tennis to win the game mode. Experienced and inexperienced tennis player, had never seen a ball hit to win the game pattern, and then ask the question to answer, such as "who has the momentum?" And "who seems to be the most active?". Although participants may have been able to draw the predicted momentum own familiar experience, but because these subjects scoring model based on their own views, the data reported reflect the momentum, and there is no prediction of the actual experience [5]. Psychological momentum clearly some issues can only be reliably convey the experience of a full appreciation of the momentum of the tennis complex participants answered the subjective experience of the actual situation that may not be connected to may have already lost the actual score (that is, a player for two or three games, but she/he play, can be improved and he/she may remain in their own ability to back the motivation and confidence). In addition, the size of the survey, while allowing those proposed by the important variables of the theoretical work of quantitative analysis, they are in a narrow (reductionism), and limit the response provided to the participants—thereby negating the experience of the participants in a comprehensive understanding. Psychological momentum (e.g., questionnaire), the concept has not yet developed systems, this problem may be further confused and not yet a comprehensive analysis

Table 73.1 The feasibility study of sports experts, questionnaire survey

Recognition of questionnaire	Entirely feasible	Basic feasible	General	Not feasible
Number	8	3	1	0

Table 73.2 The findings of data tables

j	^a j	^b j	r j
1	2.15	1.047	0.924
2	0.31	0.974	0.937
3	3.01	0.955	0.945
4	2.44	0.920	0.943
5	1.22	0.926	0.943
6	2.34	0.912	0.947
7	3.28	0.914	0.933
8	1.15	0.934	0.949
9	1.13	0.963	0.953
10	1.37	0.973	0.946
11	1.56	0.937	0.955

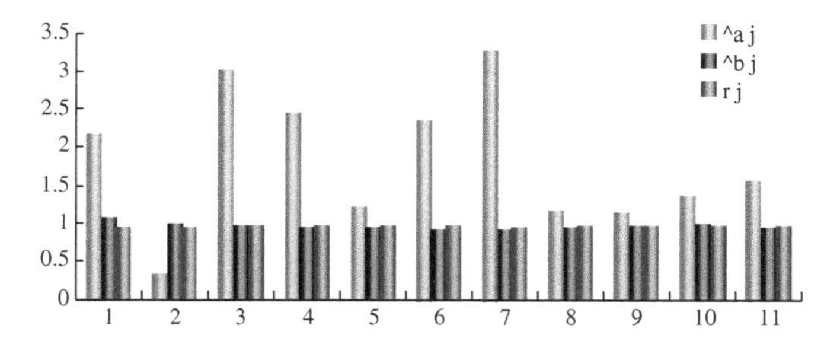

Fig. 73.1 The proportion of the model parameter data

of the experience of the athletes and coaches. The model formula is as follows [8] (Table 73.2, Fig. 73.1):

$$b_j = \frac{\sum(\overline{q_i} - \bar{x})(q_{ij} - \overline{y_j})}{\sum(\overline{q_i} - \bar{x})^2} a_j = \overline{y_j} - b_j\bar{x} \tag{73.1}$$

73.4 Research on Team's Psychological Dynamic Change

There are two main ways: in the relationship between power and performance of the test psychology, first check the archive or observational data, including closely related to the "hot hand" phenomenon (usually considered in the shooting), and test the concept following the success of string or "hot spots" temporary performance. This has been made in the shot records or assessment of individual performance test of the statistical test. Under normal circumstances, such an analysis results show that the shooting mode is not the opportunity to work with the expected. While research data from the early squash and tennis found some evidence of the effect of psychological momentum, the moment when the race to the final decision, there is no evidence that psychological momentum show. For example, has won the second game/set expected to win the deciding game/set, if the momentum effect is operating. In addition, most early studies failed to control capacity, which has been emphasized as a confounding variable.

In a more novel approach, the relationship between views and the actual performance of a 'micro' psychological dynamics of the test participants, and trained observers to employ a structure chart graphics as the representative of the "flow Volleyball game". A significant positive correlation between the momentum of the participant's survey scores and observers of the structure chart event provided some support for the existence of psychological momentum.

Secondly, the study of the relationship between power and performance of the test psychological experiments often use false feedback, as well as a variety of rating operations to create the momentum of subjective feelings and objective performance results. The inconsistent results of these findings have not been able to prove the psychological momentum; there is a causal relationship to the performance. However, despite the use of a cyclic task, requires a high level of awakening, but provide some support, only a psychological momentum-performance relationship. The questionnaire was used to measure the momentum after a false view of the circular game; the result is pre-determined by the actual performance of the participants and will not be affected. Be seen as a computer-generated visual representation of the game, the participants reviewed the completed questionnaire in four different time points on the cognitive psychological momentum. Through the performance of the average output power in four time periods were tested. Increase and decrease the momentum of the views of both sides found the view to explain the decline, increased negative convenient performance. However, this explanation is speculative, and should be given careful not to view in-depth assessment of participant's experience. In addition, previous research mission to expose the impact of psychological momentum on performance may not be conducive to fine motor.

Momentum and momentum of the psychological and cognitive/affective, cognitive and affective processes and psychological momentum, few studies have attempted to test this prediction, and those who have reported inconsistent. Test target shooting novice cortex, cognitive, emotional and behavioural responses and cognitive changes found in the psychological momentum of the relevant positive and negative feedback to manipulate the view. However, the emotional response (found by the measurement of positive and negative emotional scale) and shooting performance is not affected. A variety of studies have focused on the differences and potential advantages and limitations of qualitative and quantitative research methods (Table 73.3, Fig. 73.2).

This argument is based on the focus of their personal experience. A phenomenological approach refused to use the theory-building and is concerned only with the description of events or objects. Note for specific issues, rather than other qualitative and quantitative methods. Because it seems the phenomenon, researchers have not enough to deal with the problem of psychological momentum, seems to be the most

Table 73.3 The teamwork of the athlete's psychological evaluation parameters

		Armor		Other team members	
		Income	Cost	Income	Cost
To accept		76	88	69	93
Refuse	Convince	54	65	63	72
	Exclude	43	56	59	78

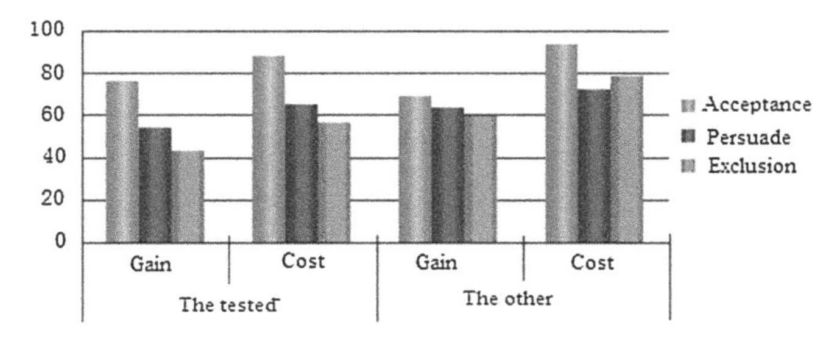

Fig. 73.2 The psychological evaluation parameters of teamwork athletes

appropriate method for the collection of descriptive information may lead to a clearer understanding of what kind of psychological momentum of what it is, would like to experience psychological momentum. Sports team spirit is the spirit of sports culture and the arts as a social cultural phenomenon in competitive sports training and competition conditions. Sports team spirit advocated inherit the Chinese athletes hard work ahead, excellent quality of glory for the country, to carry forward the tradition of sports teams collective ideology, and actively learn from the corporate culture of teamwork, and summarize the successful experience of a large number of sports practice and theoretical innovation base rather highly condensed formation. It is valuable spiritual wealth; sustainable development of China's athletics is a powerful spiritual impetus and reliable guarantee of excellence competitiveness in our athletes in the World Series.

73.5 Conclusion

Although the conceptual model there are three psychological dynamics, it is clear, quantitative analysis method of psychological momentum-driven use of limited understanding, may be added to the confusion surrounding the concept of, and doubt. In fact, the early use of "non-traditional methods of investigation in this area, the advantages of using qualitative methods is that it allows participants insight into the structure and experience of the individual, and can go beyond" macro structures, such as motivation and self-efficacy study psychological momentum, so far ignored by the researchers of micro components. There is an urgent need to re-examine the psychological power to include the conceptualization of those who are most likely to encounter psychological momentum point of view, that is, athletes. In order to make a more comprehensive understanding of the psychological momentum, future research should be qualitative, such as the existence of the phenomenon: (1) to develop a clearer concept of the psychological momentum; (2) check the athletes' experience in cognitive psychological momentum, including momentum starters; (3) investigate the specific cognitive, emotional and behavioral changes, experiencing psychological momentum and use the evidence gathered, careful assessment of the conceptual model of the psychological power; (4) analysis of the dynamic role of the research team spirit mentality.

Acknowledgments The research 'The Qualitative Research on Athletes' Psychological Dynamics in University sports' belongs to one of the performances of the humanities project of Jiujiang University in 2011. And the name of the project (No. 2011SK05) is "the value and cultivation of team spirit influenced by sports under the background of the employment and market competition". I would like to thank my colleagues that help me.

References

1. Ma Q (2008) The tension sports psychology, vol 8(3). Zhejiang Education Press, Hangzhou, pp 245–253
2. Yan G (2001) Educational psychology, vol 12(01). China Building Materials Industry Press, Beijing, pp 617–623
3. Gao Z (2004) The education theory of psychology, vol 4(22). Anhui Education Press, Hefei, pp 515-522
4. Lin Y (2010) Guopsychology of learning, vol 10(4). Police Education Press, Beijing, pp 316–319
5. Young T, DPRK (2009) Corps on the team spirit of cooperation in culture in the university physical education. Era of education (education and teaching Edition) 12(8):77-79
6. Yao DM (2005) Competition counseling and psychotherapy training, vol 5(3). People's Sports Publishing House, Beijing, pp 35–39
7. Zhao F (2007) On cultivate students the importance of teamwork in sports teaching. Sci Technol 7(4):45–47 (Academic)
8. Zhang X, Duan L (2011) The team spirit in sports. Junior Sports Training 2(3):225–230

Chapter 74
Research on Human Health Characteristics Based on Physical Exercise and Diet Mechanism

Xiaoping Xie and Jingping Min

Abstract With the demand in higher quality of life increasing, people begin to put more importance on their own health. This paper makes further research and study on physical exercise and dietary of fat person and human health. They made a large number of surveys to analyzing and comparing obese people in body quality. And then they went through sports training and establish additional strength training, and improve the quality of life, and analyzing the effect of exercise and diet measures. Through diet measures and aerobic training, the fat person health index is researched; we can use data feedback adjustment and the diet custom training content, etc. to achieve the purpose of strengthening the body quality.

Keywords Obesity · Training · Diet measures · Physical exercise

74.1 Introduction

Obesity is a common complication of pathology and in the past few decades, no matter in the developed countries or in developing countries; it has become a more important public health problem. Because overweight and related metabolic disorder soften occur in obese people, so from risk factors angle, obesity must be considered as a kind of disease. In fact, a deadly cardiovascular disease, joint disease hampered and behavioral problems will also be caused by obesity. In order to meet the need of multidisciplinary disease management, this paper studies the role of quality of life in reducing disease complications as much as possible [1].

X. Xie (✉) · J. Min
Public Sports Department, Jingchu University of Technology, Jingmen 448000, China
e-mail: Xiaoping_Xie@yeah.net

W. Du (ed.), *Informatics and Management Science VI*, Lecture Notes in Electrical Engineering 209, DOI: 10.1007/978-1-4471-4805-0_74,

74.2 Research Methods

74.2.1 Research Plan

We conducted the research of obese patients from July 2007 to November, and the analysis of their body mass index (BMI). During that period a total of 152 obese patients have received research, and participated in the training [2]. But in the end only 83 patients were included in the study analysis, the rest of the 69 people were ruled out from this research because they did not correctly complete the original training plan and be ruled out [3].

The patients were randomly divided into three groups. G1 group has 50 people; G2 group have 51 people; G3 group have 51 people. Then the training and the organization are as follows: (1) 29 patients are arranged in the G1: they have no health consultation also no doing physical exercise plan; (2) 26 patients are arranged in the G2: they can get diet and health services also carry a aerobic exercise training; (3) 28 patients are arranged in the G3: they can get diet and health services and also carry a aerobic exercise training and are supplemented with arm and leg muscle strength training [4].

74.2.2 Evaluation Parameters

All patients had two weeks of special evaluation consultation in physical medicine and functional rehabilitation department and endocrine department [5, 6].

1. Assessment of human measurement parameters. We evaluated a lot of human measurement parameters: weight, height, body mass index (BMI) and waist circumference measurement (WL). We used impedance instrument to evaluate body fat weight, quality, metabolization rate Table 74.1, Fig. 74.1.
2. Assessment of the effect of cardiovascular disease. Each obese patient uses the treadmill pressure test evaluation (ST) to assess the influence of cardiovascular

Table 74.1 Comparison of the patient's medical record and the characteristics of the human measurement

	G1	G2	G3	p
Age	38 ± 10	36.1 ± 12	37.5 ± 8	NS
Sex (male/female)	5/24	4/22	4/24	NS
Disease				
Diabetes	5	7	5	NS
Blood lipid disorders	3	4	5	NS
Spinal corrective	6	8	9	NS
Duration (years)	10 ± 5.73	9 ± 5.8	9.6 ± 6.5	NS
Weight (kg)	103.4 ± 21.6	99.2 ± 18	96.3 ± 10.9	NS
Body mass index (kg/m^2)	38.2 ± 4.9	37.3 ± 6.7	36 ± 3.86	NS
Waistline	113.69 ± 14.4	112.34 ± 20.11	110.65 ± 8.8	NS

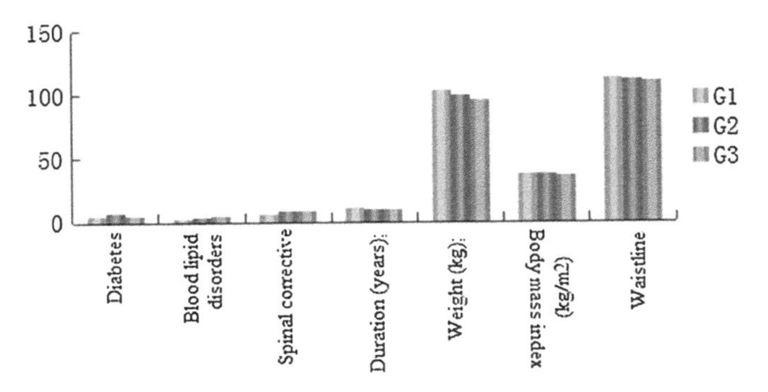

Fig. 74.1 Personal medical records and human measurement data figure

Table 74.2 Comparison of different arterial blood pressure parameters

	G1	G2	G3	p
Static heart rate	83	86.2	79	NS
Maximum power of heart rate	148.5	150.8	140.8	NS
Static arterial blood pressure contraction/relaxation	127.8/72.7	127/79.2	130.3/75.7	NS
Maximum power of contraction/diastolic blood pressure artery	178.1/87.5	168.4/87.5	172.5/90	NS
Total time of pressure test	10 min	10 min 25 s	11 min 13 s	NS

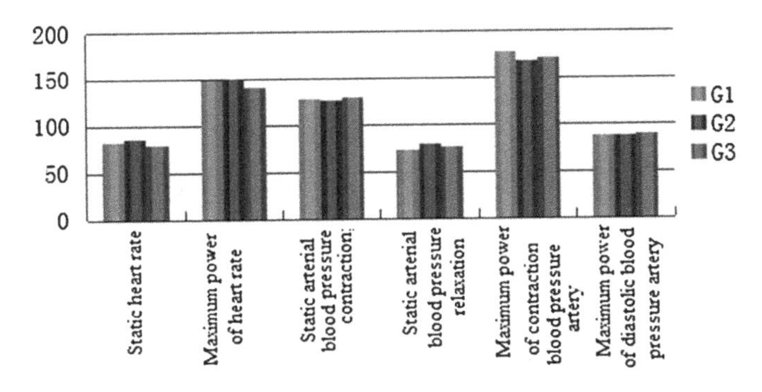

Fig. 74.2 Different arterial blood pressure stress test parameters figure

disease. At the same time make assessment for high blood pressure patient's artery (AHT) per minute, and also the arterial blood pressure measurement Table 74.2, Fig. 74.2.

3. Assessment of muscle strength. The purpose of the initial assessment and final arm and leg muscle strength is to determine maximum repeat rate (MR1). Measure muscle on the shoulder, elbow, knee gluteal muscle and related parameters Table 74.3, Fig. 74.3.

Table 74.3 Comparison of the biggest muscle strength (MR1) produced by different muscles

Muscle group	G1 (kg)	G2 (kg)	G3 (kg)	p
Shoulder pressure	8.9 ± 2.6	9.9 ± 1.5	9.4 ± 2.2	NS
Cubits flexor	8.7 ± 1.4	7.4 ± 3.6	8 ± 2.9	NS
Cubits extensor muscles	7.4 ± 1.8	6.8 ± 1.4	7.1 ± 1.7	NS
Gluteus maximus	6.7 ± 1.7	7.57 ± 2.2	7.1 ± 2	NS
Strands of muscle	7.2 ± 2	6 ± 2.5	6.5 ± 2.3	NS
Knee extensor muscles	14.1 ± 1.7	14.6 ± 1.8	14.4 ± 1.7	NS
Knee flexor	6.3 ± 1.5	6.9 ± 2	6.6 ± 1.7	NS

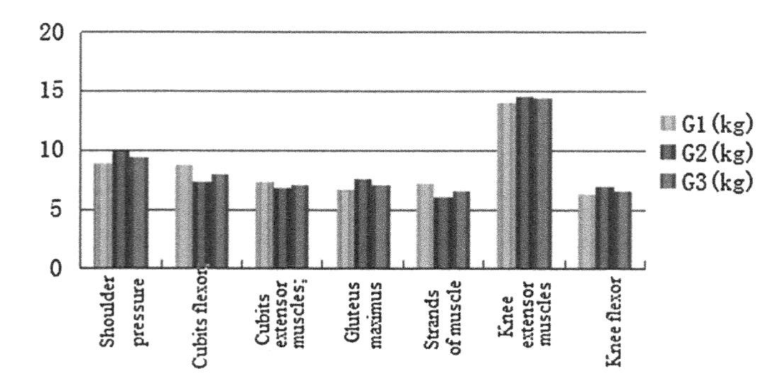

Fig. 74.3 Comparison of the biggest muscle strength (MR1) produced by different muscles

4. Assessment of metabolic disease. At the start and the last of the training, all of the patients are forbidden sugar content, total cholesterol, triglyceride and high density lipoprotein cholesterol Table 74.4, Fig. 74.4.
5. Evaluation of daily breathing difficulty in the activities. In the beginning and the end of the training, breathing difficulties in the daily life means the breathing quantity is between 0 and 100 mm (0 represents the lack of breathing problems and 100 represents the most intense breathing difficulty) Table 74.5.

74.3 Analysis of Data Results

Before developing plans, all of the patients are trying to lose 2 and 3 kg. the average of lost weight is as follows: G1 is 2.45 kg; G2 is 2.50 kg; G3 of 2.45 kg (p = NS).

1. Epidemic characteristics and obesity assessment [7].

Distribution of research groups (including 13 men (15.6 %) and 70 women (84.4 %), the average of age ± the standard deviation was 37.2 ± 10 (range: 18–60 years). These people obesity has 9.5 ± 6 years time (range: 3–20).

Table 74.4 Comparison clinical parameters of material content

	G1	G2	G3	p
Cholesterol				
Total (mmol/l)	4.75	5.15	4.96	NS
High density lipoprotein (mmol/l)	0.55	0.55	0.57	NS
Low density lipoprotein (mmol/l)	2.86	2.17	2.11	NS
Triglycerides (mmol/l)	1.35	1.12	1.45	NS
Fasting blood sugar disease	5.7	5.8	6.03	NS

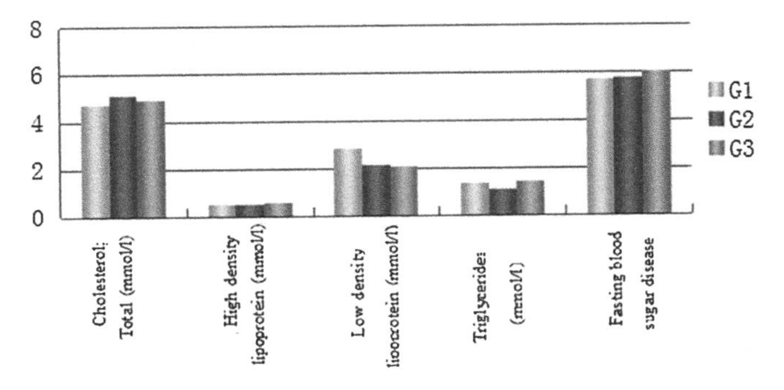

Fig. 74.4 Clinical parameters of material content

Table 74.5 Parameters of various impedance measuring method that changing with time

	G1(kg)	G2(kg)	G3(kg)
Fat content			
Before	44.4 ± 4.97	44.7 ± 7.91	39.92 ± 9.03
After	44.58 ± 4.65	40.84 ± 9.77	35.74 ± 9.41
P	NS	0.032	<0.001
Muscle quantity			
Before	52.93 ± 9.07	55.91 ± 10.6	53.19 ± 5.76
After	52.9 ± 8.98	56.68 ± 12.4	55.68 ± 8.31
P	NS	NS	0.02
Physiological metabolism			
Before	1784 ± 176	1828 ± 339	1698 ± 188
After	1726 ± 171	1842 ± 337	1793 ± 275
P	NS	NS	0.026

The average weight 99.7 + /−17.5 kg (range: 76–148). Average body mass index is 37.2 + /−5.2 kg/m^2 (range: 30–53.8 kg/m^2); body fat mass is 41.95 + /−7.7 kg (such as: 41 % of your body weight); thin constitution 53.5 + /−7.5 kg, and fundamental metabolism rate is 1742.62 + 203 kcal/d. Our research samples include 23 patients with high blood pressure, 12 years old with dyslipidemia and

17 people with diabetes. Table 74.1 summarizes and compares the three groups of epidemic and human measure characteristics.

74.4 Conclusion

Micro blog marketing has just started in China. It's just beginning. With the gradually increasing of the domestic micro blog users and the micro blog related product, micro blog marketing will be more and more gotten the attention and researched. To try to explore using micro blog marketing as soon as possible, the enterprises should integrate the skills and strategies tailored for the customers. A set of a licable marketing scheme is very important. Hope the paper can be valuable for the enterprises to improve the competitiveness of their own micro blog marketing.

This study has confirmed that diet and exercise combined can be beneficial to the management of obesity. The combination of this not only effectively control the weight, lower fat accumulation and abdominal fat, fix obesity-produced metabolic disorder, and thereby reducing the cardiovascular risk factors. Obese management plan can increase the obese patients' physical skills. And after the experiment, the mental states of the testers and the quality of life have been significantly improved. Comparing two ways of physical exercise, we found that endurance and combination of forces for some parameters produced more significant improvement, such as weight loss, muscle strength (mainly in the arm) and waist circumference. Also, we found one's state of mind and life quality to have been more significantly improved. Although containing no strength exercise didn't result in the improvement of cardiovascular parameters, it is really an effective way of physical exercise to prevent obesity.

Therefore, we believe further increasing the scope of study sample and the use of different ways will help prove the benefits of physical exercise and diet mechanism combined, and to find the most suitable combination to manage obesity and the disease it brings, in order to find a more effective and healthy life style to improve human health.

References

1. South J, Blass B (2003) Disease control department of the People's Republic of China. China diabetes prevention and control guide, vol 87. Beijing Medical University Press, Beijing, pp 12–40
2. Yang YX, Wang GZ, Pan XC (2002) Chinese food ingredients table 2002, vol 76. Beijing Medical University Press, Beijing, pp 21–220
3. Ye HQ, Yu MH, Chen HZ (2005) Practical medicine, vol 56, 12th edn. People's Medical Publishing House, Beijing, pp 1042–1043

4. Wang L, Li L (2006) Disease control department of the people's republic of China. Chinese adults overweight and obesity prevention and control guide, vol 55. People's Medical Publishing House, Beijing, pp 21–42
5. Liu MM, Feng ZY, Zhu LZ (2004) Community diet and exercise treatment strategies of metabolic syndrome patients. Chin J Gen Pract 7:1844–1846
6. Zhao Z, Liu C, Jiang QY (2005) Clinical observation of early intervention on the way of life of the patients with the metabolic syndrome. Chin J Med Univ 34:272–274
7. Li YW, Ai H, Zhang BH (2007) Treatment function of exercise and diet therapy for patients with the metabolic syndrome and people in high risk group. China Rehabil Med J 22:9–12

Chapter 75
Analysis of Physical Exercise Adjustment on Human Circulation Immune Cells and Soluble Medium

XiaoPing Xie and JingPing Min

Abstract At present, analysis of the benefits of sport exercise to human body is mainly from quality, physical endurance and power. Surveys show that the heart failure patients have circulating immune cells and excessive soluble cell apoptosis mediation factor, and this may be because they can't insist on physical exercise and clinical deterioration. This paper combines the actual investigation, through plasma level tests on the 24 bit stability of heart failure patients with tumor necrosis factor alpha (TNF alpha), soluble TNF receptor I and II, interleukin 6 (IL-6), soluble IL-6 receptor, and 12 weeks of random, cross design of physical training, and normal control project comparison before and after exercise, and through the heart of the exercise test measures to test the vo2max heart failure patients' functional state. It tries to discuss the effect of the physical stamina training cycle inflammation to human body immune cells and soluble apoptosis related factors (sFas) and Fas ligand (sFasL), to analyzing the regulatory role of physical exercise in the circulating immune cells and soluble medium.

Keywords Mathematical model human circulating immune cells · Soluble medium · Physical stamina training

75.1 Introduction

The latest surveys show that chronic heart failure (CHF) patients' immune cells and soluble cell apoptosis medium are exceptional, which may be comprehensive characterization caused by abnormal heart and endothelial function. Therefore,

X. Xie (✉) · J. Min
Public Sports Department, JingChu University of Technology, Jinmen 448000, China
e-mail: Xiaoping_Xie@yeah.net

W. Du (ed.), *Informatics and Management Science VI*, Lecture Notes in Electrical Engineering 209, DOI: 10.1007/978-1-4471-4805-0_75,

abnormal immune response seems to be an important factor of CHF syndrome and continuity. Immune cell factors, such as a tumor necrosis factor (TNF)'s and interleukin 6 (IL-6), can adjust the heart and peripheral vascular function through a variety of mechanisms to, including containing nitrogen oxide synthase (NOS) performance, oxygen free radical's excessive production of myocardial cells and abnormal rules of endothelial heart and cell apoptosis induction phenomenon. In addition, CHF patients' extraordinary performance with nitrogen oxides of skeletal muscle to induce form of chitin synthase is very likely resulted from circulating immune inflammatory cells and it seems to be responsible for muscle contraction capacity attenuation or bone myocardial cell apoptosis, or related to the exercise tolerance level limitations and the severity of the CHF [1, 2]. And there are reports that, for patients with heart failure cell apoptosis, their circulation medium, such as soluble related factors (sFas) and soluble Fas ligand (sFasL) are increased, and this is closely related to the severity of the symptoms and the forecast of heart failure patients [3].

There is no existing form of experimental data show physical exercise's influence on the immune cells in the human body circulation factor and the soluble receptor, and the same is true with CHF patients soluble receptor sFas cell apoptosis and soluble cell apoptosis induced sFasL. Therefore, we are looking for surveys of physical exercise affect immune cell factor TNF's and IL-6 and soluble receptor, soluble tumor necrosis factor receptor type 1 (sTNF-RI), soluble tumor necrosis factor receptor type II (sTNF-RII) and soluble interleukin 6 receptor (sIL-6 R) serum level, as well as the sFas/sFasL system of expression of serum level of apoptosis medium, and when these 24 patients have stable, appropriate serious heart failure in these induction training with the time of change have anything to do with exercise tolerance, oxygen consumption (vo2max) expression [4].

75.2 Research Group

These 24 patients with moderate serious CHF patients (age 55 $\pm$ 2 years old) provide information and agree to participate in our study. We are using 2-dimensional echocardiography to estimate CHF patients' average left ventricular ejection fraction of the modified formula according to Simpson 23.2 $\pm$ 1.3 %. The included standard is: stable CHF patients continue for at least 3 months; Ischemic cause (n = 11, documented myocardial infarction or coronary artery angiography and coronary artery bypass graft surgery) and idiopathic dilated cardiomyopathy (n = 13); because of breathing difficulties or fatigue for exercise restrictions and to achieve at least a single exchange rate ability; early dynamic ecg monitoring ventricular or the lack of evidence of other serious rhythm of and wrong symptoms of the heart. There are patients with infectious disease, malignant tumor, collagen protein or other inflammatory disease, and in the past 2 weeks there are patients taking anti-inflammatory or immunosuppressive, all of them will be excluded from our study [5].

We use the questionnaire survey to research people who participate in aerobic, strengthen, stretching self assessment. Exercise questionnaire is based on trans-theoretical model of behavior change, and it is different from engagement in a particular behavior preparation phase. The present study includes 5 stages: lack of deliberate practice, meditation (want to exercise), ready to (plan to exercise), action (present exercise) and maintain (continuous exercise). The first 3 stages mean that lack of exercise, and the latter two represent the current stage in the exercise. The specific standard is defined as three different sports, aerobic exercise, stretching and strength training. Aerobic exercise is a regularly plan to increase physical activity, for example, taking a walk, aerobic exercise, running, bike riding, swimming, boating, etc. Conventional strength training is to use strong body for physical activity, for example, increasing or removing free weights, using weight or resistance training machine, etc.

The results show that from beginning to the end of a sophomore year, 70 % of the 280 students' gains weight; 26 % has weight reduction, and only 3 % of the students' weight is remain the same. For those who gained weight, they had an average weight gain for 4.1 + /−3.6 kg. And for the BMI index, 69 % of the students are higher.

Data showed that the whole sports participation has not changed; aerobic exercise in number is in decline, but the number of stretching is on the increase. For freshman and sophomore students' consumption of fruits, vegetables and high fat fast food, the situation did not appear to change, and fried food consumption is on the decline. Through research and analysis and found no relationship between the changes of weight, body mass index (BMI), exercise and eating behavior of the. Significant changes include: students in maintenance stages involved in aerobic exercise are reducing, the training of merger students involved in stretch action and maintenance stages increased, but the corresponding proportion is in the fall.

Therefore, the patients and healthy controls will have 12 weeks of training plan. The plan is home-based, in a random, crossover design bicycle exercise training program to avoid conflict with other sports. Training plan includes 30 min every day a week's sports training; patients and normal control group was asked to 50 RPM exercise to maintain their heart rate monitor scope for previously identified the maximum heart rate of 60–80 %. Cardiopulmonary exercise testing: the heart patients go through sports pressure tests to assess their vo2max (ml/kg/min) of their sports ability and other drug use data testing. The results are shown in Table 75.1.

According to the data, we used statistical principle to cross test results and the results were analyzed statistically. The data show that by using proinflammatory cytokines and apoptosis vo2max medium as baseline, after the training, we conducted and analyzed repeated measurement and the variables [6] (Fig. 75.1, 75.2).

Table 75.1 Statistical data of people's all kinds of proinflammatory cytokines baseline soluble medium and clinical characteristics

Number	Age (year)	Medication	EF (%)	VO2max (ml/kg per min)	sFas (ng/ml)	sTNF-RI (ng/ml)	sTNF-RII (ng/ml)
1	65	D, ACE	27	14.8	4.57	5.2	2.185
2	61	ACE, Dig.	16	14.0	4.51	4.7	2.854
4	57	ACE, AA	40	14.8	6.57	5.3	2.236
4	50	ACE	45	12.4	4.53	5.4	3.354
5	56	ACE	24	11.8	14.33	4.9	4.823
6	58	ACE, Dig.	45	14.2	3.55	2.4	1.922
7	61	ACE	21	14.5	6.12	2.6	2.487
8	44	ACE	44	16.6	7.86	3.9	2.790
9	57	ACE	24	11.4	6.88	6.1	1.525
10	64	ACE, AA	17	12.8	9.45	4.23	3.174
11	64	ACE, AA	21	15.2	2.09	3.2	3.257
12	47	ACE	24	19.7	4.98	3.7	2.440
14	52	ACE, Dig.	19	20.1	6.43	2.78	2.322
14	57	ACE, AA	17	12.4	7.96	2.5	4.798
15	64	ACE	28	14.5	12.47	2.4	2.340
16	54	ACE, AA	22	14.6	12.38	2.3	4.469
17	58	ACE	15	15.4	8.53	2.9	1.981
18	60	ACE	17	17.1	6.12	2.65	2.033
19	40	ACE, Dig.	25	14.6	11.47	2.86	3.212
20	61	ACE, AA	18	14.7	5.55	1.92	1.790
21	46	ACE	26	16.4	3.52	4.4	2.965
22	61	ACE, AA	21	19.5	3.46	3.35	2.686
24	40	ACE, Dig.	20	26.4	2.50	2.14	1.944
24	59	ACE	40	22.8	5.28	2.22	2.538
Average	55		23.2	16.2	5.7	3.5	2.7

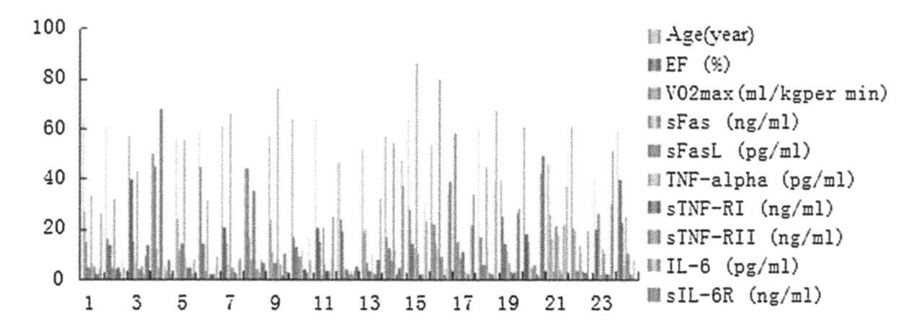

Fig. 75.1 Tionship of age on all kinds of proinflammatory cytokines baseline soluble medium

75.3 Analysis of Experimental Results

The experimental results show that: when $p < 0.01$, during the training, the induction of inflammation cell factors and apoptosis factors on CHF patients with a control group will reduce (Fig. 75.3) were fore, we observed during the training

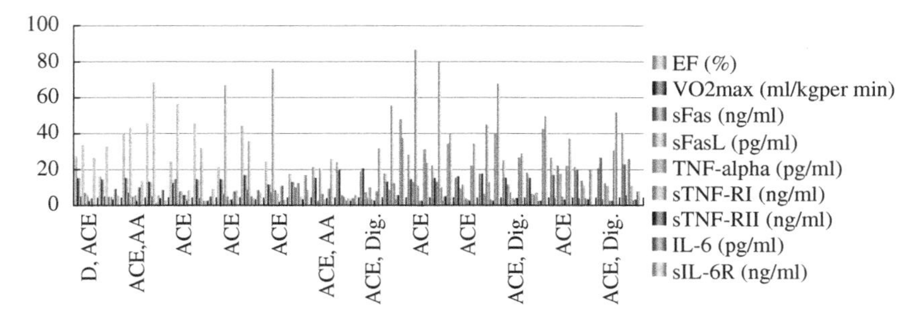

Fig. 75.2 Tionship of medication on all kinds of proinflammatory cytokines baseline soluble medium

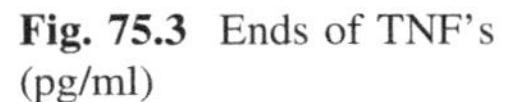
Fig. 75.3 Ends of TNF's (pg/ml)

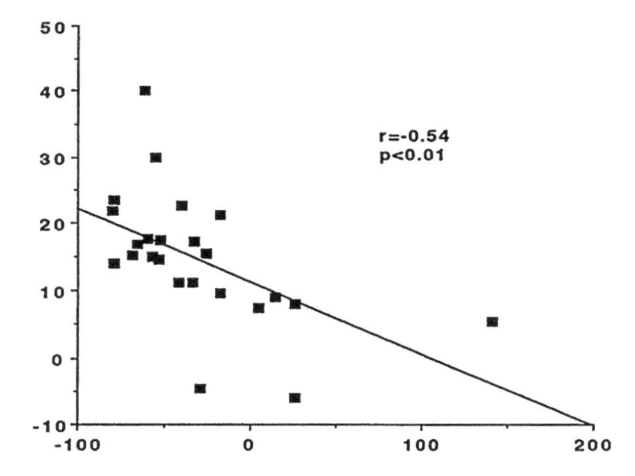

the cycle of serum concentrations cell factors reduced significantly. And for circulation cell factors IL-6, its soluble receptor sIL-6 and apoptosis medium sFas also reduced significantly [7]. The results show that sports performance, vo2max, improvement and training plan can obviously promote the vo2max and reduce serum mixture in cell factors and the percentage of cell apoptosis, which explains the activation of the attenuation immune inflammatory cells factors and sFas/sFasL systems' ability to improve the sports, it's necessary relation between realization and physical CHF patients training.

In sports training, physical exercise produced energy that has been significantly reduced ($p < 0.005$). Through the analysis of variance and the comparison of more proinflammatory cytokines and apoptosis factor, though there are no differences testing baseline and training athletic performance, data showed that the influence of the sFasL was also on a decline Fig. 75.4.

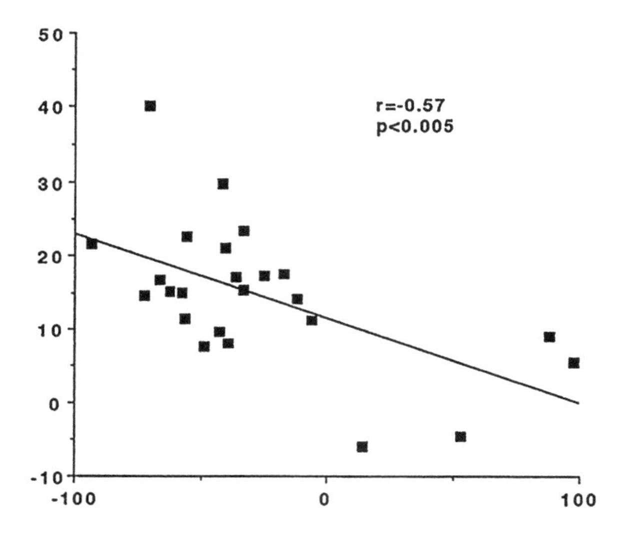

Fig. 75.4 Trends of sFasL (pg/ml)

75.4 Conclusion

We have recently shown that exercise training program can effectively reduce the inflammation markers (for example, macrophages proinflammatory cytokines, monocytes, soluble cells, dissolve the blood vessel cells, etc.) and we have reflected more glue cell interactions, as the development of the inflammatory process CHF may cause exceeding motion reduce. According to our research extension, from this original observation, we can not only find epiphenomenon, but also explain the characteristics of the syndrome of pathogenesis and the movement of intolerance. Therefore, by virtue of its anti-inflammatory effect and influence of the disease, and the positive adjustment of the surrounding of the immune response, we can improve the sports ability of human body. Analysis of proinflammatory cytokines and apoptosis inducers participation mechanism and myocardial dysfunction and myocardial cell death factors, researches through the skeletal muscle bones myocardial cell apoptosis nature and endothelial dysfunction for clinical trials have great scientific significance. Therefore, our research provides a new thinking in pathology mechanism, and gives a reference for future scientific research through analysis of symptoms effect of the complex syndrome CHF training and physical sports.

References

1. Guo L (2009) Effects of cell apoptosis in their own immune mechanism. Foreign Med Immunol 23(1):7–21
2. Yu QW, Zhang DQ, Li NL et al (2010) The significance of SFas and sFasL in autoimmune disease. Biomaterials 6:30–33

3. Li F, Ling HH, Luo JY (2005) Chronic heart failure patients' serum soluble Fas ligand and soluble Fas receptor changing levels. China J Circ 155(6):206–211
4. Nozawa K, Kayagaki N, Tokano Y et al (1997) Soluble Fas and soluble Fas ligand in rheumatic disease. Arthritis Rheum 12:1126–1129
5. Zhu XH (1996) Fas system research progress Microbe. Immunology 2:27–33
6. Suda T, Tanaka M, Miwa K et al (1996) Apoptosis of mouse naive T cells induced by recombinant soluble Fas ligand and activation induced r resistance to Fas lig. J Immunol 157:3918
7. Cascino I, Fiucci G, Papo tt G et al (1995) Three functional soluble form of the human apoptosis inducing Fas molecule are produced by alternative splicing. J Immunol 154(6):2706–2712

Chapter 76
Research on Athletes Impact of High Strength and Long-Term Aerobic Training

Xinxin Zheng, Tao Jiang and Bing Liu

Abstract The purpose of this study is to determine whether the high strength and long-term aerobic training can lead to left ventricle will have different mechanisms when increasing cardiac output at sub maximal upright bike exercise. We made ribs muscle research on 15 very competitive long-distance runners and 14 healthy sedentary adults, observing the largest peak of the four-chamber two-dimensional echocardiography at rest and during upright bicycle movement.

Keywords Frank-starling mechanism · High intensity · Long-term · Aerobic training

76.1 Introduction

Athletes engaged in long-term high-intensity aerobic training has been known as left ventricle cognitive expansion in the rest. However, the mechanism of left ventricular dilatation, the possibility of a mechanism to increase cardiac output during training is different between the sedentary and athlete, which has been a controversial topic. A description of these mechanisms is beneficial to not only adaptive sports training and understanding of the movement in cardiac

X. Zheng (✉)
Department of Physical Education, Hebei Vocational College of Foreign Languages, Qinhuangdao 066300, China
e-mail: xin2_zheng@163.com

T. Jiang
Institute of Physical Education, HeBei Normal University, Shijiazhuang 050024, China

B. Liu
P.E. Offoce, Luannan No.2 Senior Middle School, Tangshan 063500, China

W. Du (ed.), *Informatics and Management Science VI*, Lecture Notes in Electrical Engineering 209, DOI: 10.1007/978-1-4471-4805-0_76,

hypertrophy and left ventricular volume overload state. Researchers believe that exercise cardiac output is increased by the increase in left ventricular diastolic volume and stroke volume, especially in the high-intensity trained endurance athletes. These data have been cited as evidence of the mechanism of Frank-Starling. Thus, athletes and sedentary projects have priority in using it during exercise [1]. Therefore, increased cardiac output during exercise is fundamentally different mechanisms. However, there are scholars who believe that the increase in stroke volume early in the exercise of well-trained athletes is feasible, but further increasing to the maximum does not continue to increase [2].

In the past, in the process of many cardiovascular dynamics examining athletes, the exercise test is often the implementation of the supine position, and when upright exercise is always heart before and after load conditions. In addition, left ventricular volume estimation often used in technology, and this technology can take up to several minutes of data acquisition, which may mask the impact of left ventricular loading conditions and function, especially the dynamic changes in maximal exercise [3]. Therefore, it has been difficult to apply these meaningful data to the upright position of extreme sports. We hypothesized that endurance athletes during exercise will have a similar reaction. If the athletes' diastole volume reduces in high-intensity training, and the movement of cardiac output first has something to do with the increase in heart rate and myocardial contractility, and they do not use the Frank-Starling mechanism to accommodate the largest vertical movement. Therefore, the purpose of this study is the left ventricular volume changes between relatively competitive long-distance athletes and sedentary adults in the largest upright cycling [4].

76.2 Subjects and Steps

Two groups of subjects were studied. Exercise group has 15 competitive long-distance runners, and all of them are college running team members. Their age was 19 ± 1 years (mean $\pm$ 1SD); 9 men, 6 women. These athletes are an average of 4.4 ± 1 years, with 116 ± 5 km/week average training distance. All athletes have more than 70 km/week, at least in the last 8 weeks of training. The control group has 14 sedentary, healthy adults, and the average age was 28 ± 6 years, 10 men and 4 women. Each set of testers thereof in accordance with the testing requirements did not take in any inappropriate way to test. All of them in the rest and upright exercise bike have a normal cardiac physical review and normal electrocardiogram (ECG) [5].

All subjects were tested at the same time of the day. They exercise in the upright position on the bicycle ergometer and receive continuous ECG monitoring test. They have ECG and blood pressure monitor blood pressure tests at rest, exercise and peak exercise every 3 min. The training starts at zero workload and 75 (kp-m)/m and increases until this issue is due to exhaustion, and resolutely refused to further the implementation.

Echocardiography protocol. We use the advanced technology of wide-angle mechanical scanner to get the subcostal four-chamber view of two-dimensional ultrasound echocardiography. The main point of view is upright tape recorded every 3 min sports and peak exercise about 15 s when subcostal four chamber view video is in the rest. No images after exercise were used for measurement of left ventricular volume. In order to obtain repetitive video of the left ventricle, the following will be applied to the previously mentioned programs. Pocket PC sensor is placed to the left of the next approach and it is lower than the xiphoid, gently upward and pushing the sensor so that we can get the upward angle of visualization of the heart. Endometrial echo around the entire vertex clearly shows mitral valve leaflets, visualization, and intuitive left ventricle until the longest axis sensor point of view. In the course of the study, the angle of the sensor moves slightly from side to side, to ensure that the left ventricular long axis visualization. Echocardiography figures were recorded quiet or holding breath at rest and peak exercise.

Calculation of left ventricular volume. Diastolic and heart contraction (videotape frame before mitral valve opening) images using the forward and off-line analysis was carried out to determine the reverse slow motion and frame by frame analysis of video tapes. Intima is defined as the interface between the cavity and the myocardial echo, cavity between the epicardium is defined as the brightest among the pericardium echo. Ribs diastolic and systolic frame are displayed at the end of the longest axis, and all or almost all of the endocardial and epicardial are in a single one-stop-frame image visualization, which is determined to exercise at peak every 3 min for the rest of the theme. Diastolic endocardial and epicardial border manually systolic endocardial bordering XY digital converter (0.01 [0.03 cm] resolution) can track down the computer (smart D-100), so that the diagram boundary can be directly covered in the video image.

In order to detect images at rest and exercise that not through the left ventricle similar part, the quality of the left ventricle at rest and during exercise at each stage were calculated. We believe that the main body of the left ventricular mass did not change from rest to movement, making any changes in left ventricular mass in excess of 95 %, and equipment and analytical methods will be used to establish the confidence of the measurement limit (about 8 %). Therefore, the calculation of any of the cardiac cycle of left ventricular mass and the rest of the left ventricular mass difference (>10 %) will be excluded from the analysis, and the other cardiac cycle will calculate left ventricular volume. The rest of the cardiac cycle is always the first calculation to establish the baseline of the left ventricular mass, then, the subject of random sequence of the movement during the cardiac cycle and peak exercise were analyzed.

76.2.1 Statistical Methods

All data are expressed as ±1 SD of the average. We use the significance of heart rate peak in exercise and rest, systolic blood pressure, ejection analysis of variance test to test scores and exercise time. The difference between the two sets during the rest is the difference between intermediate and peak movement of left ventricular volume was tested using two repeated measures analysis of variance factors. P values <0.05 were considered statistically significant.

Relationship between cardiac work and cardiac diastolic, systolic volume index can be shown by cardiac work (W), which is obtained by each stroke volume (V) multiplied by the pressure (P), which is $w = p \times v$ (Fig. 76.1).

76.3 Statistical Results

Comparison the percentage of heart rate changes between each item at rest and peak exercise. Therefore, 0 % represents the heart rate at rest, 50 % represents at the halfway point between the rest and peak exercise heart rate, and 100 % represents the individual's peak heart rate. Athletes and sedentary heart rate differences between the rest (62 ± 10 vs 69 ± 8 beats/min) are relatively close, but did not reach statistical significance (P = 0.065). Gradual increases in heart rate for the two groups in the movement. Among 25 % heart rate (beats/min) (93 ± 10 vs 96 ± 14), 50 % (134 ± 14 vs 127 ± 16) and 75 % (166 ± 10 161 ± 18) peak exercise, during exercise peak (187 ± 7 compared with 186 ± 13), these two groups are the same. Also the peak movement between the athletes and sedentary systolic blood pressure (187 ± 16–181 ± 19 mm Hg) or two-dimensional ultrasound echocardiography, left ventricular ejection fraction (0.80 ± 0.08 vs 0.78 ± 0.08) shows no significant difference. As expected, the athletes (16.3 ± 2.6 min) has a longer exercise time than the sedentary (12.0 ± 2.6 min) (P < 0.001).

In Fig. 76.2, the athlete's cardiac output volume index was 85 ± 14 ml/m2 at rest, and early exercise will increase the index (95 ± 17 ml/m2, at 25 % [P < 0.001 correspond to at rest], and then gradually decline to 50 % of

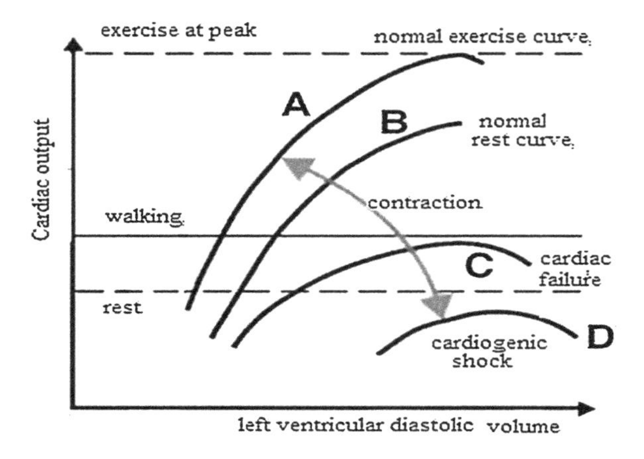

Fig. 76.1 Frank-starling mechanism

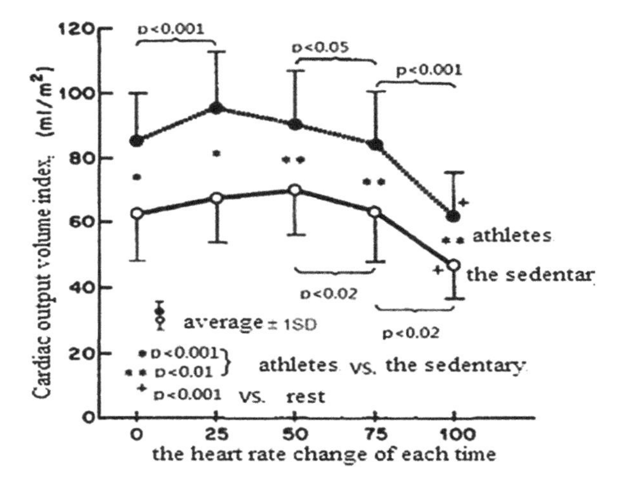

Fig. 76.2 Cardiac output volume index of the athletes and the sedentary during exercise

90 ± 17 ml/m2; 75 % of peak heart rate was 83 ± 18 ml/m2; peak training is 61 ± 14 ml/m2 ($P < 0.001$ and rest in comparison) (Fig. 76.2). Sedentary group diastolic index at rest was 62 ± 14 ml/m2, 25 % for 67 ± 14 ml/m2, 50 % for 70 ± 15 ml/m2, resting heart rate 75 % is 63 ± 15 ml/m2, and the movement reduces at the peak value of 46 ± 10 ml/m2 ($P < 0.001$ compared with at rest). Between the two groups ($P < 0.01$–0.001), there is a significant difference in cardiac output volume index at rest and during peak exercise.

Stroke volume changes are shown in Fig. 76.3. When at rest, the athletes' stroke volume index was 48 ± 15 ml/m2, and this value increased significantly ($P < 0.001$); peak heart rate was between 25 % and 66 ± 13 ml/m2; in high-movement, 50 and 75 % have no change in peak exercise levels, and decreased to 49 ± 14 ml/m2 ($P < 0.001$ than 75 % of the level). Peak exercise stroke output

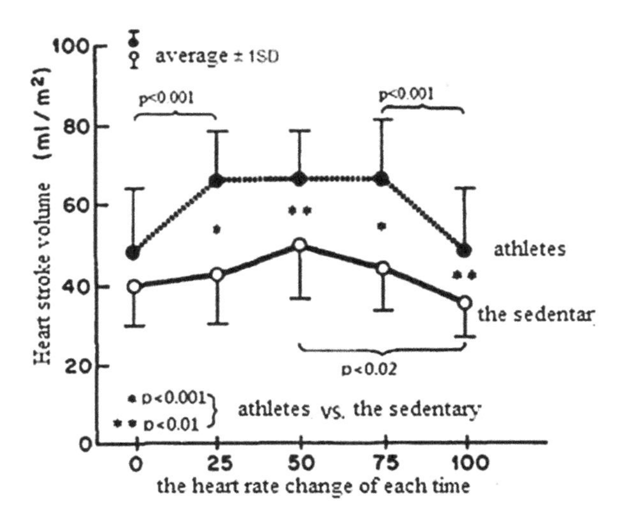

Fig. 76.3 Heart stroke volume index of the athletes and the sedentary during exercise

was not significantly different from rest, because both the cardiac output volume index and stroke volume index have a similar decline during peak exercise. Sedentary group in the rest have the stroke volume index 40 ± 10 ml/m2, and there is an increasing trend, but not significantly ($P = 0.09–0.32$). Peak exercise is 25, 50 and 75 %, and peak movement is 36 ± 9 ml/m2, which also reflects the maximal exercise diastolic and systolic volume index and the reduction of similar magnitude. Athletes at peak exercise ($P < 0.01$), stroke volume index was significantly greater than the sedentary group ($P < 0.01$ than 0.001) during the exercise.

Athletes at rest have a larger cardiac end-diastolic volume index (85 ± 14 ml/m2) (mean) than sedentary adults (62 ± 14 ml/m2), and a larger systolic volume index (37 ± 11 right 21 ± 6 ml/m2). In low-and moderate-intensity exercise, diastolic and pulse volume index of these two groups have increased, but in the high-intensity training and peak training, diastolic volume index these two groups decreased significantly below the rest value (athletes, 61 ± 14; sedentary group, 46 ± 10 ml/m2, compared with the rest to meet the $P < 0.001$). This reflects the decline in diastolic volume index; at peak training, stroke volume index decreased from moderate training value of the two groups and showing no difference in the values in the rest.

76.4 Conclusion

Long-term competitive well-trained long-distance runners have larger cardiac output volume and final shrinkage in normal rest compared with sedentary adults adhering to the moderate and has different cardiac output volume and stroke output from maximum vertical cycling (but not increase). For left ventricular dilatation of the athletes at rest, they may use the Frank-Starling mechanism to maintain the low-and moderate-intensity exercise, increase cardiac output in the normal rest. However, in the high-intensity exercise, cardiac output volume to reduce the stroke volume is unchanged between the athletes and the sedentary group. Thus, while the rest of the long distance runners have an enlarged left ventricle, they use the same mechanism as the sedentary adults to increase cardiac output in the dynamic movement of the upright. In low-and middle-level training, the mechanism of the Frank-Starling is a dominant mechanism to increase cardiac output, but at peak exercise, possibly because of lower diastolic left ventricular filling, enhanced contraction is the main mechanism to maintain stroke volume.

References

1. Li WJ et al (2009) The combined effects of high-intensity aerobic and anaerobic training on the athletes' biochemical indicators. Mod Clin Med Bioeng 22(05):5–22
2. Peng Y, Wu NY (2010) Impact of high intensity aerobic training on middle distance runner bodily functions. Phys Educ 30(2):3–20

3. Ye T (2008) Exercise physiology tutorial, vol 1(2). Higher Education Press, Beijing, pp 432–441
4. Cao HY, Shao ZP (2007) Impact of high intensity interval training on speed skating athletes aerobic endurance. Sports 22(2):40–55
5. Ji JM, Xiao GQ (2011) Research of the effects of high-intensity training on aerobic endurance. Rehabilitation 4(1):88–110

Chapter 77
Research of Complex Training of Young Athletes Explosive Power

Zhiping Wang, Dong Li, Lei Wang and Lichao Zhang

Abstract Explosive force are represented by speed and strength, which have decisive impact on the athletes performance, so how to strengthen young athletes' explosive power is the cause of general concern. The article makes a descriptive statistical analysis and is based on the perspective of complex training in young athletes tested for aerobic and anaerobic training to test their athletes muscle strength and performance enhancement degree of differentiation and forming power and rate model, further using complex training to evoke the significant growth of the maximum muscle strength, and exploring the way of young athletes' explosive power enhancement.

Keywords Complex training · Explosive power · Differentiation · Descriptive statistical analysis

---

Z. Wang (✉)
Department of Physical Education, Northeastern University at Qinhuangdao, Qinhuangdao 066004, China
e-mail: zhip_w@163.com

D. Li
Department of Physical Education, Qinhuangdao Institute of Technology, Qinhuangdao 066100, China

L. Wang
P.E. Office, Luannan No.2 Senior Middle School, Tangshan 063500, China

L. Zhang
P.E. Office, Shijiazhuang Foreign Studies School, Shijiazhuang 050000, China

W. Du (ed.), *Informatics and Management Science VI*, Lecture Notes in Electrical Engineering 209, DOI: 10.1007/978-1-4471-4805-0_77,

77.1 Introduction

In sport, athletes' explosive power plays a vital role for outstanding achievements, so we are also increasingly concerned about the athletes' explosive power, and gradually focus on young athletes explosive power analysis to further explore how to enhance young athletes explosive power. Explosive power is not only related to all the muscle groups of the body, but it needs the muscle groups to participate in it together, which is the largest outbreak of a particular instant. The explosive power has a direct impact on personal power training of young athletes' the most important factors. Therefore, this paper focuses on the young athletes explosive research, comprehensive test of aerobic and anaerobic exercise in several ways, and the sports movement and enhanced physical exercise to identify the complex training's impact on young athletes explosive power, and to enhance the explosive power of young athletes, to improve physical activity for young athletes, and to improve exercise results [1].

We use the questionnaire survey to research people who participate in aerobic, strengthen, stretching self assessment. Exercise questionnaire is based on trans-theoretical model of behavior change, and it is different from engagement in a particular behavior preparation phase. The present study includes five stages: lack of deliberate practice, meditation (want to exercise), ready to (plan to exercise), action (present exercise) and maintain (continuous exercise). The first three stages mean that lack of exercise, and the latter two represent the current stage in the exercise. The specific standard is defined as three different sports: aerobic exercise, stretching and strength training. Aerobic exercise is a regularly plan to increase physical activity, for example, taking a walk, aerobic exercise, running, bike riding, swimming, boating, etc. [2]. Conventional strength training is to use strong body for physical activity, for example, increasing or removing free weights, using weight or resistance training machine, etc.

This paper is to study the complex training's impact on young athletes' explosive power and through the combination of the physiological kinematics and physical exercise theory, conduct a comprehensive analysis of complex training and muscle performance, muscle activity, the shape and muscle structure, muscle contraction, the load and explosive indicators.

77.2 Study Groups and Steps

For young athletes, while considering enhancing their explosive power, we must be fully expressed concern of the stage of their growth and development and fully consider and analyze their physical, psychological characteristics and physiological and psychological adaptation as well as to their withstanding range. If we take high intensity training which may be inflicting bodily harm and it is not conducive to enhance the explosive force of young athletes, but just the opposite. But if we take the low load of training methods, then it's not effective; so we must find a reasonable and effective way of their training and combine a variety of training methods effectively to form a complex training program, not only in enhancing the fun exercise, but also a timely manner to enhance the physical exercise to strengthen the muscle strength of young athletes, the speed rate of athletes and their explosive power.

The research groups are primarily college young athletes, age 19 ± 3 years, male 22, female 8. The testers are divided into two sub-groups of aerobic and anaerobic training, and their muscles corresponding indicators were tested and recorded before and after training for the comparative analysis.

The complex training steps are as follows [3].

First, the high load weight training and polymeric training help to develop the strength of the movement. Athletes have trainings of bench press, squat, added weight and enhance the type of training to enhance muscle strength and development of sports performance. A weight and enhanced methods combined training is a "complex training". Involve complex training in biomechanics, such as similar high-load alternating weight training and polymeric exercises. Complex training is a simple, and may be the optimal strategy of training and development of athletes' athletic ability. Combine complex training examples such as the combination of strength loss or squatting, and depth jumps. Complex training can also be applied to the snatch, clean and jerk. Complex training, high intensity weight training is the optimal adaptation for follow-up polymeric training to supplement the movement. Complex training is suitable for a variety of groups and individual sports, and is helpful for injury rehabilitation of athletes training.

The complex training models should follow the cyclical requirements of athletes with functional strength or involved in the preparation of the basis of the strength of the circulation process. Athletes should be within the cycle range of the main low-intensity training. Ultimately athletes can combine the specific complex movements. Polymeric exercises must be in accordance with all possible strength and limited repetition to ensure that the high intensity of the work. This paper will conduct a periodic training with the increased load of weight training; repeat and allow reducing fatigue and athletes' rehabilitation. In a large number of combinations there can be weight and polymeric exercises. The combination of polymeric training and weight training provides complex adjustment programs in training programs and increase changes.

After a typical high-intensity training, they can do similar polymeric training, based on biomechanical perspective of sports training. They can also do movements such as light load on the ground after the jump squat enhanced legs vertical jump training, many possible weights and enhanced combination of exercises for complex training [4].

Complex training requires a combination of biomechanics and movement speed. Complex high-intensity training is the velocity of the request. Sports is a specific "functional training" complexes, and provides the flexibility of the generalized increase in capacity. The principle of recovery is equally important and we need to consider how to implement a complex training program. Complex training puts weight and polymeric training on the same day. Reduce the associated procedures, reduce fatigue and allow the athlete to concentrate on work performance are restored to its previous level of training cycle, so it is recommended to resume training time to not more than 96 h and to exercise the same muscle group rehabilitation. Therefore, they do complex training three times a week, and in rest intervals they should be allowed to make up the energy [5]. Complex exercise rest time is very important, part of whose cycle can affect the muscles.

77.3 Statistical Results

The first test for the testers is strength and speed test. In the beginning, the strength and speed rate of the two curves are basically the same; the rate increase, forces are increasing too. But when the rate is around 170.S-1, the increase of the strength has some difference, and the distance between the two curves gradually increase.

The testers' muscle indicators were recorded before and after training, and the results are shown in Table 77.1.

The change of the indicator parameters in Table 77.1, three test groups before and after training muscle (Fig. 77.1).

From Table 77.1 and Fig. 77.2, males in Group 1 before and after aerobic training, creatine kinase and myoglobin have relatively large changes and in the rate of increase, in addition to glucose decreased after training, and the remaining indicators are in rising trend. Males in Group 2 before and after anaerobic training, lactic acid, lactate dehydrogenase, creatine kinase and myoglobin increases are larger, but all indicators are showing an upward trend. Females in Group 3, is enzyme of creatine kinase did not change before and after anaerobic training,

Table 77.1 Changes of muscle indicator parameters of three test groups before and after training

Parameters	Group 1 (male) aerobic		Group 2 anaerobic 1 male		Group 3 anaerobic 1 female	
	Before training	After training	Before training	After training	Before training	After training
Lactic acid (m mol/L)	2.42	2.56	2.95	12.53	2.89	8.31
Lactate dehydrogenase (U/L)	270	289	251	320	221	241
Creatine kinase (U/L)	1,260	1,490	1,000	1,330	920	1,190
Creatine kinase isoenzyme (U/L)	70	80	50	70	30	30
Alkaline phosphates (U/L)	95	98	92	110	90	95
Myoglobin (μg/L)	143.7	438.6	55.54	207.1	110.3	376.5
Hemoglobin (g/L)	148	150	140	151	127	129
Glucose (m mol/L)	4.8	4.34	6.5	8.2	4.2	5.2

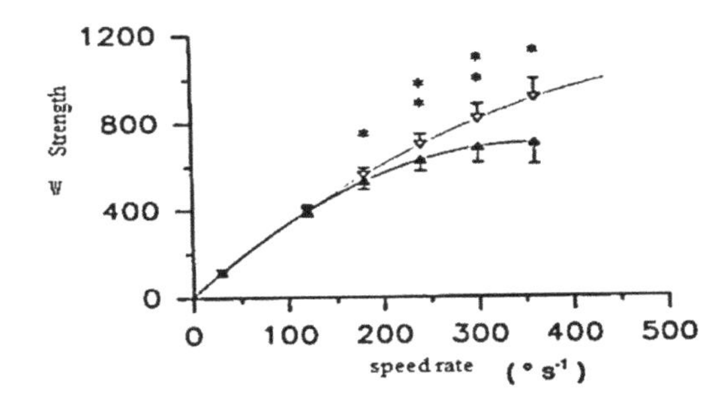

Fig. 77.1 Strength and speed rate curves of the test groups

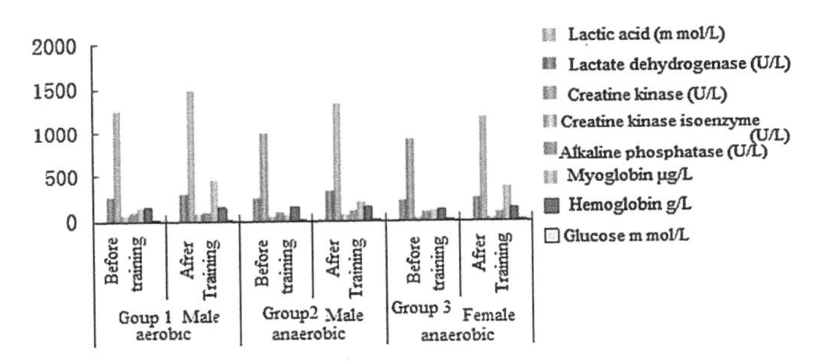

Fig. 77.2 Muscles indicators' changing trend of the three test groups before and after training

lactate, creatine kinase, and myoglobin increased. The above reflects in anaerobic training, lactic acid, lactate dehydrogenase, and other indicators will be a substantial change. But in either anaerobic or aerobic training, creatine kinase and myoglobin before and after training will increase. However, for males and females, creatine kinase isoenzyme in women before and after training did not change, but other indicators' increases are smaller than males.

For testers we only test anaerobic and aerobic training, and this is in a form of ±SD. The descriptive statistical analysis is shown in Table 77.2.

The results of combination of anaerobic and aerobic training of complex training are shown in Table 77.2 and Fig. 77.3. Before and after anaerobic training and aerobic training, the creatine kinase is enzyme after aerobic training reduced, but after anaerobic training it increased, and the remaining indicators are consistent with the trends before and after the anaerobic and aerobic training. Before and after lactic acid test is basically a slight increase. Lactate dehydrogenase increased the before and after training, but generally is about 30 U/L; creatine kinase, hemoglobin and glucose had smaller increases before and after training; alkaline phosphates and myoglobin increased greatly before and after training, which are more than 23 U/L and 50 μg/L.

Table 77.2 Muscle indicators parameters' changes in aerobic and anaerobic training groups before and after training

Parameters	Aerobic training group (n = 14)		Anaerobic training group (n = 16)	
	Before training	After training	Before training	After training
Lactic acid (m mol/L)	2.60 ± 0.44	6.44 ± 4.14	2.66 ± 0.44	11.28 ± 1.86
Lactate dehydrogenase (U/L)	214.2 ± 42.88	246.12 ± 44.6	184.6 ± 46.6	246.14 ± 44.2
Creatine kinase (U/L)	467.4 ± 14.2	474.4 ± 474.0	480.4 ± 286.4	484.7 ± 466.1
Creatine kinase isoenzyme (U/L)	84.8 ± 24.6	44.00 ± 17.06	21.78 ± 10.04	41.00 ± 14.42
Alkaline phosphates (U/L)	26.6 ± 14.6	81.42 ± 24.76	81.60 ± 44.82	104.64 ± 48.6
Myoglobin (μg/L)	44.6 ± 46.8	140.8 ± 114.6	28.11 ± 26.46	87.76 ± 86.12
Hemoglobin (g/L)	141.6 ± 11.47	144.6 ± 14.6	148.6 ± 12.66	146.24 ± 14.4
Glucose (m mol/L)	5.52 ± 0.97	5.61 ± 1.14	5.36 ± 0.45	6.89 ± 1.01

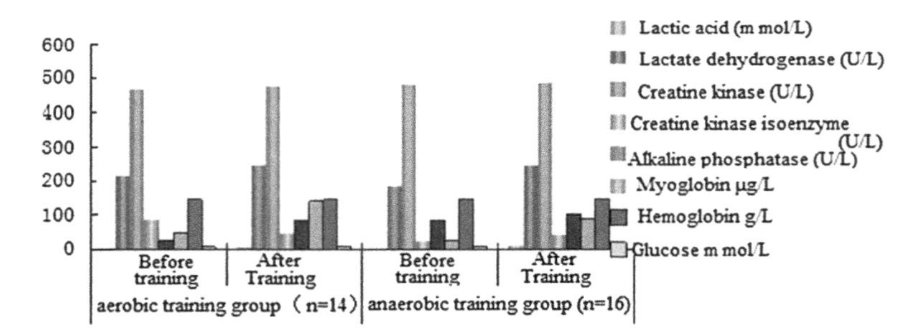

Fig. 77.3 Muscle indicators' changing trends in aerobic and anaerobic training groups before and after training

77.4 Conclusion

Complex training can be the combination of conventional sports-specific movement, enhanced training and targeted exercises. These training methods can be carried out alternately. Complex training may be a practical strategy of the best players in personnel training. Based on the physiological kinematics, for the young athletes' the maximum muscle strength is improved, and the muscle activity, morphology and muscle structure are changed, and this also enhanced the reaction force of the muscle, which have a close relationship with muscle contraction, load, and nerve reaction. This greatly enhances the explosive power of the young athletes. Complex training, from the perspective of physical exercise, perfects the movement skills, exercise, speed and intensity of exercise of athletes, which are closely related to the explosive power of the young athletes. For young athletes explosive power enhancement, we can take more complex trainings, both in forms of exercise and can make it more rational and effectively, depending on the project plans to make reasonable arrangements to enhance the explosive power of young athletes.

References

1. Feng CL (2010) The jumping movement of explosive power of high school students. Sci Technol Inf 33(19):88–90
2. Wu SJ (2009) Views on the explosive power training in the track and field events. J Fujian Inst Phys Educ 21(2):67–89
3. Tian MJ (2010) Sports training. People's Sports Press, Beijing 67:192–194
4. Li WT (2010) Research of the best load of explosive power. Sports Sci Technol 40(12):120–133
5. Liu BX (2006) Research on "maximum power training method" of the development of explosive force from the body. Sports Sci 23(5):42–47

Chapter 78
Input–Output Model of Sports Economy Based on Computer Technology

Lina Zhu

Abstract In order to accelerate the industrialization progress of sports, to promote the fast development of China's sports economy, and to realize the rise of China's sports industry, the paper explored and analyzed the problem of how to utilize computer technology to realize input–output model analysis in sports economy. Based on the data calculated by MATLAB, sports economy input–output table in future three years was formulated. Then, by drawing support from computer technology, the development trend of sports economy can be predicted, and the input and output of sports economy can be analyzed. In this way, we may be able to provide a scientific and effective method for analyses on sports economy, to better combine sports with economy and to further facilitate economic development.

Keywords Sports economy · Input–output model · Computer

78.1 Introduction

Life lies in motion. Human beings' daily life can never go without sports. Along with the situation, economy also has close connection with sports. As a burgeoning area, sports economy is being attached with increasingly more importance by governments of all levels [1]. In economic aspects, sports economy is playing an increasingly significant role, gradually becoming a new growth point promoting the development of economy. However, so far, China's sports economy has not yet

L. Zhu (✉)
Department of Physical Education, Qingdao University of Science and Technology, Qingdao 266089, China
e-mail: Zhulina2013@yeah.net

W. Du (ed.), *Informatics and Management Science VI*, Lecture Notes in Electrical Engineering 209, DOI: 10.1007/978-1-4471-4805-0_78,

formed its completed system, still staying in a chaotic situation. For example, there are many gymnasiums still remaining in field rent, paid instruction and other similar low grade business stage. In order to meet China's economic development demand, to better promote the rise of sports economy and to perfectly combine sports with economy, we need to constantly explore and analyze input and output of sports economy, so as to provide reference for our decisions [2–4].

78.2 Input–Output Model

Input-output mode is a sort of economic mathematics model, referring to linear algebra equation set for describing economic contents presented by input–output table in mathematic forms [5].

Input–output table is a chessboard-like table for reflecting input source and output of product production. The Table 78.1 shows the general form of input–output table.

Basic equation relation in input–output table: seeing vertically, intermediate input + initial input = total input; seeing horizontally, intermediate product + final product = total output.

Total input of each department = total output of the department; total volume of quadrant II = total volume of quadrant III. This is the general equation of input–output table, i.e. total initial national input = total final product.

Concretely speaking, the so-called input–output model is a linear algebra equation system based on the former two basic equation relations. When formulating the input–output table, modeling shall be conducted with relatively stable factors of previous years as the basis, so as to bring to play already-built tables, to make analysis and forecast on future tendency through the model. Here, the author will place the emphasis on product input and output model. Dividing by phase, it is comprised by static model and dynamic model. Static model is comparatively mature, with longer application history and wider range. As for dynamic model, there is still a gap before it is being practically applied, which needs to be further

Table 78.1 Input–output table

Distribution		Intermediate product Department 1 department 2 … department n	Final product Y_i	Total output X_i
Intermediate input	Department 1	$x_{11}\ x_{12}\ \ldots\ x_{1n}$	y_1	X_1
	Department 2	$x_{21}\ x_{22}\ \ldots\ x_{2n}$	y_2	X_2
		$x_{n1}\ x_{n2}\ \ldots\ x_{nn}$	$\vdots$	$\vdots$
	Department n		y_n	X_n
Added value		$v_1\ v_2\ \ldots\ v_n$	v	v
Total input X_j		$X_1\ X_2\ \ldots\ X_n$	X	X

researched both theoretically and practically. In the following, we are mainly to introduce the two sorts of models, as well as deduction of several major coefficients.

78.2.1 Static Input–Output Model

The so-called static input–output model refers to input–output model excluding time factors. Static product input–output model is the basic form of input–output analysis. As for input–output models of other types, they can be regarded as extensions of static model. Thus, in order to understand input–output principle, we have to get to know static product input–output model first.

78.2.2 Material Input–Output Model

If input–output table adopts material unit of measurement, it will consequently be regarded as a material pattern input–output table. Material input–output table is free from the influence of price, which can more directly reflect the input–output relation among departments. However, material measurement unit is restrained by product quality, so that the applicable range of material input–output table is quite limited. In material input–output tables, product is the basis of classification, and the measurement units are material objects. Simplified material input–output table is shown as the Table 78.2.

Analyzing in lines, what reflected is the distribution and usage situation of products. Some of the products are intermediate products, being consumed during production of other products. The rest could be used for investment and consumption as final products. The sum of these two parts is just the total production volume of all products in this phase.

Seeing by columns, what reflected is actually product input consumed during production process of different products. In the meanwhile, we should especially notice that, the measurement unit of products in columns may be different, so that

Table 78.2 Material pattern input–output table

	Input	Intermediate product				Final product	Total
		1	2	…	n		
Material consumption	1	q_{11}	q_{12}	…	q_{1n}	y_1	Q_1
	2	q_{21}	q_{22}	…	q_{2n}	y_n	Q_2
	…	…	…	…(I)	…	…(II)	…
	n	q_{n1}	q_{n2}	…	q_{nn}	y_n	Q_n
Labor		q_{01}	q_{02}	…(III)	q_{on}		

we shall never calculate blindly. As for this, normally, material input–output models only contain line models, rather than column models.

The equation relation of material input–output table is that, intermediate products and final products constitute total products. With symbols, it can be described as follows:

$$\begin{aligned} q_{11} + q_{12} + \cdots\cdots + q_{1n} + y_1 &= Q_1 \\ q_{21} + q_{22} + \cdots\cdots + q_{2n} + y_2 &= Q_2 \\ \cdots\cdots \\ q_{n1} + q_{n2} + \cdots\cdots + q_{nn} + y_n &= Q_n \end{aligned} \quad \text{or} \sum_{j=1}^{n} q_{ij} + y_i = Q_i (i = 1, 2, \ldots, n) \quad (78.1)$$

78.2.3 Direct Consumption Volume

This term is also referred to as technical coefficient or investment coefficient, normally indicated by a_{ij}. Its meaning is consumption volume of product i, when producing product j at certain amount.

Relational expression of direct consumption coefficient:

$$a_{ij} = \frac{q_{ij}}{Q} \quad (i, j = 1, 2, \ldots, n) \quad (78.2)$$

It has explicit meaning, easy for calculation. Moreover, it is quite important in input–output model analysis. As for this, the basic premise for a successful input–output model analysis is to correctly figure out the direct consumption coefficient. Here, we import direct consumption coefficient $a_{ij}(i, j = 1, 2, \ldots, n)$ into equation (78.1): $q_{ij} = a_{ij}Q_j(i, j = 1, 2, \ldots, n)$

$$\sum_{i=1}^{n} a_{\text{ij}}Q_j + y_i = Q_i \quad (i = 1, 2, \ldots, n) \quad (78.3)$$

This equation can also be modified into a matrix form:

$$AQ + Y = Q \quad (78.4)$$

Thus, it can also be presented as $Y = (I - A)Q$.

In the equation, "I" is the unit matrix.

Besides, we may also revise (78.4) into the following equation according to the relation between final product and total product:

$$Q = (I - A)^{-1}Y \quad (78.5)$$

On this basis, if Y is given, it will be easy for us to work out Q according to (78.5).

78.2.4 Total Consumption Coefficient

Generally speaking, during production, different products have not only all sorts of direct consumption relation, but also all sorts of indirect consumption relation. As for this, total consumption coefficient is just an overall reflect of all direct and indirect relations.

78.2.5 Final Product Coefficient

Normally, element $\overline{b}_{\mathrm{ij}}$ in matrix $(I - A)^{-1}$ is called full demand coefficient or final product coefficient, i.e. the final product coefficient shall be:

$$(I - A)^{-1} = B + I = \begin{pmatrix} \overline{b}_{11} & \overline{b}_{12} & \dots & \overline{b}_{1n} \\ \overline{b}_{21} & \overline{b}_{22} & \dots & \overline{b}_{2n} \\ \dots & \dots & \dots & \dots \\ \overline{b}_{n1} & \overline{b}_{n2} & \dots & \overline{b}_{nn} \end{pmatrix} = \begin{pmatrix} \overline{b}_{11} + 1 & \overline{b}_{12} & \dots & \overline{b}_{1n} \\ \overline{b}_{21} & \overline{b}_{22} + 1 & \dots & \overline{b}_{2n} \\ \dots & \dots & \dots & \dots \\ \overline{b}_{n1} & \overline{b}_{n2} & \dots & \overline{b}_{nn} + 1 \end{pmatrix} \tag{78.6}$$

This matrix is used to indicate the full demand volume of the ith product, when the final usage of the jth product is to be increased by one unit.

78.2.6 Distribution Coefficient

The so-called distribution coefficient (hij) refers to the ratio between the consumption volume qij of the ith product being distributed for the production of the jth product and the total domestic production volume Qi (excluding imported products) of the ith product. If so, the following equation can be deduced:

$$h_{ij} = \frac{q_{ij}}{Q_i} (i, j = 1, \dots, n;\ 0 < h_{ij} < 1) \tag{78.7}$$

78.3 Computer Implementation of Input–Output Model

Sports economy starts from the dimensionality of production and operation, combining and developing public sports and related economic activities as a special industry. Based on a simple instance, the author has established the input–output model and realized its computer implementation, so as to explain application of computer technology in sports economy.

In the following, based on an instance, we are to demonstrate how to make use of computer technology MATLAB for input–output analysis. Table 78.3 is an input–output table, in which, data in each line shows the distribution situation of all departments' products in all consumption departments' final demand. For example, in the first line, sports competition department is to provide 2,189 billion Yuan products for its own consumption, 130 million Yuan for sports facility department, 260 million Yuan for other departments, for social final product 2,874 billion Yuan (social consumption 2,147 billion Yuan, social accumulation 720 million Yuan) from the total output of 5,473 billion Yuan. Assuming that the direct consumption coefficient is constant (if there is no major changes in technical conduction, this value could keep constant within a certain period) in recent two years, the annual growth rate of sports competition department's total output is 8.4 %, sports facility department 6.7 %, other departments 9.2 %. Now, through calculation, we are able to predict and plan three departments' total output, final output, and all intermediate circulation volume among these departments by the end of the third year.

Step 1 Direct Consumption Coefficient

```
≫x1 = [21.89 1.30 2.60; 2.73 0.86 0.43; 1.37 0.42 0.86; 17.28 4.81 2.59; 11.45 1.44 2.16]; % Input xij matrix
≫y = [21.74 7.20; 4.03 0.58; 4.90 1.09]; % Input Y matrix
≫x = [54.73 8.63 8.64]; %Input X matrix
≫for i = 1:3,n a(:,i) = x1(:,i)/x(i);end; %Calculation of direct consumption coefficient
≫a %direct consumption coefficient
```

Step 2 Departments' Total Output in Three Years

```
≫t1 = sum(x1); %Sum up each column in Xij matrix
≫t1(1) = t1(1)*(1 + 0.084)^3 %Total output of sports competition department in three years
ans = 69.7002
≫t1(2) = t1(2)*(1 + 0.067)^3 %Total output of sports facility department in three years
ans = 10.4834
≫t1(3) = t1(3)*(1 + 0.092)^3 %Total output of others department in three years
ans = 11.2508
≫y1 = (eye(3)−a(1:3,1:3))*t1 ' %Final output
ans = 36.8578 5.4021 7.8760
```

Step 3 Full Consumption Coefficient

```
≫b = inv(eye(3)−a (1:3,1:3))−eye(3)
```

Table 78.3 Sports economy input–output table (Unit: RMB 100 million Yuan)

Output		Intermediate usage				Final usage			Total output
		Sports competition	Sports facility	Others	Total	Consumption	Accumulation	Total	
Intermediate	Sports competition	21.89	1.30	2.60	25.79	21.74	7.20	28.94	54.73
	Sports Facility	2.73	0.86	0.43	4.02	4.03	0.58	4.16	8.63
	Others	1.37	0.42	0.86	2.65	4.90	1.09	5.99	8.64
	Total	25.99	2.58	3.89	32.46	30.67	8.87	39.54	72.00
Added value	Labor	17.28	4.81	2.59	24.48				
	payment	11.45	1.44	2.16	15.05				
	Net social income	28.74	6.05	4.75	39.54				
	Total	54.73	8.63	8.64	72.00				
Total income									

Step 4 Data of Part I and Part III in Input–output Table in Three Years

```
≫for i = 1:3,xx(:,i) = t1(i)*a(:,i);end
≫xx
yy = sum(y ')
```

Step 5 Data of Part II in Input–output Table in Three Years

```
≫for i = 1:3
for j = 1:2
y(i,j) = y(i,j)*y1(i)/yy(i);
end
end
≫y
```

78.4 Conclusion

By drawing support from MATLAB, we are able to work of all data, i.e. sports economy input–output table for future three years. With computer technology, we are able to predict the development trend of sports economy. This would be beneficial for China's sports departments to fully bring to play the diversified function of sports, to accelerate the industrialization of sports, to better promote the fast development of China' sports economy and to realize the rise of China's sports industry.

References

1. Chen H, Sullivan HJ et al (2002) Savenye perspective on the future of computer use in China. Educ Tech Res Dev 50(1):92–101
2. Maggiolini P (2011) Information technology benefits: a framework. Emerg Themes Inf Syst Organ Stud 3(5):281–292
3. Chen Y-Q, Qi J-M, Shi L-J (2012) Analysis of college sports consumption and the sunshine sports in China. Adv Intell Soft Comput 148:1–5
4. Mitten MJ, Opie H (2012) Sports law: implications for the development of international comparative and national law and global dispute resolution ASSER international sports law series. Lex Sportiva: What is Sports Law 45:173–222
5. Gurubatham MR (2005) Understanding and interpreting the drivers of the knowledge economy. IFIP Int Fed Inf Process 161:189–202

Chapter 79
Database Establishment Scheme of Competition Results of Gymnastics in Large Sports Games

Ning Gong and Yingying Gong

Abstract Sports is the most important factor to enhance people's physique. Holding a large sports game is a huge opportunity and challenge for the hosting city or country, which is also a test and inspection of the national sports concept, sense of honor and solidarity. In large sports games, one important aspect is recording, evaluation and reward of the athletes' performance; its fairness and efficiency determines whether the sports game is successful. The study object of this paper is the competition results of gymnastics in large sports games, and informatization method is used to establish scientific and reasonable SQL database, optimize related algorithm of data structure, record and process the information and performance of the athletes, and to rank and evaluate their performance. Establishment of this database has great significance in ensuring fair and just performance evaluation, fast announcement of results and alleviating the staff's burden.

Keywords Large sports game · Gymnastics · Database

79.1 Instruction

With fast development of global economy and people's emphasis on sports, the frequency and participators of large sports games have kept growing, and its influence has also kept expanding [1]. Take China for instance, more and more international competitions have been "settled" in China. It should be noted that behind the glorious large sports games, various computer application technologies

---

N. Gong · Y. Gong (✉)
Physical Education Department, Hunan City University, Yiyang 413049, China
e-mail: Gongyinying@yeah.net

W. Du (ed.), *Informatics and Management Science VI*, Lecture Notes in Electrical Engineering 209, DOI: 10.1007/978-1-4471-4805-0_79,

such as the computer network technology, database technology and multimedia technology have played an important role as the background support, which makes significant contribution to the smooth proceeding of large sports games. Among these technologies, the fast development of the computer database technology is implemented throughout the whole game, which plays an irreplaceable role from storage of athlete information to processing of final results.

79.2 Database Technology

79.2.1 Overview of Database Technology

Database technology is the most efficient means for information resources management. Before database establishment, the most important step is database design [2]. Database design refers to construction of optimum database schema, establishment of database and other application system, effective data storage and satisfaction of the user information requirement and processing requirement in accordance with a given environment of application. The procedure of database design is shown in Fig. 79.1.

79.2.2 Selection of Mature Database Software

Before establishment of database, selection of mainstream database software is critical. In large database, common software includes SQL Server, DB2 UDB and Oracle [3], which has their own characteristics. 15 indices of these three types of

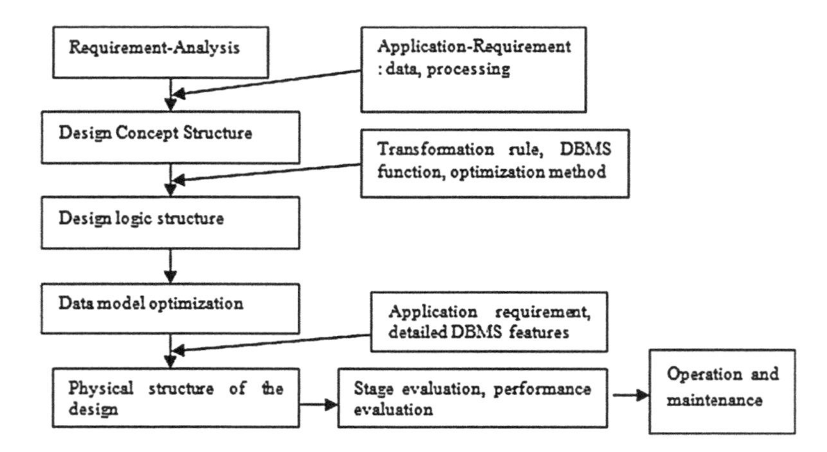

Fig. 79.1 Procedure of database design

Table 79.1 Comparison and analysis of three mainstream databases

Index	Best	Better	Good
Usability/user interface	SQL server	DB2 UDB	Oracle
Installation/configuration	DB2 UDB/SQL server		Oracle
Multi-server management	DB2 UDB/oracle		SQL server
Space management	DB2 UDB/oracle		SQL server
Backup/recovery/revocation management	DB2 UDB	Oracle	SQL server
Data loading/filing	DB2 UDB	Oracle	SQL server
Project management	Oracle	SQL server	DB2 UDB
Object management of cross-platform	DB2 UDB	Oracle	SQL server
Management of different environment	Oracle	DB2 UDB	SQL server
Parallelization/synchronization of application program	Oracle	DB2 UDB	SQL server
Security management	Oracle	DB2 UDB/SQL server	
Automatic/autonomous	SQL server	DB2 UDB	Oracle
Minimum total complexity	SQL server	DB2 UDB	Oracle
Expandability	DB2 UDB	Oracle	SQL server
Performance	DB2 UDB	Oracle/SQL server 2	

database software are compared and analyzed in the following, and classified into three levels of Best, Better and Good (Table 79.1).

79.3 Database Establishment of the Competition Results of Gymnastics

79.3.1 Rules for Gymnastics Competition

Gymnastics is one the most ancient sports. In the new rules implemented since January 1, 2006, the 10-full mark method was cancelled, and grading became more detailed [4]. Judges' scores are divided into two groups of Difficulty Score (A) and Completion Score (B). Group A starts from 0, and there is no upper limit; full mark of Group B is 10; judges are also divided into the two groups of A and B.

Athlete' valid score for single item is: valid score = starting score-average deduction-other deductions; generally speaking, in male/female team competition, the 6–5–4 method is adopted, i.e., 6 people are allowed to compete, only 5 can really compete, and best four performances are used finally. After the result of the team competition, one female and one male athlete will be chosen from the team to participate in the all-around final and various single finals. Final results of various competitions are obtained through one-time scoring, and the result of the all-around competition is the sum of the results of various single-item competitions [5].

79.3.2 Database Module Design

The performance management system can generally be divided into two modules: one is the basic information model of the athletes and judges, which includes the basic information of the athletes; the other is the performance processing module, which includes the update, inquiry and processing of the athletes' performance information [6].

79.3.3 Data Dictionary Analysis

Data item refers to the indivisible data units in the database relation, and the following table has listed the data name, data type and length and whether it can be null [7]. SQL Server was used to establish the database of "Gymnastics Performance", and it is divided into Public Table and Private Table in accordance with the actual situation of gymnastics competition. The Public Table contains information of each judge, athlete and item, and its basic list and structural description is as the following: (Table 79.2).

Data table of the basic situation of male athletes and its structure is as the following:

The Item table includes the competition time and venues of various items as well as specific information of the judges in Group A and Group B, which is as the following:

The basic situation of female athletes is that there are two items less. The coach information includes the nationality and what item he/she is coaching. Structure of the basic table is similar with that of the male athletes. Venue information includes the address and serial number of venues, and the information structure is the same with the items, so it won't be listed here.

The following is the performance table of a single competition item, and the following table is about floor exercises:

Tables for other items can be designed in accordance with this procedure, which won't be listed here (Tables 79.3, 79.4).

Table 79.2 Database public table

Database table name	Name of relational schema	Note
Athlete	Athlete	Athlete information table
Judge	Judge	Judge information table
Item	Item	Item information table
Venue	Venue	Competition venue information

Table 79.3 Basic situation of male athletes

Field name	Field type	Not null	Instruction
Athelete_sno	LONG	Primary key	Registration number
Athelete_sn	Char	Not null	Name
Athelete_sex	Char	'Male' or 'female'	Gender
Athelete_con	Char	Not null	Country
Athelete_age	Char	Not null	Age
Athelete_PE	Char	Not null	Floor exercise score
Athelete_PH		Not null	Pommel horse score
Athelete_RS		Not null	Rings score
Athelete_VT	Char	Not null	Vaulting horse score
Athelete_PB	Char	Not null	Parallel bars score
Athelete_HB	Char	Not null	Horizontal bar score
Athelete_ICESULT	Char	Not null	Total score

Table 79.4 Competition item table

Field name	Field type	Not null	Instruction
Item_ID	LONG	Primary key	Item ID
Item_NAME	Char	Not null	Item name
Item_venue	Char	Not null	Venue
Item_time	Char	Not null	Item
Item_JA	LONG	Not null	Starting score judge ID number
Item_JB1	LONG	Not null	ID number of judge 1 in group B
Item_JB2	LONG	Not null	ID number of judge 2 in group B
Item_JB3	LONG	Not null	ID number of judge 3 in group B
Item_JB4	LONG	Not null	ID number of judge 4 in group B
Item_JB5	LONG	–	ID number of judge 5 in group B
Item_JB6	LONG	–	ID number of judge 6 in group B

79.3.4 Concept Model Design

Each athlete can participate in multiple items, each item can be participated by multiple athletes, and these two aspects have a many-to-many relation; each judge scores one item, each item is scored by multiple judges, and these two aspects have a many-to-one relation; each item is held in one venue, while one venue can hold multiple items, and these two aspects have a one-to-many relation. Take athletes and competition items for example, a system E-R diagram can be established, as shown in Fig. 79.2 (Table 79.5).

79.3.5 Performance Assessment Algorithm Design

After establishment of database, algorithm for performance assessment should be designed. In accordance with the requirement analysis, performance of team

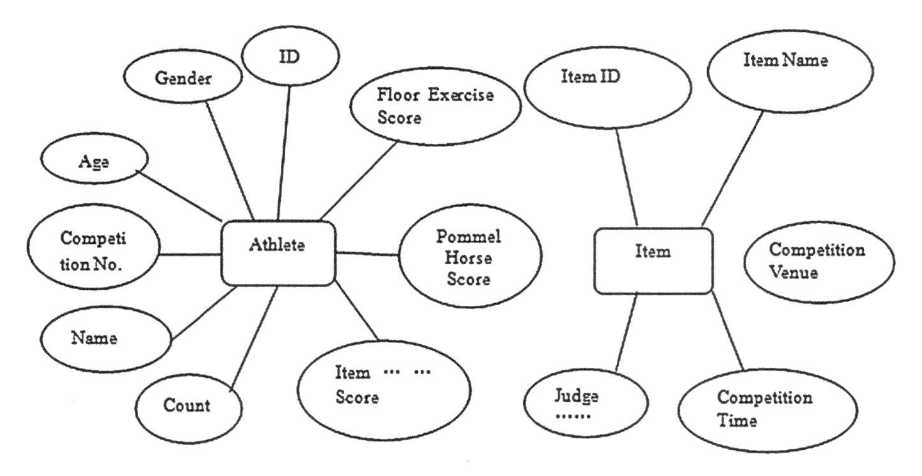

Fig. 79.2 Athletes and item E-R diagram

Table 79.5 Floor exercise performance table

Field name	Field type	Not null	Instruction
ID	LONG	Primary key	Serial number
A_ID	LONG	Not null	Athlete registration number
R_JA	Double	Not null	Scoring by the starting score judge
R_JBl	Double	Not null	Starting by judge 1 in group B
R_B2	Double	Not null	Starting by judge 2 in group B
R_JB3	Double	Not null	Starting by judge 3 in group B
R_JB4	Double	Not null	Starting by judge 4 in group B
R_JB5	Double	–	Starting by judge 5 in group B
R_JB6	Double	–	Starting by judge 6 in group B

competition is stored, each athlete in team competition will be scored, and four players with best performance in team competition will be chosen to participate in single-item competition. Players with best overall performance should be selected to participate in the all-around competition, and the selected algorithm is up-down bubble sorting algorithm.

1. First of all, compare the first and second recorded key words, if it is "reverse order" (i.e. L.r [1].key > L.r [2].key), exchange these two records, and then compare the second and third recorded key words. Proceed like this until the (n−1)th and nth recorded key words have been compared. This is the first round of bubble sorting, and its result is that the record with the biggest key word is placed in the position of the last record.

2. Then, conduct the second round of bubble sorting. Conduct the same processing for the first (n−1) records, and its result is that the record with the biggest key word is placed in the position of the (n−1) record.

   Algorithm description is as the following:

```
Void bibubble1(int r[],int n)
{
int flag = 1;
int i = 0, j;
int temp;
While (flag == 1)
{
Flag = 0;
For (j = i + 1; j < n–i–1;j++)
If (r[j] > r[j + 1])
{
Flag = 1;
Temp = r[j]; r[j] = r [j + 1]; r [j + 1] = temp;
}
i++;
}
}
```

79.4 Conclusion

In accordance with the basic procedure of database design and with the competition results of gymnastics in large sports games as the study object, this paper has studied database establishment. Through study of the rules for gymnastics competition, data requirement was analyzed, data dictionary, data logic and concept model were established, and optimized bubble sorting method was used to conduct sorting and analysis of data, which can provide certain reference to database establishment of competition results of gymnastics in large sports, and it also has great significance in ensuring fair and just performance evaluation, fast announcement of results and alleviating the staff's burden.

References

1. Schumaker RP, Solieman OK, Chen H (2010) Sports data mining methodology. Share item Integr Ser Inf Syst 12(26):15–21
2. Tjondronegoro DW, Chen YP (2005) Multi-level semantic analysis for sports video, vol 3682. Lecture notes computer science, pp 169

3. Li SG, Huo PF, Wang HJ (2012) The construction of management system for combination of sports and education. Adv Intell Soft Comput 109:683–688
4. Dowthwaite JN, Scerpella TA (2011) Distal radius geometry and skeletal strength indices after peripubertal artistic gymnastics. Osteoporos Int 22(1):207–216
5. Feng L, Ling H, Jin B (2012) The research of agent-oriented distributed active product design information database. IJACT 4(3):111–117
6. Li Z, Mahmoud S, Shamy EI, Galal T (2011) A novell security framework for web application and database. JDCTA 5(10):190–198
7. Alwan AA, Ibrahim H, Udzir NI (2011) A framework for checking and ranking integrity constraints in a distributed database. JNIT 2(1):37–48

Chapter 80
Study of Simulation Evolutionary Technology on Martial Arts

Xuntao Wang

Abstract Martial arts are obtained through killing in the history of fighting. In the present absence of actual combat, martial arts gradually lose its true mass. Fighting movement and fighting injuries can be reconstructed through the computer simulation. Moreover, we can reconstruct the genetic environment of fighting, obtain techniques by the simulation of genetic optimization, and compared with existing martial arts. We can understand the principles of the fighting techniques in order to recover the lost inheritance.

Keywords Computer · Simulation · Martial art · Inheritance · Heredity

80.1 The Simulation Requirement

The traditional Chinese martial arts leave the perfect part after experiencing the 1,000 years of cold steel age. It has reached the peak in Qing and Ming dynasty. With the firearm era development, the combat skills of cold steel age gradually lost the function. During the transmission process, the combat skill becomes vestigial [1]. It loses the spirit, and even fails to be handed down from generation to generation. The martial arts slowly step into the health protection. In addition, the fearfulness of Chinese martial arts is just the history. Without saying the traditional unique skill, the ancient books and records of martial arts are difficult to understand after the various transmissions. People have to imagine and explain the content in personal way. Chinese martial arts are facing the crisis. If we do not regain the inheritance, the martial arts will be lost [2].

X. Wang (✉)
Zhejiang Institute of Mechanical and Electrical Engineering, Zhejiang, China
e-mail: xuntao@yeah.net

W. Du (ed.), *Informatics and Management Science VI*, Lecture Notes in Electrical Engineering 209, DOI: 10.1007/978-1-4471-4805-0_80,

After carefully research the inheritance questions of Chinese martial arts, the modern society increases the body protection of the martial arts learners. However, the martial arts are the technology, which is full of kill and wound. Is has the clear practice characteristic. Base on the body protection to practice the martial arts will obtain the false and nice appearance. This is an important question to inherit the martial arts technology through fighting at close quarters after protecting the human security.

The foreign scientist utilizes the computer technology to imitate the life and give it some simple rules such as how to obtain the food, avoid natural enemy, breed the next generation, inherit the character, and mutate in the inheritance. Then the simulated lives will under the auto synthesis in the computer world. They depend on the high-speed calculation of the computer. In the very short time, the simulated lives can breed and evolve in the rapid speed. Through the 1,100 of generation evolutions, the simulated lives obtain the amazing primary intelligence in the mortal view. The simulated lives can protect themselves as well as the real lives. They can search the food, avoid enemy, and breed the excellent younger generation. There must have the new method if we use this simulation thinking in the martial arts recovery [3].

80.2 The Simulation Frame

Imagine we simulate the wrestle character in the computer world and each one has the independent characteristic. These characteristics can random distribute by the random function. Simulate various martial arts learners of the real world. Some people have the large strength, some people have the fast speed, some people have the good suppleness, some people have the strong beaten capacity, some people have the perfect balance, and some people have the good endurance. Each person can simulate the different beaten actions of the real world [4]. The physical strength, speed, direction of each action references the numeral value of the real world. The physical harm uses physical model calculation between striking objects and the objects that being strike. Each wrestle character will start to attack base on the random condition or by watching the opponent actions. After being injured, they can reduce or even remove the attack of the injured part. The inheritance possibility of injured objects that managed by the random function simulates the possibility to inherit escape and lost experience to the next generation. The winner can summary the success technology to the younger generation. Through the simulation of 1,000 wrestle characters, we can observe the inheritance characteristics of the younger generation, and test in the real martial arts in order to find out the similarities. The reasonable control of simulation parameter will obtain the similar result that compare with the real world.

80.2.1 The Simulation Model Analysis

The wrestle physical model is based on the physical collision model. It is mainly considered about the quantity, speed, rigidity, stiffness, damping factor and gravity acceleration of the model parameter. Moreover, the parameter of quantity, speed, and gravity acceleration can be obtained easier. However, the rigidity, stiffness, and damping factor can be obtained from the precision measurement. The other assistant parameters have the eruptive strength, durability, joint degrees freedom, the stability in various gestures, and the required strength vector to destroy the stability. Although these parameters are different in the real world, they are in the ascertainable quantity range.

After the parameter determination, it is necessary to do the physical model evaluation of wrestle actions. It can roughly divide into various angles, strength combination, as well as the trajectory. We can obtain all the common actions of martial arts in it. For example, in the actions of chop, hack, cut, block, prick, bar, fetch, and sweep, each term is the vector integration of one class in the range. The traditional martial arts terms are the incisive conclusion of some actions. This is convenient for the gladiator to learn the ancestors' actions without the modern help of trajectory. After clearing this point, we can translate the ancient wrestle actions into trajectory and simulate them in the computer.

80.2.2 The Balance Parameter

When simulating the trajectory, there needs to consider about the important parameter is the balance condition while the gladiator erupting the strength. Under the present balance condition and joint degrees freedom, the same action will have the great differences in the strength and direction. The physical model will calculate the difference. The force vector can divide into the 0 or 180 degrees vector with the direction of gladiator gravitational equilibrium and the vector, which parallel to the earth. Consider about the vector that correspond or adverse with the gravity direction, the worse gladiator balance will has the smaller explosive effort. This means the force from the ground in martial arts. The great power needs the support of ground. It can express the power out by the ground support. It is the tiny vector that parallel with the ground. The strength power roots in the weapon velocity inertial. The speed is the inverse ration between attack weapon quality and body quality. In the simulation, the detailed value can complete the estimation by the Newtonian mechanics formula of $m_1v_1 + m_2v_2 = 0$.

80.2.3 The Material Parameter

The destroy power of the wrestle action is the important value. The same strength through the different weapon will express the various injured conditions [5]. Therefore, there needs to simulate the injury of various weapons. It is necessary to use different materials with the various body parts to obtain damping coefficient, frictional factor measurement by cut and chop. Base on the present obtained material, silica gel can perfectly simulate human skin, muscle and the hardwood can simulate the human skeleton. Through the careful material selection, we can obtain the more correct simulated parameters for the injury.

80.3 The Weapon Simulation

Consider about the martial arts development, the weapon of each period is the restrain of last era or the opponent. When we simulate the weapon, there need to think about the sequence. If we do not simulate under the time sequence, there will have the great difference that compare with the real world.

80.3.1 The Generality Consideration

The simulated person can hold various weapons. The shape, length, weight, sharpness, rigidity, and elasticity are all reference with the real world. The weapon sequence in the virtual world simulates the real world too. The weapon sequence express the wrestle relationship of mutually generation and restriction. The first expressed is hardness bronze weapon, and then the steel weapon with perfect gravity and elasticity. Moreover, there have the good elasticity weapon of white was weapon. The weapon of both sides can be random. This is simulate the world gladiator cannot select the opponent weapon in the real world. The result of the natural weed out makes the gladiator can handle different kinds of weapons in order to avoid the disadvantage of one specific weapon. The reasonable configuration parameter makes the weapon in the virtual world should be hand arm, knife, sword, and some common weapons of the real world.

80.3.2 The Fairness Parameter

The weapon sharpness and durability can add the depreciation parameter. The gladiator through the random function can obtain the special weapon such as sharpness-treasured sword. In the simulation world, inherit the weapon from generation to generation. The gladiator will depend on the weapon that cannot obtain the fair wrestle result. Influence the simulation result and add the depreciation parameter. If the gladiator has no better technology, although some generations will obtain more win that depend on the random obtained weapon parameters, they will not win all the time. This can promote the wrestle technology inheritance cannot influence by the weapon parameter. The ordinary wrestle technology will promote the lost of inheritance. The technology to determine the technology inheritance is the real wrestle technology. In here, through the simulation result, we believe the existed is reasonable, although the reasonable might not existed.

80.3.3 The Various Types of Wrestling

The site wrestle has two types: unarmed combat and instrument combat. The main simulations will begin with these two types. The great advantage of instrument combat will influence the simulation and we will not realize it at this time. We can consider it in the further accretion. This is corresponding with the real combat. Empted-hand to catch the knife is infrequent in the real world. If there has, it is unusual to happen between the past master and ordinary people.

80.4 The Outlook of Simulation Result

Through the computer, we simulate the grapple fighting games. The great wrestle technology can entail the larger rate to the next generation. Moreover, the gaudy actions will be eliminated in the natural wrestle game. Through the 1,100 of natural selection and elimination, the inherited one is the best. It is possible to similar with the present martial arts technologies. We can find the lost martial arts actions through revolution way. Through the computer simulation, we can find one wrestle skill has force and the method it goes. Moreover, from the wrestle history, we can find out the reason that skill cannot be inherited. Maybe some wrestle movements are full of power, there must have some disadvantages that cannot be inherited. We can find out the deadly shortage, certificate the reasonable technology and the inheritance advantage.

References

1. Shan Z (2001) The review and outlook of China martial art development. J Beijing Sport Univ 11(1):21–22
2. Changle Z (2005) The calculation modeling method in the research of perspective philosophy. J Xiamen Univ (Arts Soc Sci), the 1st phase of 12(4):35–36

3. Xiaoming R, Zuoli W (2003) The research progress of artificial life philosophy. Philosophical trends the 5th phase of 2(12):27–28
4. Hailiang C, Zhongyi W (2003) The speed evaluation of car collision accident. J Chang'an Univ 8(8):84–86
5. Li L (2006) Diagnoses and simulation research of collision accident between automobile and pedestrian. Hu'nan Univ Doctoral degree memoir 22(5):20–26

Chapter 81
A Heterogenous Parallel Algorithm for Stadium Evacuation Route Assignment

Yi Liu and Bo Liu

Abstract To speed up the time-intensive, long-duration calculation process of large-scale, multi-criteria evacuation route assignment problem, this paper introduce a GPU-accelerated genetic algorithm which utilize the lightweight many-thread architecture of CUDA GPU to fasten the evaluation procedure of evacuee movement. The proposed algorithm introduces a two-level heterogeneous parallel model to harness the power of CPU and GPU separately. Experimental results show that the proposed algorithm run 6.23 times faster than sequential version counterpart and can achieve better search result within specified time unit with respect to Wuhan sports center evacuation route assignment problem.

Keywords Heterogeneous computing · Parallel evolutionary algorithm · Evacuation route assignment · Stadium evacuation

81.1 Introduction

Evacuation in an emergency is a current research trend and crucial problem for large or populous public area. The main purpose of emergency evacuation is to evacuate a great number of people from endangered area to the safe area within the shortest possible time [1]. A significant number of literatures have been dedicated to investigate the problem of emergency evacuations planning [2]. Since evacuation

Y. Liu (✉) · B. Liu
School of Computer Science and Technology, Wuhan University of Technology, Wuhan, People's Republic of China
e-mail: csliuyi@163.com

B. Liu
e-mail: whutliubo@gmail.com

W. Du (ed.), *Informatics and Management Science VI*, Lecture Notes in Electrical Engineering 209, DOI: 10.1007/978-1-4471-4805-0_81,

planning is a very complex problem with many behaviors and time-dependent factors involved [1], a complete and viable evacuation plan need to take into account an all-comprehensive aspects of assorted underling factors such as congestion degree and total path length to ensure full, fast and complete evacuation.

Providing an evacuation route plan for large-scale evacuee in short time is one of major challenge in emergency evacuation. The high complexity of current evacuation route assignment algorithm often cause the generation of an effective multi-criteria route assignment plan take several minutes or several hours to complete. The disadvantages of current evacuation route planning software systems are the rarity of software that can generate an effective plan in specified short time thus its real-application area are severely limited. Li et al. proposed a three-objective genetic algorithm [3] to address the problem of large-scale stadium evacuation route assignment problem, but the proposed MOEA is a sequential single-thread algorithm which has difficulty to handle 60,000 evacuees simultaneously. Liu et al. proposed a fine-grained parallel genetic algorithm [2] to speed-up the long iteration process in [3] and achieve significant enhancement of time efficiency and search ability. But the concurrent, many-thread calculation process of tens of thousands of single-movement of evacuee on evacuation graph is not completely fit into multi-core architecture of CPU. Compute Unified Device Architecture (CUDA) [4] is a parallel integrated platform especially design for solving large-scale data parallelism problem quickly and swiftly. It enables dramatic increases in computing performance by harnessing the power of the graphics processing unit (GPU) hence speed-up the long-duration, parallelizable computation process significantly. Consider the aforementioned points and to address this real application problem, this paper introduces a heterogeneous parallel genetic algorithm which utilizes the power of multi-core CPU and many-core GPU and applies it to solve a multi-objective stadium evacuation route selection problem. Wuhan Sport Center in Wuhan city of China was taken as the experiment scenario to test the feasibility and performance of the proposed algorithm. Experimental results show the proposed algorithm has more convergent performance and more divergent diversity compared to classic sequential NSGA-II algorithm and perform 6.23× faster on a CPU + CUDA GPU machine than sequential NSGA-II algorithm with respect to the aforementioned stadium evacuation route assignment problem.

The remainder of this paper is organized as follow: Sect. 81.2 introduces the proposed HPMOGA algorithm, Sect. 81.3 investigates experiment results and compares them with the results of NSGA-II algorithm, and Sect. 81.4 concludes the paper.

81.2 Heterogenous Parallel Model

Single instruction many threads (SIMT) CUDA architecture [5] is a data parallel hardware architecture design for concurrency of thousands of homogeneous threads. Consider the fact that data-intensive, similar single-step movement of evacuee on evacuation graph in [3] is a large-scale concurrent operation and CUDA's SIMT architecture of many similar, lightweight threads is especially suitable for this large-scale (tens of thousands) addition/multiplication operation while the coarse-grained operator of chromosome crossover, mutation and global selection is more logic-intensive and it is in accordance with CPU's SISD (single instruction single data) architecture, we divide the operator of evacuation route assignment algorithm into two levels according to granularity of parallelism. That is, the evolution level (level 1) and evaluation level (level 2).

1. Evolution Level—CPU: Crossover, mutation, global selection
2. Evaluation Level—GPU: Single-step movement of evacuee (addiation/multiplication operation of many evacuees)

Hence the task assignment between CPU and GPU is illustrated as follows.

The perturbation operations in Fig. 81.1 include crossover and mutation operation of genetic algorithm and the host thread is the controller thread of GPU. The evaluation process of single route assignment plan consist of tens of thousands single-step movement of evacuee within specified time slice thus make it especially hard for current multi-core CPU (4–8 core versus tens of thousands threads) to calculate simultaneously and that is the motivation of this two-level task assignment of operator between CPU and CUDA GPU. The MOEA (multi-objective evolutionary algorithm) based on proposed heterogeneous parallel model is referred to as HPMOGA in rest of this paper. The other implementation detail of HPMOGA is the same as in [2].

81.3 Experimental Results

To validate the time efficiency of proposed HPMOGA algorithm, we've choosen sequential NSGA-II (denoted as SGA), fine-grained Parallel genetic algorithm (PMOGA) [3] as comparison candidates in this section. The problem formulation and evacuation scenario of Wuhan sports center stadium is the same as in [3].

The running time of three different EAs is list as follows, su_ratio denotes real speed-up ratio in Amdahl's Law.

All data in this section are sampled from ten run instances of each test algorithm.

We can see from Table 81.1 that HPMOGA performs 6.23 times faster than sequential NSGA-II algorithm and runs 1.37 times faster than CPU-based PMOGA algorithm. As search performance within specified time unit is concerned, we can see form Table 81.2 that the mean HV (hyper-volume) of HPMOGA algorithm is approximately 30 % higher than sequential SGA algorithm. And the convergence performance measured by mean HV value within 120 s from Fig. 81.2 is clearly showed that the search ability within same time unit between three algorithms is very different and HPMOGA is clearly better than SGA and PMOGA.

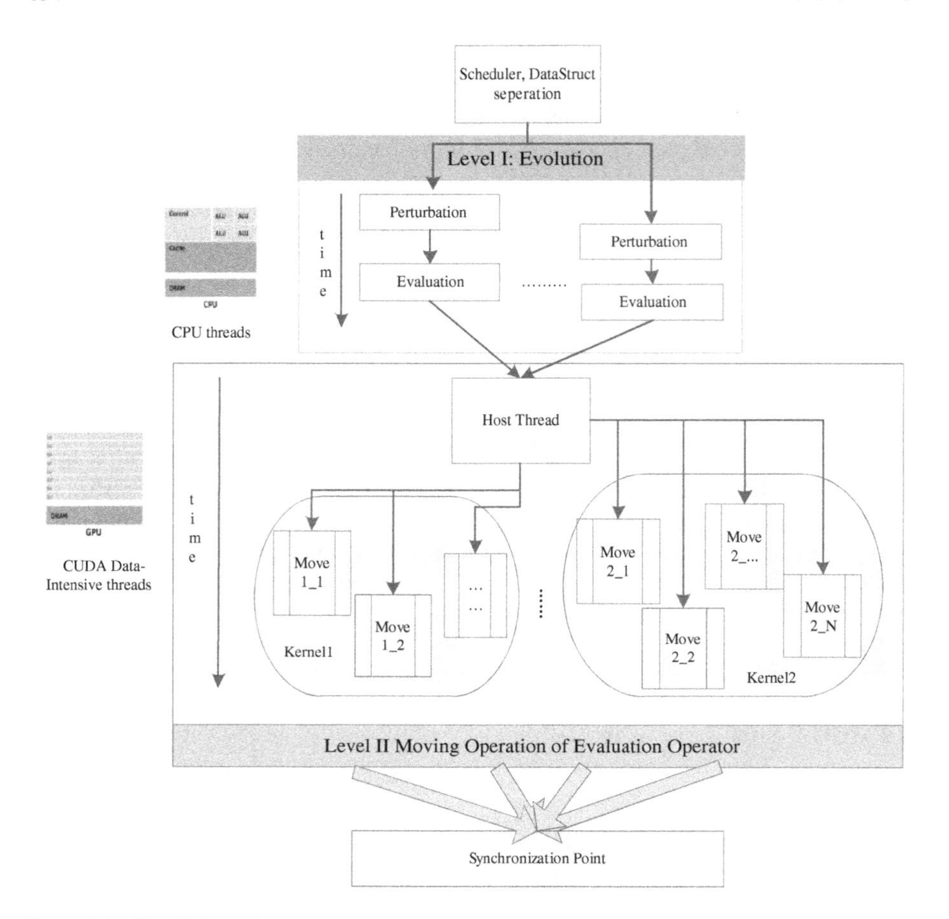

Fig. 81.1 CPU/GPU task assignment

Table 81.1 Running time comparison of different MOEAs

Algorithm	Su_ratio	Time(s)	
		Mean	Std
SGA	1	400.352	0.323456
PMOGA	4.536	88.26102	0.235737
HPMOGA	6.231	64.25165	0.302345

Table 81.2 Comparison of search ability within 300 s of different MOEAs

Algorithm	Mean iteration num.	Mean HV value
SGA	60.2	2.7522e + 006
PMOGA	353.2	2.8941e + 006
HPMOGA	422.55	3.3259e + 006

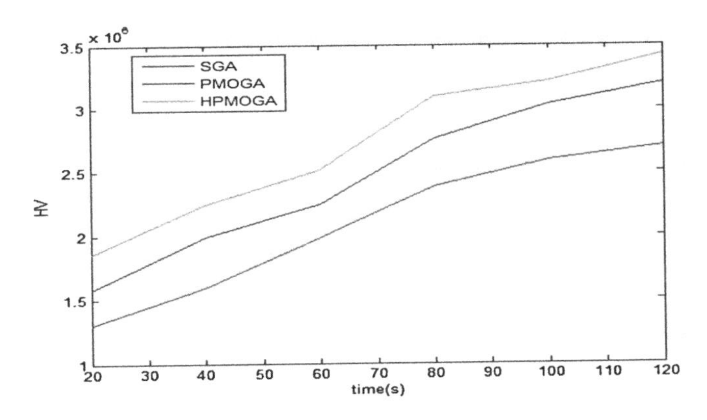

Fig. 81.2 Comparison of mean HV value within 120 s

81.4 Summaries

This paper introduces a GPU-accelerated genetic algorithm which utilizes the lightweight many-thread architecture of CUDA GPU to fasten the evaluation process of evacuee movement in [3]. The proposed algorithm introduces a two-level heterogeneous parallel model to harness the power of CPU and GPU separately. Experimental results show that the proposed algorithm run 6.23×s faster than sequential version counterpart and can achieve better search result within specified time unit. We would like to consider the possibility of accelerating the HPMOGA algorithm further with parallelization of more fine-grained operators in our future research.

Acknowledgments This work was supported in part by National Science Foundation of China (Grant No. 61170202 and No. 40971233). We would like to express our gratitude to these supports and funds.

References

1. Saadatseresht M, Mansourian A, Taleai M (2009) Evacuation planning using multi objective evolutionary optimization approach. Eur J Oper Res 198(1):305–314
2. Liu Y, Xiong S (2008) A fine-grained parallel multi-objective genetic algorithm for stadium evacuation route assignment. Int J Digit Content Technol App 32(3):43–49
3. Li Q, Fang Z, Li Q, Zong X (2010) Multiobjective evacuation route assignment model based on genetic algorithm. 18th International Conference on Geoinformatics 12(3):1–5
4. The SIMT architecture (2008) http://electronicdesign.com/article/embedded/simt-architecture-delivers-double-precision-terafl
5. Ziztler E, Deb K, Thiele L (2000) Comparison of multi objective evolutionary algorithms: empirical results. Evolutionary Computation. MIT Press, Cambridge 8(2):382–288

Chapter 82
Study on Fostering Sports Reserve Talents and Countermeasures in University

Wei Bao

Abstract In the competitive sports into just number quantity is not, talent team not listens, a first team is too little, and quality through, not the top tip; Coaches age reasonable structure, rich experience and professional spirit is good, but the degree of low level of popular dissatisfaction with treatment, scientific research ability and scientific researchers cooperation is poor, lack of theoretical knowledge, science and culture knowledge needs to be improved; Competitive sports items establish buy fewer, the optimization effect is loaded, it is not rational structure, key projects and the development of national fall point Olympic project fit don't birds, many athletes and coaches proportion of smoked disorders; The spring training funds from body bent bureau investment, training, contest funds shortage, set the varied is insufficient, the cultivation of a unit in a rightful standard make more reasonable, job training, professional athletes and to become the main quality of the college conveying goal.

Keywords Competitive sports · Reserve talents · Design and development

82.1 Background and Significance

Since 1992, our country sports system reform on the socialization, industrialization development road of its main to guiding ideology is to ask the development of competitive sports go the way of the market, to adapt to the market economic system reform, the market to configure the sports resources, fully mobilize social,

W. Bao (✉)
Physical Education Department, Harbin Engineering University, Harbin 150001, China
e-mail: baowen11@126.com

W. Du (ed.), *Informatics and Management Science VI*, Lecture Notes in Electrical Engineering 209, DOI: 10.1007/978-1-4471-4805-0_82,

enterprises and individuals in all aspects of the enthusiasm of the sport do, Form the main body of investment diversification pattern, by the government alone into a multi-channel, levels and forms of Training system up [1].

At present our province athletic sports reserve talented person cultivation [2], is based on a high investment, low output, high attrition rate.

On the basis of extensive development pattern, the athletes yield is very low. In primary school or junior high school from (including sports preach all the school) to amateur sports school [3], who a line or the national team at all levels in the selection, only very a few athletes can be selected, a large number of athletes from training units at all levels will be eliminated. Our country every 4,000 athletes can produce a world champion [4].

The sustainable development of competitive sports is a systems engineering, all countries in the world of sports contests to contemporary consciousness of the focus on reserve talented person's competition [5], so the development of supportive talents has become the key to the sustainable development of competitive sports. With the rapid development of computer network technology, how to utilize the network and the advantage of computer technology to improve the content of science and technology products and the management of the work efficiency of competitive sports reserve talented person, has become an important task. Because of intellectual property protection in a foreign country and we different cultural background, no such management system of competitive sports reserve talented person in foreign direct can use our gift. And system function of management system software is domestic at present more of a single; its main function is the basic information of the limited the management of athletes, lack of basic database system design, so it can't make an objective evaluation and comparison of the competition ability and potential competition, so it does backup talents choice. At the same time, the system is mainly to consider using the administrative personnel, and not easy to the coaches and athletes, so many of the information cannot be upgraded and feed-backed. In order to make the management of competitive sports reserve talented person more pointed, more effectively, make the management of university students more human, standardization, systematization, so as to give full play to a bigger role and produce more benefits. According to the shortages, this paper develops a kind of management system and analysis of competitive sports talent management based on the network.

82.2 System Architecture and Development Tools

82.2.1 System Architecture

The browser/server model is used in the system. Based on B/S model is three structure of the system. The first set of is the user and the whole system of the interface between. The second tier network server used to handle the request from the user's machine. The user sends request much server distribution on the network.

At the same time, the server browser is from treatment requirements and feedback the information user requirements browser. The rest of the tasks, such as data requirements, processing, the result back, a generation of the dynamic web pages, database access, and implement application program, complete network servers. The third is the database server; the mission is C/S model similar, responsible for the coordination of SQL by different network for server, and management database.

82.2.2 System Development Tools

Database Development Tool: SQL Server 2000 is selected as the back-end database management system. SQL Server is a relational database management system with Transact-SQL acting as its database query and programming language.

Database server: Microsoft SQL Server 2000

Other development tools include, Visual Studio 2005, NET Framework 2.0, IIS (server), Dreamweaver 8.0 (For the completion of webpage production and code compiling), Photoshop (For the production of pictures in the webpage), SPSS16.0 (For the data analysis and processing).

82.3 System Design and Development

82.3.1 System Overview

The system consists of three modules, namely, basic information subsystem, assessment information subsystem, and information management subsystem. The system architecture is shown in Fig. 82.1.

82.3.2 Database Selection and Design

SQL Server 2000 is applied in this system to serve as the support system for the database. OLEDB is applied in the query processor to realize the communication between Microsoft SQL Server data storage components. OLEDB provides decentralized and different types of query capabilities for SQL Server 2000 query processor. It supports the decentralized query among multiple SQL Server 2000 servers, and also supports the decentralized query for any OLEDB provider. In addition, it provides better flexibility and security if compared with the system administrator login, and the dual-duty administrator of both system and database fully feels this kind of flexibility in the aspect of security configuration.

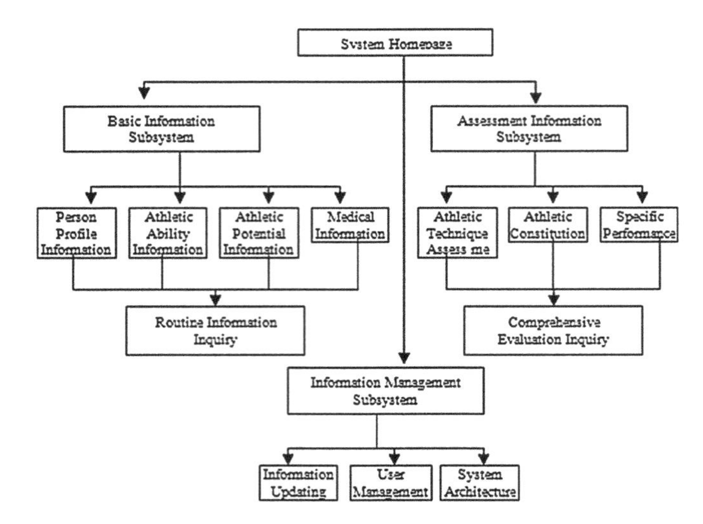

Fig. 82.1 Overall structure of information management for reserve talents of competitive sports

In this system, all lists are uniformly arranged into a database (Knowledge standard database), and the following several types are shown in the lists of this system.

Administrator situation storage list: Record the user names, passwords, and corresponding permissions of administrators at all levels.

Management user information storage list: A the time of registration, record the user's communication situations, divide and store the user information into the data list as per the sexes for the future statistical use.

Questionnaire: Store the answers and relevant data of the questionnaire.

Expert system-related list: Relevant lists in the knowledge database, reasoning database and truth database applied in the expert system.

82.3.3 System Function Features and Security Architecture

82.3.3.1 System Function Features

In consideration of the restrictions for service conditions, the system also designs a single-computer system which can run under the offline mode, and can realize good combination with the network.

82.3.3.2 System Security Architecture

System administrator is responsible for the system user authorization.

Authorized system users (all departments) are allowed to enter the system management software and carry out the relevant permitted operations.

Only the user passing through the identity validation can access to the system.

82.3.4 System Module Structure and Function Design

82.3.4.1 Functions and Contents of Basic Information Subsystem for Reserve Talents

Reserve talents Archival Information System Module

It mainly includes the personal status and the training experience contents of athletes. The personal status of athletes includes the name, origin, time of enrollment, sports events, exercise level, best results, hobbies, parental status and so on. The training experience contents of athletes mainly include the serial numbers, names, training phases, training projects, starting and ending years, training locations, coaches, main training tasks, best results in competition, and other conditions.

Reserve talents Athletic Ability Information Management Module

It mainly includes such practical information as the body shapes, physiological functions, physical and psychological qualities of the athletes. Managerial staffs and coaches can input, save, and inquire the athletic ability information of each athlete, and deploy the models in method database to predict, evaluate, grasp the current level of training, and predict the future development potential.

Reserve talents Athletic Potential Information Management Module

According to the reserve personnel evaluation index system of sports ability, this is possible; get the evaluation result calculated by selecting the fuzzy neural network model method of database. Mathematical operation results model can output level of athletes, in which all indexes are located in, integrated assessment, the size of the factors such as the contribution rate of form, function, psychological athletic performance. So, coach can decide which areas athletes also have even more potential, and concern for the greater potential in the region in the future training plan. Therefore, the athlete's sports level and are the improvement of the performance of an effective way more quickly.

Reserve talents Health Information Management Module

It mainly includes a variety of physiological and biochemical parameters, genetic and past medical history, injuries and so on over the years.

82.3.4.2 Functions and Contents of Reserve Talents Assessment Information Subsystem

Comprehensive evaluation model, physical fitness evaluation model and evaluation model, the sports movement constitution technical evaluation model, the specific performance evaluation model, and established the elite athlete's standards according to the age, sporting events. The qualitative and quantitative and qualitative and quantitative, is used in the evaluation method. The factors that can be counted as the evaluation index, it is planned to apply multiple regression analysis, gray theory method to establish the mathematical model of calculation, because of factors are difficult to quantify index, it is planned to apply expert fuzzy comprehensive evaluation and analytic hierarchy process (AHP), make qualitative factors of the quantitative analysis.

82.3.4.3 Subsystem Management

It mainly is used for the system management and maintenance, including the user management, database management and the system architecture management.

82.4 Summary and Outlook

Management system of competitive sports reserve talented person comprehensive integration of the basic information, forecasting, evaluation, statistics, decision-ma king model and all kind of elite athletes competitive capacity model, realize the information management scientific and automation, so as to provide a powerful decision support for the scientific management and implementation, for competitive sports reserve talented person.

The system is easy to use features, have strong scalability, comprehensive functions. In the future, further exploration and research will be system function expansion, system security, athlete's video information acquisition, technical analysis, judgment, etc.

The development of the management system of competitive sports reserve talented person to provide reference other software development and competitive sports management and training and other fields, and accumulated experience.

Acknowledgments This paper is supported by the Fundamental Research Funds for the Central Universities: HEUCF111603.

References

1. Du B (2008) Design and implementation of sports and scientific research information management systems master thesis from Taiyuan University of Technology, May
2. Lou M et al (2001) Design and implementation of sports training management information system for diving athlete. Comput Eng App 16(1):22–25
3. Zhang L (2001) Design and development of sports training plan management system for race walking athletes, Master thesis from Qufu normal University 4:14–17
4. Defang L (2005) Development and application of network psychological test and analysis system. J Southwest Jiaotong Univ Social Sci 6(2):75–78
5. Xu Q (2007) Research and implementation of national physical fitness detection and service information management system, master thesis from Tianjin University

Chapter 83
Primary Exploration of the Diversified Development of Sports Public Service

Hongyan Yao

Abstract The ultimate objective of sports public service is to pay attention to the development of human beings. It is the absolute trend of health conception pursuit by the growing material and cultural lives of the people. When the modern society is interpreting the concept, the connotation and the development objective of sports public services, it conforms to the scientific concept of development that takes humans as essentials. With the target of the development of three stages, it realizes the service system at which the government macro-control and all sectors of society mutually participate. In addition, it also realizes the service system in which the rural–urban sports public service develops with the characteristics of diversification, multi-subjects and multi-centers. These are to be gradually realized on the perspectives of management mode, supply mode, and operation mechanism as well as the social guarantee mechanism, etc. In this way, the sports public service in our country is ensured to develop gradually and stably.

Keywords Sports · Public service · Ultimate objective · Human development

83.1 Introduction

With the rapid development of the economy in our country, the production ability and the financial ability in our country develops continuously [1, 2]. The living level of the people in our country keeps on improving as well [3, 4]. It tries to find out the way for establishing an innovation model from the two aspects of

H. Yao (✉)
Hebi College of Vocation and Technology, Hebi, Henan 458030, China
e-mail: hongyanyaocc@126.com

W. Du (ed.), *Informatics and Management Science VI*, Lecture Notes in Electrical Engineering 209, DOI: 10.1007/978-1-4471-4805-0_83,

management organization and activity contents [1, 5, 6]. It also analyzes the existing problems and the ways for solving these problems. It is very important to consider how to perfect the connotations, the operation mechanisms and the supply guarantee system of the sports public service for they play a vital role in the promotions of the equity of the sports public service [2, 7].

This paper has primary explorations from three aspects, the current situation of the sports public service, the problems and the countermeasures.

83.2 Current Situation of Sports Public Service in Our Country

Sports public service is a service content in the fundamental public service. It has contained three basic parts, which are shown as the followings [3]: (1) guarantee the fundamental survival rights of human beings, (2) satisfy the fundamental respect and fundamental ability needs, and (3) satisfy the needs of fundamental health. The development of sports public service in our country has mainly gone through the following three stages:

83.2.1 Stage One

From the establishment of the country to the beginning of the reformation and opening up, implement the planned economy in our country.

83.2.2 Stage Two

From the reform and opening up to the beginning of the twenty first century, Public sports service is gaining growing attention during this period. Primary explorations have been made on its fundamental model and the operation guarantee mechanism. However, the development of the public sports service is relatively slow and a lot of problems have been exposed.

83.2.3 Stage Three

Since entering into the twenty first century, the concept of the sports public service is being recognized by all sectors in the society. They have accepted the attention from all sectors of the government and the sports popular projects have also gained continuous development.

The ultimate objective of sports public service is to pay attention to the development of human beings. It is the absolute trend of health conception pursuit by the growing material and cultural lives of the people. While obtaining the deserved development, we cannot ignore the potentially risks. They are shown as the following five points.

The government public service department management system is old and the management method is single.

The sports public service government department composition takes the government institution as the subject. The forms are excessively single and its socialize participation and control over the public field is very strict. They have, to a large extent, restricted the participation of the social forces into the management part.

The sports public service products are restricted in its supply. There is even a lack of theoretical guidance and technical guidance of the development of the community sports.

Regional science is a field of the social sciences concerned with analytical approaches to problems that are specifically urban, rural, or regional. Any social science analysis that has a spatial dimension is embraced by regional scientists. Regional science was founded when some economists began to become dissatisfied with the low level of regional economic analysis and felt an urge to upgrade it. The contradictory occurs, with the shortage of supply and the waste.

The community sports public service system is not perfect. The value of sport is immense. Sport has unique attributes that enable it to bring a particular value to society. It is very popular, which derives from the fact that it is fun and enjoyable, and connects people and communities, bringing together players, teams, coaches, volunteers and spectators to create community networks. This has resulted in the limitations of the social community sports form development.

83.3 The Concept, Connotation and the Development Objectives of Sports Public Service

This paper bases on the previous researches and combines the theoretical research results of the sports industry to the sports public service. It is summarized as the followings: The ultimate objective of sports public service is to pay attention to the development of human beings. It is the absolute trend of health conception pursuit by the growing material and cultural lives of the people. When the modern society is interpreting the concept, the connotation and the development objective of sports public services, it conforms to the scientific concept of development that takes humans as essentials. With the target of the development of three stages, it realizes the service system at which the government macro-control and all sectors of society mutually participate. In addition, it also realizes the service system in which the rural–urban sports public service develops with the characteristics of diversification, multi-subjects and multi-centers. The connotation construction should be shown in the following contents as the content, the types and the characteristics of the sports public service.

The entire objective of the sports public service is to satisfy the basic sports needs of the social members. It focuses on the improvement of the physical health and the life quality of the citizens. It can not only provide the citizens with basic sports cultural enjoyment, but can also provide and make sure the sports public service and produced needed under the sports environment and condition of the social survival and development. In order to realize this overall target, three stages must go through. In the first stage (short-term objective), it should pay major attention to the governmental investment and gradually turn to the stage with multiple subjects, multiple centers and socialization. In the second stage (medium objective), the sports public service system construction, the equipment construction and the service system construction are basically perfect. In the third stage (long-term objective), it tries to realize the equity of nationwide sports public service.

83.4 The Diversified Development and its Countermeasure Research into the Sports Public Service in Our Country

According to the connotations of the public sports service, this paper has put forward five countermeasures focusing on the problems of sports public service in our country.

83.4.1 Improving the Operation Mechanism of the Sports Public Service

From the perspective of the sports public service in our country, it can be found that the orientation problem of the sports public service places great emphasis on the government service function. It greatly strengthens the subject place and the leading function of the government in the sports public service. Sports public service is formed from government innovation and it belongs to public administration study. The need of public sport is the basic need of human in social life. It can not only provide the citizens with basic sports cultural enjoyment, but can also provide and make sure the sports public service and produced needed under the sports environment and condition of the social survival and development. In order to realize this overall target, three stages must go through. Otherwise, it will bring unnecessary material and spiritual loss to the country and the people. The core concept of the government is to serve the public. This is also the good manifestation of the "taking people as the essentials" in the sports public service.

83.4.2 Improving the Supply Model of the Sports Public Service

In the sports public service, the government should be taken as the leading core. When the role of macro-control and decision is giving into full play, it should at the same time break the traditional method in which the government does the unified procurement in terms of the sports public service and products. Multiple social forces should be introduced. The forms are excessively single and its socialize participation and control over the public field is very strict. They have, to a large extent, restricted the participation of the social forces into the management part. It is the absolute trend of health conception pursuit by the growing material and cultural lives of the people. Such things as the enterprises, associations, groups, communities and local government should be taken into consideration on the services and products. The living level of the people in our country keeps on improving as well. It tries to find out the way for establishing an innovation model from the two aspects of management organization and activity contents. It also analyzes the existing problems and the ways for solving these problems. It is very important to consider how to perfect the connotations, the operation mechanisms and the supply guarantee system of the sports public service for they play a vital role in the promotions of the equity of the sports public service. It should make sure the quality of the sports supply products, the technology support of the after-service and its maintenance. It should make sure the public interest of the people and gradually establish the new-type sports public service supply mechanism with the center of the government and the mutual participation of the other supply subjects.

83.4.3 Social Diversified Forces

Along with the progress of reformation and development in China, how to improve the work of marketing and service of sports service enterprise by combining the modernization instrument and advantaged method of management becomes an important problem in the presence of every handlers to adapting to the development of market economy. Gradually form an adverse market competition mechanism with diversification participation, diversified management and multiple forms supply. The reformation of the sports public service has broken through the traditional planned economic model. Through the cooperation between all industries and all sectors in the government, give full play into the adjustment and demand function of the market. In terms of the sports public service products, apply the principle of survival of the fittest so as to enable a great amount of resources to be fully used instead of being wasted.

83.4.4 Strengthen the Investment Degree

When the county makes macro-control efforts on the rural and city sports public service, the investment on finance, materials and human resources should be strengthened. At the same time, perfect the rural sports public management institutions. Cultivate the sports management skill talents and let the farmers to carry out the technology and theoretical guidance and let the farmers to do well the maintenance of the sports public products. Give full play the subject role of the people.

Give full play the forces of the community and the retire group. Perfect the community sports public service organization and do well the monitoring mechanism of the sports public service and the emergency treatment mechanism of the emergencies.

Sports public service is formed from government innovation and it belongs to public administration study. The need of public sport is the basic need of human in social life. Public sports service refers to the public goods and mixed goods provided by public organizations. Public organizations should provide public sports service, and all the citizen are the objects of public sports service. The need for public sports is plentiful and the model of public sports service is multiple.

The operation model, the supply model, the management model and the social guarantee mechanism of the sports public service cannot be accomplished at one stroke. Sports public service system in China is undertaking a new stage of development, with fair and equal, and public welfare, diversity, convenience and universal characteristics. People-oriented is he basic values to public service system construction. Centering on the rights and interests of the masses to carry out various Sports fruitful work of public services is the government's goals and mission. Gradually form an adverse market competition mechanism with diversification participation, diversified management and multiple forms supply. Realize the equity of the development of the rural sports public service. Make sure that the sports public service industry in our country develops gradually and stably.

References

1. Liu Y (2010) Connotation, character and classification framework of chinese sports public service in the period of social transformation. J Chengdu Phys Edu Inst 10:35–37
2. Bingyou F (2009) Theoretical framework and systematic structures of public sports services. J Phys Edu 6:67–71
3. Cai J, Fan B (2010) Historical evolution of public service development of sports of our country. Zhejiang Sport Sci 7:54–57
4. Brown G, Yule G (1987) Discourse analysis. Cambridge, Cambridge University Press, 3:48–54
5. Jiazu G (2000) Cross-cultural communication. Nanjing: Nanjing Normal University Press 45(5):09–14

6. Zuoliang W (ed) (1990) Writing: ELT in China. Papers presented at the international symposium on teaching English in the Chinese context. Beijing, Foreign Language Teaching and Research Press, 4(2):780–786
7. Stern HH (1992) Issues and options in language teaching London, Oxford University Press, 4:49–57

Chapter 84
Sports Games Management System Based on GIS

Yang Guo and Xiaofeng Xu

Abstract In the traditional sports games management model, there are huge organizational structure, low efficiency, slow information response, unreasonable resource configuration, trivial and miscellaneous data processing, inaccurate information analysis, and other problems. In this paper, through the analysis on the functions, structure and characteristics of Geographic Information System (GIS), the application fields of GIS are discussed and analyzed, and also the authors propose that it is feasible to apply geographic information system to the sports field and especially large sports games. According to the analysis on the feasibility of applying the geographic information system in the management systems of large sports games, an operational model of GIS management system of large sports games is constructed. With this system, hopefully, the management efficiency of China's large sports games can be improved greatly, and simultaneously the modernization, information and organizational operation of large sports games can be increased.

Keywords GIS · Sports games · Application

Y. Guo (✉)
Shaanxi Polytechnic Institute, Xianyang, China
e-mail: yangguo11112@126.com

X. Xu
Baoji Vocational Technology College, Baoji, China

W. Du (ed.), *Informatics and Management Science VI*, Lecture Notes in Electrical Engineering 209, DOI: 10.1007/978-1-4471-4805-0_84,

84.1 Introduction

Along with the rapid development of computer technology and the successful application of computer technology in the fields of remote sensing, Internet, and web of things, GIS has been penetrated in all walks of life in recent years.

In this paper, through the analysis on the technological advantages of geographic information system, the feasibility of applying geographic information system in large sports games is discussed, a model for the management of large sports games is constructed with geographic information system, and also suggestions are provided by the authors for the security of network system. Hopefully, this paper can provide a reference to improve the management and security of large sports games in China.

84.2 Summary to GIS

84.2.1 What is GIS

GIS (geographic information system) is a technical guarantee system, which is based on geographical spatial database, carries out the collection, management, operations, simulation analysis and display of relevant space data through the computers, offers a variety of space and dynamic geographic information including important positions and surface features with geographical model analysis method, and provides information for the overall planning, scientific decision-making, managements and researches. It is an emerging subject cross geographical information, information management and technology, space science, management science and other subjects, and also is a computer application system that is established for the sake of geography research and even human life, work, management and decision-making.

84.2.2 GIS Structure

Geographic information system comprises of hardware, software, data, personnel and methods. Hardware equipments include data storage equipments, data input equipments, output equipments, computer and network equipments, display equipment and other peripheral equipments, etc. Software equipments mainly include GIS system software, operating system software, database management software, and system development software. Data is the core of geographic information system, and mainly comprises of data processing and managed database data. Personnel comprise of professional technical persons in management, input, analysis and operation, and the professional technical ability of personnel plays a decisive role in the quality of the construction and operation of the system. Geographic information system is a closed ring structural system, and all of its subsystems are connected with each other from upper to lower and mutually dependent. The subsystems successively are managing object, collecting different geographic data in space, forming data source, inputting data, entering database, managing data, forming GIS, acquiring decision-making on data analysis, users use, solving problems, and serving for management objects [1].

In a geographic information system, geographical entities are abstractly expressed with three basic characteristics, namely point, line and area (polygon) [2]. There are 6 combination relationships among these three characteristics. These combinations give reflection to the proximity, connectivity, closing, inclusiveness, and consistency of the topological relationship among these characteristics, and also are an object-oriented space data structural model and a data model with four levels [3]. These four levels are (POINT, LINE AND AREA), OBJECT, ELEMENT, and CLASS. It means that a complicated entity can be composed by several simple primitives; primitives are described with a group of targets; each target is expressed with the abstract point, line and area of (GIS) geographical entity [4, 5].

84.2.3 System Functions and Characteristics of GIS

Geographic information system mainly comprises of four basic functions: (1) data acquisition and editing function; (2) drawing function; (3) spatial database management function; (4) spatial analysis function [6].

84.3 Analysis on the Feasibility of Applying GIS in the Management of Large Sports Games

84.3.1 An Effective Information Management System Necessarily Introduced to Large Sports Games

Large sports games are highly different from the general sports event because of the large number of participants, wide competition terrain distributions, high-density games, extensively-sourced game teams, high attention from people, and higher requirements on competition terrains and corresponding managements. The smooth, secured and highly-efficient development of games can give real expression to the image and management level of a city. This naturally requires the host of games to possess the ability in providing a powerfully functional management system for information security. In the mean time, this system is required to be capable of collecting information of each department and also making analysis and management on this information.

84.3.2 Highly-Efficient Operation of Large Sports Games Guaranteed by GIS

In the Sixth Asian Winter Games, public service geographic information system was introduced for services. During the period of games, this system provided public services and information such as venues information, athlete information, and games live broadcast information, result searches, and tourist routes for all athletes.

From the analysis on the structure and functions of GIS, geographic information system is a computer information system that can be used for acquiring, simulating, retrieving, analyzing and expressing geographical spatial data; GIS features large coverage, powerful data acquisition, analysis and processing abilities, and intensive information. Besides, it has the ability to transfer the information of departments participating in games in real time, make analysis and management on data, and also provide the most direct basis for the administrations of games. However, all these can speed up GIS to process the problems in all nodes more rapidly and conveniently.

84.4 Model for the Operation of GIS Management System of Large Sports Games

The application of geographic information system in the sports games has involved sports information science, sports management theory, and sports games organizations and so on. Sports information science is a subject that combines sport science, computer technology, computer graphics, and database and so on. The management of sports information is an important part of sports information science, and gives reflection to the intelligence, data, news, messages, instructions, and other data related to sports games. Sports news can be classified into internal or external news, original news or processed news, normative news or non-normative news, news reflecting the management or operation of sports games, financial news or organizational news, etc.

When a geographic information system platform is applied in the management system of large sports games, the games information, the ensured normal operation of management system and the emergencies can be reflected within the shortest time. In the mean time, it can implement an analysis and form a preliminary decision-making, thus providing a reference for the command center of games. To introduce a GIS management system to large sports games, it is necessary to convert all sports venues, athletes villages, media centers, and urban roads related games, traffic information and other security systems to attribute data. The acquired data should be screened, selected, and modified first, and then is analyzed. In the process, the most important point is analyzing all sports venues, roads and traffic information. According to the analysis of shortest path and minimum and maximum caches, it is necessary to make a reasonable plan on the positions of supporting facilities. At the same time, the models of supporting facilities can be constructed and adjusted with GIS. All these models are necessary to be able to clearly reflect the operational situation and abnormal conditions of all nodes and report them to the command center in time, promoting command center to process the problems in time through information system, as shown in Fig. 84.1.

The main functions of geographic information system comprise of data acquisition and processing, data organization, space inquires, space mapping, graphics display and so on. All these can be developed smoothly with JAVA and ActiveX. Combined with the basic components of GIS, these functions can compose a GIS-based sports games management system. The algorithms, which are commonly used in geographic information system, include the shortest path algorithm, and the shortest dynamic path algorithm, etc.

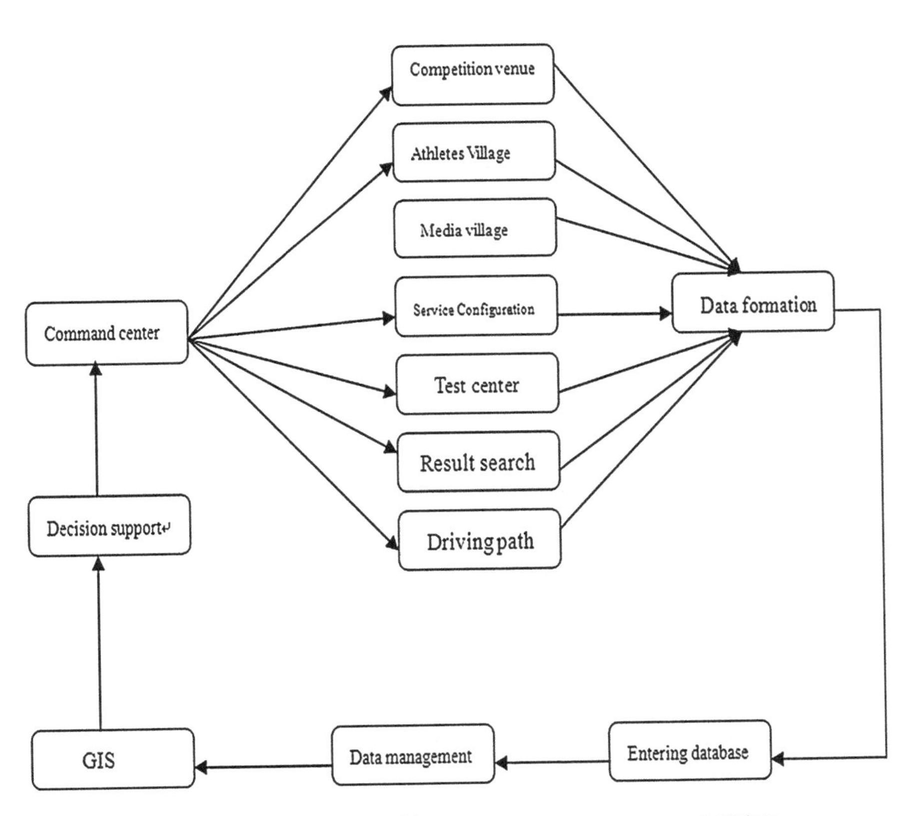

Fig. 84.1 Operation model for GIS-based large sports games management system

84.5 Conclusion

Geographic information system, after decades of development, has changed into a very mature technology with a gradual step. In China, as an emerging subject, geographic information system just develops for a relatively short time.

However, it has been applied extremely extensively in the fields of communication and transportation, water conservancy, resource investigation, public facilities management, environmental assessment, urban planning, hazard prediction, post, military and police.

In general, large sports games are with a complicated organizational and management system. For this reason, an geographic information system is very applicable to the organization and management of large sports games.

Furthermore, the operational efficiency and organizational level of large sports games can be increased highly.

At the same time, geographic information system is helpful for updating ideas and improving the information level of large sports games.

In this paper, through the introduction to geographic information system, the authors propose the feasibility of applying geographic information system in large sports games, and also establish an operation model for it.

Hopefully, this paper can provide a reference for improving the information, organizational and operational qualities of large sports games of China.

With the maturity of geographic information system technology as well as the development of WAP and GPS embedded systems, the qualities and construction of digital sports and sports information will be accelerated greatly.

References

1. Su C, Wang C (2011) Design and implementation of public facilities inspection system based on mobile GIS. Sci Surveying Mapp 28(04):061–065
2. Zhang S et al (2011) Risk analysis on high-temperature disasters in east china based on GIS. J Catastrophology 26(2):59–65
3. Fuling B (2004) China's higher education for GIS: progress. Features and research. Geomatics World 5:16–18
4. Li Y (2004) Review of market development of China's GIS. Geomatics World 8(5):40–42
5. Li X (2004) General review of GIS standardization. Geomatics World 9(5):11–15
6. Chengqi C, Guo S (2004) Discussion on the present situation and the perspective of China's geographic information system construction and management. Geomatics World 12(5):28–30

Chapter 85
Impact of Music on Comprehensive Quality of Students in Sports Dance Teaching

Xiaoyan Zhang

Abstract Music is the soul of sports dance. Sports dance can be carried out smoothly by the accompaniment of music. Therefore the two should be in harmony and unity to perform the essence of sports dance totally. This article tries analyzing the impact of music in sports dance teaching for students in physiological, psychological, creativity, imagination, aesthetic and so on to explore the important role of music in the development of students' comprehensive quality, in order to help sports dance teaching.

Keywords Music · Sports dance · Teaching · Comprehensive quality · Influence

85.1 Introduction

Sports dance is known as "international ballroom". It is the combination of art, sports, music and dance in one and also the perfect combination of power and beauty. It is a duet project that needs a male and female to participate and complete. We divide it into Ballroom and Latin dance including ten kinds of dance. Ballroom consists of Waltz, Viennese Waltz, Tango, Foxtrot, and Quickstep, Latin dance including Rumba, Cha-cha, Samba, Jeans and Paso [1]. Each dance has its own dance steps, music and unique dance style. Such as Waltz, its dance style is to show the elegance like prince and princess. The music is slow and elegant. It is the rhythm of three beats and remake on the first beat. We need to choreograph a set of complete actions for students learning on the basis of the characteristics of waltz

X. Zhang (✉)
Sports Department, North China Institute of Science and Technology, Beijing, China
e-mail: xiaoyanz120@126.com

W. Du (ed.), *Informatics and Management Science VI*, Lecture Notes in Electrical Engineering 209, DOI: 10.1007/978-1-4471-4805-0_85,

music and dance. We ask students to dance in the accompaniment of music and the beat of music. This requires that students must understand the music, feel the music, control music, and integrate music and sports dance together firmly.

85.2 Music Can Regulate Students' Psychological

Music through strength the emotion of dance and enrich the substance of dance to make the dance even more vivid and attractive and directly touch the audience. Psychologists said: "Beethoven's music can make the sad people happy and frivolous by solemn." In fact, Music for people's physical and psychological regulation has been realized by the ancients very early. The psychological effects of music on people have been experienced by many people. In today's enormous pressure and social working fatigue, people are willing to turn on the radio or put on an mp3 to listen to a wonderful song during off hours or driving time for decompression. This will help the body feel relaxed and fatigue to be reduced. For sports dance music, for 10 kinds of dance there will be 10 kinds of different music. This will stimulate the different emotions of the people [2]. Dance content is different according to different performances of different dance. It can induce inspiration when dancers are dancing, while at the same time stir the emotions of excitement. Such as the elegant waltz melody and soothing music can reduce the feeling of fatigue of students when learning, and thus can produce a refreshed, relaxed feeling. While the music of Samba which come from Brazil has a cheerful rhythm [3]. It is performed by the Brazilian people singing and dancing, and enthusiastic singing dance styles. When students hear Samba music, they should also put their emotions on the dance, and show their joy and cheerful mood with dance. Therefore in sports dance class, students have to learn the music content, style, features and so on accurately, detailed, deeply understood, and to be fully inspired when dancing. Both use the hearing to feel the dance, but also visually, so as to put themselves and the audience into a wonderful mood. If in the whole process of a sports dance class, the teacher teaches the students' actions use passwords instead of music, then students will demonstrate a boring mood and dissatisfaction to the learning process. Therefore, later in the teaching process, when teachers teach the students movements, they must often use music to exercise, so that students have an understanding of the process of music in the psychological way, so as to promote the smooth progress of teaching.

85.3 Music Can Coordinate the Students' Dance Movements

It is found that music can not only improve the students 'interest, but also coordinate students' movements, enabling them to enjoy the beauty in beautiful and rhythmic melodies to appreciate the pleasant musical accompaniment in the sports dance teaching practice, so that those actions feeling difficult to master become very interesting and mobilize the enthusiasm of their practice. If a student in practicing just by the beat of their own number without using music, encountered uncoordinated movements of students, it will make the original very graceful looking movements seem awkward, besides he will feel uncomfortable, thus they will have the mood of weariness. Strong and weak beats in the music let students know which actions are on the strong beat, which actions are on the weak beat. This will not only cultivate the students' sense of rhythm and coordination of movements, but can also help students develop imagination and expression. Developing students' delicate artistic sensibility, and enhance self-confidence. A point of sports dance in practice process is inseparable from music. Teaching music, some are cool lake elegant, some are touching and some are impassioned, so that high and low tension and relief are complementary. The music's sense of melody, rhythm, the intensity of the dancers' action and dancers' temperament affects the body's emotion, and how the practitioners' actions play. This has a great inspiring, supporting and coordinating role in teaching.

Students in sports dance lessons through music training can adjust their central nervous system to produce associative memory, and constantly deepen their experience of rhythm, thereby adjusting the rhythm of movements to coordinate the action of sports dance. Besides it can also improve the student-paced changes in capacity and resilience, so that students in the melody and rhythm, make completing technical movements more accurate and vibrant. Sports dance music can facilitate the dancer's expressive power and sense of style, so the rhythm of music for dancers is alien, benign stimulus. Practitioners in the accompaniment of music can naturally change the range of motion and strength, enhance technical movements required by the specific sense of rhythm; it will also help practitioners put speed and power in a reasonable allocation, resulting in a precise sense of time; also improving the accuracy of the action, faster formation of dynamic stereotype. When the sports dance music played, students will produce a warm atmosphere with the melody and rhythm, beat and rich atmosphere, the mood of students is mobilized to generate interest in learning sports dance, so beautiful the sound of music moving. Soothing and elegant, lively, beautiful, dance movement of the limbs, so that their bodies get the full range of exercise, so that teaching can be carried out smoothly, better accomplishing the learning task.

85.4 Music Can Develop Students' Creativity and Thinking Capabilities

Einstein said: "Much of my invention and creativity is inspired from music". Music can make people easily intoxicated with inspiration, thus one can see the role of music, creativity and imagination. The use of music during the teaching process in sports dance, can cultivate students' creativity and thinking capabilities.

Our sports dance teaching in addition to teaching students to learn the provisions of action so that they could practice their own music to orchestrate the action, so a set of dance according to the characteristics of students and their imaginative choreography. Some students lack of imagination and creativity, and only in accordance with blindly accepting the teachings of the teacher, can't think independently and play to their imagination. These kind of people are hard to take up new century modernization of the heavy responsibility, because any inventors and artists lacking in creativity and imagination can never be successful, and through self sports dance action practice can develop students' good habits of mind, and guide them to the artistic imagination and associative thinking, to achieve the purpose of enlightened wisdom, let their life and work have more imagination and creativity in the future.

85.5 The Combination of Sports Dance and Music Can Improve the Aesthetic Quality of the Students

Music is the soul of sports dance. Organic combination of music and sports dance can help students improve their aesthetic quality. Only mixing every action, every shape, and every expression change of sports dance in the music can make the sports dance produce harmony of beauty, and sports dance movements. Only with music closely integrated can one give the action life. Sports dance music allows practitioners to give full play to their emotions, they can bring the audience together to the mood of the dance performance, fascinate reverie thousands, in particular, be able to fully express, only tacit, and cannot explain in words the subtle emotion, and to achieve the perfect sports. Teachers in teaching process should focus on developing students' sense of rhythm, teach them use music's strong, weak, anxious, slow beats to fully demonstrate the ability of movement styles of art, even a small action should also be strictly required to get attention. If practitioners can put themselves into the music, it must be excellent. For example, waltz, its' performance is elegant like prince and princess. In action it requires boys with the chest raised and girls looked up with the basis on the top of the left rear chin upwards, holding the up jaw slightly. Men dressed in tuxedo, girls in big dresses that mop the floor. The two people in the process of dancing accompanied by music with the skirt flying really let the audience feel very soothed and elegant, while the dancers themselves will be into the prince and princess situation.

Of course, the action of expression is not easy, hard training is essential, and the hard to understand feelings are more important.

85.6 Conclusion

In the teaching process of sports dance, music can not only make students have better motor performance during practice, but also gives a quality improvement in all respects. It can not only develop students' psychology, but also can coordinate the students' movements, develop students' creativity and thinking skills, and even can improve the aesthetics. So we need to train students' music and rhythm, in order to stimulate the students' ability to understand dance, arouse our emotions, to inject new vitality and life to sports dance. The proposal is appropriately to adopt music in the teaching process to stimulate the student's hearing, and ask students to do self action frequently to stimulate their creativity and imagination, thereby improving their overall quality.

References

1. Xiu H, Luo X (1999) Music aesthetics theory. Shanghai Music Publishing House, Shanghai 3(4):19–25
2. Zhang R (2005) Dancesport, vol 2(3). Shanghai Education Press, Shanghai, pp 11–27
3. Fan G (2004) Dancesport foundation course. People's Education Press, China, pp 7–62

Chapter 86
A New Designed Baseball Bat Based on Sweet Spots

Pengpeng Zheng, Yaoju Huang and Jianwei Liu

Abstract This paper describes model testing of a wooden baseball bat with the purpose of finding the peak frequencies and vibration modes and their relation to the so-called "sweet spot", including the position of the "sweet spot", the effect of corking a bat and the comparison between wood and metal bats. Present an approach to determining the location of "sweet spot", which can be classified into three categories: (1) Batter feels the least vibrational sensation. (2) The batted ball is at maximum speed. (3) Maximum energy is transferred to the ball. To analysis the vibration sensation, rotational dynamics model is set. The impulse received by hands varies with different impact positions and form a curve. Through the curve, the "sweet spot" that with least vibrational sensation is found. For maximum batted ball speed, velocity model is set on the basis of the conservation of momentum, regarding the bat as a rigid body. Finally, the curve of batted ball speed of different material and impact position is obtained and the impact point with maximum speed is found. Vibrations involving beam elements model is developed to analysis the maximum energy being transferred. By using beam theory, the "sweet spot" is located. Furthermore, through the three kinds of "sweet spot" this paper designs a new baseball bat that different "sweet spots" gather in a "sweet zone".

Keywords: Sweet spot · Baseball · New baseball

P. Zheng (✉) · Y. Huang · J. Liu
School of Science, South China of Technology, Guangzhou 510000, China
e-mail: pengzheng145@126.com

W. Du (ed.), *Informatics and Management Science VI*, Lecture Notes in Electrical Engineering 209, DOI: 10.1007/978-1-4471-4805-0_86,

86.1 Introduction

Over the years, papers in the journals have addressed both experimental and theoretical issues associated with the baseball–bat collision. Brody studied the vibrational spectrum of a hand-held bat during and after the collision and showed that the bat behaves as a free body on the short time scale of the collision. Cross did an extensive study of the vibrational spectrum of free and hand-held bats and concluded that there exists a zone of impact locations on the barrel end of the bat where the impact forces on the hands due to recoil and vibration are minimized. In more recent years many important theoretical papers have appeared that go beyond the rigid approximation by treating the bat as a dynamic, flexible object.

A base bat can be viewed as an ordinary structure used to play a game. However, there is nothing simple that can define a baseball bat. Its shape produces a very complex structure, which affects the dynamic behavior and produces significant properties that govern the bat [1].

Therefore, the best way to understand the behavior of a baseball bat is to use a modal analysis approach, which helps define the parameters of a structure for all the elastic models in the frequency range of interest. The modal frequency and mode shapes parameters from a complete description of the inherent dynamic characteristics of the baseball bat [2].

Baseball bat design involves a tremendous amount of physics and engineering as well as bat-ball collision. Rotational dynamics and the conservation of momentum are the main physical methods used in this paper. The task of this paper includes locating "sweet spot", explaining how the "sweet spot" effect works, why players prefer corking bats [3], analyzing performances of bats of different materials, and the reason for prohibiting metal bats. During the analysis the parameters are changed to check the sensitivity of the models.

86.2 Models

86.2.1 Maximum Velocity Model

Firstly, this paper try to find the sweet spot based on the velocity of the ball. After hitting the ball, the ball will get a better velocity, that is to say, to get the best energy [4]. According to the physical principle, this paper analyzes the distance of the ball. Because of the difference of the hitting spot, the force is different, moreover, the angular momentum of the ball and bat are different. The difference is shown by Fig. 86.1.

From Fig. 86.1, we know that

$$\mathrm{u} = \omega H \tag{86.1}$$

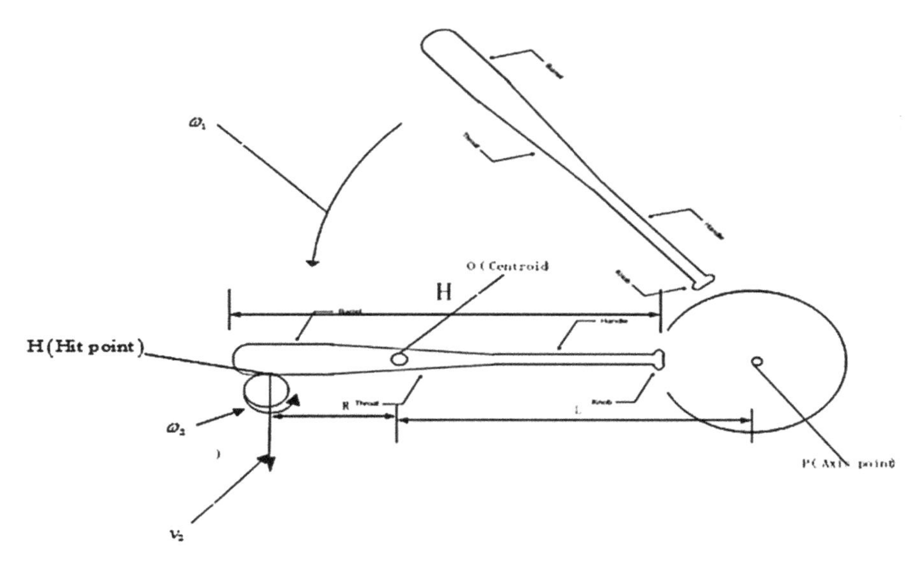

Fig. 86.1 Velicity model

$$v = \omega(H + r) = u + \omega r \tag{86.2}$$

$$m_1 v_1 + m_2 u_1 = m_1 v_2 + m_2 u_2 \tag{86.3}$$

$$m_1 v_1 r + I\omega_1 = m_1 v_2 r + I\omega_2 \tag{86.4}$$

Because the hit spot is not the center of the mass, we improve e by:

$$e = \frac{u_2 + r\omega_2 - v_2}{v_1 - u_1 - r\omega_1} \tag{86.5}$$

Because different materials make different effect on the energy loss, we introduce μ to describe the effect of energy loss, and get the final expression of velocity of the ball [5]:

$$v_2 = v_1 + \frac{m_2 I(1 + e)u_1 - 2m_2 I v_1 + m_2 I\omega_1(1 + e)r}{I(m_1 + m_2 - m_1 m_2 \mu) + m_1 m_2 r^2} \tag{86.6}$$

To discuss the question conveniently, we assume that the μ of the rigid body is 0, and the μ of wood is 0.7.

We get the final results by:

$$r = 0.1$$

After calculation, we know that the maximum speed is 67 m/s.

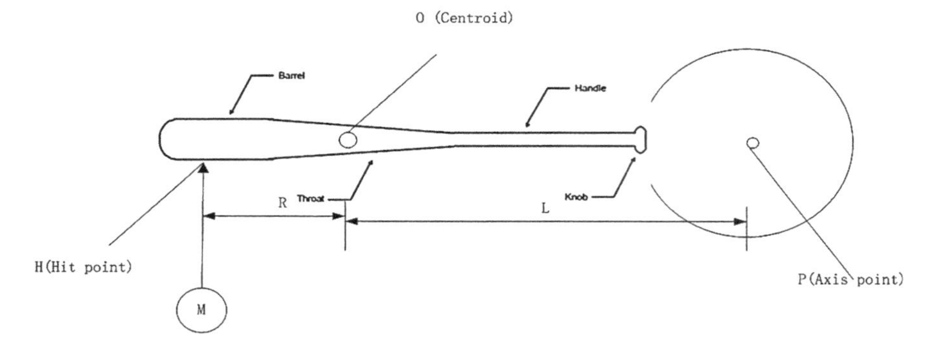

Fig. 86.2 Rotational dynamics model

86.2.2 Rotational Dynamics Model

To transfer the maximum power to the ball when hit, the "sweet spot" is pursed. An assumption based on torque might seem to identify the end of the bat as the sweet spot, but this is known to be empirically incorrect.

In this model, the sweet spot of a baseball bat is mainly defined in terms of the center of percussion. To meet the requirement, the definition that location where minimize energy is transferred to the hand is adopted (Fig. 86.2).

The length between *O* (Centroid) and *H* (Hit point) is *R* and between *O* (Centroid) and *P* (Axis point) is L. The mass of the bat is m. The acceleration of the ball is *A* and the centroid is A'. Therefore, the force the ball impacts on the bat is *F* and the force bat impacts on hands are F'. The angular acceleration of *O* is ω. (Moment of inertia is *I*).

Set axis point as reference point to develop the rotational equations of motion.

$$F' = \left(\frac{mdL}{I} - 1\right)F \tag{86.7}$$

If it is required that $F' = 0, L$ should be:

$$L = \frac{1}{md} \tag{86.8}$$

86.2.3 Vibrations Involving Beam Elements Model

Consider the baseball bat as a whirling beam about an axis. A baseball strikes the bat, causing the bat to free vibration. Regarding the bat as a long and thin stick, as Fig. 86.1 shows, in which the central shafts of every cross section are in a same plane. A collision caused by a baseball enforces the bat bended and then the bat starts to vibrate. The bat in this situation is called Bernoulli-Euler Beam.

Set the coordinate system as Fig. 86.1 shows, $y(x,t)$ represents displacement of the across section which is away from the original point at the time.

From equations given by the theory of vibration, the vibratory partial differential equation is:

$$\frac{\partial^2}{\partial x^2}\left(EJ\frac{\partial^2 y}{\partial x^2}\right) + \rho A\frac{\partial^2 y}{\partial t^2} = 0 \tag{86.9}$$

In which E represents the Young's modulus of the bat, is the density of the bat and A is the area of one differential section of the bat. $y(x,t)$ Can express like this:

$$y(x,t) = Y(x)b\sin(\omega t + \varphi) \tag{86.10}$$

So Eq. (86.11) can be expressed as:

$$(EJY'')'' - \omega^2 \rho AY = 0 \tag{86.11}$$

With the boundary conditions:

$$\left.\begin{array}{lll} Y = 0 & or & (EJY'')' = 0 \\ Y' = 0 & or & EJY'' = 0 \end{array}\right\} x = 0 \quad or \quad x = l \tag{86.12}$$

The equation combined with the boundary conditions constitute the eigenvalue problem of partial differential equation, and ω^2 is called the eigenvalue.

If the stick is uniform section, equation can be simplified as:

$$Y^{(4)} - \beta^4 Y = 0 \tag{1.12}$$

where $\beta^4 = \frac{\omega^2}{a^2}$ and $a^2 = \frac{EJ}{\rho A}$.

But in practice the baseball bat is with irregular shape, in which the expression of the transmission of the vibration is much more complex. In our model we adopt the Rayleigh-Ritz method which is one of the solving processes of finite element analysis.

According to Rayleigh-Ritz method, $Y(x)$ expands as:

$$Y(x) = \sum_{i=1}^{n} a_i\varphi_i(x) = a_1\varphi_1(x) + a_2\varphi_2(x) + \cdots + a_n\varphi_n(x) \tag{86.13}$$

In which $\varphi_i(x)$ is called the fundamental function. The meaning of the expression is to convert the continuous stick into n disperse parts.

So

$$\omega^2 = st\frac{\int_0^l EJ(Y'')^2 dx}{\int_0^l \rho AY^2 dx} \approx \frac{\sum_{i=1}^{n}\sum_{j=1}^{n} k_{ij}a_i a_j}{\sum_{i=1}^{n}\sum_{j=1}^{n} m_{ij}a_i a_j} = \frac{\alpha^T K\alpha}{\alpha^T M\alpha} \tag{86.14}$$

where $K = \begin{Bmatrix} k_{11} & \cdots & k_{1n} \\ \vdots & & \vdots \\ k_{n1} & \cdots & k_{nn} \end{Bmatrix}$, $M = \begin{Bmatrix} m_{11} & \cdots & m_{1n} \\ \vdots & & \vdots \\ m_{n1} & \cdots & m_{nn} \end{Bmatrix}$, and $\alpha = [a_1 a_2 \ldots a_n]^T$.

With the Rayleigh-Ritz method, we solve the partial differential equation and finally get the sweet zone of a standard baseball bat (83.8 cm long and 0.905 kg weight) is ranged from 64.9 to 72.4 cm (the sweet zone is defined as the area which the vibration Amplitude is minimum).

86.3 Make a New Baseball Bat

From the study conducted above, a conclusion is given that the sweet spot or sweet zone is not unique. But as for the strength model, we find another sweet spot which based on the minimize strength. Moreover, in the model of energy, another sweet spot based on energy is found. Unfortunately, these three sweet spots are different. That means, using the bat which is commonly used in the American Major League Baseball, the players can not hit the longest distance, while the strength and energy he get in hand are the lowest. As for all the initial variables of the three models, we assume the material identified, because as for random material, we can calculate the effect of hitting as we like. According to the Table 86.1.

To make our model more common, we choose while ash as the identified material to design a new bat. Because of the shape of the bat is complex, we simplify the shape of the bat. After the simplification, the shape of the bat is simplified as two cylinders and a circular truncated cone. We change the shape of the simplified bat, and try to make the three sweet spots at the same area. We fix the area of the energy of the sweet spot, and try to make the sweet spots of velocity and strength meets this area. Then we can get the best point. As for the simplified bat, we divide it into three parts,

So the volume of the bat is:

$$V = V_1 + V_2 + V_3 \tag{86.15}$$

And the mass of the bat is:

$$m = \rho V \tag{86.16}$$

To make the discussion more convenient, we fix the volume of the bat, that is to say V is a constant.

Table 86.1 Material of bats

Material	White ash	Maple	Bamboo	Hickory	Birch	Composite
Utilization	Majority	Sometimes	Seldom	Seldom	Sometimes	Seldom

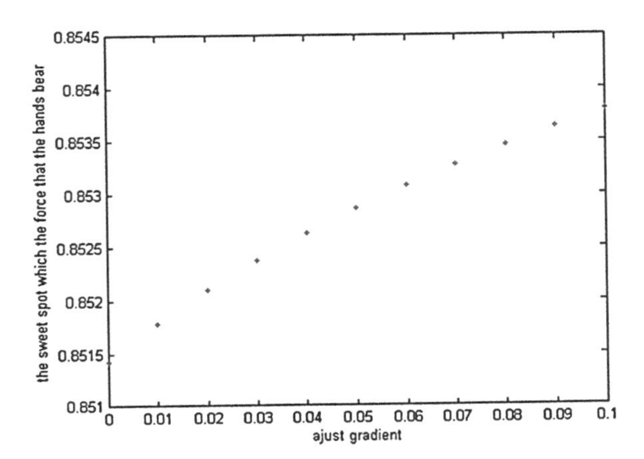

Fig. 86.3 Sweet spot of energy

$$y = kx + \frac{R_1}{2} \tag{86.17}$$

$$y = -kx + \frac{R_1}{2} \tag{86.18}$$

And the center of the mass is on x-axle. So we assume the center of the mass $C(x_0, y_0)$. Obviously $y_0 = 0$, and we can get:

$$x_0 = \frac{\int_{-L_1-L_2}^{L_3} xdm}{m} \tag{86.19}$$

Then we get moment of inertia I:

$$I = \int_{-L_1-L_2}^{L_3} x^2 dm \tag{86.20}$$

Then we get the needed variables of model 1 and model 2. Using the variables, we substitute in model 1 and model 2. And make fine tuning of k, while V is fixed, to meet the outcome of model 3.

After calculation, the sweet spot of strength is too far from the area of energy, From the Fig. 86.3.

From the figure, though r change a lot (more than 0.1 m, it is large as for the length of the bat), the sweet spot change little. And the change is far too little to approach the area of the sweet spot of the energy. So it can be concluded that the strength sweet spot cannot be adjust to the sweet spot of energy and velocity.

But the sweet spot of velocity is near the area of energy. So after the fine tuning of k, the sweet spot of velocity is in the area of energy. And when the sweet spot of velocity is in the middle of this area, the effect of hitting ball is considered to be the best. Using mat lab simulate the process of fine tuning of k and its effect, we get the final results:

$$r = 0.133 \tag{86.21}$$

After adjustment, a new bat is designed.

References

1. Zandt LL (1992) The dynamical theory of the baseball bat. Am J Phys 60(2):172–181
2. Bruel K (2009) Structural testing: part 2–modal analysis and simulation, vol 21. Denmark, pp 4–48
3. Greenwald RM, Penna LH, Crisco JJ (2001) Differences in batted ball speed with wood and aluminum baseball bats: a batting cage study. J Appl Biomech 17:241–252
4. Fleisig GS, Zheng N, Stodden DF, Andrews Article JR (2002) Relationship between bat mass properties and bat velocity. Compos Struct 23:481–488
5. Shenoy MM, Smith LV, Axtell JT (2001) Performance assessment of wood, metal and composite baseball bats. Compos Struct 52 397–404

Chapter 87
Statistic and Evaluation of Data Between Chinese and Foreign Female Freestyle Swimmers in Long Distances

Geng Du and Cheng Xiong

Abstract Comparative analysis has been given between Chinese and foreign elite female freestyle swimmers in long distance while they performed at the 29th Olympics, the 14th FINA World Championships and the 11th National Games in China. We have analyzed the race component times such as start time, mid-pool (free/clean swimming) time, turn time, final time, stroke length and stroke frequency. The results showed the advantages and disadvantages of Chinese swimmers comparing to their foreign counterparts. It can be applied to modify future training plans of Chinese swimming to improve their poor race components.

Keywords Swimming · Long distance · Freestyle · Technical analysis

87.1 Introduction

The main technical parameters in swimming contain start time, mid-pool time, turn time, final time, and stroke length and stroke frequency [1]. This paper, comparative analysis has been given among the final results time, the female athlete's age and the main technical parameters of 400 and 800 m freestyle in 29th Beijing Olympics, 14th FINA World Championships and 11th National Games in China [2]. The researchers find out the characteristics and gap of techniques and

G. Du (✉)
Department of Sports Training, Wuhan Institute of Physical Education, Wuhan 430079, China
e-mail: gengdu285@126.com

C. Xiong
Department of Postgraduate, Wuhan Institute of Physical Education, Wuhan 430079, China

W. Du (ed.), *Informatics and Management Science VI*, Lecture Notes in Electrical Engineering 209, DOI: 10.1007/978-1-4471-4805-0_87,

tactics between Chinese and foreign female athletes of 400 and 800 m freestyle. This research fills in the blank of our lagging research in female freestyle swimmers in long distances data statistic and evaluation and enriches the statistic and evaluation system.

87.2 Methods

Firstly, to conduct statistical analysis the final results of the top 8 swimmers of 400 and 800 m freestyle in 29th Beijing Olympics, 14th FINA World Championships and 11th National Games in China; then figure out the start time, mid-pool (free/clean swimming) time, turn time, final time, stroke length and stroke frequency; finally compare with the difference between Chinese and foreign swimmers and review of the literature material and to find the relevant features on the differences [3].

87.3 Results

87.3.1 Age Analysis

From 2008 Beijing Olympics final results announcement, 14th FINA World Championships results announcement, 11th National Games results and report announcement.

From Table 87.1, we find that there is an obvious gap between Chinese elite swimmers and foreign elite swimmers in both 400 and 800 m freestyle. Chinese top 3 and top 8 swimmers are slower 0.67 and 2.04 s than Beijing Olympics, 0.66 and 1.52 s than Shanghai FINA World Swimming Championships in 400 m freestyle. In 800 m freestyle, Chinese top 3 and top 8 swimmers are slower 4.70 and 4.77 s than Beijing Olympics, 4.13 and 4.16 s than Shanghai FINA World Swimming Championships. With the distance becomes long, the gap is growing more and more obvious, which reaches 4–5 s. Another phenomenon is that the average age of Chinese elite female swimmers is quite smaller than foreign elite swimmers', the top 3 and top 8 are 3–5 years younger than 29th Olympics' and 14th FINA World Swimming Championships in 400 m freestyle, at the same time the range is 2 years old from 29th Olympics to 14th FINA World Championships. In 800 m freestyle, the range contains at 3–6 years, and 14th FINA World Championships is older than Beijing's. Above all, we can see that the players in 800 m freestyle are younger than that in 400 m's. It turns out that Chinese elite players are quite younger than the foreign ones, at the same time, along with the distance growing; the gap between Chinese players and foreign players is increasing.

Table 87.1 Age analysis of top 8 women's 400 and 800 m freestyle in 29th Beijing Olympics, 14th FINA world championships and 11th national games in china

Average value		Top 3			Top 8		
		Beijing	Shanghai	Jinan	Beijing	Shanghai	Jinan
400 m	Result	4:03.34	4:03.35	4:04.01	4:04.90	4:05.42	4:06.94
	Age(y)	20	22.3	19.3	21	22.1	17.9
800 m	Result	8:19.12	8:19.69	8:23.82	8:24.42	8:25.03	8:29.19
	Age(y)	20	22.7	16.7	20.9	22.5	15.4

From the above analysis in this paper shows that our country's still present a phenomenon of small age, although the last to shorten the national games, but the gap still contains 3–6 years old [4]. Therefore, strengthening our country swimming backup talents team construction, increase talents reserves thickness and depth, and through the scientific training enhances our country swimming training level, help maintain the best young athletes competitive state, cultivate their self-confidence, improving training quality is to increase our country's female long distance swimmers and the guarantee of the fundamental conditions [5].

Comparative analysis of top 3 average split time of 400, 800 m freestyle in Beijing Olympics, Shanghai FINA World Championships and 11th National Games.

From Table 87.2, with comparative analysis, we find out Chinese swimmers are 0.73 s slower than Beijing Olympics and 1.55 s faster than Shanghai FINA World Championships in first half of 400 m freestyle, and in last half, Chinese swimmers are 0.06 faster than Beijing Olympics and 2.21 s slower than Shanghai FINA World Championships. In 800 m freestyle, Chinese swimmers are obviously slower in 0–200, 200–400, 400–600 and 600–800 m. It is also the mainly reason why Chinese female swimmers cannot catch foreign swimmers. In the women's long distance freestyle, the physical distribution plan is not careful, tactical intent is not clear and exposed a long distance swim in the women's fitness reserves is insufficient, training level is low, execute tactical ability is poor weakness.

Table 87.2 Female swimmers' top 3 average split time of 400, 800 m freestyle in Beijing Olympics, Shanghai FINA world championships and 11th national games

Average of top 3	Beijing	Difference 1	Shanghai	Difference 2	Jinan
400 m time	4:03.34	−0.67	4:03.35	−0.66	4:04.01
0–200 m time (s)	1:59.75	−0.73	2:02.03	1.55	2:00.48
200–400 m time (s)	2:03.59	0.06	2:01.32	−2.21	2:03.53
800 m time (s)	8:19.12	−4.70	8:19.69	−4.13	8:23.82
0–200 m time (s)	2:03.29	−0.86	2:02.76	−1.39	2:04.15
200–400 m time (s)	2:06.36	−0.81	2:06.94	−0.23	2:07.17
400–600 m time (s)	2:05.32	−1.35	2:06.40	−0.27	2:06.67
600–800 m time (s)	2:04.15	−1.68	2:03.59	−2.24	2:05.83

Difference 1 = 29th Olympics–11th NGM
Difference 2 = 14th FINA WCS–11th NGM

Table 87.3 Related technical parameters of top 3 of 400 m freestyle in Beijing Olympics, Shanghai FINA world championships and 11th national games

Average of top 3	Beijing	Difference 1	Shanghai	Difference 2	Jinan
400 m time	4:03.34	−0.67	4:03.35	−0.66	4:04.01
Start 15 m time (s)	7.59	0.17	7.47	0.05	7.42
7 × 15 m turn time (s)	61.98	0.87	63.37	2.26	61.11
15 m average turn time (s)	8.85	0.12	9.05	0.32	8.73
275 m mid-pool time (s)	170.87	−1.72	169.58	−3.01	172.59
Final 5 m time (s)	2.9	0.01	2.93	0.04	2.89
Stroke frequency (t/min)	51.46	5.15	47.82	1.51	46.31
Stroke length (m/t)	1.88	−0.18	2.03	−0.03	2.06
Speed (m/s)	1.61	0.02	1.62	0.03	1.59
Technical index (3)	3.02	−0.27	3.3	0.01	3.29

Difference 1 = 29th Olympics–11th NGM
Difference 2 = 14th FINA WCS–11th NGM
Technical index = stroke frequency × stroke length

Related technical parameters of top 3 of 400, 800 m freestyle in Beijing Olympics, Shanghai FINA World Championships and 11th National Games.

The world outstanding female players in long speed control ability are worth our country's learning from. Speed is not only training and body function level of concrete manifestation, but also the speed of the average speed directly determines the final match result and the ownership of the championships so can't ignore the physical distribution on the average of the speed and control. From Table 87.2, we found that our country in the 400 m freestyle players, the gap within 1 s total grade, and pay attention to detail assurance and the way will be the key to win the game, but in the 800 m, our players always follow the excellent foreign players, so get long distance freestyle, the key is that the outstanding result how to adjust the best competitive state, and constantly improve the body function level, adjust measures to local conditions to develop in its actual situation and the technical characteristics of the physical distribution plan, with the best average speed swam all the way.

In Table 87.3 shows that Chinese players in the 15 m of the technology link performance given some good players in world, than the Olympic Games, world championships and 0.05 to 0.17 s fast, our state's players out on certain advantages in possession; In turn link, the players in China still have advantages, respectively, compared with the Olympic Games, world championships and 2.26, 0.87 s fasts; And in the last 5 m on sprint, China's athletes also lead the world to a good player, but in the middle, our players and the world, there are certain gap between players, respectively, than the Olympic Games, world championships and 3.01 as 1.72 s. That difference makes our players to be lagging behind in the final result status. The reason we have to pay attention to foreign athletes in row frequency, stroke length of the performance, our country's stroke frequency are lower than the Olympic Games, world championships in 5.15, 1.51 times/min, but in stroke length on much more than the Olympic Games 0.18 m/time, more than 0.03 m/

Table 87.4 Related technical parameters of top 3 of 800 m freestyle in Beijing Olympics, Shanghai FINA world championships and 11th national games

Average of top 3	Beijing	Difference1	Shanghai	Difference 2	Jinan
800 m time	8:19.12	−4.70	8:19.69	−4.13	8:23.82
Start 15 m time (s)	7.47	−0.12	7.55	−0.04	7.59
15 × 15 m turn time (s)	133.16	−0.48	133.41	−0.23	133.64
15 m average turn time (s)	8.88	−0.03	8.89	−0.02	8.91
555 m mid-pool time (s)	355.65	−4.03	355.88	−3.80	359.68
Final 5 m time (s)	2.84	−0.07	2.85	−0.06	2.91
Stroke frequency (t/min)	48.16	4.21	47.58	3.63	43.95
Stroke length (m/t)	1.94	−0.16	1.97	−0.14	2.11
Speed (m/s)	1.56	0.02	1.56	0.02	1.54
Technical index (3)	3.03	−0.22	3.07	−0.18	3.25

Difference 1 = 29th Olympics–11th NGM
Difference 2 = 14th FINA WCS–11th NGM
Technical index = stroke frequency × stroke length

world again. According to Zhong Yu [1], the swimming stroke length of that mainly by the player's arms to show and upper limbs muscle strength of the decision. Athletes stroke rate and the athletes' physical condition in the outside, main also and athletes of the power and swimming stroke about technology [6]. From athletes delimit the declining ability of the frequency has reflected the pathway of athletes fatigue and energy consumption, make athletes fast stroke need muscle ability can't last. That is athletes the decline was more than stroke frequency stroke length declines, which is also influence that swimming speed is the main factor of stroke frequency drop. Therefore, in a long distance swim in the project to maintain high speed stroke of the frequency of ability directly related to the whole of the Speed system.

From the Table 87.4, we can see, our country's 800 m grades than the Olympic Games, world championships is slow 4.70 and 4.13 s, which lead to the main factor is the gap between the way speed slowly. If our country player to achieve better grades, the way the speed of swimming are vital, but want to progress on the one hand, will come from the following several aspects to discuss and study related technical parameters, tally up the problems need to be solved.

According to the contrast of the technical indicators can be found, our country's 15 m and the world of good players gaps; Turn a bit slow the Olympic Games time respectively players and world championships in 0.48, 0.23 s players; Sprint 5 m also at a disadvantage, slower than the Olympic Games 0.07 s, than the world championships as 0.06 s, gap is not obvious, The way our country's swim slower than the Olympics 4.03 s, world player than slow 3.80 s, the index of the gap is significant.

There is a gap between Chinese female swimmers and foreign female swimmers in the intermediate time that makes Chinese females swimmers to be lagging behind in the final results. Stroke length is mainly composed of the swimmers' arm that to show and upper limbs muscle strength of the decision. Except stroke frequency and body fitness, physical strength and technique are the most significant of all the technical parameters.

With the stroke frequency ability of female swimmers becoming decreasing, it gives an index to that the nerve conduction fatigue and energy consumption to be come out, that is the decline of stroke frequency is more than that of stroke length, which is also the major factor of influence in dropping of stroke frequency. Therefore, the ability of keeping stroke frequency in high speed in long distance freestyle is directly to be related to the speed of the whole course. The gap between Chinese and foreign females swimmers is main in the intermediate time, for the reason that prorate of stroke frequency, stroke length, fitness distribution and keeping the best average speed. Chinese are better than foreign female swimmers of the intermediate time in last half, that the combination of stroke length and frequency, delimit good relationship of scale, the appropriate improve cross frequency, so as to improve the average speed, from the technical level solve stroke length and frequency and combination problem, in order to improve the average speed providing technical support and guarantee.

87.4 Conclusion

Thought relative technique date analysis, there is no significant differences between the world excellent swimmers and ours at the start of 15 m, and even more faster than any other swimmers who joined Olympics and world championship in the 400 m, start technique not only brings time lead during the match, but also creating psychological advantage for swimmers, it will disrupt the rhythm of the opposition and will help us to exert our best rhythm in the match. Our swimmers equal to or even better than the world excellent swimmers at the first 15 m sprint, and there is not obvious difference at the final mile. Overall, comparing to the world excellent swimmers, we performed a relatively low stroke frequency in 400 and 800 m freestyle, and we did not effectively combined the stroke frequency with stroke length. Appropriate to improve stroke frequency and keep the fastest average speed to rush into the finish line is the key to get the best result. There is tiny distance for us to reach the between the world excellent swimmers level in 400 m freestyle, the main drawback is stroke frequency and speed up swimming speed at the first half. It is crucial for us to find a breakthrough point in 400 m freestyle. In the 800 m freestyle, we are behind the world excellent swimmers, to shorten the distance, we should focus on stroke frequency, and according to our technique characteristics and ability we should adjust the ratio of stroke frequency and stroke length, and pay more attention on start and turn, constantly improve training level and mid-pool.

Gradually establish and perfect the system of cultivating potential swimmers is foundation of the sustainable and healthy development of the swimming career. We should draw on foreign experience, increased investment in science and technology, vigorously promote innovation, constantly improve the scientific training conditions; equipment advanced training instrument and science technology support system. We should stick on "inviting, going out", learn from foreign advanced swimming theory, lay a solid foundation for nation team, and boost the process of nation team; those are the only way to develop the swimming career.

References

1. Zhong Y, Liu X, Lei W (2003) The research on change of speed, stroke frequency and stroke range every 10 in 50 m freestyle swimming. Chin Sports Sci Technol 2:93–96
2. Neiling W, Lingling Y (2009) The technical and tactical analysis on elite female freestyle swimmers in short distance in 27th Olympic games. Chin Sports Sci Technol 38(2):55–58
3. Lingling Y (2008) The technical and tactics analysis on world elite male freestyle swimmers in short distance participating in 27th Olympic games. Chin Sports Sci Technol 37(10):19–20
4. Xuexiong M (1991) Optimum stroke frequencies for short and middle distance swimmers. Sports Sci 22(2):45–47
5. Lingling Y (2004) Comparative analysis on technique of elite 100 m freestyle swimmer at home and abroad. Chin Sports Sci Technol 12(4):25–29
6. Peng J, Zhai F (2012) Competitive pattern of the world swimming and chinese swimming position from the 14th FINA world championships in Shanghai. Chin Sports Sci Technol 3(1):106–111

Chapter 88
Research on Urban Practitioners Participation of Martial Arts Fitness and Job Satisfaction

Mingming Guo

Abstract In today's society, people's pace of life is accelerating, especially practitioners of some larger cities. In the survival state that the increasing pressure from various quarters, the health status of this part of the population is worrying people. In this study, the relationship of urban practitioners' participation in martial arts fitness and job satisfaction is studied empirically. Using mainly literature, questionnaire survey and data management statistical method, from current situation of participating in the martial arts fitness activities, Shenyang City in membership, in-service personnel are investigated as the object of study, using random cluster sampling for the original data and statistical analysis. The survey found that job satisfaction of the city employees involved in the martial arts fitness groups was significantly higher than non-participating groups; on satisfaction of job, promotion, salary, the boss and colleagues, the martial arts fitness groups were higher than non-participating groups. In different levels of participation in martial arts fitness their job satisfaction level is different; there is very significant correlation between the frequency of participation in martial arts fitness and participation time and job satisfaction. Among them, the crowds of the high participation frequency get greater job satisfaction value than the low participation of the frequency of the crowd; when each involved in the martial arts exercise for 60–90 min, job satisfaction is in the highest level; and each involved in the martial arts fitness to moderate sweating intensity, their job satisfaction value is greatest. I hope the study of the relationship between the martial arts fitness and job satisfaction can contribute to expand the field of sports research, and increase the attention of organizations and individuals in sports participation, provide a theoretical basis for follow-up study.

Keywords Urban · Practitioners · Martial arts fitness · Job satisfaction

M. Guo (✉)
Shenyang Sports University, Shenyang, 110102 Liaoning, China
e-mail: mingmingguozx33@126.coms

W. Du (ed.), *Informatics and Management Science VI*, Lecture Notes in Electrical Engineering 209, DOI: 10.1007/978-1-4471-4805-0_88,

88.1 Introduction

The job satisfaction is satisfaction the staffs have for the job they engaged in, and has got the attention of the research in organizational behavior. From the perspective of the development of job satisfaction research, the researchers first explore the structure of job satisfaction and the preparation of the revised scale, and then developed to explore the factors that affect satisfaction. In recent years, along with the development of personality psychology research, researches on how personality factors influence job satisfaction are also increasing. Although the research field in the expanding, but the introduction of the sports participation variables appeared in few studies [1]. Kim Jee of-Young and Bynum Won-Tae (Korea) in its the impact of various entertainment activities on professionals' satisfaction with job introduced the sports participation variables, and the results show that sports activities have a positive impact on job satisfaction. The studies of Fujiwara Kin-solid (Japan) have shown that the fatigue generated by labour of sports activities group is lower than the group of non-implementation of sports activities; the satisfaction of wage income of the sports activities group is greater than the non-implementation group; the satisfaction of workplace relationships of the implementation group is also higher than the non-implementation group; physical activity can improve job satisfaction [2, 3]. In China, job satisfaction studies focused on the relationship between the demographic variable, personality characteristics, work stress and job satisfaction, and the research involved in martial arts fitness and job satisfaction has not been found. The paper uses decision theory based on individual factors, from a sociological point of view, with the survey about Shenyang City employees participating in the martial arts fitness and the status of the job satisfaction to study the relationship between participation in the martial arts fitness and job satisfaction, in order to expand the sports research areas and raise the attention of organizations and individuals in sports and provide a theoretical basis for subsequent research.

88.2 Objects and Methods of the Study

88.2.1 Object of the Study

This study investigated the random sampling in Shenyang, and the membership, serving officers in Shenyang City are chosen as the objects [4].

Check out the related research results and literature to understand the status quo of the current study and lay the theoretical basis. According to the research needs, this questionnaire selected the statistical factors involved in physical activity status factors and existing common job satisfaction (JDI) national mass sports survey listed. The scale contains include five subscales, each representing the five dimensions of job satisfaction, including: (1) Work satisfaction. (2) Promotion satisfactions. (3) Salary satisfaction. (4) Boss Steering satisfaction. (5) Colleague's satisfaction. In this study, the split-half reliability of job satisfaction scale was 0.74, with good reliability. Questionnaire using random cluster sampling, a total of 1,000 questionnaires were issued, the recovery was 94.3 %, the effective rate was 89.4 %, the total number of valid samples were 843.SPSS13.0 software for Windows Social Sciences statistical software to analyze the raw data, the concrete application of the frequency analysis of the comparison of descriptive statistics, comparison of mean and standard deviation, variance analysis, Spearman correlation analysis are used to do the data analysis.

88.3 Results and Analysis

88.3.1 Employees Basic Statistics and Analysis. Demographics Survey

Various demographic factors will have a greater impact on the measurement results. Therefore, in the course of the study, the respondent's gender, education, marital status, age, working time and other factors are included in the questionnaire, the results are shown in Table 88.1.

Practitioners involved in the Martial Arts Fitness Survey: According to the definition of sports participation by Lu Yuanzhen and the need for this research, population directly involved in the martial arts fitness, personal sports experiences, and last year participated in the martial arts exercise for at least 1 will be classified as martial arts fitness exercise participation groups.

The statistical results show, the number of people involved in martial arts fitness is 562, two times of non-participation groups. Practitioners involved in the martial arts fitness are higher than the national average (45 %, 2001). Factors that affect this result may be: first, with the development of social and economic progress, people's living standards year by year increase, and sports spending power of practitioner's increases. Second, sports in Liaoning province is developed, in the rapid development of competitive sports, mass sports and community sports has also seen an unprecedented development opportunity. From the enterprises and institutions to government agencies, from urban to rural, have set off a national fitness craze, so practitioners are necessarily affected to participate in sport. Third, the 2008 Olympic Games bid is successful, and sports propaganda of

Table 88.1 Distribution of demographic factors

Sex		degree				marriage		age				work year				
Male	female	graduate	university	high	junior and bellow	single	married	<25	26–35	36–45	>46	<5	6–10	11–20	21–30	>31
353	490	39	512	234	58	288	555	246	265	237	95	366	128	190	130	29

a variety of media publicity is a great promotion of employees to participate in martial arts fitness. Forth, employees are younger, highly educated, promoting the sport to be life and social, sports have go into people's lives as a healthy lifestyle. For deep description of the participation in martial arts fitness status, examining participation in the time, in the frequency, and in strength (mass sports survey, 2001) are necessary. In order to display participation in martial arts fitness status more visually, this study uses the percentage of statistical methods.

Groups to participate in sports 1–3 times a week are the most, accounting for 32.9 % of the total number; in participate time, each time in 30–60 min accounted for the highest percentage of 46.1 %; to participate in the strength of "sweating slightly" are the highest percentage of 27.8 %. From forms of participation, personal training is still the main form of participation of the sports crowd, amateur, voluntary, decentralized form of activity is still the first choice of participants, which is conducive to the persons involved in the fitness exercise effect. The percentage of exercising with friends, colleagues and exercising with their families is also high in sports activities; it enhanced communication with family and friends, getting the pleasure experience. Physical activity organized by unit is one of the forms of participation. In some enterprises and units of the good economic benefits and qualification, the acquisition of sports fitness equipment and organization of sports activities make their employees to enhance exchanges and cooperation in sports and increase the cohesion and team spirit, promote the development of enterprises. Few people choose to join in sports coaching stations and clubs to participate in physical exercise, so organizational, scientific and popular form of fitness and recreation remains to be further developed and popularized. In the choice of a specific martial art types, traditional martial arts are the main ones, some of the emerging competition routine did not become a daily means of fitness and recreation. Activities are mainly concentrated in the martial arts types of tai chi, tai chi fan, Mulan fan, Baji Quan, Taiji Sword. On the one hand these projects had better development and promotion and people are more familiar with them; on the other hand, these projects is simple to learn and asking not much of the space, equipment or economic investment.

88.3.2 The Relationship of Employee's Participation in Martial Arts Fitness and Job Satisfaction

This study, for job satisfaction of participation in the martial arts fitness groups and non-participating groups, made the comparison of mean and standard deviation, and made an analysis of variance [5]. Shown in Table 88.2, the average job satisfaction involved in the martial arts fitness groups is higher than non-participating groups, also is showed this feature in the five measurement dimensions, and differences were significant by the variance test. This shows that among practitioners groups, job satisfaction of participation in sports groups is higher than non-

Table 88.2 Job satisfaction variance analysis table of involved in the martial arts fitness groups and non-participating groups

	Participating groups (N = 562) M ± SD	non-participating groups (N = 281) M ± SD	P
Overall satisfaction	238.31 ± 26.28	230.08 ± 25.23	< 0.01
Job satisfaction	58.52 ± 8.67	56.06 ± 7.89	< 0.01
Promotion satisfaction	26.08 ± 5.17	25.30 ± 4.30	< 0.05
Salary satisfaction	26.87 ± 5.43	25.80 ± 4.48	< 0.01
Boss satisfaction	62.80 ± 8.07	60.70 ± 8.59	< 0.01
Collage satisfaction	63.98 ± 8.69	62.32 ± 8.67	< 0.01

participating groups. And based on the definition of participation in martial arts fitness groups, it can be inferred that, whether it is recurrent participation or occasionally involved in the martial arts fitness groups, their job satisfaction is higher than non-participating groups.

In this study, the Spearman correlation analysis method is used to analyze job satisfaction and level of participation (participation rate, participation in time to participate in strength) to reveal the relationship of the different participation and job satisfaction.

Shown in Table 88.3, there is a very significant correlation ($p < 0.01$) between participation frequency, time and job satisfaction, and participate in strength and job satisfaction relationship was not significant ($p > 0.05$). That the higher the frequency of participation in sports activities, the higher the job satisfaction; the longer to participate in, the higher job satisfaction; but participation strength and job satisfaction are not related. In the study of relationship of involved in the martial arts exercise frequency, time, strength and job satisfaction, analysis of variance was used.

With the increase in the frequency, job satisfaction increases, when it reaches more than 3 times/week which is the highest level, job satisfaction value is highest [6]. The difference was significant ($p < 0.05$) by variance test with the time changes, job satisfaction changes, participation time for "60–90 min" gets the highest satisfaction value, since with the increase of the time, satisfaction reduces, there is a significant differences by variance test. As exercise intensity increases, the satisfaction value increased, for the level of "medium sweating", satisfaction reached the highest value, and then reduces, but no significant difference by the variance test.

Table 88.3 Involvement and job satisfaction analysis table

	Overall satisfaction	job satisfaction	promotion satisfaction	salary satisfaction	boss satisfaction	Collage satisfaction
Frequency	0.142**	0.107*	0.098*	0.086*	0.024	0.156**
Time	0.127**	0.145**	0.136**	0.079	0.001	0.095*
Intensity	0.030	0.015	−0.055	0.081	0.039	−0.009

*$p < 0.05$; **$p < 0.01$; N = 562

According to the theory of sports and fitness, participate in martial arts exercise in order to achieve the purpose of enhancing physical fitness, heart and lung function, should be of moderate intensity exercise. In the moderate-intensity exercise, muscle energy is by aerobic metabolism (use of inhaled oxygen to burn body fat and glucose) way to get, people get better fitness results by aerobic exercise. For mental health, if mental health level improved seems has nothing to do with the improvement with maximal oxygen uptake or not. Fret physical exercise, even when cardiovascular function did not improve, anxiety and depression levels may also decline and can effectively improving the state of mind. Job satisfaction is an emotional response and emotional state. It follows that participate in strength did not affect corresponding impact on mental state of the fitness effects. The interpretation of internal relations of the strength and job satisfaction also require follow-up research and exploration.

Existing studies suggest that the nature of work, supervisor, income, working conditions on job satisfaction, only produced a small effect, the job satisfaction of employees to change employers and types of work, was also in steady state, the mood at work and work satisfaction are highly relevant: positive and negative emotions are strong predictor of overall job satisfaction, employee with positive emotions have high satisfaction, employee with negative emotions have low satisfaction. The Judge (2001) found that positive self-concept of individuals will have a more positive evaluation on its work, and will affect the actual perception of their work attitude, that is, individuals with positive self-concept, not only produce higher satisfaction because they are happy, but also because they are easy-to-a positive factor in the perception of work which result in higher satisfaction. Martial arts fitness is the most economical, direct and effective means to solve the negative emotions in work and enhance self-concept. Positive adjustment are achieved by affecting the hormones and the autonomic nervous system in the martial arts exercise in skeletal muscle and visceral feedback, this can also be seen as benign due to physical exercise stress induced by the decline in state anxiety after exercise. This phenomenon was also explained sports caused the interruption to the negative sentiment. Possible reasons are: to accelerate the flow of blood is beneficial to the central nervous system; depressive patients lower nor epinephrine, and physical exercise can increase nor epinephrine; of regular physical exercise can develop a sense of control, to improve depression useful; the improvement of body image is also beneficial to reduce depression.

A study of physical exercise impact on the emotion in nearly 30 years found that: (1) a one-time movement and long-term physical exercise can effectively reduce depression and anxiety state; physical exercise can reduce stress symptoms; physical exercise have emotional benefits the individuals of all ages, different genders. However, the duration and frequency of physical exercise is related to the degree of depression reduction. (2) Compared with relaxation training, and other recreational activities, physical exercise can effectively reduce depression. The modern development of productive forces and science and technology advancement have greatly reduced the monotonous and heavy degree people engaged in manual labor, but the modern labor did not reduce the workers' physical requirements. Requirements of different production conditions and labor status are different. Workers in high-tech conditions, often require the spirit to be highly centralized, prompt response and action, decisive and accurate, otherwise it will cause serious accidents or material losses. Practitioners can not only regulate bodily functions, but also adjust the sub-health state due to habits such as long-term unreasonable diet and rest, thus improving the physical health to improve the quality of life through participation in sports, getting a more high job satisfaction.

88.4 Conclusions and Recommendations

88.4.1 Conclusions

The survey found that for the city employees involved in the martial arts fitness groups, job satisfaction was significantly higher than non-participating groups. And on satisfaction in the work itself, promotion satisfaction, pay satisfaction, supervisor satisfaction and colleague satisfaction, the groups involved in martial arts fitness were higher than non-participating groups.

Different levels of participation in martial arts fitness led to different degree of job satisfaction. Frequency and time of participation in martial arts fitness and job satisfaction have a very significant correlation. For the high participation frequency of the crowd, job satisfaction value is greater than the low participation of the frequency of the crowd; when each involved in the martial arts exercise for 60–90 min, job satisfaction levels reached the highest value; and each involved in the martial arts exercise moderately sweating strength, their job satisfaction value is the highest.

88.4.2 Recommendations

Urban practitioners strengthen the cultural construction of the sports organizations according to their professional characteristics and available resources. Actively guide the members to develop more scientific habits, mobilize the enthusiasm to participate in fitness activities, create a strong fitness culture within the organization, and improve the cohesion and solidarity within the organization.

Practitioners of different occupations choose suitable participation frequency time and exercise intensity in the martial arts fitness for according to their own conditions in order to improve the psychological and physical health, regulate emotional states, work with positive state of mind, and get higher job satisfaction and life satisfaction.

Societies and the media need to pay more attention to the health problems of urban practitioners, abandon the old ideas of "disease-free is health", actively promote the new concept of health and lifestyle, and guide the group to take the initiative to participate in sports activities.

References

1. Zheng H et al (2004) Social structure of the contemporary Chinese city: status and trends, vol 12. China Renmin University Press, pp 56–59
2. Lu X (2004) Contemporary Chinese social mobility, vol 7. Social Sciences Academic Press, pp 22–25
3. Gong M (2002) Organizational behavior, vol 1. Shanghai Finance University Press, pp 15–18
4. Li P et al (2004) Social stratification in China, vol 9. Social Sciences Academic Press, pp 85–89
5. Shi J et al (2003) Organizational behavior, vol 5. Petroleum Industry Press, pp 159–162
6. Xia L et al (2002) On the job satisfaction and job performance, vol 7. Southwest China Normal University (Humanities and Social Sciences), pp 32–34

Chapter 89
Research on Inherent Laws of Taijiquan Teaching Method

Hai Yu

Abstract Chinese martial arts comprehensive towards the world, that makes the internationalization of martial arts is obvious, Taijiquan spread is the important component of martial arts internationalization spread. At present, the international to learn Taijiquan from Chinese has three main ways: one is to come to China to learn Chinese Tai Chi Boxer; second, the Chinese Tai Chi Boxer go overseas teach them; third, living overseas Chinese Tai Chi Boxer teaching them. Along with the international spread of Taijiquan, the traditional teaching method meets the bottleneck, in the international Tai Chi Boxer communication field is mainly divided into two categories: one is the professional coach and teacher, the other is folk boxer. Professional coach or teachers in teaching, technology think of the important of specification but folk boxer tend to teaching, due to the language barrier; the two methods are difficult to fit the international Tai Chi enthusiasts of Taijiquan culture and technique of the roots of appeal. In this context, based on Taijiquan self rule on teaching reform attempt. This article puts forward the two steps and four stages of teaching methods through experimental study, The two steps refers to pattern teaching and teaching style molding; The four stages is the interpretation of single movement method, the training between two students, action teaching method and in the second step of the reform of personal characteristics. Studies show that two step four sections teaching method for foreign students in the short term to correctly grasp of Taiji boxing, laying a good foundation for further study of Taijiquan culture and various schools of techniques.

Keywords Interpretation of single movement · Training between two students · Action teaching method · Reform of personal characteristics

H. Yu (✉)
Shenyang Sports University, Shenyang 110102, Liaoning, China
e-mail: haiyu284a@126.com

W. Du (ed.), *Informatics and Management Science VI*, Lecture Notes in Electrical Engineering 209, DOI: 10.1007/978-1-4471-4805-0_89,

89.1 Introduction

The martial arts routines is to play, play sports, fall, get, hammer, thorn actions, in accordance with the advance and retreat, offensive and defensive movement. Due to cultural differences, it difficult for foreigners to understand some professional terms because of the language barrier, leading to many foreign players have learned actions, but is unable to realize the inner consciousness, rigid-flexible converting and points of the actual changes, so can only like and difficult alike in spirit [1]. This is the reason why many foreign players with good physical quality can complete the high difficulty movement, but in the martial arts consciousness was very difficult and in the Chinese contestant to counterbalance. Martial arts originates in China, belongs to the world, lets foreign athletes really master Chinese martial arts is the need of history, science teaching is to make foreign players correctly grasp the first condition of technology [2]. Through the experiment, we should reform the existing methods of teaching to make it scientific and consistent with the martial arts movement, so that foreign players can be in short-term inside master martial arts techniques and form as soon as possible personal style, making the teaching quality is greatly improved.

89.2 Research Object and Method

89.2.1 Research Object

Martial Arts Internationalization Background with Taijiquan exercises their teaching pattern teaching method.

89.2.2 Experimental Subjects

Liaoning university students court of the Japan, France, South Korea, England, Australia, Sweden fifteen countries for a total of 90 students, including boys 50, girls 40 people. The 90 students were not practiced martial arts, randomization, each group of boys 25 people, girls 20 people, a group, a group of as control group.

89.2.3 Research Methods

Documentation: consulting relevant literature: for the construction of the Taijiquan teaching method and design teaching experiment preparation. Interviews with experts: visit the domestic related experts of 25 people, according to the expert proposal, constructing teaching methods, and determine the test method [3]. Teaching experiment: from 2010 August to 2011 August, a total of 128 h, 45 min each hour, each class is 2 h. The study contents for 24 Taijiquan (48 h), yang-style Taijiquan competition routines (80 h) [4]. The experimental group adopts with two steps four stages teaching method: the first stage is unitary teaching, every action teaching by the interpretation of movement, and then double feed action simulation actual combat drill teaching method, then the action method of teaching, learning and mastering the whole movement when after the second stage, the shaping of teaching style according to the characteristics of students, teachers' personality and then into plastic stage, the formation of different exercise style [5]. The control group adopted normal teaching method. After the experiment, the national employ experienced referees 2 people, the national level 3 people, to the two groups for the evaluation of examination results are records and statistics. Logic analysis method, according to the research and testing of the objective results, using the relevant knowledge of logical analysis, in order to comply with the Taijiquan exercise their teaching method theoretical scientific exploration.

89.3 Results and Analysis

89.3.1 Experimental Results

The two group of twenty-four Style Taijiquan repertoire and Style Yang Taijiquan-Style Competition routine examination results analysis, two group scored significantly, use two steps four stages teaching method in the experimental group was superior to control group routine examination results. (Two groups of routine examination results statistical comparison see Table 89.1.)

After the examination, the referee group make a summary for two groups of athletes achievement analysis, the experimental group in the quantitative evaluation of action specifications without fastening point students accounted for 92 %, the remaining 8 % students who have focused on the balance action, that reflect the experimental group has hard enough action quality, the main technical specifications, step hand type accurately, footwork practices in line with the specifications, in which the part of the students in the balance of action points, which belongs to the psychological stress caused by accidental mistake. The control group in quantifiable action specifications without penalty points for 0 students, in addition to balance the action button, the shape of the hand, footwork technique type is the point deduction, obvious gap in the movement of specifications.

Table 89.1 Experimental group and the control group results comparison n = 90

Groups	x	s	t	p
Experimental group	86.6	2.35	12.82	<0.01
Control group	75.56	3.094		

Table 89.2 Experimental group and the control group on Taijiquan connotation cognitive comparison n = 90

Groups	Like Taijiquan (%)	Continue study (%)	Learning about cultural connotation (%)	Master defense techniques (%)
Experimental group	100	87	100	100
Control group	66	45	25	15

In the qualitative performance level evaluation, students in the experimental group on the whole can reflect the action fluency, spiritual focus, unity of inside and outside, in action specifications under the overall framework can form their own style. While the control group on the whole action is connected with a hamstring phenomenon, especially the consciousness and external action in one, and is not very good to relax, action, action points are not clear. Reflect the student control group also remains in the technical movement imitation stage and don't have clear concept on Taijiquan connotation.

After the experiment, the two groups issued about Taijiquan cognitive questionnaire, statistics and find that the two groups of cognition differences are relatively large, reflecting the teaching of Science in favor of exercisers to form correct cognition, thus cultivating continue to learn Taijiquan interest. (The experimental group and the control group on Taijiquan connotation cognitive comparison see Table 89.2.)

In the interest of Taijiquan on questionnaire survey, all the students generated strong interest on the Taijiquan, in the questionnaire, the experimental group generally reflected Taijiquan for the body relax obviously, practice when enjoy them, pleasurable, through the practice, the culture of China, further understanding, produce strong interest. The control group of students after studying, who are interested in Taijiquan accounted for 66 %, other students feel dull, lost interest in further investigation, found in the control group, the Tai Chi interest most stay at the surface, there is still a mystery to Tai Chi, Tai Chi no clear cognition on.

In the future about whether to continue to learn Taijichuan questionnaires, in which the 87 % of the students said to continue to learn, especially to learn other genres of the further traditional Taiji and equipment, especially for the primal chaos sword and athletic push expressed strong interest, the remaining 13 % of the students said although no continue to learn, but made clear to keep on regular Taijiquan. The control group had 45 % students said they would further learning Taijiquan, the rest of the students are not explicitly expressed continued to practice and learn Taijiquan.

On Taijiquan culture connotation understanding the extent of the investigation, the experimental group on the cultural connotation of Taijiquan has more explicit cognition, they can understand Taiji Culture in Taijiquan techniques in guiding role, at the same time can use the related cultural knowledge on the philosophical meaning of Taiji boxing.

In the attack and defense related Taijiquan skills survey, students in the experimental group 24 Chi and 42 Style Taijiquan movement of each of the attack and defense have clear knowledge, while the control group students on the action moves defense meaning comprehension is often subjective judgment, thus reflecting the experimental group teaching method pay attention to the offensive and defensive action, and Taijiquan exercise their own laws coincide, grasp the essence of the martial arts.

89.3.2 Two Steps Four Stages of Teaching Theory

With the development of history, the needs of the society and the influence of cultural background, forming a routine, the historical progress that martial arts especially modern competitive martial arts routine to high, difficult, the United States, a new direction, to adapt to the international competition need, now largely out of martial arts category, but its basic technical material comes from the traditional martial arts kick, hit, throw, take, hammer, needling techniques, routine arrangement still comply with offensive and defensive retreat, Xu, rigid-flexible virtual movement law. The show is aesthetic of the attack and defence technology of great exaggerated art form, therefore, the principle is still attack martial arts movement of the core. Second, martial arts sports as official event, must be unified rules, routine practice must be standardized, and because the person's individual differences, set in the technical specifications of the drill, based on individual must exist style, be based on these three presents two step four stages of teaching methods.

From the point of view of teaching theory, different project teaching method in the teaching of common laws first with based, at the same time, to meet project its own movement characteristics and regularity. At present, in the international communications, Taijiquan teaching mainly has two kinds, one kind is the professional teachers and professional coach, they focus on the action from the angle of the specifications, requirements of movement quality, in the international promotion, because of learning groups are often not professional athletes, thus make students feel boring, often stop learning, because folk boxer their style is different, the heavy teaching, often make students feel there is no uniform standard, learn not know what course to take, these factors are the fundamental causes of the formation of teaching bottleneck. In fact, as long as we grasp Taijiquan exercise itself, motion law, combining the teaching of common law, our system can adapt to the international promotion of Taijiquan, break through the bottleneck, through the research, we found that Tai Chi Chuan techniques derived from martial arts fight, at the same time into Chinese culture connotation, as long as in the way of grasping the essentials of the attack cultural essence, can achieve better teaching effect, the two step four step teaching method precisely grasp the two, so the two step four stage teaching method in the international promotion of Taijiquan has positive significance.

89.3.3 Two Steps Four Stage Teaching Method in International Taijiquan Teaching in Promotion of Advantage

Two step four stage teaching method reflects the martial arts development characteristics, so that foreign students can get comprehensive system. Evolutionary learning and mastery of martial arts techniques two step four stage teaching method with the martial arts teach.

Two step four stage teaching method firstly makes the students understand the sequence of events for each move, by double exposure can correct experience martial arts boxing fighting method. In Taijiquan teaching, as the scholar is not easy to grasp the bird's tail through double feeding techniques, can recruit good students experience skill points, and push the essentials and the various parts of the body and position requirements. Moreover, handle type teaching, this one action, beginners often trick with improper, gravity transform is ineffective, when scholars through the one or two learning to understand the dynamic contains fighting method and essentials, can easily master the technical movement. And the actual situation of change is through the technique of attack and defense conversion to achieve, through double exercises, foreign students in that action meaning and experience points after the conversion, in the action criterion is very easy to grasp the actual situation of change. If teach white crane spreads its wings, which has the waist handle, then the virtual step up, beginners general practice, although with the password, but is not easy to transform strength. If you understand the waist handle is collected pick each other under the arm, with the conversion of thoracolumbar, broke a synthesis, empty step up is the other if the retracement, homeopathic set step on strike, beginners will quickly grasp the actual situation transformation, even if not immediately practice, but has formed the right concept, through subsequent practice further experience method, will be very good action. Martial arts routine about martial arts awareness training, how to correctly cultivate consciousness of Martial arts, has been attached great importance to the teaching of martial arts. Two step four stage teaching method follows the martial arts movement, through the action meaning teaching, double feed action simulation actual combat drill teaching can cultivate students consciousness of martial arts. Tai Chi requires that, to, such as Mask hand brachial beat should accomplish "livestock such as strong bow out strength, such as arrows", the students do not understand the meaning of attack, it is difficult to achieve the unity of inside and outside, through the action meaning explanation and double fed action simulation actual Combat Defense exercises, students will be experience this martial arts consciousness. Teach inserting step hook hand side kick, after the attack and defense simulation exercises, will attack and defense integration, accelerate the speed of action, fully embodies the martial defensive consciousness, we can assume in sabre play teaching, beginners learning to wrap the head wrap brain techniques, easy to make the blade away from the body too far, and through double offensive and defensive exercises, not only will quickly grasp the dynamic technology and methods, but also train martial defense conversion consciousness, make full step-by-step method of coordination, to achieve a multiplier effect.

Due to the failure to establish a correct concept of beginner, action is not easy to regulate. Two step four stage teaching method firstly the meaning of the action on the students to set up correct concept, further by having a martial art meaning of footwork and skill with the exercises, on various parts of the body, the strength of experience, especially the double simulated combat exercise allows the brain to establish the correct action mode, make the action to reach the standard. In the use of two step four stage teaching method, teachers pay attention to the appropriate language prompt and correct action demonstration, student action of standardization can be in short-term inside do.

Two steps four stages teaching method, in the initial teaching consciousness of martial arts has been influence character by environment, personalized style of the early formation has laid a good foundation, two step four stage teaching method through the progressive teaching, pay attention to the personality training, through the first three paragraph teaching, action technology basic specification, consciousness of martial arts is the initial training, on the basis of, combining the characteristics of students, stimulate students' initiative, give full play to students' creativity, as according to the stature characteristic, shape or elegant and light or heavy, speedy, such a targeted form as soon as possible personal style, improve the effect of teaching.

89.4 Summary

Two step four stags teaching method is in the modern teaching method on the basis of absorbing traditional teaching methods, and the formation of. It is based on the analysis on the offensive and defensive action, double feed action exercises in the form of the language teaching both sides out of bondage, and understand each action of the attack and defense, as soon as possible to establish the concept of action, accelerate the teaching process.

From the action meaning of teaching, double feed action simulation practice teaching, action teaching, personality and plastic teaching consists of two step four step teaching method in line with the law of education and teaching, so that foreign students can be in short-term inside master martial arts techniques, comprehend the meaning and the formation of personal style attack as soon as possible, can effectively improve the quality of teaching.

References

1. Xilian S (2008) Taijiquan international communication and China of the soft power of the ascending. J Wuhan Inst Phys Edu 4(6):57–60
2. Xuerong Y (2009) From the perspective of mass media of modern martial arts propagation. Sci Technol Innovat Herald 2(13):89–94
3. Xiandan Y (2007) Media and Chinese martial arts research in 2002–2006. China sports newspaper as a case. Soochow Univ 2(6):183–186
4. Zhilong L (2009) Taijiquan origin and development. Gansu Sci Technol 5:74–77
5. Liu S (2003) The spreading of martial arts culture obstacle factors analysis. J Tianjin Univ Sport 18(2):41–43

Chapter 90
Study on Basketball Teaching and Skills Assessment

Zhengyu Li

Abstract The completion of students' basketball technical movement is a multi-factor complex problem. To enable teachers' score work to change from being empirical to scientific, from qualitative to quantitative, from unified to standardize and from irregular to standardized, method of fuzzy mathematics should be employed to obtain accurate effect of assessment, which is a scientific method of actual needs, in line with basketball skills assessment. Therefore, this essay assesses students' basketball skills by applying the fuzzy comprehensive evaluation and achieves good results.

Keywords Fuzzy mathematics · Basketball skills · Assessment · Computers

90.1 Introduction

Evaluation is a serious and careful work. Lack of scientific factors of combining the evaluation method, it cannot objectively reflect the quality of teaching and learning [1]. In addition, it also can restrain students' learning enthusiasm and contributions of unhealthy trend. However, there is a prominent problems and the method of evaluating the score before, also is a technology assessment of a student's performance is dependent on direct experience of the teacher and the subjective impression of their due to lack of objective quantitative standard teachers to score [2]. And inevitably have certain limitation, unilateralism and optional sex identification process.

For a long time, because of the influence of traditional education thought, college basketball teaching content according to the model of initial training in

Z. Li (✉)
Henan Business College, Zhengzhou, China
e-mail: zhengyu12li@126.com

W. Du (ed.), *Informatics and Management Science VI*, Lecture Notes in Electrical Engineering 209, DOI: 10.1007/978-1-4471-4805-0_90,

nature the young athletes: basic position, basic footwork, dribbling, catch the ball, shooting, just the basic tactical cooperation and the quick break, preview fast break, defense and the people attack one another defense and other tactics [3]. Even in years of reform, the situation has not yet is the basic elements change; college physical education teaching has to continue to use the traditional teacher and students' one-way method. In this teaching methods, teachers are not only designer teaching is also executives and managers controller teaching process [4]. What to teach and how to teaching and students' learning and how to learn is to decide the teacher. This in the largely limits the students' enthusiasm and initiative, not easy to attract students' interest basketball, which affect the teaching effect. In 1995, the state council promulgated the national fitness program "points out, the foundation of the school sports fitness. Especially in August 2002, China formal education promulgated and implemented university physical education curriculum teaching guidance Outline, followed by a sports education reform tide the whole country. Value-in teaching is mentioned methods, the outline "explicitly pointed out," teaching method. It is character and diversity, and enhances the each other the activities. In the student and teacher, student and students, improve the students' enthusiasm, the maximum contact the cultivation of the students' creativity [5, 6]. It is not difficult to find that, Outline requires universities sports teaching shows on students' development of education for the ideological content, learning as the center, capacity development for core. Therefore, this paper introduces the complete teaching university special method, in order to basketball elective course really mixed the thought of "people-oriented", "health first" and the "lifetime sports" of the classroom teaching process [7].

In the traditional basketball teaching, in general, is a student should first learn basic skills, such as ready to position, footwork movement, transportation ball, shooting, side screen, and back screen etc., based on repeated practice. With this simple action, then learn the simple coordinate, such as a transfer a cut, suddenly the gap, screen, knowledge and learning to judge and tactics close to the terminal cooperation, the last few used in practice.

Human values of the humanism psychology, respect reason, believe that man is positive and rational, pursue valuable target, and have a positive attitude towards life growth. The core of humanistic thought is the person's nature, that man has the excellent natural potential role education to the potential of the inner is discussed [8]. Humanistic education workers, performance for Rogers, there is puts forward "to the learners as the center" the education view. They believe that the purpose of the foundation education is to cultivate student's individual character and exploring their potential. Education should leave the choice of self-examination ability, to students, take students' autonomous learning, self development is the goal.

Teaching should respect the four aspects: participation, autonomy, emotion, and creative, develop the students' self-concept, to improve the students' with the principle of combining the autonomy and teachers' leading, pay high pay attention to students' individual character development and personal values [9].

90.2 The Mathematic Model of Basketball Skills' Fuzzy Evaluation

90.2.1 Determining the Set of Evaluating Objects, Factors and Comments

Set the set of objects $F = \{f_1, f_2, f_3, \ldots\}$, students' basketball skills are the evaluating objects.

Set the set of factors $U = \{u_1, u_2, u_3, \ldots\}$, U is the assessment of basketball skills, u_1 is single pitching motion, u_2 is comprehensive technical movement, u_3 is the application of competing, u_4 is the application of strategies.

Set the set of comments $V = \{v_1, v_2, v_3, \ldots\}$, V stands for the comments of teachers' on the completion of students' technical movement, v_1 is D, v_2 is C, v_3 is B, v_4 is A, v_5 is A+ . According to this, the table of the relationship between the five classes and their corresponding figures in percentage see Table 90.1.

90.2.2 Establishing Weight Distributions of Evaluating Factors

After three rounds of questionnaire survey by the expert method of Delphi, weight coefficients have been determined. $A = (0.2,0.3,0.2,0.3)$, $A_1 = (0.1,0.3,0.2,0.3,0.1)$, $A_2 = (0.2,0.3,0.1,0.1,0.3)$, $A_3 = (0.2,0.3,0.1,0.2,0.2)$, $A_4 = (0.2,0.1,0.3,0.1,0.3)$.

90.3 The Result of Assessing and the Analysis

Set that the result of teachers' assessment and evaluation of a student's basketball skills is u_1 shooting movement: 3 points for lifting a ball, 4 points for balancing, 4 points for arm moving, 2 points for gimmick, 5 points for effect; u_2 comprehensive technical movement: 3 points for coordination and cohesion, 4 points for right movement, 4 points for synthetic effect, 5 points for the speed of joint, 4 points for appropriate application; u_3 the application of confrontation: 2 points for the speed,

Table 90.1 The table of the relationship between evaluation classes and their corresponding figures in percentage

Membership	40	55	70	85	100
D (1 points)	0.3	0.5	0.2		
C (2 points)		0.3	0.5	0.2	
B (3 points)			0.3	0.5	0.2
A (4 points)				0.8	0.2
A+ (5 points)				0.2	0.8

4 points for the threatening of attacks, 4 points for the effect of defense, 3 points for the power of confrontation, 5 points for the shift of attaching and defending; u_4 the application of strategies: 3 points for first pass, 3 points for receiving, 4 points for pushing, 4 points for the stage of ending, 5 points for shift.

Calculate the synthetic calculation between weight vector Ai and fuzzy relationship matrix Ri, and we can get the first—class synthetic assessment result:

Shooting movement:

$$B_{1=}A_{1o}R_1 = (0.1, 0.3, 0.2, 0.3, 0.1) = \begin{pmatrix} 0 & 0 & 0.3 & 0.5 & 0.2 \\ 0 & 0 & 0 & 0.8 & 0.2 \\ 0 & 0 & 0 & 0.8 & 0.2 \\ 0 & 0.3 & 0.5 & 0.2 & 0 \\ 0 & 0 & 0 & 0.2 & 0.8 \end{pmatrix}$$

$$\begin{aligned} &= (0.1 \wedge 0 \vee 0.3 \wedge 0 \vee 0.2 \wedge 0 \vee 0.3 \wedge 0 \vee 0.1 \wedge 0, 0.1 \wedge 0 \vee 0.3 \wedge \\ &\quad 0 \vee 0.2 \wedge 0 \vee 0.3 \wedge 0.3 \vee 0.1 \wedge 0, 0.1 \wedge 0.3 \vee 0.3 \wedge 0 \vee 0.3 \wedge 0 \vee \\ &\quad 0.3 \wedge 0.5 \vee 0.1 \wedge 0, \\ &\quad 0.1 \wedge 0.5 \vee 0.3 \wedge 0.8 \vee 0.2 \wedge 0.8 \vee 0.3 \wedge 0.2 \vee 0.1 \wedge 0.2, \\ &\quad 0.1 \wedge 0.2 \vee 0.3 \wedge 0.2 \vee 0.2 \wedge 0.2 \vee 0.3 \wedge 0 \vee 0.1 \wedge 0.8) \\ &= (0 \vee 0 \vee 0 \vee 0 \vee 0, 0 \vee 0 \vee 0 \vee 0.3 \vee 0, 0.1 \vee 0 \vee 0 \vee 0.3 \vee 0, 0.1 \\ &\quad \vee 0.3 \vee 0.2 \vee 0.2 \vee 0.1, 0.1 \vee 0.2 \vee 0.2 \vee 0 \vee 0.1) \\ &= (0, 0.3, 0.3, 0.3, 0.2) \end{aligned}$$

For the same reason we can get:

Comprehensive technical movement: B_2 = (0, 0, 0.2, 0.3, 0.2)
The application of confrontation: B_3 = (0, 0.2, 0.2, 0.2, 0.2)
The application of strategies: B_4 = (0, 0, 0.2, 0.3, 0.3)

Normalize B_1 = (0, 0.3, 0.3, 0.3, and 0.2) and we get:
B_1 = (0, 0.27, 0.27, 0.27, 0.18)
For the same reason we can get:
B_2 = (0, 0, 0.29, 0.43, 0.29)
B_3 = (0, 0.25, 0.25, 0.25, 0.25)
B_4 = (0, 0, 0.25, 0.38, 0.38)

and Bi = (40,55,70,85,100), shooting movement, comprehensive technical movement, the application of confrontation and the figures in percentage of strategy-applying: shooting movement:

$$S_1 = (0, 0.27, 0.27, 0.27, 0.18) \begin{pmatrix} 40 \\ 55 \\ 70 \\ 85 \\ 100 \end{pmatrix}$$

$$= 0 \times 40 + 0.27 \times 55 + 0.27 \times 70 + 0.27 \times 85 + 0.18 \times 100 = 74.7$$

Comprehensive technical movement: S2 = 85.9 (points)
The application of confrontation: S3 = 77.5 (points)
The application of strategies: S4 = 87.8 (points)

Calculate the synthetic calculation between weight vector A_j and fuzzy relationship matrix Rj, and we can get the second-class synthetic assessment result:

$$G = A \circ B = (0.2, 0.3, 0.2, 0.3) = \begin{pmatrix} 0 & 0.3 & 0.3 & 0.3 & .02 \\ 0 & 0 & 0.2 & 0.3 & 0.2 \\ 0 & 0.2 & 0.2 & 0.2 & 0.2 \\ 0 & 0 & 0.2 & 0.3 & 0.3 \end{pmatrix}$$

$$= (0.2 \wedge 0 \vee 0.3 \wedge 0 \vee 0.2 \wedge 0 \vee 0.3 \wedge 0,$$
$$0.2 \wedge 0.3 \vee 0.3 \wedge 0 \vee 0.2 \wedge 0.2 \vee 0.3 \wedge 0,$$
$$0.2 \wedge 0.3 \vee 0.3 \wedge 0.2 \vee 0.2 \wedge 0.2 \vee 0.3 \wedge 0.2,$$
$$0.2 \wedge 0.3 \vee 0.3 \wedge 0.3 \vee 0.2 \wedge 0.2 \vee 0.3 \wedge 0.3,$$
$$0.2 \wedge 0.2 \vee 0.3 \wedge 0.2 \vee 0.2 \wedge 0.2 \vee 0.3 \wedge 0.3)$$
$$= (0 \vee 0 \vee 0 \vee 0, 0.2 \vee 0 \vee 0.2 \vee 0, 0.2 \vee 0.2 \vee 0.2 \vee 0.2,$$
$$0.2 \vee 0.3 \vee 0.2 \vee 0.3, 0.2 \vee 0.2 \vee 0.2 \vee 0.3)$$
$$= (0, 0.2, 0.2, 0.3, 0.3)$$

The figures in percentage of basketball skills' comprehensive assessment:

The assessment and evaluation result of this student's basketball skills is 80.5 points.

We have done an evaluation of 36 students' basketball skills assessment, and the result complies with normal distribution, which suggests using fuzzy mathematics to assess basketball skills is rather scientific.

Put the names, numbers and other main information of every student and the index of basketball skills assessing into the computer, through the calculation of computer program, and the result of each student's basketball skills assessing will be got immediately. You can store information and inquire at any time. It makes the standardized and scientific management much easier.

90.4 Conclusions

Use of scientific attitude and modern means to achieve an accurate results, it has a high reliability and scientific.

Basketball technical indexes of evaluation are based on experience of doctors and experts from years of sports teaching summarized. It has proved that the index system and evaluation model, this study set up to meet the requirements of evaluation and their ultimate goal is quite high practical value.

The fuzzy comprehensive assessment model of evaluation of basketball technology not only can assess single skills but also the comprehensive quality and technology diagnosis, to provide the scientific basis of teaching and training.

The application of computer technology provides the security, make scientific evaluation of the basketball skills.

References

1. Lu J (1989) Pragmatic vague mathematics, vol 10. Science Technology Document Press, Beijing pp 101–113
2. Peizhuang W (1983) Fuzzy set theory and its application, vol 5. Shanghai Science and Technology Press, China pp 95–99
3. Kaiqi Z, Xu Y (1989) Fuzzy system and expert system, vol 12. Xi' an Jiaotong University Press, China pp 124–131
4. Thomas L, Gool J, Braphy E (2010) Seeing Through the Classroom-Teaching Mode and Method Series, vol 6(04). Light Industry Press, China pp 321–325
5. Zhenming M (2009) Searching for successful sport teaching , vol 09. Beijing Sport Coll Press, China pp 21–25
6. Zhiping Y (2001) On cooperation in the new physical education mode. Sports Sci Res (3):31–33
7. Ma L (2005) Cooperative learning, vol 9(10). Higher Education Press, Beijing pp 142–146
8. Wang T (2002) The basic idea of cooperative learning. Educational Research 2:68–71
9. Johnson DW, Johnson RW, Holubec EJ (1994) Circles of learning: cooperation in the class room, Vol 9. Interaction Book Company, Minneapolis pp 25–28

Chapter 91
Analysis of Recessive Marketing Research for Domestic and International Sporting Events

Zebo Qiao

Abstract Through the literature material, this paper expounds on the sport at home and abroad market research status of recessive, comparative analysis found that domestic market research about sports recessive and many blank fill the need, there are many drawbacks needs to improve. According to the situation of our country, the author thinks that with a longer time sports market will persist recessive behavior, market behavior of the recessive correctly and reasonable legal attitude should be reasonable and legitimate limit is not comprehensive rigid prohibited, restricted and prohibited in the legislation in the range should be clearly specified, and the time and space also stipulates range. In addition in view of the existing sports an event in a relatively short time, for marketing brings the harmful recessive needs to refer to the legal aid and perfect sports legal assistance system as soon as possible.

Keywords Sports · Ambush marketing · Sports events · Comparative study

91.1 Preface

The modern large-scale sports events held without corporate sponsorship. With the Olympics as an example, the Olympic Games as the most influential sports events, sponsors of the sponsor of the amount of the Atlanta Olympics in 1996 all sources of funds to account for 30.9 %, the 2000 Sydney Olympic Games rose to 32.7 %,

Z. Qiao (✉)
Dept of PE, Guangdong business and studies university, Guangzhou 510320, China
e-mail: zebo90qiao@126.com

W. Du (ed.), *Informatics and Management Science VI*, Lecture Notes in Electrical Engineering 209, DOI: 10.1007/978-1-4471-4805-0_91,

and the 2004 Olympic Games in Athens rose to 34.8 %, and by the 2008 Olympic Games in Beijing the rate of 40 % [1]. It is visible that sponsors on the position of the sports events have become increasingly important. In return, the sporting events for their sponsors the product or service and large sports games in contact, and business activities of monopoly exclusive rights. So as to enhance the brand recognition and reputation, improve the economic benefit of enterprise. Face the sports events with the great influence; a lot of the sponsorship enterprise also takes all kinds of means to share the benefits of the sports events. For example through its commercial marketing misleading audience, especially the consumer, and consumers that the audience with sports events that don't exist in fact there is a certain sponsored or supported relationship. This kind of marketing behavior is sports recessive market behavior.

Although at present the international and domestic market behavior of the recessive without a strict scientific definition, but sports events of the implicit market behavior will no doubt for large sporting events of the sponsors damage and deprived of the repayment, therefore sports events at all levels to this market organizers behavior struggle and boycott, such as the international Olympic committee (ioc) as well as the nocs are required for the host sports events all kinds of laws and regulations to combat this recessive market behavior. But due to market actions hidden the means and methods of many cases avoid the current legal system framework, lead to much recessive behavior in the present market the laws and regulations of the system is hard to get checked.

International sports events to the hidden market the legal nature of the action of debate also created a new trend, such as some people think that in international sports organization under pressure for all countries in the marketing activities of recessive legal system and the frame, too much focus on international sports organization, as well as the sponsors of the corresponding business interests, and not even deprived the legitimate interests of the public, So how to avoid recessive market behavior of stakeholders to power at the same time create reasonable infringement, fair competition becomes the environment of a hot and difficult [2].

91.2 Foreign Research Status

And through to the consumers and the Olympic Games the audience for sponsors such as cognitive and agree with Angle and implicit market behavior of the enterprise, this paper analyzes the successful marketing of recessive tries to get the key factor [3]. Later, in the image of the sponsors from the consumer cognitive and memory etc. Angle to evaluate whether the recessive market behavior become successful research trend. [2], separately from the point of view of consumers to research the recessive market behavior, further stress from consumers to the understanding of the sponsors such as Angle, cognitive and attitude to fight recessive market behavior. Too much emphasis on brand awareness and consumers for the recognition, and is not to associate to the actual sponsors [4]. Such as recessive market behavior threat to the sporting events honest spirit; Sponsors status of business value belittles ultimately weaken the big sporting events of the financial viability. Based on this, advances the strategy of market action against recessive, such as must be in position and with TV or sponsors the media sponsor position between seek a better balance, the last notes that official sponsors must also take corresponding measures to prevent the recessive market behavior, also can appeal of the existing legal system.

However, put forward the measures of academic circles for fighting recessive market action practice but have not mean much. Many of the measures are put forward from the law, as Hoek and Gendall [3] mentioned, this solution, or in consciousness edge, far from a reality [5]. Results from the marketing point of view the behavior of the scholars found hidden market, the path is much less than the legal branch some more important. Through the New Zealand Olympic Games of the implicit market behavior of case study, the researchers discovered that the recessive market action case are very difficult to prove its illegal legal nature, and many of the recessive market behavior is simply avoid the current legal system. So put forward, the hidden market behavior is though business motivation, if the strategy is very successful, completely is not marketing field can solve.

Lans Retsky [4] and Coulson [5] discuss the behavior of the legal regulations recessive market framework. Lans Retsky think despite the behavior of the enterprise may be hidden market is very careful to avoid illegal, but related right holder can still successfully in the United States and LanHanM under the bill law business prove its intention improper or unfair competition. Coulson also points out that the behavior of the case against the recessive market of the improper use of the importance of the defence. This reduces the European football league like these international organizations in the face when this kind of case burden of proof, as long as proof of the improper use of behavior, and don't need to prove its behavior led to tangible can measure the loss [6].

In addition, with the launch of the 2012 Olympic Games in LONDON, and many in the research of the LONDON Olympic Games of the implicit market behavior of the relevant legal is discussed in the paper, especially for sports tournament organizing committee of the right of a specific rights associated content and the corresponding relief measures are discussed, Miller [6] years in LONDON 2012 protection that the brand as well as to the value of the sponsors at the same time, also be given to the public reports and reasonable use of the Olympic logo and reasonable brand rights. According to the London games in 2006 and the London Olympics bill, Blakely [7] think that the legal system focused too much on organizing committee and the ocog sponsors and harm the interests of the public rights.

91.3 The Study Status Quo

The research in relative to foreign research to see is still in the initial stage. Main is with Beijing's bid for the 2008 Olympic Games, the success of recessive market behavior was introduced into the researcher's perspective. The lack of legal and illegal recessive market behavior the judgment standard and lack of corresponding legal remedies and liability limit [8]. But this paper and did not put forward reasonable legal recessive market behavior of the boundaries between what is law, so that to the relevant judicial practice to bring about the fuzzy [9]. By 2008, the domestic about the Olympic Games the hidden market behavior of the more common, basically adopt such research way: the definition of recessive market behavior, the hidden market behavior performance and its harm, countries to do counterattack recessive market action legal measures to fight against the Olympic Games of 2008, China recessive market behavior to provide the corresponding Suggestions. Such as about whether need to set up a unified recessive market action against the Olympics legislation, some people think that must be of the recessive market behavior type after grasp, the existing legal systems can completely regulation. And some people think that all countries in the world combined with legislation practice of China's economic situation and Suggestions related legislation to government regulations "Beijing Olympic Games recessive market behavior regulation measures (suggestion)" way to regulation of Olympic Games recessive market behavior [10].

91.4 The Comparison Analysis of Domestic and Foreign Research

Compared to the current research status of home and abroad can be seen, the domestic about sports recessive market researching the behavior of the space and need to fill, there are many drawbacks needs to improve. Such as tacit market on the legal nature of the action of the definition of the standards, namely reasonable and illegal market behavior of the legal boundaries; Second, the international fight against recessive about market behavior made a number of laws and regulations, and these laws and regulations to present the organisers, the sponsors of the event of the legitimate interests of the excessive protection, and to the general public the interests of the propaganda reasonable report has been ignored. So how to through the legal adjustment and create events during the interests of all parties held the balance of the discussion will become the hot spot. Finally, the current domestic market for the harm of the behavior of the recessive mainly from qualitative Angle discusses, from the point of view of how consumers exploration and evaluation recessive market behavior, to quantitative analysis of the characteristics of recessive market success behavior, as well as to the sponsors of the status, the extent of the harm of the sports events, and to formulate relevant laws and reasonable scale and provides the basis for the measures also become nowadays an important research Angle.

91.5 Conclusion and Suggestion

Although recessive market behavior of sports events and sponsors to cause damage, but not all of the invisible market action is illegal behavior, and the damage is not as serious as sponsors of the announcement, if too much on race organizers and sponsors, will damage most enterprise use sports events for the rights of fair competition, and may even deprived the legitimate interests of part of the public. And at present a lot of state or government has responded to the Olympic Games like, such as the World Cup international sports league pressure, take a high limit recessive market behavior legislation trend, and this legislation in our country is not suitable for tendency, we should make use of the empirical analysis, from the Angle of the audience of the sports events, the investigation on how the recessive market behavior caused extent for contact sponsors of the event this pseudo association, and the behavior of the market for regulation recessive legal measures to provide the basis. The author thinks that in quite a long period of time to our country sports recessive market behavior will exist for a long time, market behavior of the recessive correctly and reasonable legal attitude should be reasonable limit not banned. And the restricted and prohibited in the legislation in the scope of law need to determine. In addition, sports games because of the existence of a short time, and in the market from the behavior of the hidden dangers of the stop must use legal relief measures, this request we should perfect the current our country sports law relief system.

References

1. Paul H, Schmitz S, O'hare R (2009) Ambush marketing and London 2012: a golden opportunity for advertising or not. Ent L R 20(3):74–76
2. Lyberger MR, Mccarthy L (2001) An assessment of consumer knowledge of, interest in, and perceptions of ambush marketing strategies. Sport Market Q 10(2):30–137
3. Hoek J, Gendall P (2002) Ambush marketing: more than just a commercial irritant. Ent Law 1(2):72–91
4. Lans Retsky M (1996) One person's ambush is another's free speech. Market Law 30(14):12–18
5. Coulson N (2004) Ambush marketing. Brand strategy 4(179):32–37
6. Miller T (2008) London 2012: meeting the challenge of brand protection. I S L R 4:44–47
7. Blakely A-M (2006) London Olympic games and Paralympic games act 2006: less presumptive than the London Olympics bill. Ent L R 17(6):183–185

8. Hu F, Zhenyu Z (2009) Concerning the Olympic games of the legal regulations recessive market behavior. Wuhan Inst of Phys Edu 40(4):215–219
9. Xu N (2008) The Olympic recessive market behavior of the law and the rules and regulations qualitative. Grad Student Law 3:68–72
10. Xie X (2007) Regulation Olympic games market action exploring implicit legislation. Sports Sci 10:85–94

Chapter 92
Study on Short Weapon in Traditional Martial Arts

Xiao-dao Chen

Abstract Short weapons were the general name of short handheld combating and fighting weapons in ancient China, and were called by people in comparison with a variety of longer combating and fighting weapons. In ancient China, there were no strict size standards for the classification of short and long weapons, and combating and fighting cold weapons, which were not as long as the eight of a man and also could be held with a single hand, were listed as short weapons. Yue, Cheng, broadsword, dagger, sword and golden hook were five typical representatives of ancient Chinese short weapons. In this paper, starting from the culture of the traditional Chinese martial arts, the application of short weapons in the traditional martial arts is studied in depth.

Keywords Short weapons · Traditional martial arts · Fencing · Performance of martial arts · Chinese martial arts

92.1 Introduction

Short weapon is also called as Chinese fencing, which can not only help take exercises and improve physical health, but also can be utilized for self-defense and anticipating the enemy. It has changed into one of the most important items oriented attacks and defenses in the development of martial arts equipments. Short-weapon game of China was originated from the ancient short fencing. In the

X. Chen (✉)
Department of Physical Education, Guangdong University of Finance,
Guangzhou 510520, China
e-mail: xiaod33chen@126.com

W. Du (ed.), *Informatics and Management Science VI*, Lecture Notes in Electrical Engineering 209, DOI: 10.1007/978-1-4471-4805-0_92,

modern times, "short weapon" was added in examination after the National Central Martial Arts Museum as well as a "Chinese martial arts" exam system was established by Zhijiang. Short-weapon game was one of the most important parts of the traditional Chinese martial arts, and attained a very rapid development in ancient China, modern times as well as the period of the republic of China [1]. In it, profound martial arts attack and defense thoughts and historical culture are contained.

92.2 Development History of Short Weapon in the Traditional Chinese Martial Arts

Short-weapon game of China owns a very long history. The form of "sword fighting" had been in existence as early as in the period of Shang Dynasty. It became highly popular among people in the warring state period, and simultaneously the theories of fencing method emerged in ancient China.

According to Chinese written records, from Zhou and Qin Dynasties to Yuan and Ming Dynasties, games of fencing forms included a trial of strength, horning, hand wrestling, sumo, fencing, stick-based wrestling, and thorn, etc. All these games owned an extensive mass base at that time, and simultaneously playing techniques attained a very fast development [2]. However, since the theory of silence learning was advocated by neo-Confucianism in Song Dynasty, fencing game was restricted, and also the tendency of attaching importance to martial arts but ignoring literatures became increasingly more serious. For this reason, the fencing techniques failed to be handed down in China for a time. To the period of the Republic of China, Chinese martial arts went into the eyes of people more frequently in the form of sport games because of the development of society and the universal application of firearms. The National Central Martial Arts Research Institute was founded in Nanjing in 1927, and named as the "Central Martial Arts Museum" in the next year.

Domestic strife and foreign aggressions were highly intensive during the period of the Anti-Japanese War in China. For this reason, the development of short-weapon game used to be stagnated for a time. After new China was founded, under the guidance ("developing physical exercises and enhancing the physical health of people" and "letting a hundred flowers bloom and new things emerge from the old") of the Communist Party and chairman MAO, the martial arts game, as one of the national cultural heritages, can attain a rapid development as well as a great achievement, and also changes into one of the most important ways for the national people to take physical exercises [3].

92.3 Problems in Current Chinese Short-Weapon Game

92.3.1 Dull Techniques and Actions, and Insufficient Characteristics

In the current traditional martial arts games, techniques and actions that are used by players mainly comprise of chopping and cutting actions, and the utilization rates of collapsing and dotting actions are extremely low and so are hacking and pricking actions [4]. All these give rise to the weakness, poor appreciation and less attractions in attack and defense techniques as well as national characteristics of current Chinese short-weapon game.

92.3.2 Low Techniques and Weak Normalization

Short-weapon game of martial arts, as a traditional national sport time with a high value of attack and defense, has been very outstanding in the characteristics of techniques. However, a large number of players make use of chopping and cutting actions most frequently at present, and therefore their large strengths are mainly concentrated on the "bodies" of equipments but not on the first part, and also their power application part is arm but not artifice driving the whole equipment to attack the other side.

92.3.3 Seriously Imbalanced Defending System and Low Professional Level

In the application of attach and defense techniques, players mainly take advantage of aggressive attack techniques and counter-offensive techniques, while the utilization rates of passive defense techniques and counter-attack techniques are very low. Such a similar technique characteristic is embodied in repeated attacking (players attacked each other repeatedly) in game, and thus seriously affects the appreciation value of game.

92.4 Factors Affecting the Development of Chinese Short-Weapon Game

92.4.1 External Factors

First, there are limitations from different sections. Second, there are historical factors. Whether the state attaches importance to it, the national people cherish it and economic conditions are powerful will has a close tie with the development of a sport item. Third, there is ignorance in society. Up to now, there have been only a small number of people knowing well Chinese short-weapon game. Therefore, people from all walks of life are unfamiliar with short-weapon game of martial arts, and also more support and attention have not been attached in society to short-weapon game.

92.4.2 Internal Factors

First, techniques are poor, namely there are no complete techniques system and well-improved game rules. Second, cultural intensions are lacked. Short-weapon game is one of the most important parts of Chinese martial arts, and carries on and integrates sword playing techniques of ancient China. However, no intensions and philosophy of Chinese martial arts can be found in the short-weapon game.

92.5 Development Ideas and Strategies of Short-Weapon Game of Martial Arts

92.5.1 Changing Playing Rules and Bringing the Guiding Role of Rules in Techniques into Full Play

The innovation of rules is not only a magic key to guiding the development of playing techniques, but also a driving force for the development of playing techniques.

Along with the continuous deepening and improvement of the techniques and tactics of short-weapon game, Chinese short-weapon game has developed further, and simultaneously its competitions will be increased with a gradual step. This requires Chinese people to formulate a set of complete, accurate, and fair competition rules, so as to guide Chinese short-weapon game to develop towards competition, national characteristics, and high appreciation.

At present, the development of short-weapon game is confronting with a great dilemma. Therefore, the changes of the existing competition rules for Chinese short-weapon game have become an issue, which is unavoidable in China.

In 2004, the first National short-weapon game was held in China, on which there was only a small number of audiences. In the mean time, the game was very boring and far away from being fierce, which had a very close tie with the limited playing techniques. For this reason, it is highly necessary to make a change to the playing rules of Chinese short-weapon game.

Through the comparison with Japanese Kendo, it is found that many desirable features of Japanese Kendo can be introduced to Chinese short-weapon game.

Therefore, it is thought by the author that it is necessary to make an improvement to Chinese short-weapon game from clothing, equipment, etiquette, protective gear, etc.

Therefore, with the purpose of developing Chinese short-weapon game, it is necessary to constantly make an update to the ideas and emancipate the minds of Chinese people.

Thus, Chinese short-weapon game in the modern times can be emancipated from the rules of the traditional martial arts, and then playing rules can be changed in a real sense and also the guiding role of playing rules can be brought into full play. Only in such a way, new vitalities can be injected to the development of Chinese short-weapon game.

92.5.2 Introducing and Absorbing Relevant Sports Techniques and Gradually Improving and Richening Chinese Short-Weapon Game Playing Techniques System

The techniques system is one of the major factors that play an effect on the development of Chinese short-weapon game. Techniques-based playing method is necessary to the direction of the development of Chinese short-weapon game in the future.

In the modern time, it is necessary for Chinese people to abstract the typical techniques and actions of Chinese short-weapon game, and then absorb the best parts and integrate with the advantages of international short-weapon playing techniques, so as to produce a new combination of playing techniques.

As a result, modern Chinese short-weapon game is not only with the diversity of techniques and actions, but also with the variability of playing tactics. This is what Chinese people hope to see from modern Chinese short-weapon game.

According to the current situation that the attacking techniques are very dull in Chinese short-weapon game, the scores for the actions such as dotting, collapsing and pricking can be increased in a proper degree.

In Chinese short-weapon game, chopping, cutting, pricking, dotting and cleaving actions are mainly used as the attacking techniques. These actions vary frequently, and attach high importance to the integration of playing techniques. Also, the flexible wrestling, fighting, intercepting, flashing and stretching actions are mainly used as the defense techniques, in which high importance is attached to achieving victories with clever strategies and excellent skills. This is because any cruel fighting can only make the opponent seize more chances to win and the attacker will suffer more losses.

As a sport item, Chinese short-weapon game is not so fierce and cruel like Japanese Kendo. In the formulation of playing rules, some fierce offensive actions are disabled in Chinese short-weapon game because the protective gears have not been improved perfectly, thus imposing great obstacles on the practicality and appreciation of short-weapon game.

However, the security of players in Chinese short-weapon game is ensured, target positioning is to achieve victories based on points, and this has a very close tie with the game as a sport item.

92.5.3 Strengthening Theoretical Studies and Combining Scientific Researches and Trainings

It is highly necessary to make an improvement to the overall playing quality of Chinese short-weapon game; particular importance should be attached by leading departments at all levels to how to effectively increase the scientific and standardized trainings for the playing of Chinese short-weapon game.

In the mean time, it is highly necessary for all training teams to make an enhancement to theoretical studies, and also follow the road of combining scientific researches and trainings and using brains more commonly, designing methods and continuously making explorations.

For example, the Center for the Management of Chinese Martial Arts of General Administration of Sport of China as well as some physical educational colleges can make an enhancement to the trainings for scientific researchers and teaching team of martial arts, so as to promote modern Chinese short-weapon game to get rid of the constraints from the traditional martial arts training theories and then develop towards the modern sport games.

Besides, in the training practices, it is necessary for scientific researchers and teaching team as well as players of martial arts to make cooperation with each other and contact closely, which is helpful for the scientific training of modern Chinese short-weapon game. In this aspect, the National free-combating training team had made attempts twice.

For example, during the 13th Asian Games (Bangkok) and the preparation for the 5th World Martial Arts Championship, the scientific research personnel carried out monitoring on the trainings conducted at the martial arts scientific research training base of Shanghai Institute of Physical Education, and also made a very ideal training effect.

92.6 Conclusion

Today's society is completely different from the age in which the development of Chinese martial arts was under the leading of cold weapons, and simultaneously the role and value of Chinese short weapons of martial arts have been totally different from the previous time. This has proven to be an unavoidable historical fact, and also decides the inheritance and development of Chinese short-weapon game to follow the road of reform.

To survive for a long time, it is necessary to make it serve for the demands of society and people in the modern time, to promote it widely, it is necessary to represent it with a way that the people in the modern time are willing to accept.

Modern Chinese short-weapon game has not been started for a long time, and is only in the initial stage in comparison with the development of the "traditional short-weapon attack and defense" of thousands of years. If the old-fashioned thinking way and value system are blindly used for evaluating the modern Chinese short-weapon game, the development of Chinese short-weapon game will be impeded ultimately.

Therefore, only if the old thinking way can be broken and a new value system can be constructed, unique national intension can be excavated, and a good environment can be created for the survival, development and innovation of Chinese short-weapon game.

References

1. Xiao Q (1997) Institute of martial arts of general administration of sport of China. In: The history of chinese martial arts, vol 11. People's Sports Press, Beijing, pp 413–418
2. Ma X (2003) Chinese short weapons. In: Xian, vol 9. Sanqin Press, Beijing, pp 21–28
3. Ma M (2000) Sword Essays and Studies. Lanzhou Univ Press 12:45–52
4. He D (1998) Chinese short-weapon game. Beijing Sport Univ Press 4:39–43

Chapter 93
Study on Practice of Microteaching in Professional Skills Training of Tour Guides

Yuan Dong

Abstract The practice teaching of tourism in the vocational institutions of higher education in western China is still influenced by the traditional teaching mode, which impedes the improvement of students' professional skills. The author introduces the modern educational technology of Microteaching in the professional skills training of tour guides for students majoring in tourism. Good teaching results have been achieved through resetting the microteaching objectives and contents and implementing the microteaching steps. This paper aims at promoting microteaching to be applied in the higher vocational education.

Keywords Microteaching · Professional skills · Tour guide business · Higher vocational education

93.1 Introduction

At present, the training model of the professionals in tourism management in the higher vocational colleges in China has been constantly improving and attaching more and more emphasis on the practical training on the basis of sufficient knowledge. However, the existence of the traditional teaching mode limits the role and application of the contents of the practical teaching which is still carried out in the regular classrooms. It is difficult for students to generate the special feelings of the real environment by receiving the practical training in the regular theoretical

---

Y. Dong (✉)
Tourism Department, Chongqing Education College, Chongqing, China
e-mail: yuan85dong@126.com

W. Du (ed.), *Informatics and Management Science VI*, Lecture Notes in Electrical Engineering 209, DOI: 10.1007/978-1-4471-4805-0_93,

teaching environment. In addition, there are many students in a classroom, so only a small number of them can gain the practice opportunity, resulting in the poor performance of the practical training.

In order to develop the practical education and improve the connotation of the higher vocational education in tourism management, regarding the insufficiency of the practical training education, the author will introduce microteaching to the teaching of Tour Guide Business for the students majoring in tourism management. Based on the practice and comparative analysis, it fully explains that microteaching can promote students' mastery and application of the professional skills in tourism management, enhance their employability and help them grow to be talents.

93.2 Microteaching

93.2.1 The Concept of Microteaching

Microteaching is a controlled system of practice which makes the training objects probably concentrate on solving a particular behavior, or study under the controlled conditions. It is a systematic method using the modern teaching techniques to train students' professional skills.

93.2.2 The Philosophy of Microteaching

In fact, microteaching is to provide a practice environment, break the daily complex classroom teaching down into many specific skills which are easy to master, and put forward the specific training objectives and evaluation criteria for each skill. Based on the video recording of the trainees, it presents these skills to trainees in the form of picture, sound and image, helps them have a full range of perception of their skill operating behaviors, and let them repeat trainings in a short period of time, thus achieving the purpose of strengthening the trainees' professional skills.

Since microteaching breaks the overall operating skills down, makes them miniaturized, and becomes a teaching method which can be described, observed, operated and measured, thus changing the traditional training model's vague and experienced description of the practical skills and making the improvement of students' practical skills easy to realize. Literature fully proves that microteaching has more significant effects on the teaching results than the traditional teaching methods and is more conductive to improve students' ability and level of the professional skills of tour guide [1].

93.2.3 Characteristics of Microteaching

Firstly, the teaching time is short. By this, it can reduce students' stress and burden and also facilitate the teachers to concentrate on observing and evaluating the teaching.

Secondly, the teaching content is single. It just teaches a concept or a specific content, so that the students can master it well and the teachers can complete the teaching tasks in a short period of time.

Thirdly, the training objective is single. It only pays attention to one skill, so that the trainees can grasp it and the instruct can assess it easily;

Fourthly, there are many training opportunities. It adopts the advanced training room, so that every student has the opportunity of training.

93.3 The Practical Application of Microteaching in the Course of Tour Guide Business

93.3.1 The Objectives of Implementing Microteaching

The main objective of applying microteaching in the course of tour guide business for the tourism management specialty is to enhance students' professional skills, exercise students' resilience and ability to deal with things, enable them to master the skills of leading a group for tour in psychology and behavioral habits, and become the highly skilled guide talents [2].

The professional skills generally refer to these special knowledge or ability which requires the formal education or training to obtain with the most notable feature of needing the conscious and special training and mastering the special vocabulary, procedures and disciplines through memory. It especially refers to the professional knowledge and ability a guide has for tour guiding in the course of tour guide business, including the skill of tour guiding and the explaining skill.

93.3.2 The Contents of Microteaching

In order to apply microteaching scientifically, the teachers should construct the practical training contents which match the training objectives first according to the characteristics of the ability generation. Author began practice from 2009, re-divided the traditional contents of the course of tour guide business into two modules: the tour guiding skill module and the explaining skill module. As shown in Fig. 93.1.

Among them, the tour guiding skill module consists of the tour guiding procedure of the local (national) guide, disposing the incidents in the process of tour guiding, satisfying the individual demands of the tourists, the special skills in tour guiding,

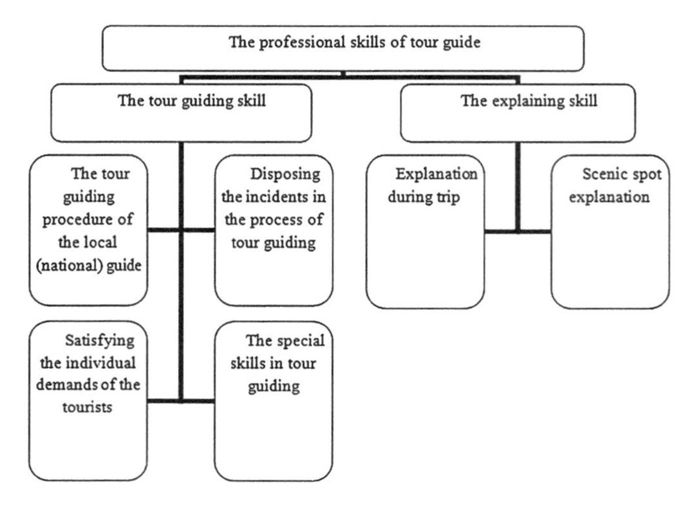

Fig. 93.1 The professional skills of tour guide

etc.; the explaining skill module consists of two programs of explanation during trip and scenic spot explanation. In each project, several tasks and many training units are designed, so that we can organize the theoretical teaching and the microteaching training one by one according to the tour guide's working procedure under the guidance of the programs and the drive of the tasks and make the teaching contents and the training programs more targeted and applicable, helping strengthening and enhancing the students' professional skills.

93.3.3 The Implementing Procedures of Microteaching

93.3.3.1 Design the Course Programs According to the Requirements of the Enterprises

During 2008 and 2009, author conducted a full investigation on a large number of tourism enterprises, famous experts in the industry, many senior tour guides and previous graduates in and out of the city. The investigation contents included the personnel structure status of the tourism industry, the development trend of the tourism professionals, talent demand condition, the demand of the job for knowledge and ability, the corresponding professional qualification and student employment, and made the scientific argumentation to grasp the talent demands of the industry and the enterprise as well as the cultivation status of the vocational schools from the middle view and thus design the course programs consistent with the tour guiding procedure for the post group on this basis; design the teaching and training methods according to the characteristics of task complexity, progressive training and skill generation of the tour guide; design the assessment program in accordance with the formation process and results of completing the task; re-integrate the original teaching contents.

93.3.3.2 Implement the Teaching Link According to the Course Programs

First of all, introduce a certain program in the tour guiding scenario; then, subdivide the program into number of tasks to lead the teaching, causing students to think, learn and prepare; finally, come to the micro-training to complete the task. It should be noticed that, to enable all students to have the micro-training in classroom time, micro-teaching must rely on the advanced training equipment and allow a number of students to have the micro-training at the same time, so as to meet the characteristics of more opportunities of the micro-teaching and training. Finally, repeatedly play the recorded training process and implement student and teacher assessment on the disadvantages in the training and improve their professional skills.

In summary, the basic links of microteaching are: program introduction—task subdivision—task preparation—micro-training, that is, task completion—assessment and improvement.

93.3.3.3 Multidimensional Assessment According to the Teaching Objectives

Multidimensional assessment is a summary of the implementation of microteaching. Combine the process evaluation with the result evaluation together, and the teacher evaluation and the student self-evaluation together. The contents include student self-assessment, assessment within and among the groups, teacher reviews, expert reviews and the comprehensive results. The ratings include the data collection, task implementation effect, guide speech creation, route design, guide speech explanation, PPT production and group cooperation. First of all, students assess the program tasks they complete; then carry out the assessment within and among the groups; finally, the teachers and the experts give the assessment. During the assessment, teachers and students make the discussion together. They fully affirm students' autonomous creativity, judge the problems in the program and discuss the ways and approaches to solve the problem. By comparing the assessment results between teachers and students, identify the reasons for the result differences.

93.3.3.4 Complete the Competitive Training According to the Higher Requirements

Penetrate microteaching into the training of all kinds of guide competitions. Through microteaching, cultivate students step by step to encourage them to achieve the best condition in the competition. In the guide speech explanation, talent show and quiz, comprehensively display the tour guide talents' high-quality skills and finally gain the good results. At the same time, it can help students get a deeper understanding of the teaching content, promote other students to learn actively, and create a better learning environment.

93.4 The Teaching Effect of Microteaching

93.4.1 Students Have a Good Mastery of the Professional Skills

As shown in Fig. 93.2, this research of the application of microteaching from 2009 to 2011 has found that students have a good mastery of the professional skills and strong operation abilities. In addition, the passing rate of the national tour guide certification exam has been increasing year by year, up to more than 90 %.

1. Tour Guiding Procedure
2. Dealing with the Accidents
3. Satisfying Customers' Demands
4. Application of the Guide Language
5. Application of the Explanation Method
6. Manner and Expression in Explanation
7. Systematic Explanation.

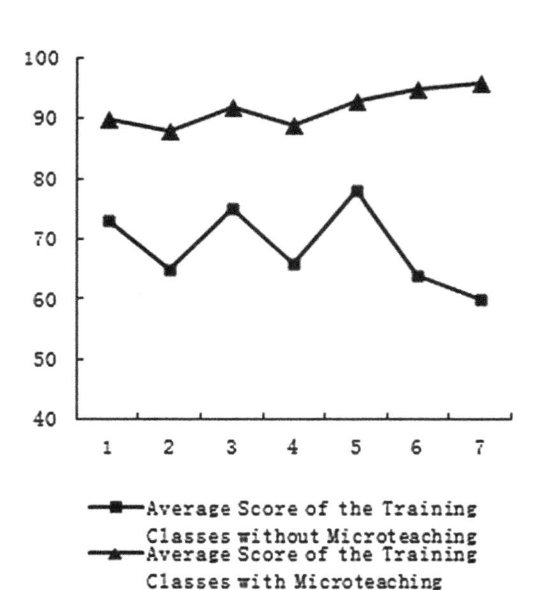

Fig. 93.2 The professional skills results comparison with and without micro-training

93.4.2 Strong Adaptability and High Competent Rate

Graduates engaged in the tour guide can quickly adapt to work and receive wide praises from the employing units. According to the follow-up survey on the graduates in 2010 and the interns in 2011, the employers generally speak very highly of the graduates. The competent rate reaches 98 % and the excellent rate 80 %.

93.4.3 The Graduates Enjoy a Rapid Development and High Evaluation

Students with the strong professional competence can quickly adapt to the posts, so they are generally welcomed by the employers. The graduates find jobs in the major travel agencies and tour guide service companies throughout the city. The graduates have a rapid development. A considerable portion of students can be quickly promoted to be the management team. The survey of CTS Group, New Century International Travel Agency Co. Ltd, Chongqing Tour Guide Service Company and others show that they speak very highly of the graduates in this major.

93.5 Conclusions

The practical application of microteaching in the professional skills training in tourism management specialty greatly enriches the opportunities for students to learn and practice and is widely welcomed by them. This method can also be extended to travel agents, hotels, hotel staffs and the training of the department heads to improve their professional skills by microteaching and achieve outstanding results in the practice, employment and skill competitions.

Acknowledgments This paper is one of the research results of the Research and Practice of Microteaching in higher vocational education, which is one of the education reformation research item of Chongqing Education College, item No. JG200918.

References

1. Cai L, Zhang H (2009) The construction of the professional skill microteaching for the tour guide and the quantitative assessment analysis. Edu Vocat 3(8):98–99
2. Chen Z (2007) The Construction of the Simulated Tour Guide Training Room Based on the Microteaching System. China Educ Technol Equip 4(11):67–78

Chapter 94
Effect of Sling-Exercise-Therapy Training on Childhood Autism

Huanxiang Ding and Rongyuan Li

Abstract To study the efficacy of conventional treatment supplemented with Sling-Exercise-Therapy on childhood Autism. 24 autistic children aging from 3 to 6 years old were randomly divided into experimental group and control group. After an observation period of two months, the two groups' scores of autism behavior checklist (ABC) and the total score were reduced. Compared with the control group, there were significant differences in the three factor scores, including sensation, body movement and communication, and the ABC total scores ($P < 0.05$). Although the factor scores of two groups' language ability and self-care ability were decreased, there was no significant difference within or between groups before and after training ($P > 0.05$). Each factor score and total scores of sensory integration were increased in varying degrees, but the experimental group's increased more ($P < 0.05$ or $P < 0.01$). After intervention, there was a significant difference between groups on two factor scores and total scores of proprioception and vestibular imbalance ($P < 0.05$). Although the score of learning ability was increased after intervention, there is no discrepancy in the statistics. The treatment method of conventional treatment combined with sling-exercise-therapy presents a satisfied effect to childhood autism.

Keywords Sling exercise · Childhood autism · Effects of treatment

H. Ding (✉)
LinYi University of Physical Education, Linyi 276005, Shandong, China
e-mail: huanxi99ding@126.com

R. Li
College of Physical Education and Health Science, Guangxi University for Nationalities, Nanning 530006, Guangxi, China

W. Du (ed.), *Informatics and Management Science VI*, Lecture Notes in Electrical Engineering 209, DOI: 10.1007/978-1-4471-4805-0_94,

94.1 Introduction

Childhood autism refers to a comprehensive mental development coagulation disease, which is characterized with social communication disorder, language developmental disorder, and mechanical and weird behavioral patterns [1]. Up to now, the causes of this disease have not been known yet. Researches and practices have proven that early education, early diagnosis and functional training intervention play a crucial role in the prognosis of childhood autism. In this paper, the effects of applying conventional treatment supplemented by sling exercise therapy on childhood autism are studied, for the purpose of providing a reference not only for the expansion of functional training of childhood autism, but also for the formulation of personalized exercise prescriptions applicable to childhood autism.

94.2 Objects and Methods

94.2.1 Objects

24 male children (3–6 years old), meeting the DSM-diagnosis standards, definitely diagnosed with childhood autism in municipal or higher-level hospitals, and receiving trainings from an autism training center, were used as research objects. Also, mental development retardation, schizophrenia and other diseases such as heart disease, epilepsy, locomotive organ disease, or significant physical disabilities disadvantageous to movements are not in these objects. Besides, objects have the ability to do simple physical exercises [2]. They were randomly divided into experimental group and control group; each groups had 12 cases; there were no significant differences between the two groups in age, course of disease, healthy condition, and autism behavior checklist (ABC) total scores comparison ($P > 0.05$), and therefore these objects were comparable.

94.2.2 Treatment Methods

94.2.2.1 Conventional Treatment

Children in both groups had received a conventional rehabilitation therapy that was centered at structured education and applied behavior analysis, etc.

94.2.2.2 Sling Exercise Therapy

In addition to conventional treatment, sling exercise therapy was also applied in training group [3]. Specifically, children (patients) with childhood autism lay prostrate on a sling, and held 10 plastic circles (diameter: 12 cm) in left hand; 10 empty water bottles of different colors were differently placed in front of these patients; when trainer gave commands and instructions, their parents pushed them to different directions, allowing them to put circles in their hands on water bottles under the shaking state; whatever these patients did in training, trainer and parents had to give applauses or language praise and encouragement. Sling exercise therapy was applied once each day; each training time was 20 min; a course of treatment was 30 days, and these patients were continuously trained for two courses.

94.2.2.3 Evaluation Tool and Method

Autism Behavior Checklist

Trainings of children were evaluated with autism behavior checklist (ABC); children's parents were guided by the same trainer to make an evaluation with ABC before and after treatment, respectively [4]. This checklist included 57 items that described children's senses, behaviors, mood, language and other abnormal findings, which could be summarized as five major factors, namely sense (9 items), communication (12 items), physical movement (12 items), language (13 items), and self-care (11 items). According to the load size of each item in checklist, 1, 2, 3, and 4 scores were respectively given; the total scores were the sum of gained scores of all items; children were diagnosed with autism if total scores were $\geq$68 scores, suspicious if $<$68 and $\geq$53, and negative if $<$53 (i.e., diagnosis was unsupported). If the total scores are higher, it suggested that autism is more serious and bad behaviors are more.

Autism Development Checklist

Autism development checklist (ADC), which was introduced by Guiying REN et al. from Institute of Mental Health of Beijing Medical University and tested in reliability and validity, was also applied [5]. Checklist was filled by children's parents; scores were given according to five grades: "never" was rated with the highest scores (5 scores), and "always" was rated with the lowest scores (1 score); after total scores were gained, the original score of each item was converted to standard T scores: low sensory integrative dysfunction (40$\sim$30 scores), average sensory integrative dysfunction (30$\sim$20 scores), and serious sensory integrative dysfunction (below 20 scores).

94.2.3 Statistical Methods

Measuring materials were expressed with mean ± standard deviation; gained data was analyzed with SPSS11.5 based on conventional mathematical methods; before and after treatment, matching samples T test was applied in analyzing differences between cases of a group, and single-factor analysis of variance was applied in analyzing differences between cases of two groups; $P < 0.05$ suggested that differences were of statistical significance.

94.3 Result

94.3.1 Comparisons Between Children with Autism in ABC Total Scores and All Factors Before and After Treatment

Result is shown in Table 94.1. After a treatment of two months, ABC total scores and each item's scores were reduced in both groups; in experimental group, sensory ability, not only physical movement and ABC total scores were highly significantly different ($P < 0.01$) and also communication ability was significantly different ($P < 0.05$) compared with conditions before experiment; in control group, just sensory ability and physical movement as well as ABC total scores were significantly different ($P < 0.05$) compared with conditions before experiment; in both groups, language ability and self-care ability were with no statistical differences ($P > 0.05$). Comparing between two groups after a treatment of two months, it was found that sensory ability, physical movement as well as ABC total scores were significantly different ($P < 0.05$).

Table 94.1 Comparison between children of two groups in ABC scores before and after treatment

Item	Experimental Group		Control Group	
	Before intervention	After intervention	Before intervention	After intervention
Sensory ability	12.13 ± 6.24	7.82 ± 3.71▲	11.36 ± 5.95	8.34 ± 4.22Δ■
Communication ability	15.62 ± 4.12	11.34 ± 3.52Δ	15.28 ± 4.32	13.74 ± 3.93■
Physical movement	17.54 ± 3.37	12.66 ± 3.27▲	18.65 ± 3.72	15.78 ± 3.35Δ■
Language ability	21.46 ± 7.92	18.57 ± 6.93	20.67 ± 8.17	18.32 ± 7.88
Self-care ability	18.65 ± 8.40	15.86 ± 7.24	18.23 ± 7.36	16.12 ± 7.50
Total scores	84.27 ± 13.38	63.24 ± 13.57▲	82.74 ± 14.56	71.47 ± 12.81Δ■

Note ▲$P < 0.01$ and Δ$P < 0.05$ compared with "before intervention"; ■$P < 0.05$ compared with experimental group

Table 94.2 Comparison between children of two groups in ADC scores before and after treatment

Item	Experimental group		Control group	
	Before intervention	After intervention	Before intervention	After intervention
Vestibular imbalance	27.12 ± 5.37	35.34 ± 6.24▲	26.95 ± 4.86	32.18 ± 5.74Δ■
Tactile defensiveness	42.42 ± 6.26	47.32 ± 6.47Δ	43.35 ± 5.67	46.32 ± 6.47Δ
Proprioception	24.31 ± 4.93	32.87 ± 5.23▲	25.75 ± 5.16	29.76 ± 5.41Δ■
Learning ability	18.27 ± 5.28	21.89 ± 4.37	17.45 ± 4.76	20.29 ± 4.83
Total scores	16.32 ± 3.57	27.33 ± 5.26▲	17.46 ± 4.23	24.25 ± 4.57Δ■

Note ▲P < 0.01 and ΔP < 0.05 compared with "before intervention"; ■P < 0.05 compared with experimental group

94.3.2 Comparisons Between Children with Autism in ADC Scores Before and After Treatment

Result is shown in Table 94.2. After an intervention of two months, sensory ability and total scores increased among all children with autism in two groups to different extent, but more highly in experimental group (($P < 0.05$ or $P < 0.01$); comparing between two groups, it was found that proprioception and vestibular imbalance as well as ADC total scores were significantly different ($P < 0.05$), and learning ability scores also increased but were of no statistical significance.

94.3.3 Effects of Sling Exercise Therapy on Sensory Integrative Dysfunction

Comparing the treatment effects of two groups, experimental group was better than control group, and there was a significant difference ($P < 0.05$) between two groups as shown in Table 94.3.

Table 94.3 Comparison between children of two groups in improvement of sensory integrative dysfunction before and after treatment

Group	N	Significant improvement (case)	Improvement (case)	No change (case)	Effective rate (%)
Experimental group	12	3	5	4	66.7
Control group	12	1	4	7	41.7

Note $\chi^2 = 16.653$; $P < 0.05$

94.4 Discussion

Neurophysiologic research results suggest that any mental activities such as cognition, thinking and reasoning and language are complex systematic functions formed by cooperative work of all areas of brain, but not the direct functions of specific cells in one area of brain; in terms of neural development pattern, learning techniques at high levels, such as cognition, language and functional behaviors, depend on the development of the processing ability of multiple sensory inputs [6].

Sling exercise therapy (SET) was originated from the rehabilitation in the Second World War. Research results show that relevant symptoms of childhood autism can be effectively improved with the application of conventional treatment supplemented by sling exercise therapy. The effects are specifically embodied in the improvement of physical movement, sensory ability, behaviors, social communication and language ability, especially in sensory ability, physical movement and social communication, indicating the application of conventional treatment supplemented by sling exercise therapy has a better treatment effect than simple conventional treatment. The reason for this result is that importance is attached by sling exercise therapy to the movements of an unstable state, making the balance control ability and stability of physical bodies in motion as well as sensory and movement coordination functions improved through the enhancement to feedback and integration functions between nerves and muscle-group.

In this study, children with autism were required to finish a series of actions by abiding by and understanding trainer's commands, searching targets, and putting circles on targets under the swaying state of lying prostrate and raising heads. Therefore, children's neck muscle tension, vestibular balance organs and joint structure as well as visual, auditory and tactile systems were greatly stimulated, and also a variety of stimulations were combined with movements, thus continuously adjusting the input of sensory information, making sensory movements integrated in many areas such as brainstem, thalamus, cerebellum, basal ganglia and brain cortex, and forcing physical bodies to appropriately response. As a result, the purposes of promoting brain development, improving brain functions, opening up new neural specialization pathways, and gradually improving the disorders of children in physical movement, sense, communication and others are achieved.

Result also show that even if the change of ABC total scores before and after training was of statistical significance and also was lower than diagnosis scores (68), but still much higher than screening boundary scores (53); there was a significant difference in sensory ability score before and after training, but children with autism are still under a middle or serious sensory integrative dysfunction state. On the one hand, this might be because the training period was too short (only two months). Similar studies made by domestic researchers on autism show that the general intervention period is six months or even longer at average. On the other hand, benchmarks for children of different ages were different, which might result in difference in training results. In addition, learning ability score increased but was of no statistical significance, and this might be because the improvement of learning ability of children was influenced by multiple factors such as physical health, intelligence and environment; children were required to initiatively search and analyze relevant information and resources in the learning process, and then their learning ability can be continuously improved through the communication with others. However, the learning ability development of children with autism is seriously hindered by the incomplete sensory information processing access, poor sensory information processing and social communication disorders that exist universally in them. Also, the study shows that although self-care and language ability scores were reduced after training, the differences between cases of one group or two groups in the two factors were of no statistical significance before or after training, suggesting that self-care and language ability of children with autism changed little after a training of two months.

In addition, there is a significant difference between female and male children in autism incidence rate: (4–5):1 is reported by many countries, but (6.5–9):1 is in China. In this study, there were few of female children receiving rehabilitation training, so female cases were not included. Therefore, further studies will be made in futures.

References

1. Cao C (2001) Psychiatric branch of Chinese medical association, China psychiatric disorders classification and diagnostic standards. Shandong Sci Technol Press 11(3):147–151
2. Cambell M, Schopler E, Cueva JE et al (1996) Treatment of autistic disorder. AM ACAD Child Adolesc Psychiatry 35(2):134–143
3. Lanting G, Chan Y (1997) Progress of clinical researches on autism. Chin Ment Health J 11(3):165–167
4. Tao G (1999) Mental disorder of children adolescents. Jiangsu Sci Technol Publishing 11:212–213
5. Tao G (1999) Mental disorder of children adolescents. Jiangsu Sci Technol Publishing 11:123–125
6. Wan Z, Qian D, Sheng X et al (2002) Analysis on childhood autism with ABC. J Clin Pediatr 20(2):80–81

Chapter 95
Study of EMG Differences in Coordinate Movement of Waist and Abdomen Muscles

Xiangxin Meng

Abstract Use Finland MuscleTesterME6,000 portable eight-guided surface EMG to record the waist and abdominal muscle electrical signal when eight shot put athletes are throwing shot puts of different quality (6, 7.26, 8 kg). The analysis results show that: when the waist and abdomen muscle mobilization order in the throwing game with the ball of 7.26 kg, the athletics are at their most theoretical status. In addition to this, the muscle discharge duration is shorter, and there are smaller IEMG values. The overall coordination capacity of the muscle is better than throwing shot puts of 6 and 8 kg. When throwing the lighter shot put (6 kg), the waist muscle IEMG value is higher with greater force. There are some meanings to the waist strength training. When throwing a heavier quality shot (8 kg), waist muscle force order is consistent to the 7.26 kg group, which has some supporting role to the study of the standard technical actions.

Keywords Surface electromyography · Coordination of waist and abdomen · Shot puts of different quality

95.1 Introduction

Coordination ability, in other words, it means under the regulation of the cerebral cortex, the nervous system dominates or regulates different muscle groups. Shot Put is strength based on speed as the core of throwing items; muscle strength is throwing the results to improve protection. Appropriate sequence and timing of the body force, throwing shot one of the core technologies. The shot put process, human

X. Meng (✉)
College of Physical Education, Linyi University, Linyi 276000, Shandong, China
e-mail: xiangx33meng@126.com

W. Du (ed.), *Informatics and Management Science VI*, Lecture Notes in Electrical Engineering 209, DOI: 10.1007/978-1-4471-4805-0_95,

motion links and equipment constitute a large part of the participation of a whole system, lower extremities, pelvis, trunk, etc. to drive a small part of the throwing arm and hand speed and power of the human body is constantly superimposed, and human kinetic energy is fully passed to the shot [1]. Such as muscle coordination, then push the shot process in all sectors can play a maximum efficacy [2]. In this study, surface electromyography and three-dimensional shooting method, and proceed from the muscle itself, and coordination of the different force of the waist and abdominal muscles to throw different quality shot in order to explore the action of the throw, "the nerve–muscle coordination of characteristics for the shot put technique. Coordination of evaluation criteria to develop a preliminary study, the theoretical basis for the throwing events designed to coordinate the quality of training methods; strength project the significance of motor skills laid the foundation for further study and coordination capacity.

95.2 Research Objects and Methods

95.2.1 Research and Measurement Objects

Eight excellent male short putters are selected as the research and measurement objects. They are from College of competitive sports in Beijing Sports University. Eight subjects were healthy. As shown in Table 95.1.

95.2.2 Research Methods

Each subject with different quality Shot Put (6, 7.26, 8 kg) to standardize the back Sliding Shot technology to complete throwing motion. EMG changes using surface electromyography test were recorded for each subject every time throwing the whole process. Test flow chart is shown in Fig. 95.1.

All test results are divided into three groups: the 6, 7.26, 8 kg group. Each shot put throwing contains $8 \times 3 = 24$ people.

95.2.2.1 Measurement Instruments

Finland MuscleTesterME6,000 portable 8 guide surface EMG, a Sony digital video camera (connect the EMG recorded simultaneously image), the electrode number of conductive paste, alcohol, cotton wool, elastic bandages and so on.

Table 95.1 Outline tables of the basic circumstances of the experimental objects

N = 6	Age (year)	Height (cm)	Weight (kg)
	20 ± 0.82	181.89 ± 2.54	103.95 ± 3.96

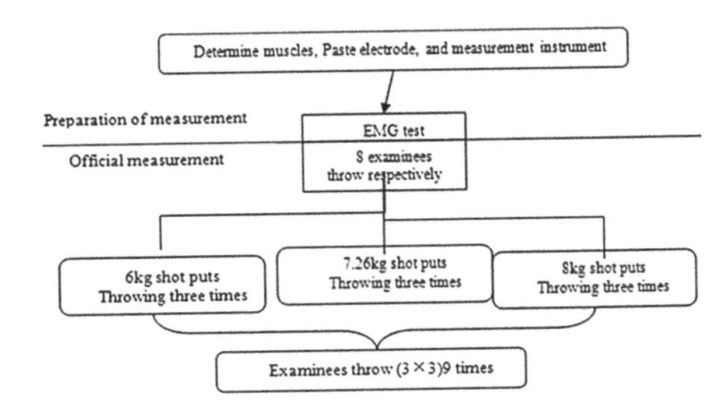

Fig. 95.1 Measurement process picture

95.2.2.2 Measurement Methods

Examinees do some preparation exercise. They do more than 15 min of jogging. When paste electrode, defatted alcohol swab to wipe the grease of the skin surface, the electrode along the direction of muscle fibers attached to the muscle belly of the measured muscle bump. Two records center connection of the electrodes and the muscle fiber direction has been, without leaving any gaps between the electrode and reference electrode attached to the vertical direction of the two recording electrodes, all electrode paste and connect the channel wire, elastic bandage fixed electrode, the requirements elastic appropriate, the electrode does not affect the subjects movement without shedding is appropriate. All wire is fixed with elastic bandages to the wires as small as possible movement.

Good subjects paste electrode connected to the instrument, according to 6, 7.26, 8 kg from light to heavy order, respectively, throwing the shot of the three qualities. Cast the shot back to specification sliding action, each of the quality of shot put throwing three times, if the foul is re-throwing.

The entire throwing process to use MuscleTesterME6,000 portable surface EMG records muscle discharge status of each throw. The MuscleTesterME6,000 the common-mode rejection ratio (CMRR) is 110 db, the absolute value of the maximum drift noise < 1.6 uV 8, 500 Hz band-pass filter range. Record frequency is 1,000 HZ.

95.2.3 Data Processing

95.2.3.1 Data Selection and Primary Processing

Surface EMG data: combined with the corresponding action of the synchronous image capture raw EMG, to use MegaWin EMG analysis software processing the raw EMG, analysis of indicators IEMG values, the discharge duration, the muscle discharge order.

95.2.3.2 Data Regular Statistic Processing

The SPSS13.0 statistical package is for routine statistical processing of all EMG variable parameters. Variable parameters are analyzed with the statistics of single factor analysis of variance. The significant level is P = 0.05.

95.3 Results and Analysis

95.3.1 Characteristics of IEMG Values of Measured Muscles when Throwing Different Quality Shot

Normal skeletal muscles are completely relaxed, there is no electrical activity, and EMG tracings for a straight line, called the power resting. When the motor nerve impulse reaches the nerve endings, the excitement passed through the nerve–muscle junction cell membrane to generate excitement in the EMG will generate action potentials [3, 4]. It can be seen that the force and action potential in a symbiotic relationship. It can be shown through the EMG picture that: which parts of a muscle or a muscle involved in activities; muscles involved in the activities measured in the movement; measured the duration of the muscle contraction and contraction strength in the movement; to complete the same technical movements of various muscles coordinates the work and so on [5].

IEMG refers to a certain period of time participating in the muscle total motor unit discharge [6]. The IEMG comprehensive reflection of the muscle discharge intensity and discharge frequency factors, and for the same muscle, the greater the tendency IEMG values the greater the muscle force [7].

The force is mutual. The quality of shot release to increase the body's load, the power of the shot weight of the body muscles is correspondingly different. Muscle strength is not simply with the increase in load increase, some muscle and even throwing the lighter shot from time to time showed a greater force, but showed a smaller force in throwing a heavier shot. It can be seen from Table 95.2.

95.3.2 Characteristics of Muscle Discharge Order and Duration of Measured Muscles when Throwing Shots of Different Quality

Technical aspects refer to the process of movement in which one or a few basic moves complete a relatively independent task in a certain period of time. The order refers not only to the technical aspects of the chronological order of position in space, but also the interface between the technical aspects of state [8].

Table 95.2 Characteristics of IEMG values of measured muscles when throwing different quality shot (uvs)

	6 kg group	7.26 kg group	8 kg group
Right musculus obliquus external abdominis	99.14 ± 45.64(−)	111.70 ± 55.72	121.91 ± 56.54(+)
Left musculus obliquus external abdominis	94.47 ± 44.17(+)	74.31 ± 17.83(−)	88.95 ± 22.13
Right rectus abdominis	288.71 ± 413.83	203.46 ± 206.53(−)	307.30 ± 336.36(+)
Left rectus abdominis	304.71 ± 537.44(+)	243.80 ± 353.10(−)	253.78 ± 376.07
Right erector muscle of spine	189.58 ± 177.30(+)	165.30 ± 100.68	147.31 ± 61.20(−)
Left erector muscle of spine	293.20 ± 300.55(+)	283.10 ± 294.50	164.90 ± 51.06(−)
Right gluteus maximus	234.09 ± 90.89(+)	208.19 ± 120.60(−)	221.95 ± 131.70
Left gluteus maximus	197.76 ± 76.38(+)	149.66 ± 45.52(−)	183.32 ± 32.42

(+): maximum value of the line (−): minimum value of the line

Muscle coordination is appropriate and reasonable muscle force [9]. In other words, the manifestation of the coordination ability to complete the action is the synergistic muscle in time of need in conjunction with rapid contraction of opposing muscle to quickly relax, in time and space need to have with the muscle to complete contraction and relaxation [10] turn.

Table 95.3 shows that throw different quality shot, with the exception of the right gluteus maximum significant difference, the remaining muscle discharge time there was no significant difference between 6 and 7.26 Kg. Three quality group compared to the 7.26 kg group of muscles discharge time generally shorter than the other two groups. Combined with muscle IEMG, muscle discharge duration and the size of the IEMG value not synchronous muscle contraction strength IEMG. Table 95.4 shows the throw different quality shot muscle mobilization order is not the same? The 8 kg group is more consistent with 7.26 kg of muscle power order, but the a 7.26 kg group of muscle mobilization is more in line with the throwing process straight—left sequence of movement of the torso. In addition to the innervations, muscle coordination and muscle structure, muscle

Table 95.3 Characteristics of muscle discharge order and duration of measured muscles when throwing shots of different quality(s)

	6 kg group	7.26 kg group	8 kg group
Right musculus obliquus external abdominis	0.25 ± 0.11(−)	0.28 ± 0.10(+)	0.27 ± 0.10
Left musculus obliquus external abdominis	0.36 ± 0.08	0.32 ± 0.03(−)	0.39 ± 0.14(+)
Right rectus abdominis	0.46 ± 0.14(+)	0.31 ± 0.15(−)	0.35 ± 0.12
Left rectus abdominis	0.38 ± 0.12	0.42 ± 0.10(+)	0.33 ± 0.17(−)
Right erector muscle of spine	0.51 ± 0.14	0.45 ± 0.16(−)	0.51 ± 0.24(+)
Left erector muscle of spine	0.43 ± 0.16(−)	0.49 ± 0.17	0.52 ± 0.12(+)
Right gluteus maximus	0.48 ± 0.07(+) #	0.34 ± 0.10(−)	0.42 ± 0.07
Left gluteus maximus	0.44 ± 0.07	0.45 ± 0.18(+)	0.40 ± 0.13(−)

(+): maximum value of the line (−): minimum value of the line
Note: Compare 7.26 kg group; #: $p < 0.05$

Table 95.4 Throwing shots of different quality, the force order of the measured muscles

	6 kg group	7.26 kg group	8 kg group
1	Right erector muscle of spine	Left gluteus maximus	Left gluteus maximus
2	Right rectus abdominis	Left erector muscle of spine	Left erector muscle of spine
3	Left gluteus maximus	Right gluteus maximus	Right gluteus maximus
4	Left erector muscle of spine	Right erector muscle of spine	Left rectus abdominis
5	Right gluteus maximus	Left rectus abdominis	Right erector muscle of spine
6	Left rectus abdominis	Right rectus abdominis	Left musculus obliquus external abdominis
7	Left musculus obliquus external abdominis	Right musculus obliquus external	Right musculus obliquus external
8	Right musculus obliquus external	Left musculus obliquus external abdominis	Right rectus abdominis

composition, the composition of the adaptability of various proprioceptors within the muscle are closely related. Enable the muscle to produce accurate and reliable coordination, the need for specialized training for a longer period of time, to the transformation of muscle structure, improve proprioception ability to adapt [11]. It cannot simply have the nerve coordination can ensure the coordination of the muscles.

95.4 Conclusion

Muscle IEMG value and the size of the load is not corresponding to the relationship. Compare muscles in throwing three qualities shot IEMG differences 7.26 kg set of values, generally the smallest. IEMG values to a certain extent reflects the size of the muscle strength, energy saving technology point of view, throwing the 7.26 kg shot, muscle coordination, force status. The 6 kg group IEMG value is of higher waist muscle strength training.

Throw different quality shot when the mobilization order, and discharge duration characteristics of muscle.

References

1. Xiaoyao W (2006) Hip the ability to put results. Shaanxi Sports Technol 2(26):135–137
2. Pingshe R, Houlin L (2006) Shot put action structure and the various stages of muscle characteristics of review. Inner Mongolia Sports Technol 2:77–78
3. Eckardt et al (1996) Facial stucture and functional findings in patients with pro-gressive muscular dystrophy. Am J Orthod Dentofacial Or-thop 110(2):185–190
4. Hongu M, Linhong J, Wang et al (2002) A study on muscle fatigue protection in rehabilitation treatment. The 6th international conference on biomendical engineering and re-habilitation engineering, Guilin 12:38–43

5. Wang J (2000) Exercise physiology research techniques, Vol 08. Zhejiang University Publishing House, Hangzhou, pp:79–85
6. West W, Hicks A, Clements L et al (1995) The relationship between voluntary EMG endurance time and intensity of effort in isometric hand grip exercise. Eur J Appl Physiol 71(4):301–305
7. Jingyi Y, Duanyuan W (1995) The EMG determination and analysis of quadriceps isokinetic concentric contraction. J Beijing Univ Phys Educ 11(4):41–43
8. Shan M (2007) A re-examination of the function of the arr. Sichuan Sports Sci 9(3):91–94
9. Ma Y, Xiaohui F (2004) Approach and method to improve the harmoniousness of the students' movements. J Shenyang Inst Phys Educ 8(4):566–567
10. Wu X, Hou W, Xiaolin Z (2000) Research on the correlation of sEMG on ECRL and hand grip force. Chin J Sci Instrum 8:21–23
11. Chu L (2004) Analysis on the nature of movement co-ordination. J Hebei Inst Phys Educ 6(2):81–82

Chapter 96
Analysis of Reaction Time Between High Performance Basketball Player and Ordinary Basketball Player

Deping Lin

Abstract Use literature material method, experimental method and mathematical statistics and other research methods and use reactive determination instrument to carry out test comparison between the simple reaction and the choice reaction of 10 high performance basketball players and 10 ordinary basketball players in College of Physical Education in Linyi University. The results have shown that there are significant differences in selective reaction between the high performance basketball players and ordinary basketball players. At the same time, there are no significant differences on the simple reaction between the high performance basketball players and ordinary basketball players. In addition to these, the effect increases with the increase of the complexity degree. Reaction time is an important indicator to make researches into the athletes' psychological characteristics. The research results carried out on 10 high performance basketball players and 10 ordinary basketball players in College of Physical Education in Linyi University have shown that there are obvious significant differences in selective reaction between the high performance basketball players and ordinary basketball players. At the same time, there are no obvious significant differences on the simple reaction between the high performance basketball players and ordinary basketball players.

Keywords Basketball player · Reaction time · Measurement and analysis

D. Lin (✉)
College of Physical Education, Linyi University, 276000 Shandong, Linyi, China
e-mail: depingl2451@126.com

W. Du (ed.), *Informatics and Management Science VI*, Lecture Notes in Electrical Engineering 209, DOI: 10.1007/978-1-4471-4805-0_96,

96.1 Introduction

Basketball is one of the competitive sports and basketball players need to have a rapid response capability. Ability to respond is the most important signs of ability for the basketball players. The strength of the reaction time happens to be the reaction time. The reaction time means the time interval to stimulate the body to respond [1].

The reaction time is divided into simple reaction time and complexity (choice) reaction time [2, 3]. In 1873, the Austrian biologist Exner was first one to propose the concept of "reaction time". After that, Wundt introduced the reaction time to his psychological laboratory.

In the 1950s, a Japanese scholar AIJ Investment Advisors Astoria put forward to divide the reaction time into two parts in the process of determination of reaction. The reaction is divided into two parts. Accordingly weiss proposes that the reaction time is divided into the former action time, and i (PMT) and after reaction time (MT). The former starts from the stimulus to the muscle action potential during this time, and the latter are known as electro-mechanical delay (EMD). In other words, it refers to the time when the muscle is excited to generate action potentials to produce shrinkage during this time [4, 5]. Fiss and other scholars have confirmed by studying the reaction time and movement, no significant correlation [6]. The general conclusion of the sports practice: the reaction speed of the speed and movement speed has nothing to do [7]. Some studies have found that the reaction time and the special requirements of the sport [6]. The reaction of athletes of different sports, even in the same sport, there are large differences in the response, which may with the Players, the division is different [4]. In general, a difficult task conditions, to improve the accuracy of the action need to reduce the reaction speed at the expense [5].

The author with the hospital response detector, the hospital two levels of basketball players in the psychological laboratory, the determination of reaction time, the purpose of determination of performance basketball player by the hospital high and ordinary post-secondary basketball player selected reaction reach appropriate conclusions, and make analysis to provide references and suggestions for teaching, training and selection of basketball beyond the hospital. There are no significant differences on the simple reaction between the high performance basketball players and ordinary basketball players. In addition to these, the effect increases with the increase of the complexity degree. Reaction time is an important indicator to make researches into the athletes' psychological characteristics. The research results carried out on 10 high performance basketball players and 10 ordinary basketball players in College of Physical Education in Linyi University have shown that there are obvious significant differences in selective reaction between the high performance basketball players and ordinary basketball players. At the same time, there are no obvious significant differences on the simple reaction between the high performance basketball players and ordinary basketball players.

96.2 Reaction Objects and Reaction Method

96.2.1 Research Objects

The research is carried out on 10 high performance basketball players and 10 ordinary basketball players in College of Physical Education in Linyi University. In terms of 10 high performance basketball players in College of Physical Education in Linyi University, the veteran of the players is more than six years; with an average height of 1.85 meters and the average age of 22.3 years. In terms of 10 ordinary basketball players in College of Physical Education in Linyi University, with an average height of 1.84 meters, the average age of 22.1 years. They all have considerable experience of the game.

In terms of 10 high performance basketball players and 10 ordinary basketball players in College of Physical Education in Linyi University, all of the 10 high performance basketball players and 10 ordinary basketball players in College of Physical Education in Linyi University are found to have ordinary ability of listening. In addition to two athletics that are found to have mild myopia, the others are tested to have ordinary vision. All subjects have physical health.

96.2.2 Research Methods

96.2.2.1 Literature Material Method

According to the research objectives and the needs of the content of the subject, the author has access to relevant information and documentation about basketball players and the reaction time.

96.2.2.2 Questionnaire Method

Questionnaire survey has been done on the 10 high performance basketball players and 10 ordinary basketball players in College of Physical Education in Linyi University. A detailed understanding of the years of competition of these two levels of basketball players and their height and weight, age, training level and competition experience are known, respectively, and extract the 10 athletes as a test object.

96.2.2.3 Experimental Method

Experimental Instrument

Take response analyzer (10 sets) as the experimental instruments. They are produced by East China Normal University, Science and Education Instrument Factory.
Experimental Process

10 high performance basketball players and 10 ordinary basketball players in College of Physical Education in Linyi University are taken as the research objects. 20 subjects were made the measurement of simple reaction time. The psychological laboratory of the Department of Physical and choice reaction time measurement, draw relevant data as accurate as possible.

96.2.2.4 Mathematical Statistics Method

Make precise statistical data research on the measured data.

96.2.2.5 Logistic Analysis Method

According to the results collated by the experimentally determined data and statistics, combined with literature data collected discussion and analysis of the scientific truth from facts, and draw relevant conclusions.

96.3 Result and Analysis

96.3.1 Variable Analysis

In the experiment, the argument is the audiovisual stimulate the feeling that the dependent variable is reaction time, the control variables have the following:

96.3.1.1 Subject Variable

The first one is the body's level of adaptation, followed by the subject's state of readiness, such as "preparation time" is also a factor affect the response time. If preparation time is too short, is the test may not have time to prepare to do the reaction; if time is too long, were prepared recession. In addition to the first one, the second one is that the attention was also affecting the reaction.

96.3.1.2 Exercise Variable

In general, exercise more, the faster the response, but progress is gradually reduced and eventually reach the lower limit of the reaction time is diminished.

96.3.2 Error Analysis

Reaction time measurement experiment is to reduce the error. Each test team has to do 160 times. Among the 160 times of experiments, 80 times choice reaction time, simple reaction times. Determined to eliminate order errors taken in strict accordance with the visual—listen—listen to—depending on the way, each unit 20, to eliminate the error caused by the different reactions in different parts of, always with one hand the same part of the key.

96.3.3 Experimental Result Analysis

High-level of basketball players and the general set of simple reaction time and choice reaction time (Table 96.1) as follows:

96.3.3.1 Differential Comparison in Simple Reaction Between High Performance Basketball Players and Ordinary Basketball Players

It can be seen from Table 96.1 that the sound of high-level of basketball players and in simple reaction time, reaction and light reaction indicators improved but did not show significant differences. This is a basketball on the neural response and the rapid ability to identify the impact of response speed is quite fast, eyes colored light reaction to a fast and close sound stimulus reaction; the majority of the basketball is the speed of ball movement, transfer of location, time difference

Table 96.1 Comparison of the simple reaction time and choice reaction time of High level basketball players and the general basketball players

Reaction time situation		High level group (M ± SD)	Ordinary group (M ± SD)	T	P
Simple reaction time	Sound reaction	0.1670 ± 0.0221	0.1728 ± 0.0218	1.614	0.828
	Light reaction	0.1285 ± 0.0165	0.1302 ± 0.0143	1.621	0.825
Choice reaction time		0.2716 ± 0.0966	0.4593 ± 0.0553	2.638	0.012

between the project of the concept of competition, the need for a long time with the eyes, your ears, and with the brain, so easy to improve the response of the central nervous system, so that the central nervous system function has been effectively improved.

Simple reaction time is single responses, that is, to test a constant single stimulation, and requires a constant response, simplify its information processing in the central nervous system, the entire reaction time is short. Fixed stimulus response mode due to the simple reaction time, the central delay of the time limit has reached a minimum, that is, increase the range of speed is very small, in such a small range to demonstrate a significant difference in speed is more difficult.

96.3.3.2 Differential Comparison in Selection Reaction Between High Performance Basketball Players and Ordinary Basketball Players

It can be seen from Table 96.1 that simple reaction time indicators in the different levels of basketball players did not show a significant difference, and in terms of choice reaction time showed significant difference. Which also confirmed this view: the central nervous system of complex reaction mechanisms may be an important factor to affect the economic capacity of basketball players. Changes in the two reactions compared athletes differences in effect size corresponding increase in the reaction complexity increases, which demonstrated significant differences in choice reaction time. Description of the economic capacity of the basketball players not only with the speed of information processing, with the complexity of the information the higher the complexity of the information, basketball player in the processing of this information the more advantage. The central delay is the main factor determining the reaction when the length of simple reaction time, complex reaction time, the central delay time range is relatively large, the time range more than twice that of the simple reaction time. It has shown significant differences in the likelihood of relatively large.

The existence of a large number of regular information in a basketball game, a lot of good basketball player by virtue of their own experience and the inherent laws of these laws, before making the actual reaction has formed one of the expected judgments. Obviously the process of this reaction is the result of high-level thinking process automation.

96.4 Result

The research results carried out on 10 high performance basketball players and 10 ordinary basketball players in College of Physical Education in Linyi University have shown that there are obvious significant differences in selective reaction between the high performance basketball players and ordinary basketball players. At the same time, there are no obvious significant differences on the simple reaction between the high performance basketball players and ordinary basketball players. The differences between the high performance basketball player and the ordinary basketball player are due to the different length of training period, the entry level of game experience; there are significant differences in choice reaction time.

96.5 Suggestions

Because there is no significant difference in simple reaction time of the basketball player a high level group and normal group, after teaching and training, basketball teachers and coaches do not simple reaction time as a decisive indicator to measure athletic ability.

The competitive ability and level of basketball players with better discernment, choice reaction time, and indicators can be used as the importance of the basketball player selection and evaluation of the level of training. When the basketball player selection, while the integrated use of genetic selection method, age, method of selection, size selection method, the movement quality of the selection method of scientific selection methods, but also attach great importance to the application of choice reaction time in the selection of basketball players, by measuring the athletes choice reaction time, the frequency of 10 s, 10 s pedaling frequency choose faster athletes in the choice reaction time.

96.6 Conclusion

The research results carried out on 10 high performance basketball players and 10 ordinary basketball players in College of Physical Education in Linyi University have shown that there are obvious significant differences in selective reaction between the high performance basketball players and ordinary basketball players. At the same time, there are no obvious significant differences on the simple reaction between the high performance basketball players and ordinary basketball players. In addition to these, it has put forward some suggestions for the teaching and training and selection of materials and some. But the response time is complex psychological and physiological phenomena. Considering the author's existing professional knowledge and experimental conditions, it is inevitable that there are some insufficient places which still need further study and improvement.

References

1. Qi-wei MA (1998) Physical psychology. Higher Education Press, Beijing 22:178–185
2. Qi-wei MA, Li-wei Z (1998) Physical sports psychology. Zhejiang Education Publishing House, Hangzhou 11:287–390

3. Bo-min Y (1997) Psychological experimental outline. Beijing University Publishing House, Beijing 8:380–384
4. Li-hong MA (1998) Research into the reaction time and the electronic mechanism delay. J Tianjin Univ Sport 3(4):26–29
5. Jin-liang LI, Li-wei Z (1995) Explorations into the reaction time problem of the athletics outline. J Beijing Univ Phys Educ 18(3):31–35
6. Zhang Q (1984) Reaction time of young athletics of different projects. Jiangsu Sports Technol 4:39–40
7. Wei-duo R, Xiao-ming W, Yun X (1993) Movement during the reaction speed accuracy trade-off studies. Psychol Technol 16(5):291–294

Chapter 97
Analysis on Chinese Women's Olympic Gold Medal Special Features and Influence Factors

Daling Shi, Yuling Song, Xinjian Luo and Aicui Hu

Abstract This paper mainly by using the method of literature, mathematical statistics, the Beijing Olympic gold medal to our country distribution of statistical analysis, aiming at improving the Chinese competitive sports scores, strengthening our country provides the reference for the development of the Olympic Games and countermeasures, make our country to create more outstanding achievement, and soon become physical power.

Keywords Women's Olympic gold medal · Special features · Influence factors

97.1 Introduction

The 29 Olympic Games, the Chinese sports delegation made 51 gold MEDALS, 100 MEDALS outstanding achievements, the first Olympic gold in the first place, creating the Chinese sports delegation to attend the Olympic Games since the best results, which in a multiple projects to the Olympic gold medal and MEDALS breakthrough. This performance greatly raised the national spirit; to the world of Chinese sports show the strength. But China still is not a physical power, in many ways need to improve, improve. This article is the games of the 29 Olympic gold

D. Shi (✉) · A. Hu
Department of Physical Education, Huazhong University of Science and Technology, Wuhan, Hubei, China
e-mail: shidaling@cssci.info

Y. Song · X. Luo
Department of Physical Education, China University of Geosciences, Wuhan, Hubei, China

W. Du (ed.), *Informatics and Management Science VI*, Lecture Notes in Electrical Engineering 209, DOI: 10.1007/978-1-4471-4805-0_97,

medal China statistical analysis of the distribution, summing up the experience and lessons for our country to help the development of the sport.

The Beijing Olympic Games, the Chinese sports delegation sent the 639 athletes, go to the Beijing Olympics all 28 sports, and a breakdown of the game. From table statistics can learn our country sports delegation in 16 of sports in the gold medal. The gold medal which mainly concentrated in traditional advantage projects, such as weight lifting, gymnastics, diving, table tennis and shooting, badminton has received 36 gold MEDALS, accounting for 70.6 % of the total number of gold MEDALS in our country. One table tennis and trampoline, our country sports delegation to the purpose all the gold medal, diving lost 1 gold MEDALS. "119" project secured four gold MEDALS, of our country 7.8 % of the total number of gold MEDALS. Women's events get 27 gold MEDALS, accounting for 52.94 % of the total number of gold MEDALS in our country, the man project get 24 gold MEDALS, accounting for 47.06 % of the total number of gold MEDALS in our country, the man project has made much progress. Won the gold medal project is mainly distributed in the skills of leading performance difficult sex, accuracy, beauty performance every other nets confrontational and combat adversarial, and physical leading such fast power sex also have the strongly strength, endurance events to improve.

In the Olympic Games, diving, weight lifting, table tennis, badminton and gymnastics were set up 46; our country has made 31 gold MEDALS, accounting for 60.8 % of the gold in China. In the game, some traditional advantage projects are getting a gold medal strength, but lost the gold medal for any variety of reasons, such as men's 10 m platform, the first five jump ZhouLv Xin always rank high on the list, the sixth jump his choice of is the difficulty of coefficient is 3.4 307 C, not the most players choose the difficulty of the coefficient is up to 3.8 5255 B, original want to ask firm, didn't want to ZhouLv Xin is planted in this movement, lost the gold medal. Chinese sports delegation in the Beijing Olympic Games, the newly set the table tennis men's and women's groups in the project, which it has the gold medal. At 27th Olympic Games in Sydney trampoline project set, the Chinese delegation made the project's full gold medal. The more and more competition, more and crueler, our country want to keep the traditional advantage in project absolute superiority, has been unlikely. Our country the potential advantages of judo and boxing, in Beijing Olympic Games made of 5 gold good grades. Our country sports delegation in canoe project continued to advantage, solid performance, obtain the Olympic gold medal. In track and field, swimming, water project this gold medal on sports, our country sports delegation has only four gold MEDALS, and other countries have a larger difference, it also shows that our progress on these items of space is very large.

In 1924, the female project officially listed in the Olympic Games. With the Olympics woman project continuously add, female the score of the athletes in a country has more and more influence athletic performance. The Beijing Olympic Games, the Chinese sports delegation has received 51 gold MEDALS, of which woman won 27 gold MEDALS, accounting for 52.94 % of the total number of gold MEDALS in our country, the man get 24 gold MEDALS, accounting for 47.06 % of the total number of gold MEDALS in our country. In our country's gold and sports, the woman is in 13 items the gold, and a man's gold project only nine. We can see that the man get finished less than women, men than women less gold project, but the gap has been in constant decrease. In the Beijing Olympic Games made brilliant achievements, and China's "the system" cloudy can be divided. Although China has been top first group, but China is still not a sports power, mass sports development slower, we should further optimization of "the system". China's traditional advantage projects play stability, increase China's Olympic Games have not finished in traditional advantage again there is great breakthrough projects, we should actively expand new project, increase the gold surface, the comprehensive development of competitive sports. Modern sports competition is not only the players' individual ability of the play, also need to corresponding scientific training and competition experience. To scientific training l practice, through the "to" bring in the practice, promoter's practice, "want to let athletes overcome" heart demons, reduce injuries, with good state leave the stadium. From the previous view, our women project develops well, mainly women's history, new project short, low penetration rate and relative competitive level is low, easier to break through. From the statistical analysis can be seen above, the Chinese male, female project level disparity is gradually narrowed. Based on this fact, we should give priority to the development of women's events at the same time, increase man project just face. Our country skills leading class project develops well, among them the performance difficult project won 18 gold MEDALS, we should summarize training rule, development of the potential advantages in the project and backward projects, such as the same inside the synchronized swimming, art gymnastic project has a further development of the space.

97.2 Research Results and Analysis

The Olympics are a host of effect with strong competition, the number and medal has obvious effect of the host. Many scholars conducted a detailed research, WuDian Ting WuYing in the 2008 Beijing Olympics gold medal China overtake America possibility in the article by using grey forecasting model GM (1, 1) preliminary forecast, all method analysis of each session of the Olympic Games all host effect, the results is: the number of host effect for 11 was 31 %. The overall strength, the host of the effect of 11 was 71 %. That is to say, from the average level of speaking, because of the existence of host effect, make each session of the Olympic Games in the host of gold MEDALS for host when jack and the average increase more than 1 %, up to 1 [1]. The 2008 Olympic Games as the Olympic host China make full use of the host nation, and the climate, the comprehensive opportunities, and in the athlete's training, management continuously strengthened, and catch up on the number of beauty Countries, the Chinese delegation won a total of 5 1 gold MEDALS, one silver 2, 2 8 bronze medal and total MEDALS 1 0 O pieces [2]. And in the 2012 Olympic Games in London, China, the host of the position will lose, instead it is away disadvantage, China's Olympic team will undoubtedly is facing a great challenge. According to the conclusion, the Chinese Olympic team finished the next Olympic Games and the average drop on the medal in theory they should achieve 1 around 1 %. That is to say, 201 2 years London Olympic Games, China won gold MEDALS in the Olympic delegation about 45-between 48 pieces, about 88–92 jack between pieces.

Look at the 28 and 29th the top five medal standings, China's most dangerous rivals is undoubtedly the United States. 28th Olympic Games the United States whether gold medal amounts or total MEDALS are in first. 29th China Beijing Olympic Games, although China as host, number 51 gold higher than the 36, that is the road, and the silver, bronze MEDALS for 38 total, respectively, 36 gold medal, 110, that is higher than the China in the 21 medal, 28 times, 100 rockets. In addition, three big ball in the six projects, the American players into the final five, won the Olympic men's and women's basketball championship, women's champion and men volleyball team champions, women's volleyball team first runner-up. In addition in swimming, athletics, shooting, aquatic sports, body such as large projects spell gold medal, the United States has strong talent advantage. And in some of the rural project has made a medal, but overall lack of Houshi, project structure a: it is not equilibrium; MEDALS distribution in a "pour pyramid", explain China in some projects of the two j has been mining gold ability to the limit, and individual project only gold medal total from individual ability to see, China's harvest 100 pieces, compared with the United States still have approximately 10 of the distance. The United States to take part in the games is very wide distribution points, obtained the medal tau is very wide also, gold, silver, bronze distribution, if their next will silver and bronze MEDALS into the gold medal, the gold medal for China is a big shock [3]. Although China team in traditional strengths continued to advantage, but our gold MEDALS has basic saturated, the value added is empty ask is not big. And in track and field, swimming and water sports in the foundation, on the strength of our country in.

But the weak, our gold MEDALS numbered. In addition the Beijing Olympic Games some China brigade of advantage project was cancelled, instead of just is we level is very low or no on the project. Boxing, rowing, and wrestling some potential advantage projects are have window, but the play is not stable, these are the first adverse factors for gold. Especially in the 201 and Olympic Games 30 th London, China, the host of the position will lose, instead it is away disadvantage, China's Olympic team can again win over the same road is the team USA, no doubt, will be faced with many severe challenges [4].

The above analysis is of the disadvantages of the Chinese team, to talk to the team have advantage: one of, nearly a few China Olympic team morale increasingly high, gold and total MEDALS soaring highs, it is an optimal situation. Instead, though the United States is still sports power, but in addition to the 1996 Atlanta Olympics besides, gold MEDALS, no obvious increase of the number, or even reduce trend. O8 Olympic Games every won gold and top, morale; Second, even if according to the next Olympic Games the Chinese delegation won gold MEDALS and medal will decline 11 % to speculate that the Chinese delegation won gold MEDALS in about 45-also between 48 pieces, medal in the 88–92 between pieces. That figure in the United States on top [5]. Third, China's Olympic team have the growing power of the motherland and 1.4 billion people's support, with strong behind; President hu jintao in Beijing for the Olympic Games conference on, the Beijing Olympic Games great success is mainly thanks to the system, the future will continue to maintain and improve the system, give full play to the athletic sports promoting our international status and prestige role. On October 16, 2009 at the opening ceremony of the first 1, the President hu jintao said again, hope sports workers carry history mission in mind, and pushed by China sports country to sports power forward. The author thinks that, although sometimes controversial sound, but in many insiders, it seems, the next few years the Chinese government on the thinking of the great force does not have the too big change, the Olympic period after not competitive sports weaken to top–down, but in the microscopic constantly strengthened, the role of the competitive sports is can't take the place of, the Chinese sports from sports country to sports power forward, and not don't athletic sports, give up the Olympic gold medal, grasping athletic sports, grasping the mass sports, and finally achieve the sports powers to the goal of sports power. To infer, the next few years our government will undoubtedly-such as always of the competitive sports support, support the London Olympics, China's Olympic delegation [6].

The London 2012 Olympic Games, China's Olympic team win over the United States, will be faced with many severe challenges. China's Olympic team won gold MEDALS and medal will drop 1 around 1 %, about 45-gold MEDALS between 48 pieces, about jack in the 88–92 between pieces. But China's Olympic team have strong support of the motherland and 1.4 billion people, with the 29th Beijing Olympic Games remaining prestige, the 56-point thrashing, serious preparation for winning is steady, and there were no big mistake, and keep the gold medal and medal first throne is totally possible. 1 take many kinds of effective measures, the greatest degree to reduce road disadvantage bring "the low effect" effect, give full play to the advantages of the Chinese Olympic team have, may first into real first. Keep today's brilliant need from the strategic make proper adjustment on; continue to play the advantage of the system, and gradually to a professional and popular with the transfer of the model. The western countries to school on the school for high level sports training mode, further perfect our country's competitive sport backup talent cultivation, doing well the sustainable development of competitive sports.

97.3 Strategic Goals

Three of Olympic Games, China won the gold medal for project for swimming (diving), weight lifting, judo and taekwondo Way, gymnastics, shooting, table tennis, badminton and sports, nine breakdowns. Kayak, women's wrestling project in the 2004 Olympics gold medal realize breakthrough, in 2008 the Olympic

Games straight gold medal. Tennis, women's volleyball, track and field in the 2004 Olympic Games won the gold medal in the 2008 Olympics gold medal not. A medal Rowing, sailing, boxing, trampoline, archery 5 projects in the Olympic Games in 2008 gold medal on realize breakthrough. Boxing, trampoline at the 2008 Olympic Games to two gold MEDALS successes out of this project came first. In addition, the Olympic Games in Beijing, China in the women's hockey, artistic gymnastics, beach volleyball, artistic gymnastics three breakdown to medal breakthrough. In our country, most of the Olympic Games by two cycle preparation for the tournament, made significant progress. But, in the last three games, our country's basic sports in the track and field swim etc and collective ball games scores around. And the world's advanced level is in the gap widening. Our country collective ball games record overall not ideal, only won the women's volleyball 1 gold MEDALS, women's hockey 1 MEDALS. Based on recent the Olympic Games every Chinese Olympic project performance variation, combined with the project development base, reserve talented person status, coaches group, organization and management level, training innovation ability, team culture construction, and other comprehensive factors suggest that Chinese Olympic projects include advantage projects, time, internal advantage projects advantage projects and project a category 4 behind. Table tennis, badminton, weight lifting, diving, gymnastics, shooting, judo (woman), trampoline, taekwondo (women) is above 9 projects (partial) in the last three Olympic Games gold medal or continuous in Beijing Olympic Games clinching the project all the gold medal: has the larger group advantage and adequate backup power reserves. Not for a certain athlete's retirement and make the whole project by the influence of the great achievements: be able to grasp regularity and a scientific, project leading training concepts and methods of training management team. Most of these projects have never won the Olympic gold medal; Track and field, the swimming events although won the gold medal, but the gold medal of the proportion of sports events in the Olympic Games Less than 5 %. Overall performance and the world advanced level disparity is large; Lack of outstanding athletes group, some individual projects are top players. But fight alone. One or two people will influence the whole project problems of record; These projects on the understanding of the project rules, training theory, training concept, training methods and mental state has a large gap, etc. The sports work, general secretary to continue to improve athletic performance level, to promote the struggle to win honor for the country, as the core of the China sports spirit, explore contemporary sports development law, improve science training level. In adhere to the system of competitive sports in China, and keep our country sports characteristics and advantages while actively explore potential, optimizing the structure, improve efficiency, promote competitive sports categories internal balanced development, and continuously enhance the competitive sports in our country the comprehensive strength and international competitiveness. General Secretary of the important speech is guide our sports programmatic document, is our system planning future strategy implementation the guiding ideology of the Olympic Games. Through to the Beijing Olympics Olympic project after world pattern analysis and classification of Chinese Olympic project again, to realize our country by sports country to sports power forward the goal, not only in the future Olympic competition keep in the gold medal of the most advantage, keep the Olympic Games before the position of the three gold, more important is Actively explore potential, optimization of the project structure, especially push track and field, swim etc and collective ball games and other international professional degree is high, influence in international sports, and that people love the rapid development of the project; Not only to pursue development quantity. The promotion project structure optimization, various balanced, improve the comprehensive strength of the Olympic Games and international competitiveness. This we put forward the future Olympic project development strategy is to keep the Olympic gold medal rank in the top Column, stable in the games were the first group of the position: advantage projects keep advantage, efforts in innovation, the training tactics ideas, training methods, organization and management, personnel training mode led the world, actively explore the pattern of development of professional advantage projects. Ensure the sustainable development of the project; Time and potential advantage projects advantage projects strive to comprehensive breakthrough, make more of the project has a certain group advantage and adequate backup power reserves, to form a profound grasp regularity and a scientific, project leading training concepts and methods of training management team; Behind the project after shame-awareness brave, striving to deepen the understanding of the law project, renew the idea, innovative methods, perfecting organization, strive to improve the track and field, swimming and other basic categories of Olympic sports events in the gold number and of the total number of gold MEDALS delegation. Improve sets Body of ball games record overall. Strive to more collective ball games gold medal, MEDALS and good places.

References

1. Beijing Olympics (2008) Baidu, Wikipedia. http://baike.baidu.com/view/50402.htm
2. Qiang Y (2010) Introduction to design arts. Chongqing University Press, Chongqing 2(8):120–125
3. Domain Z, Wang T, Xiao Y (2009) Introduction to Design Art. Tsinghua University Press, Beijing 33(4):582–589
4. Wei L, Xi L, Mao D (2010) Logo design. Southwest China Normal University Press, Chongqing 9:280–286
5. Ling C (2007) New history of body art: the journey towards integration. Tsinghua University Press, Beijing 23:394–399
6. Xiangzhong L (2010) new trend of media development in digital times. Digital Media Creative Arts / Digital Art Series Press, Beijing 2(5):101–107

Chapter 98
Research of Attention Maters in Tennis Teaching of Teenagers

Yong Yu, Chengbao Ji and Yongqi Ji

Abstract Teenagers tennis teaching is an important and complex procedure, as in a large teenagers are different from adults in physiology, psychology etc. so there are some matters needing attention. This paper introduced the matters in coaching approaches; coaches' help to select the appropriate equips for teenagers and leading teenagers to have a proper psychology in tennis.

Keywords Tennis teaching · Instructional approaches · Equip psychology

98.1 Introduction

Tennis is a sport usually played between two players (singles) or between two teams of two players each (doubles). The modern game of tennis originated in Birmingham, England in the late nineteenth century as "lawn tennis" which has close connections to various field/lawn games as well as to the ancient game of real tennis. Tennis has been always regarded as a kind of noble sports, and now more and more people including many teenagers are learning to play tennis, so as that the tennis is becoming a universal sport.

Teenagers tennis teaching is an important and complex procedure, as in a large teenagers are different from adults in physiology, psychology etc. so there are some matters needing attention. In the teaching of teenagers tennis, fitness training has become increasingly sophisticated and individualized with advances in sports science, most tennis practices continue to be intuitive, emulative and coach-led

---

Y. Yu (✉) · C. Ji · Y. Ji
Physical Education and Sports College, Beijing Normal University, Beijing, China
e-mail: yongyu121a@163.com

W. Du (ed.), *Informatics and Management Science VI*, Lecture Notes in Electrical Engineering 209, DOI: 10.1007/978-1-4471-4805-0_98,

with exiguous regard for the individual athlete [1]. Searching for an appropriate method for the teenagers seems necessary. There are two general coaching approaches: prescriptive and discovery, or variations thereof, which conceptually lie at either end of an explicit-implicit learning continuum. That is, depending on the type of instructional or practice environment, learning may occur by either explicit or implicit processes.

More than 50 % of all tennis players are afflicted with sport injuries due to their tennis activity [2]. The biggest percentage of these sport injuries are in the lower extremities and it is assumed that the suspiration during sideways movements is one of the prime factors connected with these injuries. The coaches should concentrate on these problems via help the teenagers select appropriate equip, for example, shoes and racket. The coaches also have to attention to the psychology of the teenagers to insure them performed well.

The rest of this paper is organized as follows: Sect. 98.2 provides a brief introduction to the two general coaching approaches. Section 98.3 describes the coaches' help for selecting the appropriate equips. Leading teenagers to have a proper psychology are reported in Sect. 98.4, whereas Sect. 98.5 is the conclusion.

98.2 Instructional Approaches in Tennis

Collectively the fashion in which coaches present information, structure practice, and provide feedback comprises a considerable portion of what can be considered a 'coaching approach'. In tennis, the conceptualization of different coaching approaches or philosophies has been confounded by disparate terminology and coaching parlance [1–3].

There are two general coaching approaches: prescriptive and discovery, or variations thereof, which conceptually lie at either end of an explicit-implicit learning continuum. That is, depending on the type of instructional or practice environment, learning may occur by either explicit or implicit processes. At the explicit end of the continuum (e.g., prescriptive), septic instructions are given to the learner about the rules that underlie a stimulus, or movement pattern to assist learning (e.g. lateral ball toss displacement to the right handed player's left may signal a player's intent to hit a kick serve). At the implicit end (e.g. discovery), no septic instructions are given, yet prescient learning of the underlying stimulus or movement pattern occurs.

98.2.1 Identifying Similarities and Differences

The ability to break a concept into its similar and dissimilar characteristics allows students to understand (and often solve) complex problems by analyzing them in a more simple way. Teachers can either directly present similarities and differences, accompanied by deep discussion and inquiry, or simply ask students to identify similarities and differences on their own. While teacher-directed activities focus on identifying specific items, student-directed activities encourage variation and broaden understanding, research shows. Research also notes that graphic forms are a good way to represent similarities and differences.

98.2.2 Summarizing and Note Taking

These skills promote greater comprehension by asking students to analyze a subject to expose what's essential and then put it in their own words. According to research, this requires substituting, deleting, and keeping some things and having an awareness of the basic structure of the information presented.

98.2.3 Reinforcing Effort and Providing Recognition

Effort and recognition speak to the attitudes and beliefs of students, and teachers must show the connection between effort and achievement. Research shows that although not all students realize the importance of effort, they can learn to change their beliefs to emphasize effort.

98.2.4 Homework and Practice

Homework provides students with the opportunity to extend their learning outside the classroom. However, research shows that the amount of homework assigned should vary by grade level and that parent involvement should be minimal. Teachers should explain the purpose of homework to both the student and the parent or guardian, and teachers should try to give feedback on all homework assigned.

98.2.5 Nonlinguistic Representations

According to research, knowledge is stored in two forms: linguistic and visual. The more students use both forms in the classroom, the more opportunity they have to achieve. Recently, use of nonlinguistic representation has proven to not only stimulate but also increase brain activity.

98.2.6 Cooperative Learning

Research shows that organizing students into cooperative groups yields a positive effect on overall learning. When applying cooperative learning strategies, keep groups small and don't overuse this strategy-be systematic and consistent in your approach.

98.2.7 Setting Objectives and Providing Feedback

Setting objectives can provide students with a direction for their learning. Goals should not be too specific; they should be easily adaptable to students' own objectives.

98.2.8 Generating and Testing Hypotheses

Research shows that a deductive approach (using a general rule to make a prediction) to this strategy works best. Whether a hypothesis is induced or deduced, students should clearly explain their hypotheses and conclusions.

98.2.9 Cues, Questions, and Advance Organizers

Cues, questions, and advance organizers help students use what they already know about a topic to enhance further learning. Research shows that these tools should be highly analytical, should focus on what is important, and are most effective when presented before a learning experience.

Farrow and Abernethy [4] compared explicit and implicit instructional methods in the training of anticipatory skill for the return of-serve of intermediate players. It was reported the implicit group improved their on-court prediction accuracy after the training intervention with no improvement evident in the explicit, or placebo and control groups. While evidence suggests coaching approaches reliant on explicit instruction either predispose players to conscious thought during performance, and potentially poorer performance under stress (i.e., competition) [5, 6], or to overload working memory capacity such that performance declines [7].

98.3 Selection of the Appropriate Equips

Help the teenagers select an appropriate equip including shoes and racket is necessary for a coach, as the appropriate equip is essential for the teenagers to protect them from injury and develop the best ability.

There are more than 50 % of all tennis players are afflicted with sport injuries due to their tennis activity. The biggest percentage of these sport injuries are in the lower extremities and it is assumed that the suspiration during sideways movements is one of the prime factors connected with these injuries. Shoe construction can influence the suspiration movement during sideways movement. To prevent injuries and optimize the wearer's health, it is important to consider the person's compensation strategies to the shoes [8]. The sole should be composed of a rough, nonslip surface that will maintain contact with the ground surface [9].

Following are some general rules for injury prevention no matter what sport you play. While it is impossible to prevent every injury, research suggests that injury rates could be reduced by 25 % if athletes took appropriate preventative action, including:

Be in proper physical condition to play a sport. Keep in mind the weekend warrior has a high rate of injury. If you play any sports, you should adequately train for that sport. It is a mistake to expect the sport itself to get you into shape. Many injuries can be prevented by following a regular conditioning program of exercises designed specifically for your sport.

Know and abide by the rules of the sport. The rules are designed, in part, to keep things safe. This is extremely important for anyone who participates in a contact sport. Rules of conduct, including illegal blocks and tackles are enforced to keep athletes healthy. Know them. Follow them.

Wear appropriate protective gear and equipment. Protective pads, mouth guards, helmets, gloves and other equipment is not for sissies. Protective equipment that fits you well can safe your knees, hands, teeth, eyes, and head. Never play without your safety gear.

Rest. Athletes with high consecutive days of training, have more injuries. While many athletes think the more they train, the better they'll play, this is a misconception. Rest is a critical component of proper training. Rest can make you stronger and prevent injuries of overuse, fatigue and poor judgment.

Always warm up before playing. Warm muscles are less susceptible to injuries. The proper warm up is essential for injury prevention. Make sure your warm up suits your sport. You may simply start your sport slowly, or practice specific stretching or mental rehearsal depending upon your activity.

Avoid playing when very tired or in pain. This is a set-up far a careless injury. Pain indicates a problem. You need to pay attention to warning signs your body provides.

In previous investigations, Caslon and Ruggeri [10] analyses the mutual dynamic relations between ball and racket. Leigh and Lu [11] investigated the dynamic of the interactions between ball-strings and racket in tennis. A suitable racket is the premise of playing tennis.

98.4 The Impact of Psychology During Tennis Education

98.4.1 Help Teenagers Plan Precisely with Clear Goals

T Goal orientations predict effort, enjoyment, and psychological skill [12]. Task orientation positively predicted improvement in training. Previous research has also shown that the goal positively predicted players' perceived improvement in handball. Research extends this work to tennis and indicates that task orientation is promoted if one wishes to enhance perceptions of improvement in training. So the coach should help the teenagers to plan precisely with clear goals to improve themselves. Clear goals specializes in the implementation of Integrated Marketing Management (IMM), and Interactive Marketing solutions that enables medium, and large marketing organizations to leverage the power of their information assets to achieve their marketing objectives and goals by continually reaching the right return on investment when it comes to their EMM infrastructure, Interactive Marketing, Multi-Channel Marketing, and related business processes. As a result, organizations can benefit from a faster and more efficient decision-making process, thus resulting in enhanced marketing performance, increased sales effectiveness, improved customer loyalty and ultimately, increasing the customer value to an organization.

98.4.2 Exploit the Impact of Parents

Parents exhibited many positive behaviors that facilitated development including various forms of support, emotionally intelligent discussions, and developing the child psychologically and socially through tennis [13]. Negative behaviors that inhibited development included being negative and critical, over pushing, over emphasizing winning and talent development over other domains of the child's life, and using controlling behaviors to reach tennis goals. Thus if the coach exploit the impact of parents fully, the result will be better. One of the biggest problems with children in today's society is youth apathy. Parental involvement in school can help solve this problem by emphasizing the importance of a good education, and getting their children excited about learning. "[F]or most children to succeed in school, their parents' interest in their learning is of paramount importance. But this interest ought to be with what happens on a daily basis, because this is how the child lives, and this is how he understands his life. The essential ingredient in most children's success in school is a positive relation to his parents". (Bettelheim 55) Parents' personal educational backgrounds and economic backgrounds have a significant effect on their children's education. However, if parents are a positive influence in their children's everyday lives, and most importantly in their everyday education, the future of our society will look brighter and brighter every day.

98.5 Conclusion

This paper introduced some matters that need attention in tennis teaching for teenagers. In the teaching, coaches should select an compatible method for teenagers, help teenagers to choose equips, and combine the education with their psychology to get better results.

Acknowledgments This research is supported by "the Fundamental Research Funds for the Central Universities".

References

1. Marcher R, Miguel C, Brendan L (2007) A Library on Regional Authority. J Sci Med Sport 10:1–10
2. Baker J, Cote J, Abernethy B (2003) Sport-specific practice and the development of expert decision-making in team ball sports. J Apply Sport Psycho 15:12–25
3. Abernethy B, Farrow D, Berry J (2003) Constraints and issues in the development of a general theory of expert perceptual motor performance: a critique of the deliberate practice frame work. Expert Perform Sport 3:49–69
4. Farrow D, Abernethy B (2002) Can anticipatory skills be learned through implicit video-based perceptual training. J Sports Sic 20(6):471–485
5. Masters R (1992) Knowledge, nerves and know-how: the role of explicit versus implicit knowledge in the breakdown of a complex motor skill under pressure. Br J Psycho 8(3):343–358
6. Liao C, Masters R (2001) Analogy learning: a means to implicit motor learning. J Sports Sic 19(3):07–19
7. Essence W (1988) Models of memory: information processing. Psychopharmacol Ser 6:3–11
8. Brecht J, Chang M, Price R (1995) Decreased balance performance in cowboy boots compared with tennis shoes. Arch Phys Med Rehab 76:940–946
9. Finlay E (1986) Footwear management in the elderly care programmed. Physiotherapy 72:172–178
10. Caslon F, Ruggeri G (2006) Dynamic analysis of the ball-racket impact in the game of tennis. Meccanica 6:67–73
11. Pepin K. van de P, Maria K (2011) Achievement goals and motivational responses in tennis: Does the context matter. Psycholo Sport Exerc 12:176–183
12. Ellen S, Retort E (2004) An is kinetic profile of trunk rotation strength in elite tennis players. Med Sic Sports Exec 36(11):1959–1963
13. Larry L, Daniel G, Nathan R (2010) Parental behaviors that affect junior tennis player development. Psychol Sport Exerc 11:487–496

Chapter 99
Research of Attitude of Chinese Learners on Culture Education in EFL Classroom

Dai Nalian

Abstract This paper analyses the attitude of Chinese learners towards the teaching of target culture in EFL classroom. It focuses on the attitude of the students at Xiaogan University, Hubei, China. With the help of a questionnaire which involves several aspects of target culture, we measured participants' responses. The questionnaire was composed of different sections concerning customs, beliefs, social organization, gestures, arts and notions of privacy of the target culture. Results revealed that learners have an obvious negative attitude towards the teaching of culture in EFL classroom. These responses are similar to those of other studies outside China and implications for the results are discussed.

Keywords Culture of target language · EFL classroom · Attitude · Chinese learners

99.1 Introduction

Attitude plays a vital role in the formation of our view of the world. It affects our perceptions of the people and things around us and determines how we respond to them. In view of the important role of attitude in our lives, it is not surprising that attitude studies have a long historical background. Researchers have defined and discussed it from different perspectives.

Attitude is usually defined along mentalist and behaviorist paradigm. Behaviorists describe it as a social and accordingly observable product, while Fasold [1] portrays the mentalist view of attitude as based on cognition. Hence it is supposed

D. Nalian (✉)
School of Foreign Languages, Xiaogan University, Xiaogan 432000, Hubei, China
e-mail: dainalin230@126.com

W. Du (ed.), *Informatics and Management Science VI*, Lecture Notes in Electrical Engineering 209, DOI: 10.1007/978-1-4471-4805-0_99,

to be analyzed on the basis of learners' reports about their mental attitudes. Mentalists usually define it as mental response to a given situation. People may interpret it at cognitive, ideational and experiential level [2] or discusses its model upon the basis of three factors: cognitive, affective and behavioral. Hence a dominant aspect of attitude is emotional response to particular topics. Attitude can also be described as abstract unit realized in the form of behavior. It not only predicts behavioral patterns, but also elicits its different manifestations, one of which is realized through culture [3].

Culture and attitude is mutually dependant. Paige et al. [4] define the former in terms of positive or negative. However, we should not simply delimit culture to attitude only as culture is composed of various aspects. Classifies culture as "forms of speech acts, notions of personal space, social organizations and proper gestures" while Kramsch [5] summarizes it as "membership in a discourse community". Each discussion of culture is in a sense correlated with certain aspect of attitude.

99.1.1 Definition of Culture

Previous theories view culture as a "static entity made up of accumulated, classifiable, observable, thus teachable and learnable facts" [4] while more recent ones see it as dynamic and ever changing. It is connected with beliefs and knowledge and defined as a set of attitudes based on our responses and beliefs. Ilter and Güzeller [6] use culture to refer to the belief system of a community and include knowledge and values in its definition. This view is supported by Kachru et al. [7] who defines culture as "socially acquired knowledge". Ariffin [8] believes that culture differs from country to country and community to community. The diverse nature of culture elicits various responses. Hofstede [9] divides the differences into attitudes at regional, national, generation, gender and social class levels. Robatjazi and Mohanlal [10] categorize it into Culture with capital C and culture with small c. This division complicates the definition.

99.1.2 Relationship Between Culture and Language Teaching

Attitude to language and the culture related to it is of vital importance in language teaching. Ammon [11] claims that the way learners respond to culture of the target language affects their attitude towards the language. Khuwaileh [12] holds the same view and portrays language classroom as various "cultural variables". It suggests that learning foreign language essentially involves learning its culture. Ilter and Güzeller [6] point out the positive effect of using culture on learners' cognitive attitude and claim that the teaching of culture in EFL instruction results in improved social attitude and positive view to the community of target language.

Ho [13] relates culture teaching with positive attitudes towards higher level of motivation for learning target language.

The teaching of culture in EFL classroom also elicits negative responses. Adaskou et al. [14] found that most of the teachers argue against the teaching of target culture in EFL classroom and assert that learners are more motivated when they compare target language with their own one. Fredricks [15] carried out research on an EFL class of Tajik students and maintains that learners are more likely to adopt positive attitude towards the target language if teaching material is nearer to their own culture. Shafaei and Nejati [16] holds that it is of vital importance to know learners' culture as it predicts their behavioral patterns of language learning and that any EFL teaching practice excluding learners' culture may be ineffective and exert negative influences on language learning. Ariza [17] points out that abandoning one's own culture is like forgetting oneself. Hence it becomes exceedingly important to use learners' own culture as a source to teaching EFL.

Other researches explored learners' response to the culture of the target language. Most of them advocate that the incorporation of target culture in EFL classroom should be minimized or abandoned. Abed and Smadi [18] studied Saudi students' response to target language culture and found the subjects' attitude to language learning more positive where "cultural loading" was avoided or at least kept to the minimum. Prodromou [19] found that Greek students' interest in target culture is merely based on the fallacy that its knowledge can be a source of securing their better grades in testing. In order to resolve the controversy concerning culture teaching in EFL classroom, Robatjazi and Mohanlal [10] propose that teachers take context into consideration and create an combination of target and local culture according to learners' demands. They emphasize teacher's role in creating a balance between the teaching of culture in EFL classroom and facilitating learners' communicative requirements.

All the above discussion lead us to conclude that learners show different attitudes towards the teaching of culture in EFL classroom. Thus this paper aims to investigate Chinese learners' response to the issue. The following questions are to be addressed: (1) what is Chinese learners' attitude towards the teaching of culture in EFL classroom? (2) Does Chinese learners' economic status influence their attitude towards the teaching of culture in EFL instruction? (3) Does Chinese learners' educational background affect their attitude towards the teaching of culture in EFL classroom? (4) Do Chinese learners' learning objectives affect their attitude towards the teaching of culture in EFL instruction?

99.2 Methodology

This is a case study of Xiaogan University. Case study is an intensive study of a single case which is recommended by as a valid tool for educational research. The reason for choosing Xiaogan University is access and convenience.

We used representative sampling which aims to choose a target population sample and generalizes the results of the research [20]. The population selected is experimentally accessible as it consists of 354 Chinese EFL learners at Xiaogan University who learn English language as a compulsory course at undergraduate level. Since it is impossible to include the responses of all the EFL learners from Xiaogan University, we have taken a representative sample from three different schools where English is taught as a four credit course. Of the 354 participants, 53 % are males and 47 % are females. All respondents have finished three semesters of their degrees. The three schools were chosen randomly and for the sake of convenience.

In order to find the specific answers to the four research questions, a questionnaire has been designed based on Cohen et al. [21]. It is composed of two parts. One focuses on participants' personal information and the other aims to elicit their responses to the four research questions. We used a number of question types such as rank ordering and multiple choices to trigger participants' personal information. Family income and education variables are divided into five multiple elements. Participants may choose one of them to mark the group to which they belong. As for the rank ordering, subjects are urged to rank seven learning objectives according to their preference.

The second part is divided into various aspects concerning learners' responses to cultural customs, beliefs, speech acts, social organizations, gestures, arts and notions of privacy. Each of the aspects is arranged in sub-categories and supported by at least two or more questions. Every question is an attitude statement which refers to learners' attitude towards an aspect of culture. Choices are divided into anchor statements devised on the basis of 5 Likert scales and offer a range of responses from strongly agree, agree, disagree, strongly disagree to neutral. The pretesting of the questionnaire had been conducted in the previous week of March, 2011. The questionnaire was then modified and certain changes were made based on the results. Its reliability coefficient was 0.85. The modified version was used to collect the data needed for the research.

The results of the questionnaire were analyzed by means of SPSS 17.0. We studied the correlation between learners' responses to the teaching of target culture and variables such as educational experiences and family income. The relationship between these variables may be inverse or direct. In addition, the effect of learning objectives on learners' attitude towards the teaching of target culture is also evaluated by cross tabulating learning objectives with learners' responses. Cross tabulation is described as a statistical tool of measuring correlation between sets of scores which also indicates the strength of correlation between factors. These two tools enabled us to interpret the data in a quantitative way and also added to the validity of the findings.

99.3 Data Analysis and Discussion

Descriptive statistics revealed that 52 % learners consider acquisition of higher education as the top ranking objective for learning English. As higher education seems to be the most important objective, we cross tabulated this objective with learners' responses. Results indicated that 84 % of the respondents have a negative attitude towards the teaching of target culture in EFL classroom. It is interesting that one objective for learning English language is the desire to know western culture, especially American one. Accordingly this is the objective of lowest ranking with 59 % of the participants assigning it the last.

Data analysis also shows that 65 % of the participants fall within the family income group of five thousand RMB per month. Hence the majority of the participants belong to low and lower middle class. It can be seen that there is a strong correlation between family income and learners' attitude towards the teaching of target culture. Respondents belonging to the low income group have an obviously positive attitude towards the teaching of target culture. But as family income increases, respondents tend to show negative attitude. Despite of that, the situation is not that simple or clear–cut. Respondents who adopted positive attitude in general deem some aspects of target culture negatively and those who have an overall negative attitude have no objections when certain aspects of culture are taught in EFL classroom. Hence as for the teaching of beliefs and values, 81 % of the participants responded positively. This result is somewhat surprising in the context of the present situation in China where lack of tolerance for conflicting viewpoints or values is often regarded as a sign of extremism. Besides, 79 % responded positively to the teaching of social organization of target culture in EFL classroom. It can be seen that apart from these two aspects of culture, respondents revealed strongly negative attitude to the teaching of target culture, which is similar to overall reaction of other countries against culture-loaded teaching material. For example, Abed and Smadi's [18] research arrived at the same conclusion.

In our study, the correlation is also strong between respondents' educational background and their attitude towards the teaching of target culture in EFL classroom. 65 % of the participants used to study at ordinary high schools while 35 % in key ones. Correlation analysis indicated that most our population adopted negative attitude towards the teaching of target culture and that learners respond positively only to the teaching of beliefs and social organization of target culture. They respond negatively if exposed to any of the other cultural aspects. Thus we may conclude that income and education influence learners' attitudes and those ordinary high schools do not emphasize the promotion of intercultural tolerance and critical thinking. So their students may not be expected to have positive views of the learning of target culture. Hence learners' educational background also contributes to their attitude towards target culture in EFL classroom.

99.4 Conclusion and Implications

The above discussion reveals that our participants adopted predominantly negative attitude to the teaching of target culture in EFL classroom. Both education background and family income contributed to the result. Besides, participants' learning objectives also play a significant role in their attitude.

The findings may be of great help for policy makers and course designers since it is essential for them to take learners' responses into consideration when designing curriculum and prescribing syllabi for students. Textbook writers are supposed to be aware of learners' attitudes towards different aspects of culture in EFL classroom. Ariza's [17] research focused on the same issue and proved the importance of learners' attitudes while making educational policies.

Another important aspect is that learners' negative attitude towards the teaching of target culture may result in the rejection of the language itself. So policy makers have to decide whether to insist on teaching culture at the cost of language or to revise and reanalyze teaching approach. Those who argue in favor of the combination theory of culture teaching supported by Ariffin [8] should also reconsider their position. As if both target and local culture are incorporated into EFL classroom, what will be the criteria for selection, combination and exclusion or which aspects of the target culture will prove to be acceptable to learners? Although our findings reveal that some aspects of target culture are acceptable in EFL classroom, the overall response of learners is so negative that the few tolerated elements had better be avoided or kept to the minimum. Ariffin [8] proposed that target culture should be taught to language learners who intend to visit the target language country some time in future. Our data indicates that respondents ranked this learning objective as the second lowest in the list of rankings. Hence we suggest that language classes for those learners should be arranged separately. For most of the learners, target language should be taught in the local cultural context which will lead to their positive response and active act to the language and learning act itself.

References

1. Fasold R (1985) The sociolinguistics of society. Wiley 45(44):335–345
2. Halliday M (2005) On grammar. Continuum International Publishing Group, New York 9(4):94–103
3. Speilberger CD (ed) (2004) Encyclopedia of applied psychology. Academic, London 42(15):95–106
4. Paige RM et al (2003) Culture learning in language education: a review of the literature. In: Paige DL (ed) Culture as the core: perspectives on culture in second language learning. IAP 74(32):75–89
5. Kramsch CJ (2008) Language and culture (8th edn). In: Widdowson HG ed Oxford University Press, Oxford 47(42):844–853

6. Ilter BG, Güzeller CO (2005) Cultural problems of Turkish students while learning English as a foreign language. Modern Lang J 89(2):456–461
7. Kachru BB, Kachru Y, Nelson C (2009) The handbook of world Englishes. Wiley 8(4):93–103
8. Ariffin S (2006) Culture in EFL teaching: issues and solutions. TESL Working Paper Series 4(1):75–78
9. Hofstede G (1997) Cultures and organizations: softwares of the mind. McGraw Hill, New York 34(5):82–95
10. Robatjazi MA, Mohanlal S (2007) Culture in second and foreign language teaching. Language in India 7(2):59–65
11. Ammon U (2004) Sociolinguistics: an international handbook of the science of language and society (2nd edn). Walter de Gruyter, Berlin 85(6):77–84
12. Khuwaileh AA (2000) Cultural barrier of language teaching: a case study of classroom cultural obstacles. Comput Assist Lang Learn 13(3):281–290
13. Ho M-C (1998) Culture studies and motivation in foreign and second language learning in Taiwan. Lang, Cult Curriculum 11(2):86–91
14. Nation ISP, Macalister J (2009) Language curriculum design. Taylor and Francis 64(43):965–1013
15. Fredricks L (2007) A rationale for critical pedagogy in EFL: the case of Tajikistan. Read Matrix 7(2):22–28
16. Shafaei A, Nejati M (2008) Global practices of language teaching. In: Proceedings of the 2008 international online language conference. Universal-Publishers 36(31):73–84
17. Ariza D (2007) Culture in the EFL classroom in Universidad de la Selle: An innovation project. Actualidades Padagogicas 15(2):9–17
18. Abed F, Smadi O (1996) Spread of English and westernization in Saudi Arabia. World Englishes 15(3):307–315
19. McKay SL (2004) Western culture and the teaching of English as an international language. English Teaching Forum 65(41):10–15
20. Perry FL (2008) Research in applied linguistics, becoming a discerning consumer. Taylor and Francis 24(13):35–42
21. Cohen L et al (2007) Research methods in education. Routledge, London 93(33):725–738

Chapter 100
Study on Sports Training Method Based on 3D Structure

Minggang Yang

Abstract The time structure, content structure and control structure during sports training are interrelated and interacted on each other, which constitutes the Three-dimensional structure of sports training method, and provide theoretical basis and macro frame for core questions of the training method, such as dividing of training stages, the deploying of training contents and the monitoring training method, and also provide the important theoretical guidance and practical basis for practical work of sports training at the same time.

Keywords Sports training · 3D structure · Method

100.1 Preface

Sports training method is a dynamic and complex system structure by a multidimensional structure system, do a lot of difficulties, theoretical research and practical application training quality monitoring system. It is very important to understand the basic rules of the, promote the method of sports training the scientific level of modern sports training structure system; fully understand the method of sports training.

M. Yang (✉)
Shandong University of Technology, Zibo, Shandong, China
e-mail: yangminggang@guigu.org

W. Du (ed.), *Informatics and Management Science VI*, Lecture Notes in Electrical Engineering 209, DOI: 10.1007/978-1-4471-4805-0_100,

100.2 The Concepts and Elements of Training Method

100.2.1 The Analysis of the Concepts of the Sports Training Method

Nowadays, there are many analyses to "sports training method": Wan showed that the sports training method is an educational method that the athletes, according to the requirements of the specific competition, improve their specific performance and full development of individual character in the guidance of coach, and also with the cooperation of Scientific researcher, the medical personnel [1]. While, Dachao thinks "sports training method" has two meanings, one is narrowed meaning, and the other is generalized meaning. From the narrowed meaning, the sports training method is every training sessions that athletes who are the subject of the sports training participate in the guidance of coach or the accumulation of this method. And from the generalized meaning, the sports training method is the whole training Period that athletes engages in practice, includes the duration of training session and all the time outside it [2]. The book "Sports Training Study—Physical Education Colleges Correspondence Course" notices that sports training method is a method that reflects the physical education and sports practice for sports talent. Yihai thinks, sports training method is a method that the athletes improve their contests abilities and the special sports performance, as a result of accepting the systematical training [3].

100.2.2 The Analysis of the Factors of Sports Training Method

From the concept of different sports training method, we can find that one of the movement of athletes training methods; improve the process of competition athletes and sports ability appearance. This program has two basic elements: the time element and space elements. The time factor is inevitable method as development of everything, such as tarring, development and end, for example, a player just started specific training; Then, the improvement of the specific performance; Finally, the end of the training career. However, just like the space structure of the element content space necessary for the development of each object, such as: at every stage of the athlete's training methods, including not only entity-sports training activities, and other space training management, training content-design (plans), training monitoring, research activities, and so on. A complete training method is a complex space and time, all kinds of space development content elements with the change of time.

100.3 The Three-Dimensional Structure of Modern Sports Training Method

The so-called "structure" is refers to the mutual contact way or combination of every part of the link [4]. Therefore, "sports training method of structure" reflects the overall; the stable combination and eternal connected the various components in the whole training method. Sports training methods have two basic elements-time and space. In addition, these two elements don't change and natural development, but also by human factors, they are direct and control of human beings. So it lead to structure sports training method obviously has three main structural features of the structure, content structure and control structure [5].

100.3.1 The Time Structure and its Composition of Sports Training Method

Time structure refers to the structure of sports training methods must through different stages in the whole training athletes training progress, different period have different training objectives and tasks, the sport ability also have different. Time time-tier structure reflects the composition of sports training method [6].

The terms of the basic method, things, an excellent athlete complete training methods should include at least five basic stage (stage): the basic initiated training cycle; Primary specific training period, improve specific training period; Has the advanced specific training period; Sports life maintain period (Fig. 100.1). For example, a good athlete, contact with the physical exercise, he or she is young, and then add in the junior team, and youth team, the next international (professional team), finally retire, like this [7, 8].

In addition, each training phase is by years of training, they can also be divided into several levels of training modal verbs, such as years of training, circulating training, stage training, week training, a day of training, training courses, and so on.

100.3.2 The Content Structure and its Composition of Sports Training Method

A content of sports training method is structure refers to various elements and their relationship between athletes in training method. This reflects the space sports training method.

Summarizes the theoretical circle sports training is the training content as follows: the body form, physiological function, the sports quality, health quality, athletic technology, sports strategy, sports psychology, sports ability of intelligence

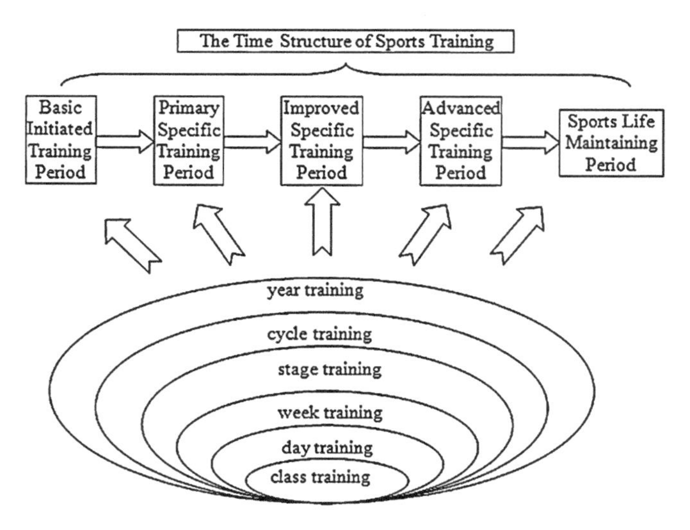

Fig. 100.1 The time structure of sports training

and so on. At the same time, some scholars summarize the training content into the eight physical stamina training and sports skill training (Fig. 100.2). The content of the sports training method the structure of two contents constitute a system and eight factors influence each other and interaction of content. In the influence factors of physical fitness, body means the external form, athletes, such as height, weight, limb proportion, such as the ratio of the perimeter. Physiological function means the athletes' physical function, including energy system, muscle system, cardiopulmonary system and the nervous system. The sports quality is including strength, speed, endurance, flexibility and agility. The main points of the health quality athlete's ability the ability of adapting to climate change and resist disease and injury methods [9].

In all sorts of elements of the sports skills is training, athletic technology including technology, technical details and technical basis [10]. Athletic strategy is including strategy, tactics, concept, tactical awareness and tactical principles, tactical form and tactical operations. Mainly includes the spirit of sports ability of the psychological characteristics of the personality and psychological method characteristics and athlete's management system, etc. Intelligence campaign is mainly refers to the athletes observation, the memory, thinking ability, attention and imagination.

So the sports training method is a kind of method that training, athletes targeted system with two aspects and eight aspects above character and actual needs of the various sports, in order to improve the performance of the standard competition and sports.

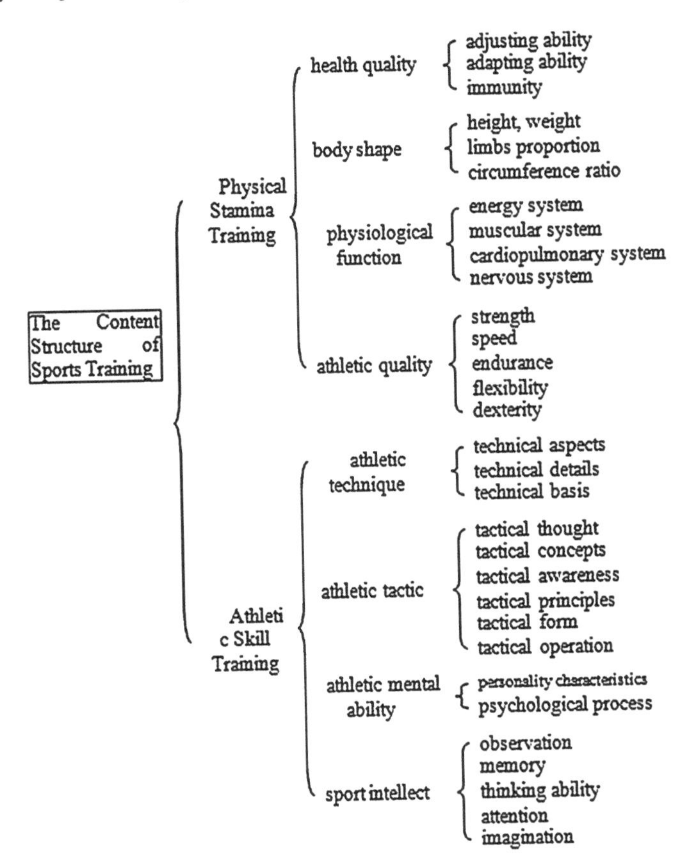

Fig. 100.2 The content structure of sports training

100.3.3 The Control Structure and its Composition of Sports Training Method

Control of all kinds of sports training structure control method is related to the task of training personnel, of inherent characteristics, the elements of there. The understanding of the training a competitive goal is by six basic control part as follows: diagnosis initial condition, establish professional training target, design training plan and implement training plan, supervise and evaluate training, achieving the training goal. And the parts control system consists of the training method. It reflects the space–time structure of comprehensive sports training method.

Control structure reflects three function systems in sports training methods, planning system, implement system, monitoring system. And planning system's main function is to: diagnosis the initial state, different stage, sure series goals, planning project training methods, etc. And implementation of the system of main

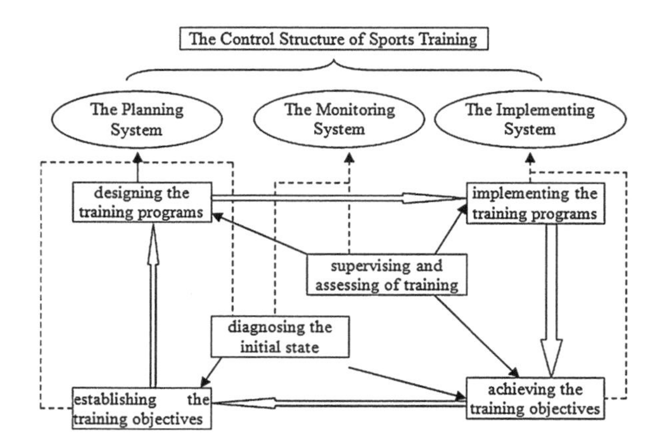

Fig. 100.3 The control structure of sports training

functions are: put forward the planning stage of training, and set up the hardware's training ground, implement concrete Labor different coach, implement the different stages of the specific objectives, etc. And monitoring system's main function is to: analysis of the initial training, training plan of the inspection and supervision of the implementation of the training course, sport ability of the evaluation of training effect (Fig. 100.3).

100.4 Conclusion

Through the analysis of the concept of sports and form a part of the training process, that the "three dimensional structure of modern sports training process is composed of the time system structure, content structure and control structure, each man has content system and internal relations".

References

1. Wan J (2000) Comment of the definition and the method of sports training. J Qingdao Edu College 6(3):76–87
2. Dachao Z (2007) The basic theoretical system of monitoring of sports training method. J Wuhan Inst Phys 12(11):367–381
3. Meng M, Zhang DC (2000) Sports training study—physical education college correspondence course, vol 8(2). People's Sports Publishing House, Beijing, pp 287–293
4. Dong LL (1999) Sports training—sports colleges general teaching, vol 203. People's Sports Publishing House, Beijing, pp 1010–1016
5. Yihai H (2005) The training theory and method of athletic sports, vol 45(32). Hubei Province People's Publishing House, Wuhan, pp 673–684
6. Yihai H (2005) Competitive sport training theory and method, vol 34(23). Hubei People Press, pp 422–435

7. Wan J (2000) Discuss on the definition and constitution of sport training process. J Qingdao Inst Edu 6(2):42–53
8. Dachao Z (2007) Establishment of basic theory system of sports training process monitoring. J Wuhan Inst Phys Edu 12(5):38–46
9. Dachao Z (2007) The basic theoretical system of monitoring of sports training process. J Wuhan Inst Phys 12(3):83–102
10. Yihai H (2005) The training theory and method of athletic sports, vol 15(4). Hubei Province People's Publishing House, pp 93–110

Chapter 101
Study on Reform of Traditional Chinese Opera

Ke Liu

Abstract Traditional Chinese opera has been in a stagnant state in the modern times, changing into a great a matter of debate among Chinese people. In this paper, the reasons, why it divorces from mass people, loses the market, and falls into a dilemma, are analyzed starting from traditional Chinese opera itself, namely whether the objective laws of its development are ignored by people in multiple years of reform. Therefore, the author calls on relevant parties to draw up a conclusion on historical experience with a serious attitude, pay high attention to the objective development laws of traditional Chinese opera, and make an adjustment to thinking ways. As a result, the development of the prosperity of traditional Chinese opera can be really promoted in the future.

Keywords Tradition · Modern · Development

101.1 Introduction

Traditional Chinese opera is known as one of the world's three major drama schools [1, 2]. It is an outstanding traditional culture of Chinese nation.

Therefore, it is irresponsible if Chinese people see such a resplendent traditional culture to wither away gradually without taking any measures, or only think superficially of its recession.

Therefore, how should Chinese people do? In this paper, it is believed by the author that it is necessary for us to think from a different perspective and find out a lesson from the experience in decades of development.

K. Liu (✉)
Conservatory of Music, Hunan City University, Yiyang, Hunan, China
e-mail: liu203bke@126.com

W. Du (ed.), *Informatics and Management Science VI*, Lecture Notes in Electrical Engineering 209, DOI: 10.1007/978-1-4471-4805-0_101,

101.2 Aesthetic Characteristics of Traditional Chinese Opera Art

Wang Guowei, master of Chinese culture, said in his book Study on the Source of Traditional Chinese Opera that opera was performing stories with songs and dances [3–5]. From the perspective of art, western drama is a "reappearance art", and Traditional Chinese opera is a "performance art".

In traditional Chinese opera, importance is attached to a high level of performance and an abstract expression [6], because it has a close, natural connection with Zhuangzi philosophy and Confucianism.

Since the local classical opera emerging in South China, traditional Chinese opera has owned a history of 800 years. The literature plays in the early times, such as the poetic dramas set to music during the Jin and Yuan dynasties, were with an open structure though only one person comprised of the lead singer, which was known as "one theme went through the whole performance".

In general, there are two climaxes in traditional Chinese opera: one is an event climax, and the other one is an emotional climax [7–9]. However, in fact emotional climax is superior to event climax in dramatics, such as "Baoyu cries before the dead" in A Dream in Red Mansions, and "Tell revolutionary family stories dearly" in the Legend of the Red Lantern.

Besides, traditional Chinese opera was influenced by the "Grand Unified Theory". For this reason, traditional Chinese opera works themed with anything were designed with a very good ending in them even if they were oriented at tragedies.

For example, Number One Scholar Zhang Xi was supposed to be a tragedy, but the couple in it was finally reunited though passing through all the obstacles. The complexes, such as "one good turn deserves another", "wickedness does not go altogether unrequited", and "the chickens have come home to roost", have universally existed in the minds of Chinese people for a long time.

Overall, the performances in traditional Chinese opera are virtualized and stylized regularly, such as "three or five persons are equivalent to a large number of mounted and foot soldiers and six or seven step are like all corners of the country", "mountains and clear water seem to really exist even though seen from a remote place", and "circumstances change with the passage of time".

Like performances, music in traditional Chinese opera is stylized. In terms of the structure and pattern of common systematic tunes, "the names of the tunes to which Qu are composed" and "the names of the tunes to which Qu are composed of plate change" are two major structures in the music of traditional Chinese opera. The tunes in these two structures are also stylized as a whole, which are helpful for the use of clusters and the inheritance of teaching.

101.3 Conditions of Traditional Chinese Opera Before and After Reform

Reform of traditional Chinese opera has been not a new subject at present. It goes through the whole process of the formation and development of the opera art in a manner of speaking.

In the early 1950s, high importance was attached to the reflection of the modern life in Chinese opera. This made actors feel very embarrassed, because character events of historic and costume opera were still very far away from them.

In traditional Chinese opera, whether male characters with a painted face or long mustache or else, all were exaggerated artistic images which were thought indifferently by the audiences, making actors give performance with high proficiency.

However, character events of the modern life commonly happen right away by us. Therefore, it is necessary for Chinese opera performances to bring a realistic sense to people.

Moreover, there is a combination between Chinese opera and Stanislavsky's System. This feature helps Chinese opera find out a very excellent way for giving reflection to the modern life.

Through the unremitting efforts made by a great number of opera workers, the artistic characteristics of traditional Chinese opera were formed in the historical period, namely legendary dramatic literatures, authentic stage environments, rich songs and dances performed by actors, and symphonic music accompaniments.

The reform of traditional Chinese opera in common systematic tunes is in fact very successful. The principle of the stylized single frame of Chinese opera music is strictly followed by reformers.

Concerning about the reform of common systematic tunes of traditional Chinese opera, Mei Lanfang once said: "Changing the shape instead of changing the essence". Cheng Yanqiu had also said: "changing short interludes with systematic tunes unchanged". In 1964, during the period of the national Peking Opera Festival, it was consistently agreed by many people that Beijing opera was always necessary to attach importance to its local characteristics, namely the attracting audiences and expressing performances with systematic tunes.

From 1980, another drama school, namely Brecht de-familiarization, actively spread to the dramatic community through school of drama, or all sorts of drama lectures. This artistic technique was applied widely by many people at that time. At this point, desalinating contradiction was very popular in drama literatures; the majorities of the settings at stages were oriented at pattern symbols; performances were completely virtualized, etc.

Drama art systematization is an expression given to the mature development of dramatic culture. Modern people can't make use of excuses to carry out evaluations on any drama schools.

Whether Stanislavsky, Brecht or Mei Lanfang, all of them are not only masters of art, but also gems of the world culture.

However, absorption and introduction are a combination of different cultures.

Therefore, it is necessary to carry out in-depth studies on two sides, for the ultimate purpose of finding out basic elements, which can be combined together.

However, due to the influence of various thoughts, especially the influence from the pop music, a great number of reformers in Chinese opera wrongly introduce pop music as the only modern music technique.

For example, jazz drum and electronic organ are introduced by some theatres. Music from keyboard instruments is with equal temperament, but folk music is just of intonation. Therefore, how such a combination is made by them?

The introduction of jazz has completely broken the most fundamental rhythms of opera music. Obviously, this is a blind introduction based on an insufficient analysis of two music systems. From that time, the reform in common systematic tunes of many varieties of Chinese operas became very aggressive.

In the music of a new opera, almost all traditional systematic tunes were removed by the composer, and only a few of tones were selected for developing motivated musical phrases; an opera was only performed exclusively by only one actor. Such a kind of opera is in shortage of the traditional opera characteristics, is difficult to sign fluently, and also is with long and multifarious singing tunes. Therefore, it is hard to be popular among actors, not to mention being recognized by audiences.

In recent years, modernism is discussed frequently by a large number of people. In fact, modern times are supposed to make traditions continued, but not ended, even though the new contents can't be really reflected by the traditional forms to some extent.

The "XiPi allegro" in the Sect. 101.4 (Li indignantly denounces Wang) of the Legend of the Red Lantern was actually the "XiPi allegro" Zhu Geliang sang in Miss JieTing, which is a traditional opera. Many famous opera works source from the reform of traditions, just like "achieving the great with doing little".

In the reform of opera music, it is highly necessary to give consideration to the establishment of a sound tunes system for opera, aiming to create new primary singing tunes and increase new traditions.

For example, "Henan wooden clappers" were introduced by Mei Lanfang to Chinese opera; hand accompaniment with XiPi chord became a new traditional tune (Southern wooden clappers).

If the tunes are incapable of attracting audiences and making opera passed on, the result will certainly be that audiences will feel stranger and stranger to Chinese opera, market will go to deeper and deeper depression until vanished.

101.4 Discussion on the Popularization and Enhancement of Chinese Opera

About the popularization and enhancement of Chinese opera, MAO Zedong used to make a systematic discussion in the Speech of Literature and Art Symposium in Yan'an 69 years ago.

In fact there is a problem about the popularization and enhancement of art at any time. The enhancement on the basis of popularization and achieving popularization under the guidance of enhancement are a dialectical relationship between the popularization and enhancement of Chinese opera. However, it should always be remembered that popularization of Chinese opera is the foundation, and the purpose of the enhancement is for wide popularization at any time.

Not all operas are ignored by people in popularization. There are still a good many operas popularized by Chinese people, such as Beijing Opera, Henan opera, Huangmei opera and Shaoxing opera.

In the reform of common systematic tunes of Chinese operas, it is necessary to not only start from contents, and also abide by the laws of opera music without losing the styles of traditional Chinese operas.

In the Speech of Literature and Art Symposium in Yan'an, MAO Zedong pointed out: "improvement can be realized only based on mass people".

With the purpose of promote Chinese opera, all kinds of activities such as art festival and opera festival are held by the state, provinces and even municipalities in many years. The subjective desire is good, and also there are some fine works to emerge.

However, on the whole, these activities seem to play no important role in popularizing local operas, cultivating audiences and wining market share. These programs, from hundreds to millions, have entered a great number of luxurious theaters, and seem to be elegant, majestic and outstanding. In some joint performances, actors were even invited from anywhere.

For example, an opera troupe somewhere, in order to participate in a provincial joint performance, only bought script from others, and also employed director, composer, dancing personnel and major actors with a lot of money, but finally acquired only a third prize, which was awarded to all troupes.

101.5 Conclusion

The author thinks that whether local operas are qualified to apply for intangible cultural heritage should be not only based on the current prevalence situation and audiences. If importance is attached only to this point, why Kunqu opera has been listed by United Nations Educational, Scientific and Cultural Organization (UNESCO) as the world's intangible cultural heritage to be protected though it has a small number of audiences and also has been popularized widely?

The policy of Chinese central government for intangible cultural heritage has been specific, and the protection is not equivalent to museum, but needs to be "dynamic". Generally, application is only a method, but the purpose should be inheritance.

Acknowledgments This paper is supported by the General Project of Department of Education of Hunan Province 2010 (No. 10C0520).

References

1. Liu W et al (1984) Theory of artistic characteristics, vol 8(4). Culture and Arts Publishing House, pp 399–407
2. Zhang L (1984) Study on Bertolt Brecht. China Social Sciences Press, Beijing
3. Ouyang Y (1984) Collection of Ouyang Yuqian drama papers. Shanghai art and literature press, Shanghai
4. Zhang Z (1982) Aristotle poetics, vol 23(4). The People's Literature Publishing House, pp 289–297
5. Zhai W (2010) Know all Chinese learning knowledge. China overseas Chinese press, China
6. Zhou Y (1995) Chinese opera culture, vol 3. China Friendship Publishing Company, pp 370–376
7. Liu K (1970) Score of the legend of the red lantern of modern revolutionary peking opera, vol 38. The People's Publishing House, pp 802–808
8. Wang WW (1974) Music score of the theme of cuckoo mountains of modern revolutionary Opera, vol 20(2). The People's Music Publishing House, pp 378–385
9. Zhaolong T (2002) Talk at random, without reference to reality. Arts Sea 29(4):77–85

Chapter 102
A Comprehensive Evaluation of the Academic Influence of Chinese Core Journals in Dramatic Art

Rui Zou

Abstract Based on the statistical data of "Chinese Journal Citation Reports-Expanded (CJCRE)", the basic characteristics and general rules of Chinese core journals in dramatic art are discussed by the method of principal components analysis (PCA). In the PCA framework, eight important and efficient evaluating indictors are selected, i.e. total cites, impact factor, cited rate, diffusion index, subject impacting index, subject diffusion index, rate of funded papers and H index. This comprehensive evaluation method can overcome the shortages of instability and unreliability, when PCA is used based on the statistical data in only one year. Meanwhile, the corresponding comprehensive evaluation results can not only provide a guide for the scientific researchers, but also improve the journal quality.

Keywords Core journal · Dramatic art · Comprehensive evaluation · Principal components analysis

102.1 Introduction

Academic journals are important carriers of academic papers and scientific research achievements publishing, which reflect the national development level of science and technology, and play an important role in social progress. The drama art periodicals of our country is developing rapidly, but compared to the number, the quality is bad and has a substantial gap with the international journal ones [1]. Therefore, how to appraise the academic level and influence of the drama art

R. Zou (✉)
Academy of Art, Qingdao Technological University, Qingdao 266033, China
e-mail: lwsunlight@gmail.com

W. Du (ed.), *Informatics and Management Science VI*, Lecture Notes in Electrical Engineering 209, DOI: 10.1007/978-1-4471-4805-0_102,

journal correctly and promote the improvement of the quality of journals is a very important research subject. Currently, most domestic agencies are lack of fair and reasonable evaluation system. Generally, they choose the influence factor as the index of offprint quality and performance evaluation, which is clearly biased. So, it is rather necessary to introduce the idea of comprehensive evaluation to evaluate the Chinese journals in dramatic art. At present, domestic scholars have used factor analysis and principal component analysis methods to comprehensively evaluate the management and agricultural scientific journals, and put forward practical methods and suggestions to related units [2, 3].

Aiming at solving the lagging behind problem of comprehensive evaluation research of drama art core journal, this paper choose the principal component analysis method and select a number of measurement indicators, and then study the academic quality level and the development rules of Chinese core journals in drama art from 2008 to 2011, on the basis of statistical data from Chinese journal citation reports-expanded. Then, the comprehensive results fully illustrate the developing quality level and discipline influence of different academic journals in the same field.

102.2 The Research Process and Comprehensive Evaluation Model

102.2.1 Select the Statistical Analysis of Indexes

In this paper, my objective is to evaluate the developing quality of the Chinese core journals in dramatic art from 2008 to 2011. Then nine kinds of journals are selected, i.e. (1) Drama; (2) Theatre Art; (3) Chinese Theatre; (4) Hundred Schools in Art; (5) Chinese Theatre Arts; (6) Dramatic Literature; (7) Contemporary Drama; (8) Jingju of China and (9) Sichuan Theatre [4]. In order to improve the precision of evaluation results, eight positive indexes are also chosen according to the statistical data from the Chinese journal citation reports-expanded, for instance, total cites (TC), impact factor (IF), cited rate (CR), diffusion index (DI), subject impacting index (SI), subject diffusion index (SD), rate of funded papers (RF) and H index. Furthermore, the following comprehensive evaluation model is calculated based on the aforementioned statistical data.

102.2.2 Construction of the Comprehensive Evaluation Model

In our paper, both the principal component analysis and dimensionality reduction ideas are introduced to translate the multiple indexes into a fewer indexes. Furthermore, we use multivariate statistical analysis to solve the problem well. In this way, the numbers of measurable indexes are reduced effectively, and the academic standards and status of the various journals in different research areas are evaluated objectively [5–8]. Table 102.1, there are obvious differences in the number of specific indexes, so that would make big dispersion of each variables' value. At the same time, the total variances of the variables are controlled by the variables with larger variance. If the principal component analysis is carried on based on the covariance matrix of the original statistics, it will give priority to the variable with larger variance, and not only make the Interpretation of principal component variables difficult, and sometimes result in an unreasonable result. Therefore, before the start of principal component analysis, we should make the original statistics normalized. Established in this paper there are m samples, each sample has n indicator data, denoted by

$$X = \begin{bmatrix} x_{11} & \cdots & x_{1n} \\ \vdots & \ddots & \vdots \\ x_{m1} & \cdots & x_{mn} \end{bmatrix} = (x_1, x_2, \ldots, x_m), \tag{102.1}$$

The steps of constructing the model of the comprehensive evaluation with principal component analysis method as follows [9, 10]:

Normalized Data. The selected indicators in this article are positive type, and then the normalized formulas are followed:

$$\bar{x}_{i,j} = \frac{x_{i,j} - \min_i\{x_{i,j}\}}{\max_i\{x_{i,j}\} - \min_i\{x_{i,j}\}}, (i = 1, 2, \ldots, m, j = 1, 2, \ldots, n) \tag{102.2}$$

where the standardization matrix $\bar{X}$ is represented by

Table 102.1 The original statistical data and its comprehensive evaluation results for the nine kinds of Chinese core journals in dramatic art in 2011

Journals	(1)	(2)	(3)	(4)	(5)	(6)	(7)	(8)	(9)
TC	89	159	130	589	112	118	45	40	147
IF	0.113	0.153	0.031	0.305	0.048	0.000	0.045	0.000	0.092
CR	0.96	0.97	0.98	0.92	0.93	0.94	1.00	1.00	0.91
DI	0.000	0.023	0.000	0.080	0.010	0.018	0.045	0.002	0.021
SI	0.16	0.16	0.2	0.43	0.21	0.17	0.09	0.07	0.23
SD	3.32	1.68	1.18	4.70	1.00	1.64	0.57	0.32	1.34
RF	0.122	0.151	0.014	0.521	0.250	0.036	0.159	0.000	0.159
H Index	3	3	8	4	2	7	8	3	6
Factor 1	0.646	0.641	0.118	3.157	0.601	0.425	0.039	−0.558	0.986
Factor 2	−0.347	−0.365	−0.891	0.139	0.153	−0.570	−0.834	−0.647	−0.161
Factor 3	−0.122	−0.140	0.594	0.151	−0.070	0.622	0.597	−0.190	0.564
Factor 4	0.125	0.218	0.038	0.275	0.142	0.06	0.406	0.153	0.086
Score	0.302	0.353	−0.141	3.722	0.826	0.538	0.208	−1.242	1.475
Rank	6	5	8	1	3	4	7	9	2

$$\bar{X} = \begin{bmatrix} \bar{x}_{11} & \cdots & \bar{x}_{1n} \\ \vdots & \ddots & \vdots \\ \bar{x}_{m1} & \cdots & \bar{x}_{mn} \end{bmatrix}. \tag{102.3}$$

Calculating Correlation Coefficient Matrix. In order to eliminate the impact of the big differences in every variances of original variable, we usually make principal component analysis based on the correlation matrix R of the normalized data.

$$R = \left[r_{ij}\right]_{m\times n} = \left[s_{ij}/\sqrt{s_{ii}s_{jj}}\right]_{m\times n} \tag{102.4}$$

And,

$$\begin{cases} s_{jk} = \dfrac{1}{m-1}\sum_{i=1}^{m}\left(\bar{x}_{ij} - \tilde{x}_j\right)\left(\bar{x}_{ik} - \tilde{x}_k\right) \\ \tilde{x}_j = \dfrac{1}{m}\sum_{i=1}^{m}\bar{x}_{ij}, (j,k = 1,2,\ldots,n) \end{cases} \tag{102.5}$$

Calculation Eigenvalues and Eigenvectors. After getting the correlation coefficient matrix R in Step-Calculating Correlation Coefficient Matrix, we calculate the eigenvalues λ_i and eigenvectors $\alpha_i(i = 1,2,\ldots,n)$ of the correlation coefficient matrix R by MATLAB's function svd.m.

Determining the Number of Main Component. Based on the eigenvalues λ_i of the correlation coefficient matrix R computed in Step-Calculation Eigenvalues and Eigenvectors, we could obtain the contribution rate of the kth principal component which is $\beta_k = \lambda_k/\sum_{i=1}^{n}\lambda_i$, then the cumulative contribution rate of the first pth principal components can be calculated as $\sum_{i=1}^{p}\lambda_i/\sum_{i=1}^{n}\lambda_i$. As the cumulative contribution rate is greater than 85 %, we select the first pth principal components instead of the original indictors $n(p<n)$ to achieve the purpose of dimensionality reduction.

Computing Principal Component Loading The factor loading is $\eta_i = \sqrt{\lambda_i}\alpha_i$, and the score of each factor is $F_i = \eta_i\bar{x}_i$.

Computing Principal Component Score According to the factor score F_i and the size of the contribution rate, we calculate the composite scores that are as follows $F = \beta_1F_1 + \beta_2F_2 + \cdots + \beta_pF_p$, and sort the statistical sample based on the composite scores.

102.2.3 Computational Results and Analysis

In our paper, the aforementioned principal component analysis (PCA) method can be easily calculated using MATLAB software. Meanwhile, the eigenvalue contribution rate of PCA method is used as a factor weight in our comprehensive evaluation method. According to the scores of this evaluation method, the Chinese core journals in dramatic art can be ranked via these scores from largest to smallest. Based on the statistical data in Chinese journal citation reports-expanded (2011) [5], we can obtain four eigenvalues accordingly, i.e. 4.0977, 1.2936, 1.0298 and 0.6591, and guarantee that the cumulative variance contribution rate of these values is greater than 85 %. Furthermore, these factor score formulas of the four principal components are:

$$\begin{cases} \hat{y}_{i,1} = 0.1887\bar{x}_{i,1} + 0.2216\bar{x}_{i,2} + \cdots + 0.0196\bar{x}_{i,8} \\ \hat{y}_{i,2} = 0.0413\bar{x}_{i,1} - 0.0609\bar{x}_{i,2} + \cdots - 0.2955\bar{x}_{i,8} \\ \hat{y}_{i,3} = 0.0044\bar{x}_{i,1} - 0.2029\bar{x}_{i,2} + \cdots + 0.8835\bar{x}_{i,8} \\ \hat{y}_{i,4} = -0.7680\bar{x}_{i,1} + 0.1900\bar{x}_{i,2} + \cdots - 0.0080\bar{x}_{i,8} \end{cases} \tag{102.6}$$

At the same time, in accordance with the above analysis, the available function of comprehensive evaluation of the principal component is:

$$\tilde{y}_i = 0.5122\hat{y}_{i,1} + 0.1617\hat{y}_{i,2} + 0.1287\hat{y}_{i,3} + 0.0824\hat{y}_{i,4} \quad i = 1, 2, \ldots, 9. \tag{102.7}$$

According to the formulas (102.5) and (102.6), we could calculate the factor score and composite score of each journal. The journals also can be rank on the basis of these scores. The specific circumstances are shown in Table 102.1.

As shown in Table 102.1, Hundred Schools in Art is ranked first in nine kinds of Chinese core journals in dramatic art, and Jingju of China is the last one. Hundred Schools in Art has a very good performance in the right chosen measurement indexes, and it outperforms the other eight kinds of journals. The performance of Sichuan Theatre which ranks second is not outstanding but very stable in the measurement of all statistical indexes. This indicates that the journal is in the steady process of development, but is more difficult to outperform Hundred School in Art. Jingju of China ranks last has not good performance on each statistical index, which illustrates that the journal is now in a more hard time. To some extent, it also reflects that in China the development of the art of Jingju—current traditional quintessence is in a depressed state. And some young people, whose acceptance of it is lower than other readily acceptable form of art or “Exotic”. Due to its lack of attractive highlights and innovative forms, people have been gradually forgotten it. And the numbers of scholars entering the research field are very small, which directly determine that academic influence of Jingju of China is lower than other similar disciplines.

According to the statistical data about nine kinds of journals provided by the Chinese journal citation reports-expanded (2008 Edition), (2009 Edition) and (2010 Edition), We will compare them longitudinally in this article on the four years of dynamic development trends between 2008 and 2011 to provide reference for editors to improve the running level. As shown in Table 102.2, the evaluation result between 2008 and 2010 was calculated from principal component analysis model. According to the changes of the nine kinds of journal ranking, Hundred Schools in Art and Sichuan Theatre maintained a high level of journal, remained

Table 102.2 The PCA-based comprehensive results and ranking for the Chinese core journals in dramatic art from 2008 to 2011

Journals	Comprehensive results				Comprehensive ranking			
	2008	2009	2010	2011	2008	2009	2010	2011
1 Drama	0.784	−0.403	−0.204	0.302	5	5	5	6
2 Theatre arts	−0.183	−0.671	−0.315	0.353	6	6	6	5
3 Chinese theatre	−0.383	−2.065	−1.461	−0.141	7	9	8	8
4 Hundred schools in art	3.792	3.453	3.833	3.722	1	1	1	1
5 Chinese theatre arts	1.724	0.769	0.861	0.826	4	4	3	3
6 Dramatic literature	2.252	0.804	0.735	0.538	3	3	4	4
7 Contemporary drama	−1.073	−0.95	−0.568	0.208	8	7	7	7
8 Jingju of China	−1.294	−1.277	−1.749	−1.242	9	8	9	9
9 Sichuan theatre	2.819	1.148	1.267	1.475	2	2	2	2

the first or second place in the 4 years. On the contrary, both Chinese Theatre and Jingju of China stay at a low level. These ranking results are affected by the diffusion index and the rate of funded papers. In order to improve the journal quality and subject influence, the journals should attract funded scientific papers as more as possible, and encourage more papers to be cited in a short time. Meanwhile, it is also a effective way to publish the papers in advance via the Internet, then the researchers can download the papers expediently and then to cite these papers fast in their published papers.

Arrange the impact factor from high and low, we can kwon the journals' quality and academic impact. In many areas, people tend to directly equate the impact factor and quality of academic journals. This article, took the data from the Chinese journal citation reports-expanded (2011 edition) for example, as shown in Table 102.3, studied the coupling relationship between impact factor and publication. In two different rankings, Hundred Schools in Art is on the top of all, and Jingju of China still stays at a low level. For the middle-ranking journals, there exist differences in ranking position, but the degree of difference is very small. Moreover, it is efficient and effective for us to utilize Impact Factor to evaluate the developing quality of the corresponding journals. Nevertheless, the Impact Factor is not always effective to evaluate the journal quality due to the shortage of stability and credibility, because only one high-cited paper could bring high Impact Factor for the journal in a short time. However, the peer-reviewed feedback information about this journal may be unsatisfactory, contrary to the above-mentioned incorrect Impact Factor. Therefore, the principal component analysis

Table 102.3 The comparison between impact factor and PCA-based comprehensive ranking in 2011

Journal		(1)	(2)	(3)	(4)	(5)	(6)	(7)
Evaluation results	Impact factor	0.113	0.153	0.031	0.305	0.048	0.000	0.045
	Comprehensive	0.302	0.353	−0.141	3.722	0.826	0.538	0.208
Ranking	Impact factor	3	2	7	1	5	8(9)	6
	Comprehensive	6	5	8	1	3	4	7

model has a strong robustness and credibility to evaluate the journal quality based on selected multiple measurement indexes.

102.3 Conclusion

The development of scientific research closely depends on scientific assessment. As an important carrier of the scientific achievements, evaluation on journals plays a component of academic appraisal present. This paper aims at solving the problem of the academic influence of the dramatic arts core journals and dynamic comprehensive evaluates the academic influence of the nine Chinese core journals from 2008 to 2011 by using principal component analysis model, on the basis of the journal information in the Chinese journal citation reports-expanded and General contents of Chinese core journal. Through the league tables of academic influence above, you can see the competition in the Chinese core journals in dramatic art becomes a trend of polarization: Hundred Schools in Art, Sichuan Theatre, Chinese Theatre Arts and Dramatic Literature stay at a high level. Meanwhile, the ranking of the Chinese Theatre and Contemporary Drama stays at a low level. The paper further analyzes the reason of this problem, then proposes practical methods and recommendations for the relevant units to improve the journal quality, and also provides a reference for the positive development of other non-core journals to promote the entire healthy development of the Chinese journals in dramatic art.

References

1. Yue W, Lei L, Li S et al (2010) The comprehensive evaluated model of the weight number of probability theory of journals information and its application. J Intell 29:82–85
2. He Y (2007) the integrated evaluation about the academic influence of Chinese academic journals in management science (2001–2004). China Soft Sci Mag 1:107–112
3. Han GX, Xin DQ (2011) Comprehensive evaluation on agricultural sciences core journals based on principle component analysis. J Anhui Agri Sci 39:20247–20248
4. Dai JL, Zhang QS, Cai RH (2011) General contents of Chinese core journal, vol 264. Beijing University Press, Beijing, pp 135–136
5. ISTIC, Wanfang Data Co. Ltd (2011) Chinese journal citation reports-expanded, vol 24. Scientific and Technical Documentation Press, China, pp 61–66
6. ISTIC, Wanfang Data Co. Ltd (2010) Chinese journal citation reports-expanded, vol 562. Scientific and Technical Documentation Press, China, pp 419–425
7. ISTIC, Wanfang Data Co. Ltd (2009) Chinese journal citation reports-expanded, vol 42. Scientific and Technical Documentation Press, China, pp 156–159
8. ISTIC, Wanfang Data Co. Ltd (2008) Chinese journal citation reports-expanded, vol 164. Scientific and Technical Documentation Press, China, pp 43–45
9. Yu YS, Liu ZX (2010) The study of comprehensive evaluation method on S&T journal: principal component analysis. J Chongqing Univ Social Sci Ed 16:119–123
10. Yu LP, Pan YT, Wu YS (2009) Research on data normalization methods in multi-attribute evaluation. Libr Inf Serv 53:136–139

Author Index

W. Du (ed.), *Informatics and Management Science VI*, Lecture Notes in Electrical Engineering 209, DOI: 10.1007/978-1-4471-4805-0,

CPSIA information can be obtained
at www.ICGtesting.com
Printed in the USA
LVHW080317270420
654475LV00020B/1670